Machine Vision: Theory, Algorithms, Practicalities

2nd Edition

Signal Processing and its Applications

Machine Vision: Theory, Algorithms, Practicalities

2nd Edition

E.R. DAVIES

Department of Physics
Royal Holloway
University of London
Egham, Surrey, UK

ACADEMIC PRESS
San Diego   London   Boston
New York   Sydney   Tokyo   Toronto

Academic Press, Inc.
525 B Street, Suite 1900, San Diego, California 92101-4495, USA
http://www.apnet.com

Academic Press Limited
24–28 Oval Road, London NW1 7DX, UK
http://www.hbuk.co.uk/ap/

ISBN 0-12-206092-X

Library of Congress Cataloging-in-Publication Data

A catalogue record for this book is available from the British Library

Typeset by Mathematical Composition Setters Ltd, Salisbury
Printed in Great Britain by The University Press, Cambridge

96 97 98 99 00 01 EB 9 8 7 6 5 4 3 2 1

Series Preface

Signal processing applications are now widespread. Relatively cheap consumer products through to the more expensive military and industrial systems extensively exploit this technology. This spread was initiated in the 1960s by the introduction of cheap digital technology to implement signal processing algorithms in real-time for some applications. Since that time semiconductor technology has developed rapidly to support the spread. In parallel, an ever increasing body of mathematical theory is being used to develop signal processing algorithms, the basic mathematical foundations, however, have been known and well understood for some time.

Signal Processing and its Applications addresses the entire breadth and depth of the subject with texts that cover the theory, technology and applications of signal processing in its widest sense. This is reflected in the composition of the Editorial Board, who have interests in:

(i) Theory – The physics of the application and the mathematics to model the system;

(ii) Implementation – VLSI/ASIC design, computer architecture, numerical methods, systems design methodology, and CAE;

(iii) Applications – Speech, sonar, radar, seismic, medical, communications (both audio and video), guidance, navigation, remote sensing, imaging, survey, archiving, non-destructive and non-intrusive testing, and personal entertainment.

Signal Processing and its Applications will typically be of most interest to postgraduate students, academics, and practising engineers who work in the field and develop signal processing applications. Some texts may also be of interest to final year undergraduates.

Richard C. Green
The Engineering Practice,
Farnborough, UK

Contents

Part 2   Intermediate-Level Processing

11 Ellipse Detection   271

12 Hole Detection   291

13   Polygon and Corner Detection   309

Part 3   Application-Level Processing

14   Abstract Pattern Matching Techniques   347

15   The Three-Dimensional World   373

16   Tackling the Perspective *N*-Point Problem   417

22   Texture   561

23   Image Acquisition   583

Part 4   Perspectives on Vision

Appendices

Appendix B: Mathematical Morphology   647

Appendix C: Image Transformations and Camera Calibration   663

Appendix D: Robust statistics   677

Preface to the First Edition

Over the past 30 years or so, machine vision has evolved into a mature subject embracing many topics and applications: these range from automatic (robot) assembly to automatic vehicle guidance, from automatic interpretation of documents to verification of signatures, and from analysis of remotely sensed images to checking of fingerprints and human blood cells. Currently, automated visual inspection is undergoing very substantial growth, necessary improvements in quality, safety and cost-effectiveness being the stimulating factors. With so much ongoing activity, it has become a difficult business for the professional to keep up with the subject and with relevant methodologies: in particular, it is difficult for him to distinguish accidental developments from genuine advances. It is the purpose of this book to provide background in this area.

The book was shaped over a period of 10–12 years, through material I have given on undergraduate and postgraduate courses at London University, and contributions to various industrial courses and seminars. At the same time, my own investigations coupled with experience gained while supervising PhD and postdoctoral researchers helped to form the state of mind and knowledge that is now set out here. Certainly it is true to say that if I had had this book 8, 6, 4 or even 2 years ago, it would have been of inestimable value to myself for solving practical problems in machine vision. It is therefore my hope that it will now be of use to others in the same way. Of course, it has tended to follow an emphasis that is my own—and in particular one view of one path towards solving automated visual inspection and other problems associated with the application of vision in industry. At the same time, although there is a specialism here, great care has been taken to bring out general principles—including many applying throughout the field of image analysis. The reader will note the universality of topics such as noise suppression, edge detection, principles of illumination, feature recognition, Bayes' theory, and (nowadays) Hough transforms. However, the generalities lie deeper than this. The book has aimed to make some general observations and messages about the limitations, constraints and tradeoffs to which vision algorithms are subject. Thus there are themes about the effects of noise, occlusion, distortion and the need for built-in forms of robustness (as distinct from less successful *ad hoc* varieties and those added on as an afterthought); there are also themes about accuracy, systematic design and the matching of algorithms and architectures. Finally, there are the problems of setting up lighting schemes which must be addressed in complete systems, yet which receive scant attention in most books on image processing and analysis. These remarks will indicate that the text is intended to be read at various levels—a factor that should make it of more lasting value than might initially be supposed from a quick perusal of the contents.

Of course, writing a text such as this presents a great difficulty in that it is necessary to be highly selective: space simply does not allow everything in a subject of this nature and maturity to be dealt with adequately between two covers. One solution might be to dash rapidly through the whole area mentioning everything that comes to mind, but leaving the reader unable to understand anything in detail or to achieve anything having read the book. However, in a practical subject of this nature this seemed to me a rather worthless extreme. It is just possible that the emphasis has now veered too much in the opposite direction, by coming down to practicalities (detailed algorithms, details of lighting schemes, and so on): individual readers will have to judge this for themselves. On the other hand, an author has to be true to himself and my view is that it is better for a reader or student to have mastered a coherent series of topics than to have a mish-mash of information that he is later unable to recall with any accuracy. This, then, is my justification for presenting this particular material in this particular way and for reluctantly omitting from detailed discussion such important topics as texture analysis, relaxation methods, motion and optical flow.

As for the organization of the material, I have tried to make the early part of the book lead into the subject gently, giving enough detailed algorithms (especially in Chapters 2 and 6) to provide a sound feel for the subject—including especially vital, and in their own way quite intricate, topics such as connectedness in binary images. Hence Part 1 provides the lead-in, although it is not always trivial material and indeed some of the latest research ideas have been brought in (e.g. on thresholding techniques and edge detection). Part 2 gives much of the meat of the book. Indeed, the literature of the subject currently has a significant gap in the area of intermediate-level vision; while high-level vision (AI) topics have long caught the researcher's imagination, intermediate-level vision has its own difficulties which are currently being solved with great success (note that the Hough transform, originally developed in 1962, and by many thought to be a very specialist topic of rather esoteric interest, is arguably only now coming into its own). Part 2 and the early chapters of Part 3 aim to make this clear, while Part 4 gives reasons why this particular transform has become so useful. As a whole, Part 3 aims to demonstrate some of the practical applications of the basic work covered earlier in the book, and to discuss some of the principles underlying implementation: it is here that chapters on lighting and hardware systems will be found. As there is a limit to what can be covered in the space available, there is a corresponding emphasis on the theory underpinning practicalities. Probably this is a vital feature, since there are many applications of vision both in industry and elsewhere, yet listing them and their intricacies risks dwelling on interminable detail, which some might find insipid; furthermore, detail has a tendency to date rather rapidly. Although the book could not cover 3-D vision in full (this topic would easily consume a whole volume in its own right), a careful overview of this complex mathematical and highly important subject seemed vital. It is therefore no accident that Chapter 16[*] is the longest in the book. Finally, Part 4 asks questions about the limitations and constraints of vision algorithms and answers them by drawing on information and experience from earlier chapters. It is tempting to call the last chapter the Conclusion. However, in such a dynamic subject area any such temptation has to be resisted, although it has still been possible to draw a good number of lessons on the nature and current state of the subject. Clearly, this chapter presents a personal view but I hope it is one that readers will find interesting and useful.

---

[*] Chapter 15 in the second edition

Preface to the Second Edition

The first edition came out in 1990, and was welcomed by many researchers and practitioners. However, in the space of six years the subject has inevitably moved on, and topics which hardly deserved a mention in the first edition had to be considered very seriously for the second edition. It seemed particularly important to bring in significant amounts of material on artificial neural networks, mathematical morphology, motion, invariance, texture analysis, X-ray inspection and foreign object detection, camera calibration and robust statistics. There are thus new chapters or appendices on these topics, and these have been carefully integrated with the existing material. The greater proportion of this new material has been included in Part 3 Application-Level Processing, though robust statistics appears as an appendix under Part 4 Perspectives on Vision, as it has some quite profound implications. In fact, it is difficult to design a logical ordering for all the chapters and appendices in Parts 3 and 4, as the topics interact with each other at a variety of different levels—theory, algorithms, methodologies, practicalities and so on. However, this should not matter, as the reader will be exposed to the essential richness of the subject, and his/her studies should be amply rewarded by increased understanding and capability.

A typical final-year undergraduate course on vision for Electronic Engineering or Computer Science students might include much of the work of Chapters 1–8, Chapters 12 and 20, and a selection of sections from other chapters, according to requirements. For MSc or PhD research students, a suitable lecture course might go on to cover Part 2 in depth and then apply this knowledge in detail to Chapters 14–17 in Part 3, with many practical exercises being undertaken at the same time on an image analysis system.

Here much will depend on the research programme being undertaken by each individual student. At this stage the text will have to be used more as a handbook for research, and indeed, one of the prime aims of the volume is to act as a handbook for the researcher and practitioner in the field of Machine Vision.

As mentioned above, this book leans heavily on experience I have gained from working with postgraduate students: in particular I would like to express my gratitude to Barry Cook, Mark Edmonds, Simon Barker, Max van Daalen, Daniel Celano and Darrel Greenhill, all of whom have in their own ways helped to shape my view of the subject. In addition, it is a special pleasure to recall many rewarding discussions with my colleagues Ian Hannah, Zahid Hussain, Dev Patel, Adrian Johnstone and Piers Plummer, the last two named having been particularly prolific in generating hardware systems for implementing my research group's vision algorithms. Finally, I am indebted to Kate Brewin of Academic Press for her help and encouragement, without which this second edition might never have been completed.

Royal Holloway,   E.R. DAVIES
University of London

About the Author

Professor E. R. Davies graduated from Oxford University in 1963 with a First Class Honours degree in Physics. After 12 years' research in solid state physics, he became interested in vision and now leads the Machine Vision Group at Royal Holloway, University of London, where he is Professor of Machine Vision. He has worked on signal recovery, noise suppression, thinning, edge detection, Hough transforms, robust pattern matching, and artificial neural networks; he specializes in algorithm design for automated visual inspection, particularly for the food industry. He has published well over 100 papers, most of them on Machine Vision. Professor Davies is a member of the Executive Committee of the British Machine Vision Association, a Fellow of the Institute of Physics and of the Institution of Electrical Engineers, and a Senior Member of the IEEE. He is on the Editorial Board of *Real-Time Imaging* as European Co-ordinating Editor. In June 1996 he was awarded a Doctor of Science by the University of London for his work on the Science of Measurement, including signal processing and machine vision.

This book is dedicated to my family:
—to my late mother, Mary Davies, to record her never-failing love and devotion
—to my late father, Arthur Granville Davies, who passed on to me his appreciation of the beauties of mathematics and science
—to my wife, Joan, for love, patience, support and inspiration
—to my children, Elizabeth, Sarah and Marion, the music in my life

Glossary of Acronyms and Abbreviations

ACM	Association for Computing Machinery (USA)
ADC	analogue-to-digital converter
AI	artificial intelligence
ANN	artificial neural network
ASCII	American Standard Code for Information Interchange
BMVA	British Machine Vision Association
CAD	computed-aided design
CAM	computer-aided manufacture
CCD	charge-coupled device
CIM	computer integrated manufacture
CLIP	cellular logic image processor
CPU	central processor unit
DEC	Digital Equipment Corporation
DET	Beaudet's (1978) determinant operator
DG	differential gradient
DN	Dreschler and Nagel (1981) corner detector
DSP	digital signal processing
ECL	emitter-coupled logic
FFT	fast Fourier transform
FOE	focus of expansion
GA	genetic algorithm
GHT	generalized Hough transform
HT	Hough transform
IDD	integrated directional derivative
IEE	Institution of Electrical Engineers (UK)

IEEE	Institute of Electrical and Electronics Engineers (USA)
KR	Kitchen and Rosenfeld (1982) corner detector
k-NN	k-nearest neighbour
LFF	local-feature-focus method (Bolles and Cain, 1982)
LSB	least significant bit
LUT	lookup table
MIMD	multiple instruction stream, multiple data stream
MIPS	millions of instructions per second
MISD	multiple instruction stream, single data stream
MLP	multilayer perception
MSB	most significant bit
NN	nearest neighbour
OCR	optical character recognition
PCB	printed circuit board
PE	processing element
PR	pattern recognition
PSF	point spread function
RAM	random access memory
RHT	randomized Hough transform
RMS	root mean square
SIAM	Society of Industrial and Applicative Mathematics
SIMD	single instruction stream, multiple data stream
SISD	single instruction stream, single data stream
S/N	signal-to-noise ratio
SPIE	Society of Photo-optical Instrumentation Engineers
SPR	statistical pattern recognition
TM	template matching
TTL	transistor–transistor logic
TV	television
ULUT	universal lookup table
VLSI	very large scale integration
WSI	wafer-scale integration
ZH	Zuniga and Haralick (1983) corner detector
1-D	one dimension/one-dimensional
2-D	two dimensions/two-dimensional
3-D	three dimensions/three-dimensional
3DPO	3-D part orientation

Acknowledgements

The author would like to credit the following sources for permission to reproduce tables, figures and extracts of text from earlier publications:

The Committee of the Alvey Vision Club for permission to reprint portions of the following paper as text in Chapter 14 and as Figs 14.1, 14.2, 14.6:

Davies, E.R. (1988g)

Butterworth (Elsevier Science–NL) for permission to reprint portions of the following papers from Image and Vision Computing as text in Chapter 5, as Tables 5.1, 5.6–5.10 and as Figs 3.20, 5.3:

Davies, E.R. (1984b, 1987e)

CEP Consultants Ltd (Edinburgh) for permission to reprint portions of the following paper as text in Chapter 19:

Davies, E.R. (1987a)

Elsevier Science–NL for permission to reprint portions of the following papers from *Pattern Recognition Letters and Signal Processing* as text in Chapters 3, 5, 8–13, 20, as Tables 3.2, 5.2–5.5, 9.3, 10.1, 10.2, 11.1, 13.1 and as Figs 3.6, 3.8, 3.10–3.17, 5.1, 5.2, 5.4, 8.1–8.4, 9.4–9.15, 10.1, 10.3–10.12, 11.5–11.10, 12.1–12.5, 13.1–13.3, 13.6–13.10:

Davies, E.R. (1986a,c, 1987c,d,g,k,l, 1988b,c,e,f, 1989a–d)

IEE for permission to reprint portions of the following papers from the *IEE Proceedings and Colloquium Digests* as text in Chapters 13, 19, 24 and as Figs 13.15–13.17, 19.2–19.6, 21.11:

  Davies, E.R. (1985b, 1988a)
  Davies, E.R. and Johnstone, A.I.C. (1989)
  Greenhill, D. and Davies, E.R. (1994b)

IEEE for permission to reprint portions of the following papers as text in Chapter 13 and as Figs 3.4, 3.5, 3.7, 3.11, 13.4, 13.5:

  Davies, E.R. (1984a, 1986d)

IFS Publications Ltd for permission to reprint portions of the following paper as text in Chapter 19 and as Figs 9.1, 9.2, 19.7:

  Davies, E.R. (1984c)

The Council of the Institution of Mechanical Engineers for permission to reprint portions of the following paper as Tables 24.1, 24.2:

  Davies, E.R. and Johnstone, A.I.C. (1986)

MCB University Press for permission to reproduce Plate 1 of the following paper as Fig. 19.8:

  Patel, D., Davies, E.R. and Hannah, I. (1995)

Pergamon (Elsevier Science Ltd) for permission to reprint portions of the following articles as text in Chapters 6, 7 and as Figs 3.19, 6.3, 6.14:

  Davies, E.R. and Plummer, A.P.N. (1981)
  Davies, E.R. (1982)

Royal Swedish Academy of Engineering Sciences for permission to reprint portions of the following paper as text in Chapter 19 and as Fig. 7.3:

  Davies, E.R. (1987f)

Springer-Verlag (Heidelberg) for permission to reprint portions of the following paper as text in Chapter 13 and as Figs 13.12, 13.14:

  Davies, E.R. (1988d)

Unicom for permission to reprint portions of the following paper as text in Chapter 19 and as Figs 8.2, 9.3, 11.3, 13.17, 14.8:

Davies, E.R. (1988h)

The help of S. May in preparing photographs for Figs 19.8 and 22.1 is gratefully acknowledged.

The author would also like to acknowledge that much of his own research (including that outlined in this book) has been supported by grants from the UK Engineering and Physical Sciences Research Council. Broadly, the papers under the author's name in the list of references from 1981 to 1983 were supported by SERC grant GR/B/26374, those from 1984 to 1986 by SERC ACME grant GR/C/09333, those from 1987 to 1989 by SERC ACME grant GR/D/88526, and those from 1992 to 1995 by EPSRC CDP grant GR/G/53590.

1

Vision, the Challenge

1.1 Introduction—man and his senses

Of the five senses—vision, hearing, smell, taste and touch—vision is undoubtedly the one that man has come to depend upon above all others, and indeed the one that provides most of the data he receives. Not only do the input pathways from the eyes provide megabits of information at each glance but the data rates for continuous viewing probably exceed 10 megabits per second. However, much of this information is redundant and is compressed by the various layers of the visual cortex, so that the higher centres of the brain have to interpret abstractly only a small fraction of the data. Nonetheless, the amount of information the higher centres receive from the eyes must be at least two orders of magnitude greater than all the information they obtain from the other senses.

Another feature of the human visual system is the ease with which interpretation is carried out. We see a scene as it is—trees in a landscape, books on a desk, widgets in a factory. No obvious deductions are needed and no overt effort is required to interpret each scene: in addition, answers are effectively immediate and are normally available within a tenth of a second. Just now and again some doubt arises—e.g. a wire cube might be "seen" correctly or inside out. This and a host of other optical illusions are well known, although for the most part we can regard them as curiosities—irrelevant freaks of nature. Somewhat surprisingly, it turns out that illusions are quite important, since they reflect hidden assumptions that the brain is making in its struggle with the huge amounts of complex visual data it is receiving. We have to pass by this story here (though it surfaces

1

now and again in various parts of this book). However, the important point is that we are for the most part unaware of the complexities of vision. Seeing is not a simple process: it is just that vision has evolved over millions of years, and there was no particular advantage in evolution giving us any indication of the difficulties of the task (if anything, to have done so would have cluttered our minds with worthless information and quite probably slowed our reaction times in crucial situations).

Thus there is much ignorance of the process of human vision. However, man is inventive, and he is now trying to get machines to do much of his work for him. For the simplest tasks there should be no particular difficulty in mechanization but for more complex tasks the machine must be given man's prime sense, that of vision. Efforts have been made to achieve this, sometimes in modest ways, for well over 30 years. At first such tasks seemed trivial, and schemes were devised for reading, for interpreting chromosome images, and so on; but when such schemes were confronted with rigorous practical tests, the problems often turned out to be more difficult. Generally, researchers react to finding that apparent "trivia" are getting in the way by intensifying their efforts and applying great ingenuity, and this was certainly so with some early efforts at vision algorithm design. Hence it soon became plain that the task really is a complex one, in which numerous fundamental problems confront the researcher, and the ease with which the eye can interpret scenes turned out to be totally deceptive.

Of course, one of the ways in which the human visual system gains over the machine is that the brain possesses some 10^{10} cells (or neurons), some of which have well over $10\,000$ contacts (or synapses) with other neurons. If each neuron acts as a type of microprocessor, then we have an immense computer in which all the processing elements can operate concurrently. Most man-made computers still contain less than 100 million processing elements, so the majority of the visual and mental processing tasks that the eye–brain system can perform in a flash have no chance of being performed by present-day man-made systems. Added to these problems of scale, there is the problem of how to organize such a large processing system, and also how to program it. Clearly, the eye–brain system is partly hard-wired by evolution but there is also an interesting capability to program it dynamically by training during active use. This need for a large parallel processing system with the attendant complex control problems illustrates clearly that machine vision must indeed be one of the most difficult intellectual problems to tackle.

So what are the problems involved in vision that make it apparently so easy for the eye, yet so difficult for the machine? In the next few sections an attempt is made to answer this question.

1.2 The nature of vision

1.2.1 The process of recognition

This section illustrates the intrinsic difficulties of implementing machine vision, starting with an extremely simple example—that of character recognition. Consider the set of patterns shown in Fig. 1.1(a). Each pattern can be considered as a set of 25 bits of information, together with an associated class indicating its interpretation. In each case imagine a computer learning the patterns and their classes by rote. Then any new pattern may be classified (or "recognized") by comparing it with this previously learnt "training set", and assigning it to the class of the nearest pattern in the training set. Clearly, test pattern (1) (Fig. 1.1(b)) will be allotted to class U on this basis. Chapter 20 shows that this method is a simple form of the nearest-neighbour approach to pattern recognition.

The scheme outlined above seems straightforward and is indeed highly effective, even being able to cope with situations where distortions of the

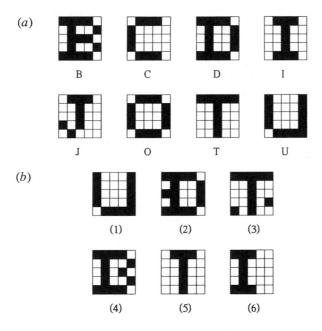

Fig. 1.1 Some simple 25-bit patterns and their recognition classes used to illustrate some of the basic problems of recognition: (a) training set patterns (for which the known classes are indicated); (b) test patterns.

test patterns occur or where noise is present: this is illustrated by test patterns (2) and (3). However, this approach is not always foolproof. First, there are situations where distortions or noise are excessive, so errors of interpretation arise. Second, there are situations where patterns are not badly distorted or subject to obvious noise, yet are misinterpreted: this seems much more serious, since it indicates an unexpected limitation of the technique rather than a reasonable result of noise or distortion. In particular, these problems arise where the test pattern is displaced or misorientated relative to the appropriate training set pattern, as with test pattern (6).

As will be seen in Chapter 20, there is a powerful principle that indicates why the unlikely limitation given above can arise: it is simply that there are *insufficient training set patterns*, and that those that are present are *insufficiently representative* of what will arise during tests. Unfortunately, this presents a major difficulty, since providing enough training set patterns incurs a serious storage problem, and an even more serious search problem when patterns are tested. Furthermore, it is easy to see that these problems are exacerbated as patterns become larger and more real (obviously, the examples of Fig. 1.1 are far from having enough resolution even to display normal type-fonts). In fact, a combinatorial explosion* takes place. Forgetting for the moment that the patterns of Fig. 1.1 have familiar shapes, let us temporarily regard them as random bit patterns. Now the number of bits in these $N \times N$ patterns is N^2, and the number of possible patterns of this size is 2^{N^2}: even in a case where $N = 20$, remembering all these patterns and their interpretations would be impossible on any practical machine, and searching systematically through them would take impracticably long (involving times of the order of the age of the universe). Thus it is not only impracticable to consider such brute-force means of solving the recognition problem, it is effectively also impossible theoretically. These considerations show that other means are required to tackle the problem.

1.2.2 Tackling the recognition problem

An obvious means of tackling the recognition problem is to normalize the images in some way. Clearly, normalizing the position and orientation of any 2-D picture object would help considerably: indeed this would reduce the number of degrees of freedom by three. Methods for achieving this

---

*This is normally taken to mean that some parameter produces a fast-varying (often exponential) effect, which "explodes" as the parameter increases by a modest amount.

involve centralizing the objects—arranging that their centroids are at the centre of the normalized image—and making their major axes (deduced by moment calculations, for example) vertical or horizontal. Next, we can make use of the order which is known to be present in the image—and here it may be noted that very few patterns of real interest are indistinguishable from random dot patterns. This approach can be taken further: if patterns are to be nonrandom, isolated noise points may be eliminated. Ultimately, all these methods help by making the test pattern closer to a restricted set of training set patterns (although care has to be taken to process the training set patterns initially so that they are representative of the processed test patterns).

It is useful to consider character recognition further. Here we can make additional use of what is known about the structure of characters—namely, that they consist of limbs of roughly constant width. In that case the width carries no useful information, so the patterns can be thinned to stick figures (called skeletons—see Chapter 6); then, hopefully, there is an even greater chance that the test patterns will be similar to appropriate training set patterns (Fig. 1.2). This process can be regarded as another instance of reducing the number of degrees of freedom in the image, and hence of helping to minimize the combinatorial explosion—or, from a practical point of view, to minimize the size of the training set necessary for effective recognition.

Next, consider a rather different way of looking at the problem. Recognition is necessarily a problem of discrimination—i.e. of discriminating between patterns of different classes. However, in practice, considering the natural variation of patterns, including the effects of noise and distortions (or even the effects of breakages or occlusions), there is also a problem of generalizing over patterns of the same class. In practical problems there is a tension between the need to discriminate and the need to generalize. Nor is this a fixed situation. Even for the character recognition task, some classes are so close to others (n's and h's will be similar) that less generalization is

Fig. 1.2   Use of thinning to regularize character shapes. Here character shapes of different limb widths—or even varying limb widths—are reduced to stick figures or skeletons. Thus irrelevant information is removed and at the same time recognition is facilitated.

possible than in other cases. On the other hand, extreme forms of generalization arise when for example an *A* is to be recognized as an *A* whether it is a
capital or small letter, or in italic, bold, suffix or other form of font—even if
it is handwritten. The variability is determined largely by the training set
initially provided. What we emphasize here, however, is that generalization
is as necessary a prerequisite to successful recognition as is discrimination.

At this point it is worth considering more carefully the means whereby
generalization was achieved in the examples cited above. First, objects were
positioned and orientated appropriately; second, they were cleaned of noise
spots; and third, they were thinned to skeleton figures (although the latter
process is relevant only for certain tasks such as character recognition). In
the last case we are generalizing over characters drawn with all possible
limb widths, width being an irrelevant degree of freedom for this type of
recognition task. Note that we could have generalized the characters further
by normalizing their size and saving another degree of freedom. The
common feature of all these processes is that they aim to give the characters
a high level of standardization against known types of variability before
finally attempting to recognize them.

The standardization (or generalization) processes outlined above are all
realized by image processing, i.e. the conversion of one image into another
by suitable means. The result is a two-stage recognition scheme: first,
images are converted into more amenable forms containing the same
numbers of bits of data; and second, they are classified, with the result that
their data content is reduced to very few bits (Fig. 1.3). In fact, recognition
is a process of data abstraction, the final data being abstract and totally
unlike the original data. Thus we must imagine a letter *A* starting as an array
of perhaps 20×20 bits arranged in the form of an *A*, and then ending as the
7 bits in an ASCII representation of an *A*, namely 1000001 (which is
essentially a random bit pattern bearing no resemblance to an *A*).

The last paragraph reflects to a large extent the history of image analysis.
At one stage (prior to about 1970), a good proportion of the image analysis
problems being tackled were envisaged as consisting of an image

Fig. 1.3 The two-stage recognition paradigm: C, input from camera; G, grab
image (digitize and store); P, preprocess; R, recognize (i, image data; a, abstract
data). The classical paradigm for object recognition is that of (i) preprocessing
(image processing) to suppress noise or other artefacts and to regularize the image
data, and (ii) applying a process of abstract (often statistical) pattern recognition to
extract the very few bits required to classify the object.

"preprocessing" task carried out by image processing techniques, followed by a recognition task undertaken by statistical pattern recognition methods (Chapter 20). These two topics—image processing and statistical pattern recognition—consumed much research effort and effectively dominated the subject of image analysis, so much so that certain other approaches such as the Hough transform were not given due recognition and hence for a time were not researched or optimized adequately. Even now, many commercial vision systems tend to concentrate overmuch on certain less powerful processing methods, ignoring approaches that do not approximate to the image processing–pattern recognition paradigm. One of the aims of this book is to redress this balance, considering carefully what might be termed "intermediate-level" processing and perhaps de-emphasizing the more traditional low-level–high-level schema.

In passing, it is worth noting that much of what has been touched on above comes under the purview of optimization. If image analysis has been constrained to a low-level followed by a high-level approach, then maybe relaxing the constraint will permit a more efficient means of extracting the data. This sentiment underlies some of the work studied in the following chapters.

1.2.3 Object location

The problem that was tackled above—that of character recognition—is a very constrained one. In a great many practical applications it is necessary to search pictures for objects of various types, rather than just interpreting a small area of a picture.

Search is a task that can involve prodigious amounts of computation and which is also subject to a combinatorial explosion. Imagine the task of searching for a letter E in a page of text. An obvious way of achieving this is to move a suitable "template" of size $n \times n$ over the whole image, of size $N \times N$, and to find where a match occurs (Fig. 1.4). A match can be defined as a position where there is exact agreement between the template and the local portion of the image but, in keeping with the ideas of Section 1.2.1, it will evidently be more relevant to look for a best local match (i.e. a position where the match is locally better than in adjacent regions) and where the match is also good in some more absolute sense, indicating that an E is present.

One of the most natural ways of checking for a match is to measure the Hamming distance between the template and the local $n \times n$ region of the image, i.e. to sum the number of differences between corresponding bits. This is essentially the process described in Section 1.2.1. Then places with a low

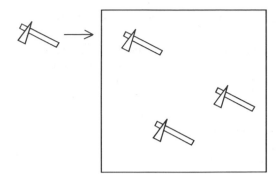

Fig. 1.4 Template matching, the process of moving a suitable template over an image to determine the precise positions at which a match occurs—hence revealing the presence of objects of a particular type.

Hamming distance are places where the match is good. These template matching ideas can be extended to cases where the corresponding bit positions in the template and the image do not just have binary values but may have intensity values over a range 0–255. In that case the sums obtained are no longer Hamming distances but may be generalized to the form:

$$\mathcal{D} = \sum_t |I_i - I_t| \tag{1.1}$$

I_t being the local template value, I_i being the local image value, and the sum being taken over the area of the template. This makes template matching practicable in many situations: the possibilities are examined in more detail in subsequent chapters.

We referred above to a combinatorial explosion in this search problem too. The reason this arises is as follows. First, when a 5×5 template is moved over an $N \times N$ image in order to look for a match, the number of operations required is of the order of $5^2 N^2$, totalling some 1 million operations for a 256×256 image. The problem is that when larger objects are being sought in an image, the number of operations increases as the square of the size of the object, the total number of operations being $N^2 n^2$ when an $n \times n$ template is used[*]. For a 30×30 template and a 256×256

---

[*] Note that, in general, a template will be larger than the object it is used to search for, because some background will have to be included to help demarcate the object.

image, the number of operations required rises to some 60 million and the time taken to achieve this on any conventional computer will be many seconds—well away from the requirements of real-time[*] operation.

Next, recall that in general, objects may appear in many orientations in an image (*E*'s on a printed page are exceptional). If we imagine a possible 360 orientations (i.e. one per degree of rotation), then a corresponding number of templates will in principle have to be applied in order to locate the object. This additional degree of freedom pushes the search effort and time to enormous levels, so far away from the possibility of real-time implementation that it is clear that new approaches must be found for tackling the task. Fortunately, many researchers have applied their minds to this problem and there are now a good many ideas for tackling it. Broadly, however, there is only one means of saving effort on this sort of scale: that of two-stage (or multistage) template matching. The principle is to search for objects via their features. For example, we might consider searching for *E*'s by looking for characters that have horizontal line segments within them. Similarly, we might search for hinges on a manufacturer's conveyor by looking first for the screw holes they possess. In general it is useful to look for small features, since they require smaller templates and hence involve significantly less computation, as demonstrated above. This means that it may be better to search for *E*'s by looking for corners instead of horizontal line segments.

Unfortunately, noise and distortions give rise to problems if we search for objects via small features—there is a risk of missing the object altogether. Hence it is necessary to collate the information from a number of such features. This is the point where the many available methods start to differ from each other. How many features should be collated? Is it better to take a few larger features than a lot of smaller ones? And so on. Also, we have not answered in full the question of what types of feature are the best to employ. These and other questions are considered in the following chapters.

Indeed, one could say, in a sense, that these questions are the subject of this book. Search is one of the fundamental problems of vision, yet the details and the application of the basic idea of two-stage template matching give the subject much of its richness: to solve the recognition problem the dataset needs to be explored carefully. Clearly, any answers will tend to be data-dependent but it is worth exploring to what extent there are generalized solutions to the problem.

---

[*] A commonly used phrase meaning that the information has to be processed as it becomes available: this contrasts with the many situations (such as the processing of images from space probes) where the information may be stored and processed at leisure.

1.2.4   Scene analysis

The last subsection considered what is involved in searching an image for objects of a certain type: the result of such a search is likely to be a list of centroid coordinates for these objects, although an accompanying list of orientations might also be obtained. This subsection considers what is involved in scene analysis—the activity we are continually engaged in as we walk around, negotiating obstacles, finding food, and so on. Scenes contain a multitude of objects, and it is their interrelationships and relative positions that matter as much as identifying what they are. There is often no need for a search *per se* and we could in principle passively take in what is in the scene. However, there is much evidence (e.g. from analysis of eye movements) that the eye–brain system interprets scenes by continually asking questions about what is there. For example, we might ask the following questions: Is this a lamp-post? How far away is it? Do I know this person? Is it safe to cross the road? And so on. It is not the purpose here to dwell on these human activities or introspection about them but merely to observe that scene analysis involves enormous amounts of input data, complex relationships between objects within scenes and, ultimately, descriptions of these complex relationships. The latter no longer take the form of simple classification labels, or lists of object coordinates, but have a much richer information content: indeed, a scene will, to a first approximation, be better described in words than as a list of numbers. It seems likely that a much greater combinatorial explosion is involved in determining relationships between objects than in merely identifying and locating them. Hence all sorts of props must be used to aid visual interpretation: there is considerable evidence of this in the human visual system, where contextual information and the availability of immense databases of possibilities clearly help the eye to a very marked degree.

Note also that scene descriptions may initially be at the level of factual content but will eventually be at a deeper level—that of meaning, significance and relevance. However, we shall not be able to delve further into these areas in this book.

1.2.5   Vision as inverse graphics

It has often been said that vision is "merely" inverse graphics. There is a certain amount of truth in this. Computer graphics is the generation of images by computer, starting from abstract descriptions of scenes and a knowledge of the laws of image formation. Clearly it is difficult to quarrel with the idea that vision is the process of obtaining descriptions of sets of objects, starting from sets of images and a knowledge of the laws of image formation (indeed, it is good to see a definition that explicitly brings in the need to know the laws of image formation, since it is all too easy to forget that this is a prerequisite when building descriptions incorporating heuristics that aid interpretation).

However, this similarity in formulation of the two processes hides some very fundamental points. First, graphics is a "feedforward" activity, i.e. images can be produced straightforwardly once sufficient specification about the viewpoint and the objects, and knowledge of the laws of image formation, has been obtained. True, considerable computation may be required but the process is entirely determined and predictable. The situation is not so straightforward for vision because search is involved and there is an accompanying combinatorial explosion. Indeed, certain vision packages incorporate graphics (or CAD) packages (see Chapter 15) which are inserted into feedback loops for interpretation: the graphics package is then guided iteratively until it produces an acceptable approximation to the input image, when its input parameters embody the correct interpretation (there is a close parallel here with the problem of designing analogue-to-digital converters by making use of digital-to-analogue converters). Hence it seems inescapable that vision is intrinsically more complex than graphics.

We can clarify the situation somewhat by noting that, as a scene is observed, a 3-D environment is compressed into a 2-D image and a considerable amount of depth and other information is lost. This can lead to ambiguity of interpretation of the image (both a helix viewed end-on and a circle project into a circle), so the 3-D to 2-D transformation is many-to-one. Conversely, the interpretation must be one-to-many, meaning that there are many possible interpretations, yet we know that only one can be correct: vision involves not merely providing a list of all possible interpretations but providing the most likely one. Hence some additional rules or constraints must be involved in order to determine the single most likely interpretation. Graphics, in contrast, does not have these problems, as the above ideas show it to be a many-to-one process.

1.3 Automated visual inspection

So far we have considered the nature of vision but not what man-made vision systems may be used for. There is in fact a great variety of applications for artificial vision systems—including, of course, all of those for which man employs his visual senses. Of particular interest in this book are forensic, medical, space, military, commercial and industrial applications of vision. By way of example, fingerprint analysis and recognition have long been important applications of computer vision, as have the counting of red blood cells, signature verification and character recognition, and aeroplane identification (both from aerial silhouettes and from ground surveillance pictures taken from satellites). Face recognition is becoming a practical possibility and vehicle guidance by vision will in principle soon be sufficiently reliable for urban use[*]. However, the main applications of vision considered in this book are those of manufacturing industry—particularly, automated visual inspection and vision for automated assembly.

In the latter two cases, much the same manufactured components are viewed by cameras: the difference lies in how the resulting information is used. In assembly, components must be located and orientated so that a robot can pick them up and assemble them. For example, the various parts of a motor or brake system need to be taken in turn and put into the correct positions, or a coil may have to be mounted on a TV tube, an integrated circuit placed on a printed circuit board, or a chocolate placed into a box. In inspection, objects may pass the inspection station on a moving conveyor at rates typically between 10 and 30 items per second and it has to be ascertained whether they have any defects. If any defects are detected, the offending parts will usually have to be rejected: that is the feedforward solution. In addition, a feedback solution may be instigated—i.e. some parameter may have to be adjusted to control plant further back down the production line (this is especially true for parameters that control dimensional characteristics such as product diameter). Inspection also has the potential for amassing a wealth of information that is useful for management, on the state of the parts coming down the line: the total number of products per day, the number of defective products per day, the distribution of sizes of products, and so on. The important feature of artificial vision is that it is tireless and that *all* products can be scrutinized and measured: thus quality control can be maintained to a very high standard. In automated assembly too, a considerable amount of on-the-spot inspection can be performed and this may help to avoid the problem of complex assemblies being rejected, or having to be subjected to expensive repairs, because (for example) a proportion of screws were threadless and could not be inserted properly.

An important feature of most industrial tasks is that they take place in real time: if applied, machine vision must be able to keep up with the manufac-

---

[*] Whether the public will accept this is another matter, although it is worth noting that radar blind-landing aids for aircraft have been in wide use for some years.

turing process. For assembly, this may not be too exacting a problem, since a robot may not be able to pick up and place more than one item per second—leaving the vision system a similar time to do its processing. For inspection, this supposition is rarely valid: even a single automated line (for example, one for stoppering bottles) is able to keep up a rate of 10 items per second (and, of course, parallel lines are able to keep up much higher rates). Hence, visual inspection tends to press computer hardware to the limits of its capabilities. Note in addition that many manufacturing processes operate under severe financial constraints so it is not possible to employ expensive multiprocessing systems or supercomputers. Hence very great care must be taken in the design of hardware accelerators for inspection applications. Chapter 24 aims to give some insight into these hardware problems.

1.4   What this book is about

The foregoing sections have examined something of the nature of machine vision and have briefly considered its applications and implementation. It is already clear that implementing machine vision involves considerable practical difficulties but, more important, these practical difficulties embody some very substantial fundamental problems: these include various mechanisms giving rise to excessive processing load and time. Practical problems may be overcome by ingenuity and care: however, by definition, truly fundamental problems cannot be overcome by *any* means—the best that we can hope for is that they must be minimized following a complete understanding of their nature.

Understanding is thus a cornerstone for success in machine vision. It is often difficult to achieve, since the dataset (i.e. all pictures that could reasonably be expected to arise) is very variegated. Indeed, much investigation is required to determine the nature of a given dataset, including not only the objects being observed but also the noise levels, degrees of occlusion, breakage, defect and distortion that are to be expected, and the quality and nature of lighting schemes. Ultimately, sufficient knowledge might be obtained in a useful set of cases so that a good understanding of the milieu can be attained. Then it remains to compare and contrast the various methods of image analysis that are available. Some methods will turn out to be quite unsatisfactory for reasons of robustness, accuracy or cost of implementation, or other relevant variables: and who is to say in advance what a relevant set of variables is? This, too, needs to be ascertained and defined. Finally, among the methods that could reasonably be used, there will be competition: tradeoffs between parameters such as accuracy, speed, robustness and cost will have to be worked out first theoretically and then in numerical detail to find an optimal solution. This is a complex and long process in a situation where workers have usually aimed to find solutions for their own particular (often short-term) needs. Clearly, there is a need to raise practical machine vision from an art to a science. Fortunately this process has been developing for some years (a relevant turning point was 1980), although it does not seem to have been reviewed adequately up to now: one of the aims of this book is to make clear statements about the situation.

Before proceeding further, there are one or two more pieces to fit into the jigsaw. First, there is an important guiding principle: *if the eye can do it, so can the machine.* Thus, if an object is fairly well hidden in an image, yet the eye can see it, then it should be possible to devise a vision algorithm that can also find it. Next, although we can expect to meet this challenge, should we set our sights even higher and aim to devise algorithms that can beat the eye? There seems no reason to suppose that the eye is the ultimate vision machine: it has been built through the vagaries of evolution, so it may be well adapted for finding berries or nuts, or for recognizing faces, but ill-suited for certain other tasks. One such task is that of measurement. The eye probably does not need to measure the sizes of objects, at a glance, to better than a few per cent accuracy. However, it could be distinctly useful if the robot eye could achieve remote size measurement, at a glance, and with an accuracy of say 0.001%. Clearly, it is worth being aware of the possibility of the robot eye having capabilities superior to those of biological systems. Again, this book aims to point out such possibilities where they exist.

1.5   The following chapters

On the whole the early chapters of the book (Chapters 2–4) cover rather simple concepts, such as the nature of image processing, how image processing algorithms may be devised, and the restrictions on intensity thresholding techniques. The next three chapters (Chapters 5–7) discuss edge detection and some fairly traditional binary image analysis techniques. Then, Chapters 8–13 move on to intermediate-level processing, which has developed significantly in the past decade, particularly in the use of transform techniques to deduce the presence of objects. Intermediate-level processing is important in leading on to the inference of complex objects, both in 2-D (Chapter 14) and in 3-D (Chapter 15). It also enables automated inspection to be undertaken efficiently (Chapter 19). Chapter 20 enlarges on the process of recognition that is fundamental to many inspection and other processes—as outlined earlier in this chapter. Chapters 23 and 24 respectively outline the enabling technologies of image acquisition and vision hardware design and finally, Chapter 25 reiterates and highlights some of the lessons and topics covered earlier in the book.

To help give the reader more perspective on the 20 or so chapters, the main text has been divided into four parts: Part 1 (Chapters 2–7) is entitled Low-Level Processing, Part 2 (Chapters 8–13) is called Intermediate-Level Processing, Part 3 (Chapters 14–24) is headed Application-Level Processing, and Part 4 (Chapter 25) is called Perspectives on Vision. The heading "application-level processing" is used instead of "high-level processing", or other possible titles, since the emphasis here is on interpretation in real-world terms, and the performance of useful functions—often using methods (as in 2-D inspection) that are not especially high-level or abstract. The purpose of Part 4 is to draw together and systematize the topics already encountered.

Although the sequence of chapters follows the somewhat logical order just described, the ideas outlined in the previous section—understanding of the visual process, constraints imposed by realities such as noise and occlusion, tradeoffs between relevant parameters, and so on—are mixed into the text at relevant junctures, as they reflect all-pervasive issues.

Finally, there are many topics that would ideally have been included in the book, yet space did not permit this. The chapter bibliographies, the main list of references and the indexes are intended to make good some of these deficiencies.

1.6 Bibliographical notes

The purpose of this chapter has been to introduce the reader to some of the problems of machine vision, showing the intrinsic difficulties but not at this stage getting into details. For references the reader should consult the later chapters. Meanwhile, some background into human vision can be obtained from the two delightful books by Gregory, *The Intelligent Eye* (1971) and *Eye and Brain* (1972).

Part 1
Low-Level Processing

This part of the book introduces images and image processing, and then proceeds to show how image processing may be developed in order to start the process of image analysis. By the end of Chapter 7 image analysis has been taken, via this "traditional" route, to a reasonably useful level. In fact Chapter 7 forms a bridge with Part 2, although it was felt better to retain it in Part 1 because it develops the lines of thought already outlined in the early chapters.

2

Images and Imaging Operations

pĭx' ĕllātĕd, a. picture broken into a regular tiling
pĭx' ĭlātĕd, a. pixie-like, crazy, deranged

2.1 Introduction

This chapter is concerned with images and simple image processing operations. It is intended to lead on to more advanced image analysis operations that are of use for machine vision in an industrial environment. Perhaps the main purpose of the chapter is to introduce the reader to some basic techniques and notation that will be of use throughout the book. However, the image processing algorithms introduced here are of value in their own right in disciplines ranging from remote sensing to medicine, and from forensic to military and scientific applications.

This chapter deals with images that have already been obtained from suitable sensors: the latter are covered in a later chapter. Typical of such images is that shown in Fig. 2.1(a). This is a grey-tone image which at first sight appears to be a normal "black and white" photograph. However, on closer inspection it may be seen that it is composed of a large number of individual picture cells, or pixels. In fact, the image is a 128 × 128 array of pixels. To get a better feel for the limitations of such a digitized image, Fig. 2.1(b) shows a section that has been enlarged so that the pixels can be examined individually. A three-fold magnification factor has been used and thus the visible part of Fig. 2.1(b) contains only a 42 × 42 array of pixels.

(a)

(b)

Fig. 2.1 Typical grey-scale images: (a) grey-scale image digitized into a 128×128 array of pixels; (b) section of image shown in (a) subjected to three-fold linear magnification: the individual pixels are now clearly visible.

It is not easy to see that these grey-tone images are digitized into a grey-scale containing just 64 grey levels. To some extent high spatial resolution compensates for lack of grey-scale resolution, and as a result we find difficulty in seeing the difference between an individual shade of grey and the shade it would have had in an ideal picture. In addition, when we look at the magnified section of image in Fig. 2.1(b), it is difficult to understand the significance of the individual pixel intensities—the whole is becoming lost in a mass of small parts. In practice, images from TV (vidicon) cameras typically give a grey-scale resolution which is accurate only to about one part in 50; any additional grey-scale resolution we try to obtain above this level will include a certain amount of noise and so it is seldom worth digitizing more accurately than 6–7 bits of information per pixel. Solid-state cameras commonly give less noise and may allow 8 or even 9 bits of information per pixel. However, there are many occasions when it is not worthwhile to aim for such high grey-scale resolutions, particularly when the result will not be visible to the human eye, and when for example there is an enormous amount of other data that a robot can use to locate a particular object within the field of view. Note that if the human eye can see an object in a digitized image of particular spatial and grey-scale resolution, it is in principle possible to devise a computer algorithm to do the same thing.

Nevertheless, there is a range of applications for which it is valuable to retain good grey-scale resolution, so that highly accurate measurements can be made from a digital image. This is the case in many robotic applications, where high-accuracy checking of components is critical. More will be said about this later. In addition, it will be seen in Part 2 that certain techniques for locating components efficiently require local edge orientation to be estimated to better than 1°, and this can be achieved only if at least 6 bits of grey-scale information are available per pixel.

2.1.1   Grey scale versus colour

Returning now to the image of Fig. 2.1(a), we might reasonably ask whether it would be better to replace the grey scale with colour, using an RGB colour camera and three digitizers for the three main colours. In principle, the answer to this question is affirmative, since it is plain that the human eye responds more quickly and accurately to what is happening in a scene if it is in colour. Colour is helpful in making many objects "stand out" when they would be subdued or even hidden in a grey-tone image. However, in many applications such as robotics the additional expense is frequently not justified. Not only is the colour camera more expensive, but three digitizers and three image frame stores are needed to hold the image. More serious, however, is the large amount of additional processing that is required to interpret colour images fully. With grey-scale images the bits representing the grey scale are all of the same type and take the form of a number representing the pixel intensity: they can thus be processed as a single entity on a digital computer. This is not the case with colour information, as three sets of bits—three individual numbers—are required to represent the information. Simple linear additions and subtractions of these numbers may help to make sense of the picture data but a large amount of processing will normally be required before it is known exactly how the information should be combined to obtain the best results in a particular case, and even then the situation may vary over the whole image. This makes for considerable extra complication in devising suitable algorithms, which will often impose considerable load on the host computer. No wonder, then, that in practical real-time applications short-cuts are taken which enable the computer interpreting the data to obtain the required information from a single grey-scale image of the type shown in Fig. 2.1(a). However, it should be pointed out that colour is of immense importance in the analysis of multispectral remote sensing imagery obtained from satellites encircling the earth or from interplanetary space probes.

2.2   Image processing operations

In what follows the images of Figs 2.1(a) and 2.3(a) are considered in some detail, examining some of the many image processing operations that can be performed on them. The resolution of these images reveals a considerable amount of detail and at the same time shows how it relates to the more "meaningful" global information. This should help to make it clear how simple imaging operations contribute to image interpretation.

When performing image processing operations, we start with an image in one storage area and generate a new processed image in another storage area. In practice these storage areas may either be in a special hardware unit called a frame store that is interfaced to the computer, or else they may be in the main memory of the computer or on one of its discs. In the past a special frame store was required to store images, since each image contains a good fraction of a megabyte of information and this amount of space was not available for normal users in the computer main memory. Nowadays this is less of a problem but for image acquisition a frame store is still required. However, we shall not worry about such details here. Instead it will be assumed that all images are inherently visible and that they are stored in various image "spaces" P, Q, R, etc. Thus we might start with an image in space P and copy it to space Q, for example.

2.2.1 Some basic operations on grey-scale images

Perhaps the simplest of imaging operations is that of clearing an image or setting the contents of a given image space to a constant level. We need some way of arranging this, and accordingly the following Pascal routine may be written for implementing it[*]:

```
for j := 0 to 127 do
    for i := 0 to 127 do
        P[i, j] := alpha;                          (2.1)
```

In this routine the local pixel intensity value is expressed as P[i, j], since P-space is taken to be a two-dimensional array of intensity values. In what follows it will be advantageous to rewrite such routines in the more succinct form:

```
SETP:  [[ P0 := alpha ]];                          (2.2)
```

since it is then much easier to understand the image processing operations (see Appendix A). In effect, we are attempting to expose the fundamental imaging operation by removing irrelevant programming detail. The double square bracket notation is intended to show that the operation it encloses is applied at every pixel location in the image, for whatever size of image is currently defined. The reason for calling the pixel intensity P0 will become clear later.

Another simple imaging operation is to copy an image from one space to another. This is achieved, without changing the contents of the original space P, by the routine:

```
COPY:  [[ Q0 := P0 ]];                             (2.3)
```

Again, the double bracket notation indicates a double loop which copies each individual pixel of P-space to the corresponding location in Q-space.

---

[*] Readers who are unfamiliar with Pascal should refer to Appendix A.

A more interesting operation is that of inverting the image, as in the process of converting a photographic negative to a positive. This process is represented as follows:

$$\text{INVERT:} \quad \text{[[Q0 := 255 - P0]];} \qquad\qquad (2.4)$$

In this case it is assumed that pixel intensity values lie within the range 0–255, as is commonly true for frame stores which represent each pixel as one byte of information. Note that such intensity values are commonly unsigned and this is assumed generally in what follows.

There are many operations of these types. Some other simple operations are those that shift the image left, right, up, down or diagonally. They are easy to implement if the new local intensity is made identical to that at a neighbouring location in the original image. It is evident how this would be expressed in the double suffix notation used in the original Pascal routine. In the new shortened notation it is necessary to name neighbouring pixels in some convenient way, and we here employ a commonly used numbering scheme:

P4	P3	P2
P5	P0	P1
P6	P7	P8

with a similar scheme for other image spaces (see Appendix A). With this notation, it is easy to express a left shift of an image as follows:

$$\text{LEFT:} \quad \text{[[Q0 := P1]];} \qquad\qquad (2.5)$$

Similarly, a shift down to the bottom right is expressed as:

$$\text{DOWNR:} \quad \text{[[Q0 := P4]];} \qquad\qquad (2.6)$$

(note that the shift appears in a direction opposite to that which one might at first have suspected).

It will now be clear why P0 and Q0 were chosen for the basic notation of pixel intensity: the "0" denotes the central pixel in the "neighbourhood" or "window", and corresponds to zero shift when copying from one space to another.

There is a whole range of possible operations associated with modifying images in such a way as to make them more satisfactory for a human viewer. For example, adding a constant intensity makes the image brighter:

$$\texttt{BRIGHTEN:} \quad \texttt{[[QO := PO + beta]];} \qquad (2.7)$$

and the image can be made darker in the same way. A more interesting operation is to stretch the contrast of a dull image:

$$\texttt{STRETCH:} \quad \texttt{[[QO := PO * gamma + beta]];} \qquad (2.8)$$

where gamma > 1. In practice (as for Fig. 2.2) it is necessary to ensure that intensities do not result which are outside the normal range, e.g. by using an operation of the form:

```
STRETSH:   [[   QQ := PO  * gamma + beta;
                QO := if    QQ < 0        then 0
                      else if QQ > 255    then 255
                                          else QQ ]];      (2.9)
```

Most practical situations demand more sophisticated transfer functions — either nonlinear or piecewise linear—but such complexities are ignored here. Note that the above example deviates from normal Pascal notation in the interests of compactness and clarity (see Appendix A).

A further simple operation which is often applied to grey-scale images is that of thresholding to convert to a binary image. This topic is covered in more detail later, since it is widely used to detect objects in images. However, our purpose here is to look on it as another basic imaging operation. It can be implemented using the routine[*]:

```
THRESH:   [[ AO := if PO > thresh then 1 else 0 ]];   (2.10)
```

If, as very often happens, objects appear as dark objects on a light background, it is easier to visualize the subsequent binary processing

---

[*] The first few letters of the alphabet $(A, B, C, ...)$ are used consistently to denote binary image spaces, and later letters $(P, Q, R, ...)$ to denote grey-scale images (see Appendix A). In software these variables are assumed to be predeclared, and in hardware (e.g. frame store) terms they are taken to refer to dedicated memory spaces containing only the necessary 1 or 8 bits per pixel. The intricacies of data transfer between variables of different types are important considerations which are not addressed in detail here: it is sufficient to assume that both $AO := PO$ and $PO := AO$ correspond to a single-bit transfer, except that in the latter case the top 7 bits are assigned the value 0.

Fig. 2.2 Contrast-stretching: effect of increasing the contrast in the image of Fig. 2.1(a) by a factor of two and adjusting the mean intensity level appropriately. The interior of the jug can now be seen more easily. Note, however, that there is no additional information in the new image.

operations by inverting the thresholded image using a routine such as:

$$\text{INV:} \quad [[\text{ A0 } := 1 - \text{A0}]]; \qquad\qquad (2.11)$$

However, it would be more usual to combine the two operations into a single routine of the form:

$$\text{INVTHRESH:} \quad [[\text{ A0:} = \text{if P0} > \text{thresh then 0 else 1}]]; \quad (2.12)$$

To display the resulting image in a form as close as possible to the original, it can be re-inverted and given the full range of intensity values (intensity values 0 and 1 being scarcely visible):

$$\text{INVDISPLAY:} \quad [[\text{ R0 } := 255 * (1 - \text{A0})]]; \qquad (2.13)$$

Figure 2.3 shows the effect of these two operations.

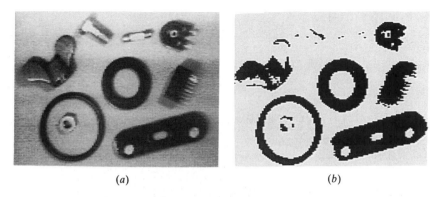

(a)   *(b)*

Fig. 2.3 Thresholding of grey-scale images: (a) 128 × 128 pixel grey-scale image of a collection of parts; (b) effect of thresholding the image.

2.2.2 Basic operations on binary images

Once the image has been thresholded, a wide range of binary imaging operations become possible. Only a few such operations are covered here, with the aim of being instructive rather than comprehensive. With this in mind, a routine may be written for shrinking dark thresholded objects (Fig. 2.4(a)) which are here represented by a set of 1's in a background of 0's:

```
SHRINK:  [[ sigma := A1 + A2 + A3 + A4 + A5 + A6 + A7 + A8;
            B0 := if  A0 = 0             then 0
                  else if sigma < 8  then 0
                                 else 1 ]];               (2.14)
```

In fact, the logic of this routine can be simplified to give the following more compact version:

```
SHRINK:  [[ sigma := A1 + A2 + A3 + A4 + A5 + A6 + A7 + A8;
            B0 := if sigma < 8 then 0 else A0 ]];          (2.15)
```

Note that the process of shrinking[*] dark objects also expands light objects, including the light background. It also expands holes in dark objects. The opposite process, that of expanding[*] dark objects (or shrinking light ones),

---

[*] The processes of shrinking and expanding are also widely known by the respective terms "erosion" and "dilation". (See also Appendix B.)

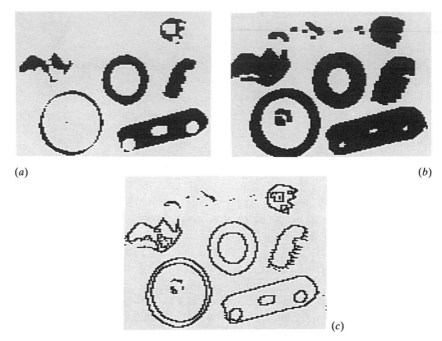

(a)   (b)

(c)

Fig. 2.4 Simple operations applied to binary images: (a) effect of shrinking the
dark-thresholded objects appearing in Fig. 2.3(b); (b) effect of expanding these dark
objects; (c) result of applying an edge location routine. Note that the SHRINK,
EXPAND and EDGE routines are applied to the *dark* objects: this implies that the
intensities are initially inverted as part of the thresholding operation and then
reinverted as part of the display operation (see text).

is achieved (Fig. 2.4(b)) with the routine:

```
EXPAND:   [[ sigma := A1 + A2 + A3 + A4 + A5 + A6 + A7 + A8;
              B0 := if sigma > 0 then 1 else A0 ]];           (2.16)
```

Each of these routines employs the same technique for interrogating
neighbouring pixels in the original image: as will be apparent on
numerous occasions in this book, the sigma value is a useful and
powerful descriptor for 3×3 pixel neighbourhoods. Thus "**if** sigma > 0"
can be taken to mean "**if** next to a dark object" and the consequence can
be read as "**then** expand it". Similarly, "**if** sigma < 8" can be taken to
mean "**if** next to a light object" or "**if** next to light background", and the

consequence can be read as "**then** expand the light background into the dark object".

The process of finding the edge of a binary object has several possible interpretations. Clearly, it can be assumed that an edge point has a sigma value in the range 1−7 inclusive. However, it may be defined as being within the object, within the background or in either position. Taking the definition that the edge of an object has to lie within the object (Fig. 2.4(c)), the following edge-finding routine for binary images results:

```
EDGE:  [[   sigma := A1 + A2 + A3 + A4 + A5 + A6 + A7 + A8;
            B0 := if sigma = 8 then 0 else A0 ]];                    (2.17)
```

This strategy amounts to cancelling out object pixels that are not on the edge. For this and a number of other algorithms (including the SHRINK and EXPAND algorithms already encountered), a thorough analysis of exactly which pixels should be set to 1 and 0 (or which should be retained and which eliminated), involves drawing up tables of the form:

ACTION   RECOGNITION

EDGE		sigma	
		0 − 7	8
A0	0	0	0
	1	1	0

SECTIONS (COLS)
S|B REVERSED
AS RECOGNITION
COMES BEFORE
ACTION

This reflects the fact that algorithm specification includes a recognition phase and an action phase, i.e. it is necessary first to locate situations within an image where (for example) edges are to be marked or noise eliminated, and then action must be taken to implement the change.

Another function that can usefully be performed on binary images is the removal of "salt and pepper" noise, i.e. noise which appears as a light spot on a dark background or a dark spot on a light background. The first problem to be solved is that of recognizing such noise spots; the second is the simpler one of correcting the intensity value. For the first of these tasks the sigma value is again useful. To remove salt noise (which has binary value 0 in our convention), we arrive at the following routine:

```
SALT:  [[   sigma := A1 + A2 + A3 + A4 + A5 + A6 + A7 + A8;
            B0 := if sigma = 8 then 1 else A0 ]];                    (2.18)
```

which can be read as leaving the pixel intensity unchanged unless it is proven to be a salt noise spot. The corresponding routine for removing pepper noise (binary value 1) is:

```
PEPPER:  [[ sigma := A1 + A2 + A3 + A4 + A5 + A6 + A7 + A8;
            B0 := if sigma = 0 then 0 else A0 ]];          (2.19)
```

Combining these two routines into one operation (Fig. 2.5(a)) gives:

```
NOISE:  [[  sigma := A1 + A2 + A3 + A4 + A5 + A6 + A7 + A8;
            B0 := if  sigma = 0       then 0
                 else if sigma = 8  then 1
                                     else A0 ]];             (2.20)
```

The routine can be made less stringent in its specification of noise pixels, so that it removes spurs on objects and background: this is achieved (Fig. 2.5(b)) by a variant such as:

```
NOIZE:  [[  sigma := A1 + A2 + A3 + A4 + A5 + A6 + A7 + A8;
            B0 := if  sigma < 2       then 0
                 else if sigma > 6  then 1
                                     else A0 ]];             (2.21)
```

As before, if there is any doubt about the algorithm, its specification should be set up rigorously—in this case with the following table:

NOIZE		sigma		
		0–1	2–6	7–8
A0	0	0	0	1
	1	0	1	1

There are many other simple operations that can usefully be applied to binary images and some of them are dealt with in Chapter 6. Meanwhile, it is worth considering the use of common logical operations such as NOT, AND, OR, EXOR. The NOT function is related to the inversion operation

(a)   *(b)*

Fig. 2.5 Simple binary noise removal operations: (a) result of applying a "salt and pepper" noise removal operation to the thresholded image in Fig. 2.3(b); (b) result of applying a less stringent noise removal routine: this is effective in cutting down the jagged spurs that appear on some of the objects.

described earlier, in that it produces the complement of a binary image:

$$\text{NOT:} \quad [[\text{ B0 } := \text{ not A0 }]]; \qquad\qquad (2.22)$$

This has the same action as

$$\text{INV:} \quad [[\text{ B0 } := 1 - \text{A0 }]]; \qquad\qquad (2.23)$$

The other logical operations operate on two or more images to produce a third image. This may be illustrated for the AND operation:

$$\text{AND:} \quad [[\text{ C0 } := \text{A0 and B0 }]]; \qquad\qquad (2.24)$$

This has the effect of masking off those parts of A-space that are not within a certain region defined in B-space (or locating those objects which are visible both in A-space and in B-space).

2.2.3   Noise suppression by image accumulation

The AND routine combines information from two or more images to generate a third image. It is frequently useful to average grey-scale images in a similar way in order to suppress noise. A routine for achieving this is:

$$\text{AVERAGE:} \quad [[\text{ R0 } := (\text{P0} + \text{Q0}) \text{ div } 2]]; \qquad\qquad (2.25)$$

This idea can be developed so that noisy images from a TV camera or other input device are progressively averaged until the noise level is low enough for useful application. It is, of course, well known that it is necessary to average 100 measurements to improve the signal-to-noise ratio by a factor of 10: i.e. the signal-to-noise ratio is multiplied only by the square root of the number of measurements averaged. Since whole-image operations are computationally costly, it is usually practicable to average only up to 16 images. To achieve this the routine shown in Table 2.1 is applied. Note that overflow from the usual single byte per pixel storage limit will prevent this routine from working properly if the input images do not have low enough intensities. Sometimes, it may be permissible to reserve more than one byte of storage per pixel. If this is not possible, a useful technique is to use the "learning" routine of Table 2.2. The action of this routine is somewhat unusual, as it retains one-sixteenth of the last image obtained from the camera, and progressively lower portions of all earlier images. The effectiveness of this technique can be seen in Fig. 2.6.

Table 2.1   Routine for averaging a set of input images.

```
ADDAVG:   begin
            [[ P0 := 0 ]];
            for i := 1 to 16 do
              begin
                input(Q); (* obtain new image in Q-space *)
                [[ P0 := P0 + Q0 ]];
              end;
            [[ P0 := P0 div 16 ]];
          end;
```

Table 2.2   Alternative 'learning' routine for averaging images.

```
LEARNAVG:   begin
              input(P); (* obtain new image in P-space *)
              for i := 1 to 15 do
                begin
                  input(Q); (* obtain new image in Q-space *)
                  [[ P0 := (P0 * 15 + Q0 + 8) div 16 ]];
                end;
              end;
```

ROUND OFF ERROR
PRE-ADD 1/2 OF FINAL $INT\left(\dfrac{\#}{16}\right)$ $5/8$ $INT\left(\dfrac{\#+8}{16}\right)$
UNIT TO PREVENT A
ROUND OFF ERROR

Fig. 2.6 Noise suppression by averaging a number of images using the LEARNAVG routine of Table 2.2. Note that this is highly effective at removing much of the noise and does not introduce any blurring (clearly this is only true if, as here, there is no camera or object motion). Compare this image with that of Fig. 2.1(a) for which no averaging was used.

Note that in this last routine 8 is added before dividing by 16 in order to offset rounding errors. If accuracy is not to suffer, this is a vital feature of most imaging routines which involve small-integer arithmetic. It is a surprising fact that this problem is seldom referred to in the literature of the subject, although it must account for many attempts at perfectly good algorithm strategies which mysteriously turn out not to work well in practice!

2.3 Convolutions and point spread functions

Convolution is a powerful and widely used technique in image processing and other areas of science. It appears in many applications throughout this book and it is therefore useful to introduce it at an early stage. We start by defining the convolution of two functions $f(x)$ and $g(x)$ as the integral:

$$f(x) * g(x) = \int_{-\infty}^{\infty} f(u)g(x-u)\,du \qquad (2.26)$$

The action of this integral is normally described as the result of applying a point spread function $g(x)$ to all points of a function $f(x)$ and accumulating the contributions at every point. It is significant that if the point spread function (PSF) is very narrow[*], then the convolution is identical to the original function $f(x)$. This makes it natural to think of the function $f(x)$ as having been spread out under the influence of $g(x)$. This argument may give the impression that convolution necessarily blurs the original function but this is not always so if, for example, the PSF has a distribution of positive and negative values.

When convolution is applied to digital images, the above formulation changes in two ways: (i) a double integral must be used in respect of the two dimensions; and (ii) integration must be changed into discrete summation. The new form of the convolution is:

$$F(x, y) = f(x, y) * g(x, y) = \sum_{i}\sum_{j} f(i,j)g(x-i, y-j) \qquad (2.27)$$

where g is now referred to as a spatial convolution mask. When convolutions are performed on whole images, it is usual to restrict the sizes of masks as far as possible in order to save computation. Thus convolution masks are not often larger than 15 pixels square, and 3×3 masks are typical. The fact that the mask has to be inverted before it is applied is inconvenient for visualizing the process of convolution. In this book we therefore present only preinverted masks of the form:

$$h(x, y) = g(-x, -y) \qquad (2.28)$$

Convolution can then be calculated using the more intuitive formula:

$$F(x, y) = \sum_{i}\sum_{j} f(x+i, y+j)h(i,j) \qquad (2.29)$$

---

[*] Formally, it can be a delta function, which is infinite at one point and zero elsewhere while having an integral of unity.

Clearly this involves multiplying corresponding values in the modified mask and the neighbourhood under consideration. Re-expressing this result for a 3×3 neighbourhood and writing the mask coefficients in the form:

$$\begin{bmatrix} h4 & h3 & h2 \\ h5 & h0 & h1 \\ h6 & h7 & h8 \end{bmatrix}$$

the algorithm can be obtained in terms of our earlier notation:

```
CONVOLV:  [[ Q0 := P0*h0 + P1*h1 + P2*h2 + P3*h3 + P4*h4
                  + P5*h5 + P6*h6 + P7*h7 + P8*h8 ]];
```
$$(2.30)$$

We are now in a position to apply convolution to a real situation. At this stage we merely extend the ideas of the previous section, attempting to

Fig. 2.7 Noise suppression by neighbourhood averaging achieved by convolving the original image of Fig. 2.1(a) with a uniform mask within a 3×3 neighbourhood. Note that noise is suppressed only at the expense of introducing significant blurring.

suppress noise by averaging not over corresponding pixels of different images, but over nearby pixels in the same image. A simple way of achieving this is to use the convolution mask:

$$\frac{1}{9}\begin{bmatrix} 1 & 1 & 1 \\ 1 & 1 & 1 \\ 1 & 1 & 1 \end{bmatrix}$$

where the number in front of the mask weights all the coefficients in the mask and is inserted to ensure that applying the convolution does not alter the mean intensity in the image. As hinted above, this particular convolution has the effect of blurring the image as well as reducing the noise level (Fig. 2.7). More will be said about this in the next chapter.

It will be clear from the above discussion that convolutions are linear operators. In fact, they are the most general spatially invariant linear operators that can be applied to a signal such as an image. Note that linearity is often of interest in that it permits mathematical analysis to be performed that would otherwise be intractable.

2.4   Sequential versus parallel operations

It will be noticed that most of the operations defined so far have started with an image in one space and finished with an image in a different space. Usually, P-space was used for an initial grey-scale image and Q-space for the processed image, or A-space and B-space for the corresponding binary images. This may seem somewhat inconvenient, since if routines are to act as interchangeable modules that can be applied in any sequence, it will be necessary to interpose restoring routines[*] between modules:

$$\text{RESTORE:} \quad [[\text{ PO } := \text{ QO }]]; \qquad\qquad (2.31)$$

Unfortunately, many of the operations will not work satisfactorily if we do not use separate input and output spaces in this way. This is because they are inherently "parallel processing" routines. This term is used as these are

---

[*] Restoring is not actually necessary within a sequence of routines if the computer is able to keep track of where the currently useful images are stored and to modify the routines accordingly: nonetheless, it is a particular problem when it is wished to see a sequence of operations as they occur, using a frame store with only one display space.

the types of process that would be performed by a parallel computer possessing a number of processing elements equal to the number of pixels in the image, so that all the pixels are processed simultaneously. If a serial computer is to *simulate* the operation of a parallel computer, then it must have separate input and output image spaces and rigorously work in such a way that it uses the original image values to compute the output pixel values. This means that an operation such as the following cannot be an ideal parallel process:

$$\text{BADSHRINK:} \quad [[\text{sigma} := A1 + A2 + A3 + A4 + A5 + A6 + A7 + A8;$$
$$A0 := \text{if sigma} < 8 \text{ then } 0 \text{ else } A0]]; \qquad (2.32)$$

This is so because, when the operation is half completed, the output pixel intensity will depend not only on some of the unprocessed pixel values but also on some that have already been processed. For example, if the computer makes a normal (forward) TV raster scan through the image, the situation at a general point in the scan will be

where the ticked pixels have already been processed and the others have not. As a result, the BADSHRINK routine will in fact shrink all objects to nothing!

A much simpler illustration of this is obtained by attempting to shift an image to the right using the following routine:

$$\text{BADSHIFT:} \quad [[P0 := P5]]; \qquad (2.33)$$

In fact, all this achieves is to fill up the image with values corresponding to those off its left edge[*], whatever they are assumed to be. Thus we have shown that the BADSHIFT process is inherently parallel.

---

[*] Note that when the computer is performing a 3×3 (or larger) window operation, it has to assume some value for off-image pixel intensities: usually whatever value is selected will be inaccurate, and so the final processed image will contain a border that is also inaccurate. This will be so whether the off-image pixel addresses are trapped in software or in specially designed circuitry in the frame store.

It will be seen later that there are some processes that are inherently sequential—i.e. in which the processed pixel has to be returned immediately to the *original* image space. Meanwhile, note that not all of the routines described so far need to be restricted rigorously to parallel processing. In particular, all single-pixel routines (essentially, those which only refer to the single pixel in a 1×1 neighbourhood) can validly be performed as if they were sequential in nature. Such routines include the following inverting, intensity adjustment, contrast stretching, thresholding, and logical operations:

INVERT:	`[[ P0 := 255 - P0 ]];`	(2.34)
BRIGHTEN:	`[[ P0 := P0 + beta ]];`	(2.35)
STRETCH:	`[[ P0 := P0 * gamma + beta ]];`	(2.36)
THRESH:	`[[ P0 := if P0 > thresh then 1 else 0 ]];`	(2.37)
NOT:	`[[ A0 := not A0 ]];`	(2.38)
AND:	`[[ A0 := A0 and B0 ]];`	(2.39)

These remarks are here intended to act as a warning. It may well be safest to design algorithms that are exclusively parallel processes unless there is a definite need to make them sequential. It will be seen later how this need can arise.

2.5 Concluding remarks

This chapter has introduced a compact notation for representing imaging operations and has demonstrated some basic parallel processing routines. The following chapter extends this work to see how noise suppression can be achieved in grey-scale images. This leads on to more advanced image analysis work that is directly relevant to machine vision applications. In particular, Chapter 4 studies in more detail the thresholding of grey-scale images, building on the work of Section 2.2.1, while Chapter 6 studies object shape analysis in binary images.

2.6 Bibliographical and historical notes

Since the aim of this chapter was not to cover the most recent material but to provide a succinct overview of basic techniques, it will not be surprising that most of the topics discussed were discovered well over a decade ago and have been used by a large number of workers in many areas. For example, thresholding of grey-scale images was first reported at least as long ago as 1960, while shrinking and expanding of binary picture objects dates from a similar period. Averaging of a number of images to suppress noise has been reported many times; although the learning routine described here has not been found in the literature, it has long been used in the author's laboratory and it is probable that it has been developed independently elsewhere. Discussion of the origins of other techniques is curtailed: for further detail the reader is referred to the texts by (for example) Duda and Hart (1973), Pratt (1978), Castleman (1979), Hall (1979) and Rosenfeld and Kak (1981). More specialized texts will be referred to in the following chapters.

2.7 Problems

1. Derive an algorithm for finding the edges of binary picture objects by applying a SHRINK operation and combining the result with the original image. Is the result the same as that obtained using the edge-finding routine (2.17)? Prove your statement rigorously by drawing up suitable algorithm tables as in Section 2.2.2.
2. In a certain frame store, each off-image pixel can be taken to have either the value 0 or the intensity for the nearest image pixel. Which will give the more meaningful results for (a) shrinking, (b) expanding and (c) a blurring convolution?
3. Suppose the NOISE and NOIZE routines of Section 2.2.2 were reimplemented as sequential algorithms. Show that the action of NOISE would be unchanged whereas NOIZE would produce very odd effects on some binary images.

3

Basic Image Filtering Operations

3.1 Introduction

Chapter 2 was concerned with simple imaging operations, including such problems as thresholding grey-scale images and suppressing noise in binary images. In this chapter the discussion is extended to noise suppression and enhancement in grey-scale images. Although these types of operation can for the most part be avoided in industrial applications of vision, it is useful to examine them in some depth because of their wide use in a variety of other image processing applications and because they set the scene for much of what follows. In addition, some fundamental issues come to light which are of vital importance.

It has already been seen that noise can arise in real images and it is hence necessary to have sound techniques for suppressing it. Commonly, in electrical engineering applications, noise is removed by means of low-pass or other filters which operate in the frequency domain (Davies, 1993a). Applying these filters to 1-D time-varying analogue signals is straightforward, since it is necessary only to place them at suitable stages in the sequence of black boxes through which the signals pass. For digital signals the situation is more complicated, since the frequency transform of the signal must first be computed, then the low-pass filter applied, and finally the signal obtained from the modified transform by converting back to the time domain. Thus two Fourier transforms have to be computed, though modifying the signal while it is in the frequency domain is a straightforward task (Fig. 3.1). It is by now well known that the amount of processing involved in computing a discrete Fourier

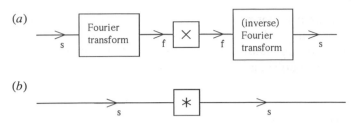

Fig. 3.1 Low-pass filtering for noise suppression: s, spatial domain; f, spatial frequency domain; ×, multiplication by low-pass characteristic; ∗, convolution with Fourier transform of low-pass characteristics. (a) Low-pass filtering achieved most simply, by a process of multiplication in the (spatial) frequency domain; (b) low-pass filtering achieved by a process of convolution. Note that (a) may require more computation overall, because of the two Fourier transforms that have to be performed.

transform of a signal represented by N samples is of order N^2 (we shall write this as $O(N^2)$), and that the amount of computation can be cut down to $O(N \log_2 N)$ by employing the fast Fourier transform (FFT) (Gonzalez and Wintz, 1987). This then becomes a practical approach for the elimination of noise.

When applying these ideas to images we must first note that the signal is a spatial rather than a time-varying quantity and must be filtered in the spatial frequency domain. Mathematically this makes no real difference but there are nevertheless significant problems. First, there is no satisfactory analogue short-cut and the whole process has to be carried out digitally (we here ignore optical processing methods despite their obvious power, speed and high resolution, because they are on the whole inflexible, bulky and difficult to marry cheaply with current digital computer technology; however, there is much ongoing research in this area and the situation will change radically over the next decade). Second, for an $N \times N$ pixel image, the number of operations required to compute a Fourier transform is $O(N^3)$ and the FFT only reduces this to $O(N^2 \log_2 N)$, so the amount of computation is quite considerable (here it is assumed that the 2-D transforms are implemented by successive passes of 1-D transforms: see Gonzalez and Wintz, 1987). Note also that two Fourier transforms are required for the purpose of noise suppression (Fig. 3.1). Nevertheless, in many imaging applications it is worth proceeding in this way, not only so that noise can be removed but also so that TV scan lines and other artefacts can be filtered out. This situation applies particularly in remote sensing and space technology. However, in industrial applications the emphasis is always on real-time processing, so in many cases it is not practicable to remove noise by spatial frequency domain operations. A secondary problem is that low-pass filtering is suited to removing Gaussian noise but distorts the image if it is used to remove impulse noise, as will be seen below.

3.2   Noise suppression by Gaussian smoothing

Low-pass filtering is normally thought of as the elimination of signal components with high spatial frequencies and it is therefore natural to carry it out in the spatial frequency domain. Nevertheless, it is possible to implement it directly in the spatial domain. That this is possible is due to the well-known fact (Rosie, 1966) that multiplying a signal by a function in the spatial frequency domain is equivalent to convolving it with the Fourier transform of the function in the spatial domain (Fig. 3.1). If the final convolving function in the spatial domain is sufficiently narrow then the amount of computation involved will not be excessive: in this way a satisfactory implementation of the low-pass filter can be sought. It now remains to find a suitable convolving function.

If the low-pass filter is to have a sharp cut-off, then its transform in image space will be oscillatory: an extreme example of this is the sinc (sin x/x) function, which is the spatial transform of a low-pass filter of rectangular profile (Rosie, 1966). It turns out that oscillatory convolving functions are unsatisfactory since they can introduce halos around objects, hence distorting the image quite grossly. Marr and Hildreth (1980) suggested that the right types of filter to apply to images are those which are well-behaved (nonoscillatory) both in the frequency and in the spatial domain. Gaussian filters are able to fulfil this criterion optimally: they have identical forms in the spatial and spatial frequency domains. In 1-D these forms are

$$f(x) = [1/(2\pi\sigma^2)^{1/2}] \exp [-x^2/(2\sigma^2)] \qquad (3.1)$$

$$F(\omega) = \exp (-\sigma^2\omega^2/2) \qquad (3.2)$$

Thus the type of spatial convolving operator required for the purpose of noise suppression by low-pass filtering is one which approximates to a Gaussian profile. Many such approximations appear in the literature: these vary with the size of the neighbourhood chosen and in the precise values of the convolution mask coefficients.

One of the most common is the following mask, first introduced in Chapter 2, which is used more for simplicity of computation than for its fidelity to a Gaussian profile:

$$\frac{1}{9}\begin{bmatrix} 1 & 1 & 1 \\ 1 & 1 & 1 \\ 1 & 1 & 1 \end{bmatrix}$$

Another commonly used mask, which approximates more closely to a Gaussian profile, is the following:

$$\frac{1}{16}\begin{bmatrix} 1 & 2 & 1 \\ 2 & 4 & 2 \\ 1 & 2 & 1 \end{bmatrix}$$

In both cases the coefficients that precede the mask are used to weight all the mask coefficients: as mentioned in Section 2.2.3, these weights are chosen so that applying the convolution to an image does not affect the average image intensity. These two convolution masks probably account for over 80% of all discrete approximations to a Gaussian. Notice that as they operate within a 3×3 neighbourhood, they are reasonably narrow and hence incur a relatively small computational load.

Let us next study the properties of this type of operator, deferring for now consideration of Gaussian operators in larger neighbourhoods. First, imagine such an operator applied to a noisy image whose intensity is inherently uniform. Then clearly noise is suppressed, as it is now averaged over 9 pixels. This averaging model is obvious for the first of the two masks above but in fact applies equally to the second mask, once it is accepted that the averaging effect is differently distributed in accordance with the improved approximation to a Gaussian profile.

Although this example shows that noise is suppressed, it will be plain that the signal is also affected. This problem arises only where the signal is initially nonuniform: indeed, if the image intensity is constant, or if the intensity map approximates to a plane, there is again no problem. However, if the signal is uniform over one part of a neighbourhood and rises in another part of it, as is bound to occur adjacent to the edge of an object, then the object will make itself felt at the centre of the neighbourhood in the filtered image (see Fig. 3.2). As a result, the edges of objects become somewhat blurred. Looking at the operator as a "mixing

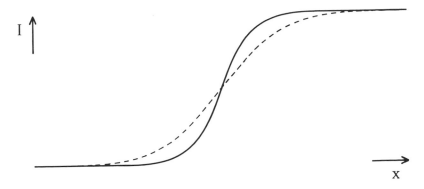

Fig. 3.2 Blurring of object edges by simple Gaussian convolutions. The simple Gaussian convolution can be regarded as a grey-scale neighbourhood "mixing" operator, hence explaining why blurring arises.

operator" which forms a new picture by mixing together the intensities of pixels fairly close to each other, it is intuitively obvious why blurring occurs.

It is also apparent from a spatial frequency viewpoint why blurring should occur. Basically, we are aiming to give the signal a sharp cut-off in the spatial frequency domain, and as a result it will become slightly blurred in the spatial domain. Clearly, the blurring effect can be reduced by using the narrowest possible approximation to a Gaussian convolution filter but at the same time the noise suppression properties of the filter are lessened. Assuming that the image was initially digitized at roughly the correct spatial resolution, it will not normally be appropriate to smooth it using convolution masks larger than 3×3 or at most 5×5 pixels (here we ignore methods of analysing images that use a number of versions of the image with different spatial resolutions: see for example, Babaud *et al.*, 1986).

Overall, low-pass filtering and Gaussian smoothing are not appropriate for the applications considered here, because of the blurring effects they introduce. Notice also that where interference occurs which can give rise to impulse or "spike" noise (corresponding to a number of individual pixels having totally the wrong intensities), merely averaging this noise over a larger neighbourhood can make the situation worse, since the spikes will be smeared over a sizeable number of pixels and will distort the intensity values of all of them. This consideration is important as it leads naturally to the concepts of limit and median filtering.

3.3 Median filtering

The idea explored here is to locate those pixels in the image which have extreme and therefore highly improbable intensities and to ignore their actual intensities, replacing them with more suitable values. This is akin to drawing a graph through a set of plots and ignoring those plots that are evidently a long way from the best-fit curve. An obvious way of achieving this is to apply a "limit" filter which prevents any pixel having an intensity outside the intensity range of its neighbours:

```
LIMIT:  [[ minP := min(P1,P2,P3,P4,P5,P6,P7,P8);
           maxP := max(P1,P2,P3,P4,P5,P6,P7,P8);
           Q0 := if  P0 < minP        then minP
                  else if P0 > maxP  then maxP
                                       else P0 ]];        (3.3)
```

To develop this technique it is necessary to examine the local intensity distribution within a particular neighbourhood. Points at the extremes of the

Fig. 3.3 Effect of applying a 3 × 3 median filter to the image of Fig. 2.1(a). Note the slight loss of fine detail and the rather "softened" appearance of the whole image.

distribution are quite likely to have arisen from impulse noise. Not only is it sensible to eliminate these points, as in the limit filter, but it is reasonable to try taking the process further, removing equal areas at either end of the distribution and ending with the median. Thus we arrive at the median filter, which takes all the local intensity distributions and generates a new image corresponding to the set of median values. As the preceding argument indicates, the median filter is excellent at impulse noise suppression and this is amply confirmed in practice (see Fig. 3.3).

In view of the blurring caused by Gaussian smoothing operators, it is pertinent to ask whether the median filter also induces blurring. In fact, Fig. 3.3 shows that any blurring is only marginal, although there is some slight loss of fine detail which can give the resulting pictures a "softened" appearance. Theoretical discussion of this point is deferred for now; the lack of blur makes good the main deficiency of the Gaussian smoothing filter and results in the median filter being almost certainly the single most widely used filter in general image processing applications.

There are many ways of implementing the median filter: Table 3.1 reproduces only an obvious algorithm which essentially implements the above description. The notation of Chapter 2 is used but is augmented suitably to permit the 9 pixels in a 3×3 neighbourhood to be accessed in turn with a running suffix (Appendix A).

Table 3.1   An implementation of the median filter.

```
MEDIAN:   begin
            for i := 0 to 255 do hist[i] := 0;
            [[ for m := 0 to 8 do hist[ P[m] ] := hist[ P[m] ] + 1;
              i := 0; sum := 0;
              while sum < 5 do
                begin
                  sum := sum + hist[i];
                  i := i + 1;
                end;
              Q0 := i - 1;
              for m := 0 to 8 do hist[ P[m] ] := 0 ]];
            end;
```

The operation of the algorithm is as follows: first, the histogram array is cleared and then the image is scanned, generating a new image in Q-space; for each neighbourhood, the histogram of intensity values is constructed; then the median is found; finally, points in the histogram array that have been incremented are cleared. This last feature eliminates the need to clear the whole histogram and hence saves computation. Further savings could be made by finding the extreme intensity values and limiting the search for the median to this range. Unlike the general situation when the median of a distribution is being located, only one (half-) scan through the distribution is required, since the total area is known in advance (in this case it is 9).

As is clear from the above, methods of computing the median involve pixel intensity sorting operations. If a bubble sort (Gonnet, 1984) were used for this purpose, then up to $O(n^4)$ operations would be required for an $n \times n$ neighbourhood, compared with some 256 operations for the histogram method described above. Thus sorting methods such as the bubble sort are faster for small neighbourhoods where n is 3 or 4 but not for neighbourhoods where n is greater than about 5, or where pixel intensity values are more restricted.

Much of the discussion of the median filter in the literature is concerned with saving computation (Narendra, 1978; Huang et al., 1979; Danielsson, 1981; Davies, 1992a). In particular, it has been noticed that, on proceeding from one neighbourhood to the next, relatively few new pixels are encountered: this means that the new median value can be found by updating the old value rather than starting from scratch (Huang et al., 1979).

3.4 Mode filtering

Having considered the mean and the median of the local intensity distribution as candidate intensity values for noise smoothing filters, it also seems relevant to consider the mode of the distribution. Indeed, we might imagine that this is if anything more important than the mean or the median, since the mode represents the most probable value of any distribution.

However, a tedious problem arises as soon as we attempt to apply this idea. The local intensity distribution is calculated from relatively few pixel intensity values (Fig. 3.4). This means that instead of a smooth intensity distribution whose mode is easily located, we are almost certain to have a multimodal distribution whose highest point does not indicate the position of the *underlying* mode. Clearly the distribution needs to be smoothed out considerably before the mode is computed. Another tedious problem is that the width of the distribution varies widely from neighbourhood to neighbourhood (e.g. from close to zero to close to 256), so that it is difficult to know quite how much to smooth the distribution in any instance. For these reasons, it is likely to be better to choose an indirect measure of the position of the mode rather than to attempt to measure it directly.

Fig. 3.4 The sparse nature of the local intensity histogram for a small neighbour-hood. This situation clearly causes significant problems for estimation of the mode. It also has definite implications for rigorous estimation of the underlying median, assuming that the observed intensities are only noisy samples of the ideal intensity pattern (see Section 3.5.4).   © *1984 IEEE.*

The author has found that the position of the mode can be estimated with reasonable accuracy once the median has been located (Davies, 1984a, 1988c). To understand the technique it is necessary to consider how local intensity distributions of various sorts arise in practical situations. At most positions in an image, variations in pixel intensity are generated by steady

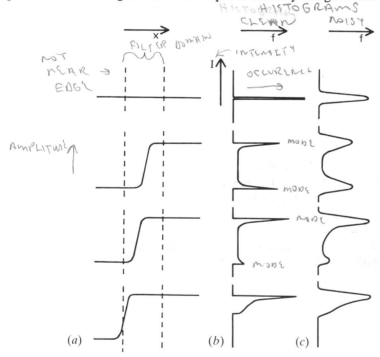

Fig. 3.5 Local models of image data near the edge of an object: (a) cross-sections of an edge falling in the vicinity of a filter neighbourhood; (b) corresponding local intensity distributions when very little image noise is present; (c) situation when the noise level is increased.   © *1984 IEEE.*

changes in background illumination, or by steady variations in surface orientation, or else by noise. Thus a symmetrical unimodal local intensity distribution is to be expected. It is well known that the mean, median and mode are coincident in such cases. More problematic is what happens to the intensity variation near the edge of an object in the image. Here the local intensity distribution is unlikely to be symmetrical and, more important, it may not even be unimodal. In fact, near an edge the distribution is in general inherently *bimodal*, since the neighbourhood contains pixels with intensities corresponding to the values they would have on either side of the edge (Fig. 3.5). Considering the image as a whole, this will be the most likely alternative to a symmetrical unimodal distribution, any further possibilities such as trimodal distributions being rare and of varied causes (e.g. odd glints on the edges of metal objects) which are outside the scope of the present discussion[*].

If the neighbourhood straddles an edge and the local intensity distribution is bimodal, the larger peak position should clearly be selected as the most probable intensity value. A good strategy for finding the larger peak is to

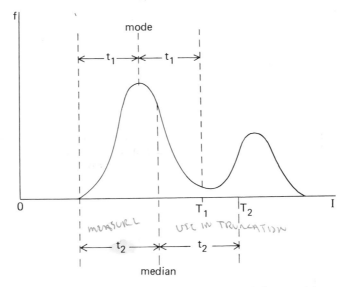

Fig. 3.6 Rationale for the method of truncation. The obvious position at which to truncate the distribution is T_1. Since the position of the mode is not initially known, it is suboptimal but safe to truncate instead at T_2.

---

[*] Here we are ignoring the effects of noise and just considering the underlying image signal.

eliminate the smaller peak. If we knew the position of the mode we could find where to truncate the smaller peak by first finding which extreme of the distribution was closer to the mode, and then moving an equal distance to the opposite side of the mode (Fig. 3.6). Since we start off *not* knowing the position of the mode, one option is to use the position of the median as an estimator of the position of the mode, and then to use that position to find where to truncate the distribution. Since it invariably happens that the three means take the order mean, median and mode (see Fig. 3.7), except when distributions are badly behaved or multimodal, it turns out that this method is cautious in the sense that it truncates less of the distribution than the required amount: this makes it a safe method to use. When we now find the median of the truncated distribution, the position is much closer to the mode than the original median was, a good proportion of the second peak having

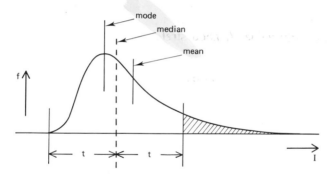

Fig. 3.7   Relative positions of the mode, median and mean for a typical unimodal distribution. This ordering is unchanged for a bimodal distribution, as long as it can be approximated by two Gaussian distributions of similar width.   © *1984 IEEE*.

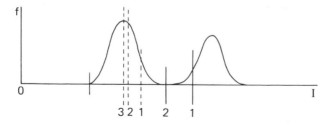

Fig. 3.8   Iterative truncation of the local intensity distribution. Here the median converges on the mode within three iterations of the truncation procedure. This is possible since at each stage the mode of the new truncated distribution remains the same as that of the previous distribution.

Fig. 3.9   Effect of a single application of 3×3 truncated median filter to the image of Fig. 2.1(a).

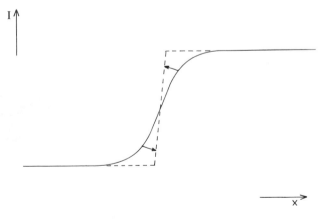

Fig. 3.10   Image enhancement performed by the mode filter. Here the onset of the edge is pushed laterally by the action of the mode filter within one neighbourhood; since the same happens from the other side within an adjacent neighbourhood, the actual position of the edge is unchanged in first order. The overall effect is to sharpen the edge.

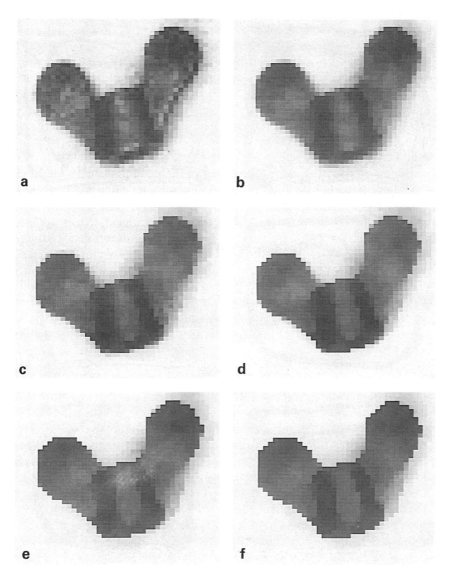

Fig. 3.11   Results of repeated action of the truncated median filter: (a) the original, moderately noisy picture; (b) effect of a 3×3 median filter; (c)–(f), effect of 1–4 passes of the basic truncated median filter, respectively.   © *1984 IEEE*.

been removed (Fig. 3.8). Iteration could be used to find an even closer approximation to the position of the mode. However, the results show that the method gives a marked enhancement in the image even when this is not done (Fig. 3.9).

It is worth enquiring more closely what has been achieved by the above procedure. Comparing the results of applying a median filter and the new filter (which we call a "truncated median filter" as it is not a true mode filter—see below), whereas the median filter is highly successful at removing noise the new filter not only removes noise but also enhances the image so that edges are made sharper. It is quite easy to see why this should happen, by reference to Fig. 3.10. Basically, at a location even very slightly to one side of an edge, a majority of the pixel intensities contribute to the larger peak and the filter ignores the pixel intensities contributing to the smaller peak. Thus the filter makes an informed binary choice about which side of the edge it is on. At first this seems to mean that it pushes a nearby edge further away. However, it must be remembered that it actually "pushes the edge away" from both sides, and the result is that its sides are made sharper and object outlines are crispened up. Particularly striking is the effect of applying this filter to an image a number of times, when objects start to become segmented into regions of fairly uniform intensity (Fig. 3.11). The complete algorithm for achieving this is outlined in Table 3.2.

This problem has been dealt with at some length for a number of reasons. First, it seems that the mode filter has not hitherto received the attention it deserves. Second, the median filter seems to be used fairly universally, often without very much justification or thought. Third, all these filters show what markedly different characteristics are available merely by analysing the contents of the local intensity distribution, and ignoring totally where in the neighbourhood the different intensities appear: it is perhaps remarkable that there is sufficient information in the local intensity distribution for this to be possible. All this shows the danger of applying operators that have been derived in an *ad hoc* manner without first making a specification of what is required and then designing an operator with the required characteristics. In fact, the situation appears to be that if we want a filter which has maximum impulse noise suppression capability, then we should use a median filter; and if we want a filter which enhances images by sharpening edges, then we should use a mode or truncated median filter (note that the truncated median filter should in some ways be an improvement on the mode filter in that it is more cautious very close to an edge transition, where noise prevents an exact judgement being made as to which side of the edge a pixel is on: see Davies, 1984a, 1988c).

Table 3.2   Outline of algorithm for implementing the truncated median filter.

```
repeat (* as many passes over image as necessary *)
  repeat (* for each pixel *)
    compute local intensity distribution;
    repeat (* iterate to improve estimate of mode *)
      find minimum, median and maximum intensity values;
      decide from which end local intensity distribution should be
        truncated;
      deduce where local intensity distribution should be truncated;
      truncate local intensity distribution;
      find median of truncated local intensity distribution;
    until median sufficiently close to mode of local distribution;
    transfer estimate of mode to output image space;
  until all pixels processed;
until sufficient enhancement of image;
```

Comments:
  (i) The outermost and innermost loops can normally be omitted (i.e. they need to be executed once only).
  (ii) The *final* estimate of the position of the mode can be performed by simple averaging instead of computing the median: this has been found to save computation with negligible loss of accuracy.
  (iii) Instead of the minimum and maximum intensity values, the positions of the outermost octiles (for example) may be used to give more stable estimates of the extremes of the local intensity distribution.

While considering enhancement, attention has been restricted to filters based on the local intensity distribution: there are many filters that enhance images without the aid of the local intensity distribution (Lev *et al.*, 1977; Nagao and Matsuyama, 1979) but they are not within the scope of this chapter. Note that the method of "sharp–unsharp masking" (Section 3.8) performs an enhancement function, although its main purpose is to restore images that have inadvertently become blurred, e.g. by a hazy atmosphere or defocused camera.

3.5   Bias generated by noise suppression filters

Despite knowing the main characteristics of the different types of filter there are still some unknown factors. In particular, it is often important (especially when making precision measurements on manufactured components) to ensure that noise is removed in such a way that object locations and sizes are unchanged. Clearly, it is necessary to examine how the various filters satisfy this criterion. In fact we seem to have answered this question already, in pointing out that the mode filter operates by pushing back edges equally from both sides. Not only does this filter not (to first order) move the centre of an edge but also it sharpens the edge so that it is actually easier to measure its position accurately. However, there are two problems with this simple answer.

First, it has been assumed that the intensity profile of an edge is symmetrical. If this is so then the mean, median and mode of the local intensity distribution will be coincident and there will clearly be no overall bias for any of them. However, when the edge profile is asymmetrical it will not be obvious in the absence of a detailed model of the situation what the result will be for any of the filters. The situation is even more involved when significant noise is present (Yang and Huang, 1981; Bovik *et al.*, 1987). Since the problem is so data-dependent it is not profitable to consider it further here.

The other problem concerns the situation for a curved edge. In this case there is again a variety of possibilities, and filters employing the different means will modify the edge position in ways that depend markedly on its shape. In robot vision applications the median filter is the one we are most likely to use, because its main purpose is to suppress noise without introducing blurring. Hence it is worth considering in some detail the bias produced by this type of filter: this is done in the next subsection.

3.5.1   Theory of edge shifts caused by median filters in binary images

This section takes the case of a continuous image (i.e. a nondiscrete lattice), assuming first (i) that the image is binary, (ii) that neighbourhoods are exactly circular, and (iii) that images are noise-free. To proceed we notice that binary edges have symmetrical cross-sections, while straight edges extend this symmetry into 2-D: hence applying a median filter in a (symmetrical) circular neighbourhood cannot pull a straight edge to one side or the other.

Now consider what happens when the filter is applied to an edge that is not straight. If, for example, the edge is circular, the local intensity distribution will contain two peaks whose relative sizes will vary with the precise position of the neighbourhood (Fig. 3.12). At some position the sizes of the two peaks will be identical. Clearly, this happens when the centre of the neighbourhood is at a unique distance from the centre of a circular object: this is the position at which the output of the median filter changes from dark to light (or vice versa). Thus the median filter produces an

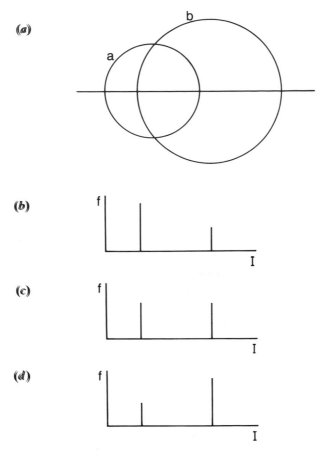

Fig. 3.12 Variation in local intensity distribution with position of neighbourhood: (a) neighbourhood of radius a overlapping a dark circular object of radius b; (b)–(d) intensity distributions I when the separations of the centres are respectively less than, equal to, or greater than the distance d for which the object bisects the area of the neighbourhood.

inwards shift towards the centre of a circular object (or the centre of curvature), whether the object is dark on a light background or light on a dark background.

Next suppose that the edge is irregular with several "bumps" (i.e. prominences or indentations) within the filter neighbourhood: clearly the filter will now tend to average out the bumps and straighten the edge, since it acts in such a way as to form a boundary on which the amounts of dark

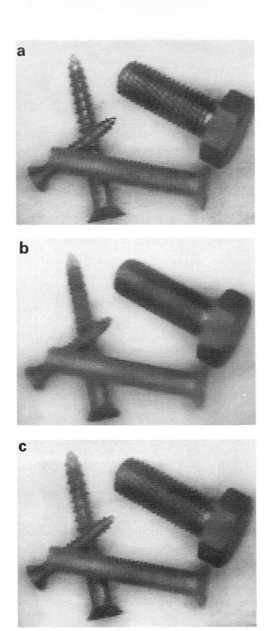

Fig. 3.13 Edge smoothing property of the median filter. (a) Original image, (b) median filter smoothing of irregularities, in particular those around the boundaries (notice how the threads on the screws are virtually eliminated although detail larger in scale than half the filter area is preserved), using a 21-element filter operating within a 5×5 neighbourhood on a 128×128 pixel image of 6-bit grey-scale; (c) effect of a new type of "detail-preserving" filter (see Section 3.5.3).

and light within the neighbourhood are equalized (Fig. 3.13). This means that the edge will be locally biased but only by a reduced amount, since the various bumps will tend to pull the final edge in opposite directions. On the other hand, if there is one gross bump within the neighbourhood—i.e. if the curvature has the same sign and is roughly constant at all points on the edge within the filter neighbourhood—then all these parts of the edge will act in concert and it will be pulled sideways a significant amount by the filter. Thus a circular section of the boundary constitutes a worst-case situation, for which the filter produces the largest bias in the position of the edge. It is clearly worth finding the size of the worst-case shift and for this reason we concentrate attention below on circular objects, in the knowledge that all other shapes will give less serious shifts and distortions.

The worst-case calculation is a matter of elementary geometry: we need to find at what distance d from the centre of a circular object (of radius b) the area of a circular neighbourhood (of radius a) is bisected by the object boundary.

From Fig. 3.14 the area of the sector of angle 2β is βb^2, while the area of the triangle of angle 2β is $b^2 \sin \beta \cos \beta$. Hence the area of the segment shown shaded is:

$$B = b^2(\beta - \sin \beta \cos \beta) \qquad (3.4)$$

Making a similar calculation of the area A of a circular segment of radius a and angle 2α, the area of overlap (Fig. 3.14) between the circular neighbourhood of radius a and the circular object of radius b may be deduced as:

$$C = A + B \qquad (3.5)$$

For a median filter this is equal to $\pi a^2/2$.

Hence:

$$F = a^2(\alpha - \sin \alpha \cos \alpha) + b^2(\beta - \sin \beta \cos \beta) - \pi a^2/2 = 0 \qquad (3.6)$$

where

$$a^2 = b^2 + d^2 - 2bd \cos \beta \qquad (3.7)$$

and

$$b^2 = a^2 + d^2 - 2ad \cos \alpha \qquad (3.8)$$

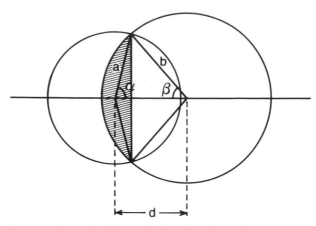

Fig. 3.14   Geometry for calculating neighbourhood and object overlap.

To solve this set of equations, we take a given value of d, deduce values of α and β, calculate the value of F and then adjust the value of d until $F = 0$. Since d is the modified value of b obtained after filtering, the shift produced by the filtering process is

$$D = b - d \qquad (3.9)$$

The results of doing this computation numerically have been found by Davies (1989b). As expected, $D \rightarrow 0$ as $b \rightarrow \infty$ or $a \rightarrow 0$. Conversely, the shift becomes very large as a first approaches and then exceeds b. Note, however, that when $a > \sqrt{2}b$ the object is ignored, being small enough to be regarded as irrelevant noise by the filter: beyond this point it has no effect at all on the final image. The maximum edge shift before the object finally disappears is $(2 - \sqrt{2})b \approx 0.586b$.

It is instructive to approximate the above equations for the case when edge curvature is small, i.e. $a \ll b$. Under these conditions, β is small, $a \approx \pi/2$ and $d \approx b$. Hence we find:

$$\beta \approx a/b \qquad (3.10)$$

After some manipulation the edge shift D is obtained in the form:

$$D \approx a^2/6b = \kappa a^2/6 \qquad (3.11)$$

$\kappa = 1/b$ being the local curvature. In Chapter 13 this equation is found to be useful for estimating the signals from a median-based corner detector.

3.5.2   Edge shifts caused by median filters in grey-scale images

To extend these results to grey-scale images, first consider the effect of applying a median filter near a smooth step edge in 1-D. Here the median filter gives zero shift, since for equal distances from the centre to either end of the neighbourhood there are equal numbers of higher and lower intensity values and hence equal areas under the corresponding portions of the intensity histogram. Clearly this is always valid where the intensity increases monotonically from one end of the neighbourhood to the other—a property first pointed out by Gallagher and Wise (1981) (for more recent discussions on related "root" (invariance) properties of signals under median filtering, see Fitch *et al.*, 1985; Heinonen and Neuvo, 1987).

Next, it is clear that for 2-D images, the situation is again unchanged in the vicinity of a straight edge, since the situation remains highly symmetrical. Hence the median filter gives zero shift, as in the binary case.

For curved edges, it again turns out that circular boundaries constitute a worst case that should be considered carefully. However, grey-scale edges are unlike binary edges in that they have finite slope. This means that it is necessary to take account of the exact form of the intensity function within the neighbourhood.

When boundaries are roughly circular, contours of constant intensity often appear as in Fig. 3.15. To find how a median filter acts we merely need to identify the contour of median intensity (in 2-D the median intensity value labels a whole contour) which divides the area of the neighbourhood into two equal parts. The geometry of the situation is identical to that already examined in Section 3.5.1: the main difference here is that for every position of the neighbourhood, there is a corresponding median contour with its own particular value of shift depending on the curvature. Intriguingly, the formulae already deduced may immediately be applied for calculating the shift for each contour. Figure 3.15 shows an idealized case in which the contours of constant intensity have similar curvature, so that they are all moved inwards by similar amounts. This means that, to a first approximation, the edges of the object retain their cross-sectional profile as it becomes smaller.

For grey-scale images the shifts predicted by this theory (with certain additional corrections: see Davies, 1989b) agree with experimental shifts within approximately 10% for a large range of circle sizes in a discrete lattice (see Fig. 3.16). Paradoxically, the agreement is less perfect for binary images, since circles of certain sizes show stability effects (akin to median root behaviour): these effects tend to average out for grey-scale images, owing to the presence of many contours of different sizes at different grey

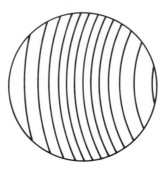

Fig. 3.15 Contours of constant intensity on the edge of a large circular object, as seen within a small circular neighbourhood.

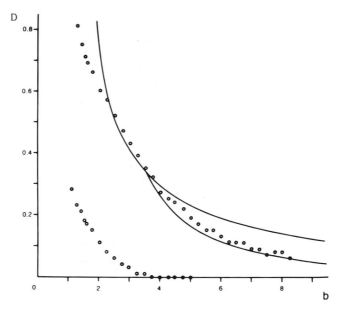

Fig. 3.16 Edge shifts for 5×5 median filter applied to a grey-scale image. The upper set of plots represents the experimental results and the upper continuous curve is derived from the theory of Section 3.5.1. The lower continuous curve is derived from a more accurate model (Davies, 1989b). The lower set of plots represents the much reduced shifts obtained with the detail-preserving type of filter (see Section 3.5.3).

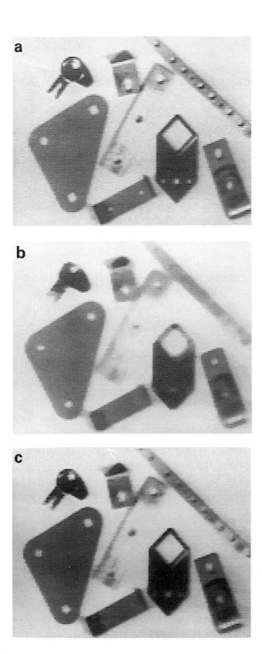

Fig. 3.17 Circular holes in metal objects before and after filtering: (a) original 128 × 128 pixel image with 6-bit grey-scale; (b) 5 × 5 median-filtered image: the diminution in size of the hole is clearly visible and such distortions would have to be corrected for when taking measurements from real filtered images of this type; (c) result using a detail-preserving filter: some distortions are present although the overall result is much better than in (b).

levels. Overall, it appears that the edge shifts obtained with median filters are now quite well understood. Figures 3.13 and 3.17 give some indication of the magnitudes of these shifts in practical situations. Note that once image detail such as a small hole or screw thread has been eliminated by a filter, it is not possible to apply any edge shift correction formula to recover it, although for larger features such formulae are useful for deducing true edge positions.

3.5.3 Edge shifts arising with hybrid median filters

Although median filters preserve edges in digital images, they are also known to remove fine image detail such as lines. For example, 3×3 median filters remove lines 1 pixel wide, and 5×5 median filters remove lines 2 pixels wide. In many applications such as remote sensing and X-ray imaging this is exceedingly important and efforts have been made to develop filters that overcome the problem. In 1987 Nieminen *et al.* reported a new class of "detail preserving" filters: these employ linear subfilters whose outputs are combined by median operations. There is a great variety of such filters, employing different subfilter shapes and having the possibility of several layers of median operations. Hence it is not possible to describe them fully here in the space available. Although these filters are aimed particularly at retention of line detail, and are readily understood in this context, they turn out to have some corner preserving properties and to be resistant to the edge shifts that arise when there is a nonzero curvature.

Perhaps the best of the filters in the new class, from the point of view of preserving edge position, is the two-level "bidirectional" linear-median hybrid filter termed "2LH+" (Nieminen *et al.*, 1987). Its operation in a 5×5 neighbourhood may be illustrated as follows. It employs the subfilters A–I in the 5×5 region:

E	—	D	—	C
—	E	D	C	—
F	F	A	B	B
—	G	H	I	—
G	—	H	—	I

pixels marked as being in the same subfilter having their intensities averaged, and dashed pixels being ignored. Nonlinear filtering then proceeds using two levels of median filtering, the final centre-pixel intensity being taken as:

$$A' = \text{med}(A, \text{med}(A, B, D, F, H), \text{med}(A, C, E, G, I))   (3.12)$$

We here ignore the line preserving properties of this filter and concentrate on its corner preserving, low-edge-shift characteristics. It is quite easy to see that the 5×5 regions

```
0  0  0  0  0        0  0  0  0  0
0  0  0  0  0        0  0  0  0  0
0  0  1  1  1        0  0  1  0  0
0  0  1  1  1        0  1  1  1  0
0  0  1  1  1        1  1  1  1  1
```

are preserved by this filter, although these examples represent limiting cases which could be disrupted by minor amounts of noise or slight changes of orientation. Thus the filter seems *guaranteed* to preserve corners only if the internal angle is greater than $135°$. This figure should be compared with the $180°$ obtained using similar arguments for the normal median filter in 5×5 regions such as

```
0  0  0  1  1
0  0  0  1  1
0  0  1  1  1
0  0  1  1  1
0  0  1  1  1
```

Figure 3.16 shows plots obtained with this filter under the same conditions as for the 5×5 median filters. Clearly, it always gives at least a four-fold improvement in edge shift over that for the median filter, and this performance improves with increasing radius of curvature b until there is zero shift for $b > 4$ (note that $b = 4$ is approximately the figure that would be expected from the corner angle of $135°$ noted above, within a 5×5 neighbourhood). Hence such detail preserving filters improve the situation dramatically but do not completely overcome the underlying problem described earlier. In addition, this improvement may not have been obtained without cost, since in some cases the filter seems to *insert* structure where none exists (Davies, 1989b). The result is to cast some doubt on the usefulness of this type of filter in all possible situations. Nevertheless, its effect on real images appears to be generally very good (see Figs 3.13(c) and 3.17(c)).

3.5.4 Problems with statistics

Thus far it has been seen that computations of the position of the mode are made more difficult because of the sparse statistics of the local intensity distribution. In fact this also affects the median calculation. Suppose the median value happens to be well spaced near the centre of the distribution (Fig. 3.4). Then a small error in this one intensity value is immediately reflected in full when calculating the median: i.e. the poor statistics have biased the median in a particular way. Ideally, what is required is a stable estimator of the median of the underlying distribution. Thus the distribution should be made smoother before arriving at a specific value for the median. In practice this procedure adds significant computational load to the filter calculation and is commonly not carried out. As a consequence the median filter tends to result in runs of constant intensity, thereby giving images the "softened" appearance noted earlier. This is apparent on studying the following 1-D example:

original: 0 0 1 0 0 1 1 2 1 2 2 1 2 3 3 4 3 2 2 3

filtered: ? 0 0 0 0 1 1 1 2 2 2 2 2 3 3 3 3 2 2 ?

Although histogram smoothing is not commonly carried out, some workers have felt it necessary to adjust the relative weights of the various pixels in the neighbourhood according to their distance from the central pixel (Akey and Mitchell, 1984). This mimics what happens for a Gaussian filter and is theoretically necessary although it is not generally implemented. On the whole workers have been happy to use the median filter without considering such problems: much more effort has been devoted to devising fast implementations of the filter rather than making it obey better theoretical models.

3.6 Reducing computational load

Significant efforts have been made to speed the operation of the Gaussian filter since implementations in large neighbourhoods require considerable amounts of computation (Wiejak *et al.*, 1985). For example, smoothing images using a 30×30 Gaussian convolution mask in an image of 256×256

pixels involves 64 million basic operations. For such a basic operation as smoothing this is unacceptable. However it is possible to cut down the amount of computation drastically, since a 2-D Gaussian convolution can be factorized into two 1-D Gaussian convolutions which can be applied in turn:

$$\exp\,(-r^2/2\sigma^2) = \exp\,(-x^2/2\sigma^2)\,\exp\,(-y^2/2\sigma^2) \qquad (3.13)$$

It is important to realize that the decomposition is rigorously provable and is not an approximation: we shall refer to this below in the context of the median filter. Meanwhile, the decompositions for the two 3×3 Gaussian filters we met earlier are:

$$\frac{1}{9}\begin{bmatrix} 1 & 1 & 1 \\ 1 & 1 & 1 \\ 1 & 1 & 1 \end{bmatrix} = \frac{1}{3}\begin{bmatrix} 1 \\ 1 \\ 1 \end{bmatrix}\frac{1}{3}\begin{bmatrix} 1 & 1 & 1 \end{bmatrix} \qquad (3.14)$$

and

$$\frac{1}{16}\begin{bmatrix} 1 & 2 & 1 \\ 2 & 4 & 2 \\ 1 & 2 & 1 \end{bmatrix} = \frac{1}{4}\begin{bmatrix} 1 \\ 2 \\ 1 \end{bmatrix}\frac{1}{4}\begin{bmatrix} 1 & 2 & 1 \end{bmatrix} \qquad (3.15)$$

Overall, this approach replaces a single $n \times n$ operator whose load is $O(n^2)$ with two operators of load $O(n)$, and for $n > 3$ there is always a worthwhile saving. Ignoring scanning and other overheads the situation is summarized in Table 3.3, the final column of which represents the saving factor.

Table 3.3 Saving achieved by factorizing an $n \times n$ Gaussian operator.

n	$2n$	n^2	$n/2$
1	2	1	0.5
3	6	9	1.5
5	10	25	2.5
7	14	49	3.5
9	18	81	4.5
11	22	121	5.5
13	26	169	6.5
15	30	225	7.5

It turns out that it is not possible to decompose the median filter in the same way without making approximations. However, it is quite common to try to perform a similar function by applying two 1-D median filters in turn (Narendra, 1978). Although the effect is similar in its outlier rejection properties to the corresponding 2-D median filter, it is simple to show by appealing to specific examples of image data that the two are not formally equivalent. An instance in which the same effect is obtained by a 3×3 median filter or by 3×1 and 1×3 median filters applied in either order is the following:

	0	0	0	0	0	0
	0	0	1	0	0	0
original image	0	0	0	1	0	0
segment:	0	1	0	0	1	0
	0	0	0	0	0	0

In this case all the 1's immediately disappear after applying a 3×3 or a 3×1 or a 1×3 median filter. In the following instance the situation is slightly more complex:

	0	0	0	0	0	0
original image	0	0	1	1	2	0
segment:	0	0	2	2	0	0
	0	0	0	0	0	0

	0	0	0	0	0	0
after applying a 3×3	0	0	0	1	0	0
filter:	0	0	0	1	0	0
	0	0	0	0	0	0

	0	0	0	0	0	0		0	0	0	0	0	0
after applying a 3×1	0	0	1	1	1	0		0	0	1	1	0	0
and then a 1×3 filter:	0	0	2	2	0	0		0	0	1	1	0	0
	0	0	0	0	0	0		0	0	0	0	0	0

	0	0	0	0	0	0		0	0	0	0	0	0
after applying a 3×1	0	0	1	1	0	0		0	0	1	1	0	0
and then a 1×3 filter:	0	0	1	1	0	0		0	0	1	1	0	0
	0	0	0	0	0	0		0	0	0	0	0	0

The following case tests the situation more thoroughly, and is one of the simplest for which the three final image segments are all different:

	0	0	0	0	0	0
	0	0	0	1	0	0
original image	0	0	1	1	1	0
segment:	0	0	1	0	0	0
	0	0	0	0	0	0

	0	0	0	0	0	0
	0	0	0	0	0	0
after applying a 3×3	0	0	0	1	0	0
filter:	0	0	0	0	0	0
	0	0	0	0	0	0

	0	0	0	0	0	0	0	0	0	0	0	0
after applying a 3×1	0	0	0	0	0	0	0	0	0	0	0	0
and then a 1×3 filter:	0	0	1	1	1	0	0	0	0	0	0	0
	0	0	0	0	0	0	0	0	0	0	0	0
	0	0	0	0	0	0	0	0	0	0	0	0

	0	0	0	0	0	0	0	0	0	0	0	0
after applying a 1×3	0	0	0	1	0	0	0	0	0	0	0	0
and then a 3×1 filter:	0	0	1	1	0	0	0	0	1	1	0	0
	0	0	1	0	0	0	0	0	0	0	0	0
	0	0	0	0	0	0	0	0	0	0	0	0

In general, spots and streaks not more than one pixel wide are eliminated quite effectively by the original or by the separated forms of the filter. Larger filters should effectively eliminate wider spots and streaks, although exact functional equivalence between the original and its separated forms is not to be expected, as has been indicated.

Finally, the problem of inexact decomposition is not an exclusive property of nonlinear filters: many linear filters cannot be decomposed exactly either, as is clear because the number of independent mask coefficients in an $n \times n$ mask is $n^2 - 1$, which is greater than the number $(n-1)$ in an $n \times 1$ or $1 \times n$ component mask.

3.6.1 A bit-based method for fast median filtering

A number of other means have been employed to speed the operation of the median filter. Of especial interest is the bit-based method of Ataman *et al.*

(1980), since it is surprisingly simple yet highly effective. Space does not permit a full account of the method to be given here, so a simplified version retaining the same main underlying principle is given.

The basic idea derives from the observation that only the most significant bit (MSB) of each of the original numbers is relevant for computing the MSB of the median. This suggests that it ought to be possible to determine the median in a number of operations of the order of the number of bits required to represent the original numbers. Since pixel intensities are frequently represented by just 6–8 bits of data, this makes the idea a particularly attractive one for further development in image processing applications.

To proceed, the discussion is specialized to the case where the numbers involved are the N pixel intensities within a given neighbourhood, each represented by L bits, and t is the number of the median in the ordered sequence of the pixel intensities (so that in Pascal notation, `t := N div 2 + 1`). Bits are here labelled from the MSB to the LSB in the order $1, 2, ..., L$ in which they will be examined. Finally, $p(r)$ denotes the number of pixels for which the rth bit is 1 while the first $r - 1$ bits are identical to those of the median (as so far determined during the processing).

In this notation, the MSB of the median will be 1 if $p(1)$ is at least t and 0 otherwise:

$$\text{median[1]} := \text{if } p(1) >= t \text{ then } 1 \text{ else } 0; \qquad (3.16)$$

It is relevant to what follows to note that if $p(1) < t$, this identifies $p(1)$ pixel intensities as being greater than the median: otherwise none are so identified.

This simple result is somewhat tricky to generalize: in fact, it is now necessary to restate the problem. At any stage, let us denote by the variable G the number of pixel intensities that have already been shown to be greater than the median. Let us also imagine that we are making a trial to determine whether, after considering the rth bit, this number should be increased. In fact, if $G + p(r)$ is at least t, then the additional pixels which contribute to the value of $p(r)$ (having rth bit equal to 1) ensure that the rth bit of the median is 1, but we are prevented from knowing any more about the number of pixels that have intensities greater than the median. However, if $G + p(r)$ is less than t, then (i) we know that median[r] is 0, and (ii) we have identified additional $p(r)$ pixel intensities that are larger than the median. This can be written algorithmically in the form:

```
if G + p(r) >= t then median[r] := 1
      else begin median[r] := 0; G := G + p(r) end;   (3.17)
```

Clearly the processing terminates when all bits have been examined.

The overall algorithm now appears as in Table 3.4. Note that $p(r)$ will have to be a (Pascal) function requiring separate declaration, in accordance with the definition given earlier. If the whole algorithm including $p(r)$ is expressed in Pascal, then it is unlikely to be significantly faster than other relevant methods. However, if it is written in machine code or implemented partly in suitable hardware, then it should achieve its potential speed. This is ultimately due to the problem of bit packing and unpacking which certain high-level languages are unable to handle efficiently.

Table 3.4   The complete bit-based median algorithm.

```
t := N div 2 + 1;
G := 0;
r := 0;
repeat
  r := r + 1;
  if G + p(r) >= t then median[r] := 1
    else begin median[r] := 0; G := G + p(r) end;
until r = L;
(* It is assumed that p(r) has already been implemented as a
software or hardware function *)
```

3.6.2   VLSI implementation of the median filter

Some very subtle schemes have been devised for implementing the median filter in a computationally efficient manner in VLSI microcircuits (Offen, 1985). Here the problem is inherently one of partitioning functionality so that useful portions of the task can be solved in a single VLSI chip. This problem turns out to be as much a hardware problem as one of devising an ingenious breakdown of the fundamental algorithm. The simplest VLSI implementations depend on cascading devices for implementing a bubble sort (Oflazer, 1983). As such they should be suitable for implementing the truncated median filter as well as the median filter.

3.7   The role of filters in industrial applications of vision

It has been shown above how the median filter can successfully remove noise and artefacts such as spots and streaks from images. Unfortunately, many useful features such as fine lines and important points and holes are effectively indistinguishable from spots and streaks. In addition, it has been seen that the median filter "softens" pictures by removing fine detail. It is also found to clip corners of objects—another generally undesirable trait (but see Chapter 13). Finally, although it does not blur edges, it can still shift them slightly. In fact, shifting of curved edges seems likely to be a general characteristic of noise suppression filters (but see Section 3.5.3).

Such distortions are quite alarming and mitigate against the indiscriminate use of filters. If applied in situations where accurate measurements are to be made on images, particular care must be taken to test whether the data are being biased in any way. Although it is possible to make suitable corrections to the data, it seems a good general policy to employ noise removal filters only where they are absolutely essential for object visibility. The alternative is to employ edge detection and other operators that automatically suppress noise as an integral part of their function. This is the general approach taken in subsequent chapters: indeed, it is one of the principles underlined in this book that algorithms should be "robust" against noise or other artefacts which might upset measurements. There is quite significant scope for the design of robust algorithms, since images contain so much information that it is usually possible to arrange for erroneous information to be ignored.

3.8   Sharp–unsharp masking

When images are blurred either before or as part of the process of acquisition, it is frequently possible to restore them to substantially their ideal state. Properly, this is achieved by making a model of the blurring process and applying an inverse transformation which is intended exactly to cancel the blurring. This is a complex task to carry out rigorously but in some cases a rather simple method called sharp–unsharp masking is able to produce significant improvement (Gonzalez and Wintz, 1987). This technique involves first obtaining an even more blurred version of the image (e.g. with the aid of a Gaussian filter) and then subtracting this image from the original:

```
UNSHARP:
  begin
    [[ Q0 := (P0 *4 + (P1 + P3 + P5 + P7) *2 + P2 + P4 + P6 + P8)
          div 16 ]];
    [[ P0 := P0 * (1 + alpha) - Q0 * alpha ]];
  end;                                                            (3.18)
```

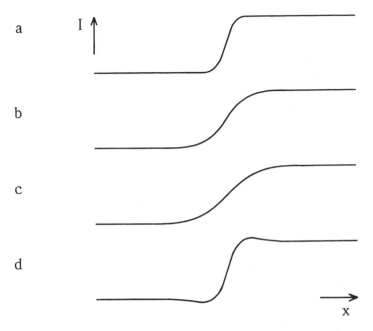

Fig. 3.18 The principle of sharp–unsharp masking: (a) cross-section of an idealized edge; (b) observed edge; (c) artificially blurred version of (b); (d) result of subtracting a proportion of (c) from (b).

The principle of the technique is illustrated in Fig. 3.18 and its effect is seen in Fig. 3.19. Unfortunately, this particular technique is difficult to set up rigorously, since the amount of artificial blurring to apply and the proportion of the blurred image to subtract are rather arbitrary quantities and are usually adjusted by eye (the value of alpha is typically $1/4$). Thus the method is in practice better categorized under the heading "enhancement" than "restoration", as it is not the precise mathematical technique normally understood by the latter term. Of such enhancement techniques Hall (1979) states: "Much of the art of enhancement is knowing when to stop".

3.9 Concluding remarks

Although this chapter has dwelt on the implementation of noise suppression and image enhancement operators based on the local intensity distribution, it has made certain other points. In particular, it has shown the need to make a specification of the required imaging process and only then to work out the

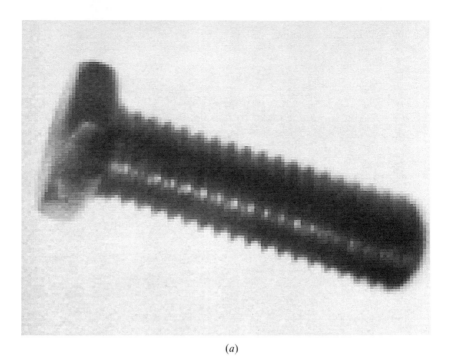

(a)

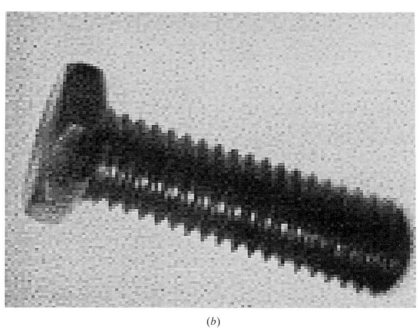

(b)

Fig. 3.19 Sharp–unsharp masking: (a) original, slightly blurred image; (b) improvement after sharp–unsharp masking.

algorithm design strategy. Not only does this ensure that the algorithm will perform its function effectively, but also it should make it possible to optimize the algorithm for various practical criteria including speed, storage and other parameters of interest. In addition, this chapter has demonstrated that any undesirable properties of the particular design strategy chosen (such as the inadvertent shifting of edges) should be sought out. Finally, it has demonstrated a number of fundamental problems to do with imaging in discrete lattices—not least being problems of statistics that arise with small pixel neighbourhoods.

The next chapter moves on to a particularly vital problem in machine vision—that of segmenting images in order to find where objects are situated. This work builds on what has been learnt in the present chapter about edge profiles and how they are "seen" by neighbourhood operators.

3.10   Bibliographical and historical notes

Much of the work of this chapter has built on a paper by the author (Davies, 1988c) which itself rests on considerable earlier work on Gaussian, median and other rank order filters (Hodgson et al., 1985; Duin et al., 1986). It should be noticed that the edge shifts that occur for median filters are not limited to this type of filter but apply almost equally to mean filters (Davies, 1991b). In addition, other inaccuracies have been found with median filters and methods have been found to correct them (Davies, 1992f).

The literature hardly mentions mode filters, presumably because of the difficulty of finding simple mode estimators that are not muddled unduly by noise and which still operate rapidly. Indeed, only one overt reference has been found (Coleman and Andrews, 1979). Other work referred to here is that on decomposing Gaussian and median filters (Narendra, 1978; Wiejak et al., 1985), and the many papers on fast implementation of median filters, both in software and in hardware (e.g. Narendra, 1978; Huang et al., 1979; Ataman et al., 1980; Danielsson, 1981; Oflazer, 1983; Davies, 1992a).

Considerable efforts have been devoted to studying the "root" behaviour of the median filter, i.e. the result of applying median filtering operations until no further change occurs. In fact, most of this work has been carried out on 1-D signals, including cardiac and speech waveforms, rather than on images (Gallagher and Wise, 1981; Fitch et al., 1985; Heinonen and Neuvo, 1987). Root behaviour is of interest as it relates to the underlying structure of signals, although its realization involves considerable amounts of processing.

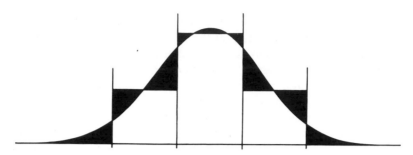

Fig. 3.20 Approximating a discrete to a continuous Gaussian. This diagram shows how a balance needs to be struck between subpixel errors and those arising from the truncated part of the function.

Some of the latest work on filtering aims to improve on rather than to emulate the median filter. Work of this type includes that of Heinonen and others (Nieminen *et al.*, 1987). See also the neural network approach to this topic discussed in Chapter 21.

Finally, the author has reported methods of optimizing linear smoothing filters in small neighbourhoods by minimizing the total error in fitting them to a continuous Gaussian function (Davies, 1987e): it turns out that a balance has to be struck between subpixel errors within the neighbourhood and errors that arise from the proportion of the distribution that lies outside the neighbourhood (Fig. 3.20).

3.11 Problems

1. Draw up a table showing the numbers of operations required to implement a median filter in various sizes of the neighbourhood. Include in your table (a) results for a straight bubble sort of all n^2 pixels, (b) results for bubble sorts in separated $1 \times n$ and $n \times 1$ neighbourhoods, and (c) results for the histogram method of Section 3.3. Discuss the results, taking account of possible computational overheads.

2. Show how to perform a median filtering operation on a binary image. Show also that if a set of binary images is formed by thresholding a grey-scale image at various levels, and each of these binary images is median filtered, then a grey-scale image can be reconstructed which is a median filtered version of the original grey-scale image. Consider to what extent the reduced amount of computation in filtering a binary image compensates for the number of separate thresholded images to be filtered.

3. An "extremum" filter is an image-parallel operation which assigns to every pixel the intensity value of the closer of the two extreme values in its local intensity distribution. Show that it should be possible to use such a filter to enhance images. What would be the *disadvantage* of such a filter?

4. Under what conditions is a 1-D signal that has been filtered once by a median filter a root signal? What truth is there in the statement that a straight edge in an image is neither shifted nor blurred by a median filter, whatever its cross-section?

4

Thresholding Techniques

4.1 Introduction

One of the first tasks to be undertaken in vision applications is to segment objects from their background. When objects are large and do not possess very much surface detail, segmentation is often imagined as splitting the image into a number of regions each having a high level of uniformity in some parameter such as brightness, colour or texture. Hence it should be straightforward to separate objects from one another and from their background, and also to discern the different facets of solid objects such as cubes.

Unfortunately, the concept of segmentation presented above is an idealization which is sometimes reasonably accurate but more often in the real world it is an invention of the human mind, generalized inaccurately from certain simple cases. This problem arises because of the ability of the eye to understand real scenes at a glance, and hence to segment and perceive objects within images in the form they are known to have. Introspection is not a good way of devising vision algorithms, and it must not be overlooked that segmentation is actually one of the central and most difficult practical problems of machine vision.

Thus the common view of segmentation as looking for regions possessing some degree of uniformity is to a large extent invalid. There are many examples of this in the world of 3-D objects: one is a sphere lit from one direction, the brightness in this case changing continuously over the surface so that there is no distinct region of uniformity; another is a cube where the direction of the lighting may lead to several of the facets having equal brightness values, so that it is impossible from intensity data alone to segment the image completely as desired.

Nevertheless, there is sufficient correctness in the concept of segmentation by uniformity measures for it to be worth pursuing for practical applications. The reason is that in many (especially industrial) applications only a very restricted range and number of objects are involved, and in addition it is possible to have almost complete control over the lighting and the general environment. The fact that a particular method may not be completely general need not be problematic, since by employing tools that are appropriate for the task in hand, a cost-effective solution will have been achieved in that case at least. However, in practical situations there is clearly a tension between the simple cost-effective solution and the general-purpose but more computationally expensive solution; this tension must always be kept in mind in severely practical subjects such as machine vision.

4.2   Region-growing methods

The segmentation idea outlined above leads naturally to the region-growing technique (Zucker, 1976b). Here pixels of like intensity (or other suitable property) are successively grouped together to form larger and larger regions until the whole image has been segmented. Clearly there have to be rules about not combining adjacent pixels that differ too much in intensity, while permitting combinations for which intensity changes gradually because of variations in background illumination over the field of view. However, this is not enough to make a viable strategy, and in practice the technique has to include the facility not only to merge regions together but also to split them if they become too large and inhomogeneous (Horowitz and Pavlidis, 1974). Particular problems are noise and sharp edges and lines which form disconnected boundaries, and for which it is difficult to formulate simple criteria to decide whether they form true region boundaries. In remote sensing applications, for example, it is often difficult to separate fields rigorously when hedges are broken and do not give continuous lines: in such applications segmentation may have to be performed interactively, with a human operator helping the computer. Hall (1979) found that in practice regions tend to grow too far[*], so that to make the technique work well it is necessary to limit their growth with the aid of edge detection schemes.

---

[*] The danger is clearly that even one small break will in principle join two regions into a single larger one.

Thus the region-growing approach to segmentation turns out to be quite complex to apply in practice. In addition, region-growing schemes usually operate iteratively, gradually refining hypotheses about which pixels belong to which regions. The technique is complicated because, carried out properly, it involves global as well as local image operations. Thus each pixel intensity will in principle have to be examined many times, and as a result the process tends to be quite computationally intensive. For this reason it is not considered further here, since we are particularly interested in methods involving low computational load which are amenable to real-time implementation.

4.3 Thresholding

If background lighting is arranged so as to be fairly uniform, and we are looking for rather flat objects that can be silhouetted against a contrasting background, segmentation can be achieved simply by thresholding the image at a particular intensity level. This possibility was apparent from Fig. 2.3. In such cases the complexities of the region-growing approach are bypassed. The process of thresholding has already been covered in Chapter 2, the basic result being that the initial grey-scale image is converted into a binary image in which objects appear as black figures on a white background, or as white figures on a black background. Further analysis of the image then devolves into analysis of the shapes and dimensions of the figures: at this stage object identification should be straightforward. Chapter 6 concentrates on such tasks. Meanwhile there is one outstanding problem—how to devise an automatic procedure for determining the optimum thresholding level.

4.3.1 Finding a suitable threshold

One simple technique for finding a suitable threshold arises in situations such as optical character recognition (OCR) where the proportion of the background that is occupied by objects (i.e. print) is substantially constant in a variety of conditions. A preliminary analysis of relevant picture statistics then permits subsequent thresholds to be set by insisting on a fixed proportion of dark and light in a sequence of images (Doyle, 1962). In practice a series of experiments is performed in which the thresholded image is examined as the threshold is adjusted, and the best result ascertained by eye: at that stage the proportions of dark and light in the image are measured. Unfortunately, any changes in noise level following the original measurement will upset such a scheme, since they will affect the relative amounts of dark and light in the image. However, this is frequently a useful technique in industrial applications, especially when particular details within an object are to be examined: typical examples of this are holes in mechanical components such as brackets (notice that the mark-space ratio for objects may well vary substantially on a production line, but the proportion of hole area *within* the object outline would not be expected to vary).

The technique that is most frequently employed for determining thresholds involves analysing the histogram of intensity levels in the digitized image (see Fig. 4.1): if a significant minimum is found, it is interpreted as the required threshold value (Weska, 1978). Clearly, the assumption being made here is that the peak on the left of the histogram corresponds to dark objects, and the peak on the right corresponds to light background (i.e. it is assumed that, as in many industrial applications, objects appear dark on a light background).

This method is subject to the following major difficulties:

(i)   the valley may be so broad that it is difficult to locate a significant minimum;

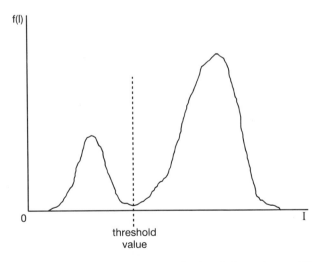

Fig. 4.1   Idealized histogram of pixel intensity levels in an image. The large peak on the right results from the light background; the smaller peak on the left is due to dark foreground objects. The minimum of the distribution provides a convenient intensity value to use as a threshold.

(ii)   there may be a number of minima because of the type of detail in the image, and selecting the most significant one will be difficult;

(iii)   noise within the valley may inhibit location of the optimum position;

(iv)   there may be no clearly visible valley in the distribution because noise may be excessive or because the background lighting may vary appreciably over the image;

(v)   either of the major peaks in the histogram (usually that due to the background) may be much larger than the other and this will then bias the position of the minimum;

(vi)   the histogram may be inherently multimodal, making it difficult to determine which is the relevant thresholding level.

Perhaps the worst of these problems is the last: if the histogram is inherently multimodal, and we are trying to employ a single threshold, we are applying what is essentially an *ad hoc* technique to obtain a meaningful result. In general such efforts are unlikely to succeed and this is clearly a case where full image interpretation must be performed before we can be satisfied that the results are valid. Ideally, thresholding rules have to be formed after many images have been analysed. In what follows such problems of meaningfulness are eschewed and attention is concentrated on how best to find a genuine single threshold when its position is obscured as suggested by problems (i)–(v) above (which can be ascribed to image "clutter", noise and lighting variations).

4.3.2 Tackling the problem of bias in threshold selection

This subsection considers problem (v) above—that of eliminating the bias in the selection of thresholds which arises when one peak in the histogram is larger than the other. First, note that if the relative heights of the peaks are known this effectively eliminates the problem, since the "fixed proportion" method of threshold selection outlined above can be used. However, this is not normally possible. A more useful approach is to prevent bias by weighting down the extreme values of the intensity distribution and weighting up the intermediate values in some way. To achieve this, note that the intermediate values are special in that they correspond to object edges. Hence a good basic strategy is to find positions in the image where there is a significant intensity gradient—corresponding to pixels in the regions of edges—and to analyse the intensity values of these locations while ignoring other points in the image.

To develop this strategy, consider the "scattergram" of Fig. 4.2. Here pixel properties are plotted on a 2-D map with intensity variation along one axis

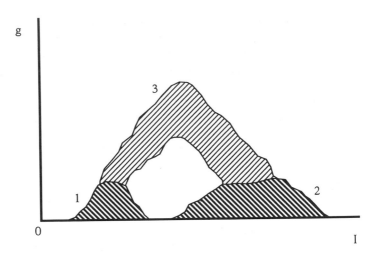

Fig. 4.2 Scattergram showing the frequency of occurrence of various combinations of pixel intensity I and intensity gradient magnitude g in an idealized image. There are three main populated regions of interest: (1) a low-I, low-g region; (2) a high-I, low-g region; and (3) a medium-I, high-g region. Analysis of the scattergram provides useful information on how to segment the image.

and intensity gradient magnitude variation along the other. Broadly speaking, there are three main populated regions on the map: (i) a low-intensity, low-gradient region corresponding to the dark objects; (ii) a high-intensity, low-gradient region corresponding to the background; and (iii) a medium-intensity, high-gradient region corresponding to object edges (Panda and Rosenfeld, 1978). These three regions merge into each other, with the high-gradient region that corresponds to edges forming a path from the low-intensity to the high-intensity region. The fact that the three regions merge in this way gives a choice of methods for selecting intensity thresholds. The first choice is to consider the intensity distribution along the zero gradient axis but this clearly has the same disadvantage as the original histogram method— namely, introducing a substantial bias. The second choice is that at high gradient, which means looking for a peak rather than a valley in the intensity distribution. The final choice is that at moderate values of gradient, which was suggested above as a possible strategy for avoiding bias.

It is now apparent that there is likely to be quite a narrow range of values of gradient for which the third method will be a genuine improvement; in addition, it may be necessary to construct a scattergram rather than a simple intensity histogram in order to locate it. If instead we try to weight the plots on a histogram with the magnitudes of the intensity gradient values, we are likely to end with a peaked intensity distribution (i.e. the second method) rather than one with a valley (Weska and Rosenfeld, 1979). A mathematical model of the situation is presented in Section 4.3.3 in order to explore the situation more fully.

4.3.2.1 Methods based on finding a valley in the intensity distribution

This section considers how to weight the intensity distribution using a parameter other than the intensity gradient, in order to locate accurately the valley in the intensity distribution. A simple strategy is first to locate all pixels that have a significant intensity gradient, and then to find the intensity histogram not only of these pixels but also of nearby pixels. This means that the two main modes in the intensity distribution are still attenuated very markedly and hence the bias in the valley position is significantly reduced. Indeed, the numbers of background and foreground pixels that are now being examined are very similar, so the bias from the relatively large numbers of background pixels is virtually eliminated (note that if the modes are modelled as two Gaussian distributions of equal widths and they also have equal heights, then the minimum lies exactly half-way between them).

Though obvious, this approach clearly includes the edge pixels themselves, and from Fig. 4.2 it can be seen that they will tend to fill the valley between the two modes. For the best results the points of highest gradient must actually be removed from the intensity histogram. A well-attested way of achieving this is to weight pixels in the intensity histogram according to their response to a Laplacian filter (Weska et al., 1974). Since such a filter gives an isotropic estimate of the second derivative of the image intensity (i.e. the magnitude of the first derivative of the intensity gradient), it is zero where intensity gradient magnitude is high: hence it gives such locations zero weight but it nevertheless weights up those locations on the shoulders of edges. It has been found that this approach is excellent at estimating where to place a threshold within a wide valley in the intensity histogram (Weska et al., 1974). In fact, Section 4.3.3 shows that this is far from being an ad hoc solution to a tedious problem: rather, it is in many ways ideal in giving a very exact indication of the optimum threshold value.

4.3.2.2 Methods which concentrate on the peaked intensity distribution at high gradient

The other approach indicated earlier is to concentrate on the peaked intensity distribution at high gradient values. In fact it seems possible that location of a peak will be less accurate (for the purpose of finding an effective intensity threshold value) than locating a valley between two distributions, especially when edges do not have symmetrical intensity profiles. However, an interesting result was obtained by Kittler *et al.* (1984, 1985), who suggested locating the threshold directly (i.e. without constructing an intensity histogram or a scattergram) by calculating the statistic:

$$T = \frac{\sum_i I_i g_i}{\sum_i g_i} \tag{4.1}$$

where I_i is the local intensity and g_i is the local intensity gradient magnitude. This statistic was found to be biased by noise, since it resulted in noisy background and foreground pixels contributing to the average value of intensity in the threshold calculation. This effect occurs only if the image contains different numbers of background and foreground pixels and is a rather different manifestation of the bias between the two histogram modes which was noted earlier. In this case a formula *including* bias was obtained in the form

$$T = T_0 + C(0.5 - b) \tag{4.2}$$

where T_0 is the inherent (i.e. unbiased) threshold, C is the image contrast and b is the proportion of background pixels in the image.

To understand the reasoning behind this formula, denote the background and foreground intensities as B and F respectively, and the *proportions* of background and foreground pixels as b and f, where $f + b = 1$ (here we are effectively assuming that a negligible proportion of pixels are situated on object edges: this is reasonable as the numbers of foreground and background pixels will be of order N^2 for an $N \times N$ pixel image, whereas the number of edge pixels will be of order N). Assuming a simple model of intensity variation, it is found that

$$T_0 = (B + F)/2 \tag{4.3}$$

and

$$C = F - B \tag{4.4}$$

Hence:

$$T = (B + F)/2 + (F - B)(f - b)/2$$
$$= B(1 - f + b)/2 + F(1 + f - b)/2 \qquad (4.5)$$
$$= Bb + Ff$$

This equation can be interpreted as follows. The threshold T is here calculated by averaging the intensity over image pixels, after weighting them in proportion to the local intensity gradient. Within the background the pixels having appreciable intensity gradient are noise pixels and these are proportional in number to b and have mean intensity B; similarly, within the foreground, pixels with appreciable intensity gradient are proportional in number to f and have mean intensity F. As a result, the mean intensity of all these noise pixels is $Bb + Ff$, in agreement with the above equation for T. On the other hand, if pixels with low intensity gradient are excluded, this progressively eliminates normal noise pixels and weights up the edge pixels (not represented in the above equation), so that the threshold reverts to the unbiased value T_0. By applying the right corrections, Kittler et al. were able to eliminate the bias and to obtain a corrected threshold value T_0 in practical situations, even when pixels with low intensity gradient values were not excluded.

This method is useful since it results in unbiased estimates of the threshold T_0 without the need to construct histograms or scattergrams. However, it is probably somewhat risky not to obtain histograms as they give so much information—for example, on the existence of multiple modes. Thus the statistic T may in complex cases amount to a bland average which means very little. Ultimately it seems that the safest threshold determination procedure must involve seeking and identifying *specific* peaks on the intensity histogram for relatively high values of intensity gradient, or *specific* troughs for relatively low values of intensity gradient.

4.3.3 A convenient mathematical model

To understand more clearly the problems outlined above, it is useful to construct a theoretical model of the situation. It is assumed that a number of dark, fairly flat objects are lying on a light worktable and are subject to soft, reasonably uniform lighting. In the absence of noise and other perturbations, the only variations in intensity occur near object boundaries (Fig. 2.3). With little loss of generality the model can be simplified to a single dimension: it is easy to see that the distribution of intensities in the image is then as shown in Fig. 4.3 and is closely related to the shape of the edge profile. In fact, the intensity distribution is effectively the derivative of the edge profile:

$$f(I) = c \, |\, \mathrm{d}x/\mathrm{d}I \,| \tag{4.6}$$

since for I in a given range ΔI, $f(I) \, \Delta I$ is proportional to the range of distances Δx that contribute to it. Taking account now of the fact that the entire image may contain many edges at specific locations, the distribution $f(I)$ can in principle be calculated for all values of I including the two very large peaks at the maximum and minimum values of intensity.

So far the model is incomplete in that it takes no account of noise or other variations in intensity. Essentially, these act in such a way as to broaden the intensity values from those predicted from edge profiles alone. This means that they can be allowed for by applying a broadening convolution $b(I)$ to $f(I)$:

$$F(I) = \int_{-\infty}^{\infty} b(u - I) f(u) \, \mathrm{d}u \tag{4.7}$$

This gives the more realistic distribution shown in Fig. 4.3(c). These ideas confirm that a high histogram peak at the highest or lowest value of intensity can pull the minimum one way or the other and result in the original automatic threshold selection criterion being inaccurate.

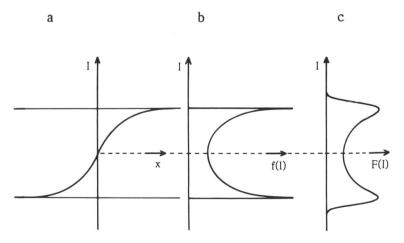

Fig. 4.3   Relation of edge profile to intensity distribution: (a) shape of a typical edge profile; (b) resulting distribution of intensities; (c) effect of applying a broadening convolution (e.g. resulting from noise) to (b).

In what follows the analysis is simplified by returning to the unbroadened distribution $f(I)$. We make the following definitions of intensity gradient g and rate of change of intensity gradient g':

$$g = dI/dx \tag{4.8}$$

$$g' = dg/dx = (dg/dI)(dI/dx) = g(dg/dI) \tag{4.9}$$

Applying equation (4.6) it is found that

$$f(I) = c\,|\,g\,|^{-1} \tag{4.10}$$

If we wish to weight the contributions to $f(I)$ using some power p of $|\,g\,|$, the new distribution becomes:

$$h(I) = f(I)\,|\,g\,|^p = c\,|\,g\,|^{p-1} \tag{4.11}$$

This is clearly not a sharp unimodal distribution as required by the earlier discussion, unless $p > 1$. We could conveniently set $p = 2$, although in practice workers have sometimes opted for a nonlinear, sharply varying weighting which is unity if $|\,g\,|$ exceeds a given threshold and zero otherwise (Weska and Rosenfeld, 1979). This is not pursued further here. Instead we consider the effect of weighting the contributions to $f(I)$ using $|\,g'\,|$ and obtain the new distribution:

$$k(I) = f(I)\,|\,g'\,| \tag{4.12}$$

Applying equations (4.6) and (4.9) now gives

$$k(I) = c\,|\,dg/dI\,| \tag{4.13}$$

To proceed further an exact mathematical model is needed. A convenient one is obtained using the tanh function:

$$I = I_0 \tanh x \tag{4.14}$$

with I being restricted to the range $-I_0 \leqslant I \leqslant I_0$ (note that we have simplified the model by making I zero at the centre of the range). This leads to:

$$\begin{aligned} dI/dx &= I_0 \operatorname{sech}^2 x \\ &= I_0(1 - \tanh^2 x) \\ &= I_0(1 - I^2/I_0^2) \end{aligned} \tag{4.15}$$

Applying equations (4.8) and (4.9) now gives

$$g = I_0(1 - I^2/I_0^2) \tag{4.16}$$

and

$$g' = -2I(1 - I^2/I_0^2) \tag{4.17}$$

Note that the form of g given by the model, as a function of x, is:

$$g = I_0 \operatorname{sech}^2 x = 4I_0/(e^x + e^{-x})^2 \tag{4.18}$$

and for high values of $|x|$

$$g \simeq 4I_0 e^{-2|x|} \tag{4.19}$$

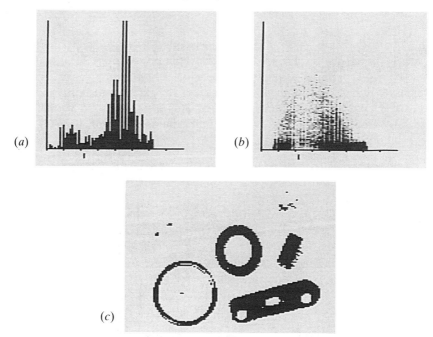

(a)

(b)

(c)

Fig. 4.4 (a) Histogram and (b) scattergram for the image shown in Fig. 2.3(a). In neither case is the diagram particularly close to the ideal form of Figs 4.1 and 4.2. Hence the threshold obtained from (a) (indicated by the short lines beneath the two scales) does not give ideal results with all the objects in the binarized image (c). Nevertheless, the results are better than for the arbitrarily thresholded image of Fig. 2.3(b).

While it is possible that a Gaussian variation for g might fit real data better, the present model has the advantage of simplicity and is able to provide useful insight into the problem of thresholding.

Applying equations $(4.10)-(4.13)$ and (4.16) it may be deduced that

$$f(I) = c \, |I_0(1 - I^2/I_0^2)|^{-1} \qquad (4.20)$$

$$h(I) = c \, |I_0(1 - I^2/I_0^2)|^{p-1} \qquad (4.21)$$

$$k(I) = c \, |2I/I_0| \qquad (4.22)$$

We have now arrived at an interesting point: starting with a particular edge profile we have calculated the shape of the basic intensity distribution $f(I)$, the shape of the intensity distribution $h(I)$ resulting from weighting pixel contributions according to a power of the intensity gradient value, and the distribution $k(I)$ resulting from weighting them according to the spatial

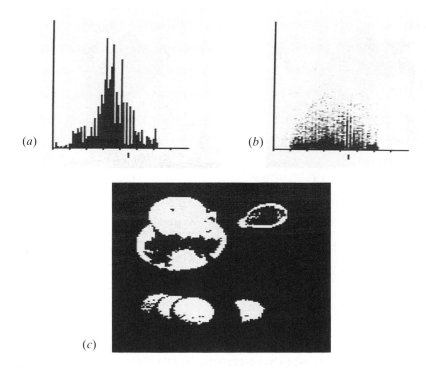

Fig. 4.5 (a) Histogram and (b) scattergram for the image shown in Fig. 2.1(a). In neither case is the diagram at all close to the idealized form, and the results of thresholding (c) are not a particularly useful aid to interpretation.

derivative of the intensity gradient, i.e. according to the output of a Laplacian operator. The $k(I)$ distribution has a very sharp valley with linear variation near the minimum. By contrast, $h(I)$ varies rather slowly and smoothly near its maximum. This shows that the Laplacian weighting method should provide the most sensitive indicator of the optimum threshold value and thereby provides strong justification for the use of this method. However, it must be made absolutely clear that the advantages of this approach may well be washed away in practice if the assumptions made in the model are invalid. In particular, broadening of the distribution by noise, variations in background lighting, image clutter and other artefacts will affect the situation adversely for all of the above methods. Further detailed analysis seems pointless, in part because these interfering factors make choice of the threshold more difficult in a very data-dependent way, but more importantly because they ultimately make the concept of thresholding at a single level invalid (see Figs 4.4–4.6).

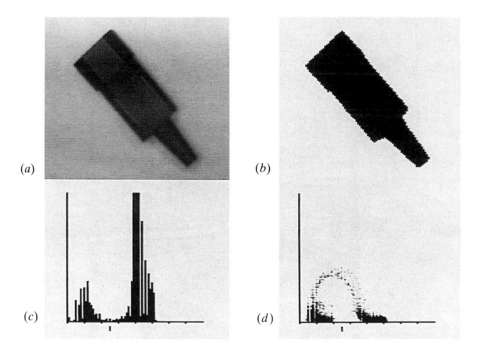

(a) (b)

(c) (d)

Fig. 4.6 A picture with more ideal properties. (a) Image of a plug which has been lit fairly uniformly. The histogram (c) and scattergram (d) approximate to the ideal forms, and the result of thresholding (b) is acceptable. However, much of the structure of the plug is lost during binarization.

4.3.4 Summary

It has been shown that available techniques are able to provide values at which intensity thresholding can be applied but they do not themselves solve the problems caused by uneven lighting. They are even less capable of coping with glints, shadows and image clutter. Unfortunately, these artefacts are common in most real situations and are only eliminated with difficulty in practice. Indeed, in industrial applications where shiny metal components are involved, glints are the rule rather than the exception, while shadows can seldom be avoided with any sort of object. Even flat objects are liable to have quite a strong shadow contour around them because of the particular placing of lights. Lighting problems are studied in detail in Chapter 23. Meanwhile it may be noted that glints and shadows can only be allowed for properly in a two-stage image analysis system, where tentative assignments are made first and these are firmed up by exact explanation of all pixel intensities. We now return to the problem of making the most of the thresholding technique, by finding how variations in background lighting can be allowed for.

4.4 Adaptive thresholding

The problem that arises when illumination is not sufficiently uniform may be tackled by permitting the threshold to vary adaptively (or "dynamically") over the whole image. In principle there are several ways of achieving this. One involves modelling the background within the image. Another is to work out a local threshold value for each pixel by examining the range of intensities in its neighbourhood. A third approach is to split the image into subimages and deal with them independently. Though "obvious", this last method will clearly run into problems at the boundaries between subimages, and by the time these problems have been solved it will look more like one of the other two methods. Ultimately all such methods must operate on identical principles. The differences arise in the rigour with which the threshold is calculated at different locations and in the amount of computation required in each case. In fact, in real-time applications the problem amounts to finding how to estimate a set of thresholds with a minimum amount of computation.

The problem can sometimes be solved rather neatly in the following way. On some occasions—such as in automated assembly applications—it is possible to obtain an image of the background in the absence of any objects. This appears to solve the problem of adaptive thresholding in a rigorous manner, since the tedious task of modelling the background has already been carried out. However, some caution is needed with this: objects bring with them not only shadows (which can in some sense be regarded as part of the objects) but also an additional effect due to the reflections they cast over the background and other objects. This additional effect is nonlinear, in the sense that it is necessary to add not only the difference between the object and the background intensity in each case but also an intensity which depends on the products of the reflectances of pairs of objects. These considerations mean that using the no-object background as the equivalent background when several objects are present is ultimately invalid. However, as a first approximation it is frequently possible to assume an equivalence. If this proves impracticable, there is no option but to model the background from the actual image to be segmented.

On other occasions the variation in background intensity may be rather slowly varying, in which case it may be possible to model it by the following technique (this is a form of Hough transform—see Chapter 8). First, an equation is selected which can act as a reasonable approximation to the intensity function, e.g. a linear or quadratic variation:

$$I = a + bx + cy + dx^2 + exy + fy^2 \tag{4.23}$$

Next, a parameter space for the six variables a, b, c, d, e, f is constructed; then each pixel in the image is taken in turn and all sets of values of the parameters that could have given rise to the pixel intensity value are accumulated in parameter space. Finally, a peak is sought in parameter space which represents an optimal fit to the background model. So far it appears that this has been carried out only for a linear variation, the analysis being simplified initially by considering only the differences in intensities of pairs of points in image space (Nixon, 1985). Note that a sufficient number of pairs of points must be considered so that the peak in parameter space resulting from background pairs is sufficiently well populated. This implies quite a large computational load for quadratic or higher variations in intensity: hence it is unlikely that this method will be generally useful if the background intensity is not approximated well by a linear variation.

4.4.1 The Chow and Kaneko approach

As early as 1972 Chow and Kaneko introduced what is widely recognized as the standard technique for dynamic thresholding: it performs a thorough-going analysis of the background intensity variation, making few compromises to save computation (Chow and Kaneko, 1972). In this method the image is divided into a regular array of overlapping subimages and individual intensity histograms are constructed for each one. Those that are unimodal are ignored since they are assumed not to provide any useful information which can help in modelling the background intensity variation. However, the bimodal distributions are well suited to this task: these are individually fitted to pairs of Gaussian distributions of adjustable height and width and the threshold values are located. Thresholds are then found, by interpolation, for the unimodal distributions. Finally, a second stage of interpolation is necessary to find the correct thresholding value at each pixel.

One problem with this approach is that if the individual subimages are made very small in an effort to model the background illumination more exactly, the statistics of the individual distributions become worse, their minima become less well defined and the thresholds deduced from them are no longer statistically significant. This means that it does not pay to make subimages too small and that ultimately only a certain level of accuracy can be achieved in modelling the background in this way. Clearly, the situation is highly data-dependent but it might be expected that little would be gained by reducing the subimage size below 32×32 pixels. Chow and Kaneko employed 256×256 pixel images and divided these into a 7×7 array of 64×64 pixel subimages with 50% overlap.

Overall, this approach involves considerable computation and in real-time applications it may well not be viable for this reason. However, this type of approach is of considerable value in certain medical, remote sensing and space applications. In addition, note that the method of Kittler *et al.* (1985) referred to earlier may be used to obtain threshold values for the subimages with rather less computation.

4.4.2   Local thresholding methods

In real-time applications the alternative approach mentioned earlier is often more useful for finding local thresholds. It involves analysing intensities in the neighbourhood of each pixel to determine the optimum local thresholding level. Ideally, the Chow and Kaneko histogramming technique would be repeated at each pixel but this would significantly increase the computational load of this already computationally intensive technique. Thus it is necessary to obtain the vital information by an efficient sampling procedure. One simple means of achieving this is to take a suitably computed function of nearby intensity values as the threshold: often the mean of the local intensity distribution is taken, since this is a simple statistic and gives good results in some cases. For example, in astronomical images stars have been thresholded in this way. Niblack (1985) reported a case in which a proportion of the local standard deviation was added to the mean to give a more suitable threshold value, the reason (presumably) being to help suppress noise (clearly, addition is appropriate where bright objects such as stars are to be located, whereas subtraction will be more appropriate in the case of dark objects).

Another statistic that is frequently used is the mean of the maximum and minimum values in the local intensity distribution. The justification for this is that whatever the sizes of the two main peaks of the distribution, this statistic often gives a reasonable estimate of the position of the histogram minimum. The theory presented earlier shows that this method will only be accurate if (i) the intensity profiles of object edges are symmetrical, (ii) noise acts uniformly everywhere in the image so that the widths of the two peaks of the distribution are similar, and (iii) the heights of the two distributions do not differ markedly. Sometimes these assumptions are definitely invalid—for example, when looking for (dark) cracks in eggs or other products. In such cases the mean and maximum of the local intensity distribution can be found and a threshold deduced using the statistic

$$T = \text{mean} - (\text{maximum} - \text{mean}) \qquad (4.24)$$

where the strategy is to estimate the lowest intensity in the bright background assuming the distribution of noise to be symmetrical (Fig. 4.7): use of the mean here is realistic only if the crack is narrow and does not affect the value of the mean significantly. If it does, then the statistic can be adjusted by use of an *ad hoc* parameter:

$$T = \text{mean} - k(\text{maximum} - \text{mean}) \qquad (4.25)$$

where k may be as low as 0.5 (Plummer and Dale, 1984).

This method is essentially the same as that of Niblack (1985) but the computational load in estimating the standard deviation is minimized. Each of the last two techniques relies on finding local extrema of intensity. Using these measures helps save computation but they are clearly somewhat unreliable because of the effects of noise. If this is a serious problem, quartiles or other statistics of the distribution may be used. The alternative of prefiltering the image to remove noise is unlikely to work for crack thresholding, since cracks will almost certainly be removed at the same time as the noise. A better strategy is to form an image of T-values obtained using equation (4.24) or (4.25): smoothing this image should then permit the initial image to be thresholded effectively.

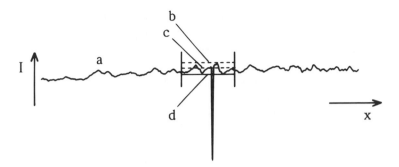

Fig. 4.7 Method for thresholding the crack in an egg. (a) Intensity profile of an egg in the vicinity of a crack: the crack is assumed to appear dark (e.g. under oblique lighting); (b) local maximum of intensity on the surface of the egg; (c) local mean intensity. Equation (4.24) gives a useful estimator T of the thresholding level (d).

Unfortunately, all these methods work well only if the size of the neighbourhood selected for estimating the required threshold is sufficiently large to span a significant amount of foreground and background. In many practical cases this is not possible and the method then adjusts itself erroneously, for example so that it finds darker spots within dark objects as well as segmenting the dark objects themselves. However, there are certain applications where there is little risk of this occurring. One notable case is that of OCR. Here the widths of character limbs are likely to be known in advance and should not vary substantially. If this is so, then a neighbourhood size can be chosen to span or at least sample both character and background, and it is thus possible to threshold the characters highly

Table 4.1 A simple algorithm for adaptively thresholding print.

```
minrange := 255/5; (* minimum likely difference in intensity
  between print and background: this parameter can be preset
  manually or "learnt" by a previous routine *)
repeat (* for each pixel *)
find minimum and maximum of local intensity distribution;
range := maximum − minimum;
if range > minrange
  then T := (minimum + maximum)/2 (* print visible in
      neighbourhood *)
    else T := maximum − minrange/2; (* all white neighbourhood *)
  if P0 > T then Q0 := 255 else Q0 := 0; (* now binarize print *)
until all pixels processed;
```

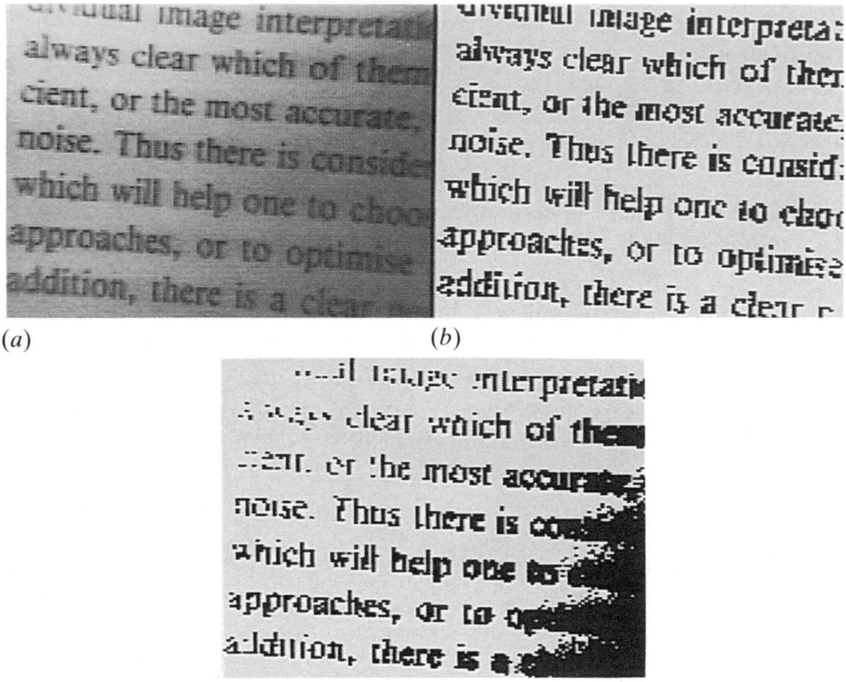

(a)   (b)

(c)

Fig. 4.8 Effectiveness of local thresholding on printed text. Here a simple local thresholding procedure (Table 4.1), operating within a 3 × 3 neighbourhood, is used to binarize the image of a piece of printed text (a). Despite the poor illumination, binarization is performed quite effectively (b). Note the complete absence of isolated noise points in (b), while by contrast the dots on all the i's are accurately reproduced. The best that could be achieved by uniform thresholding is shown in (c).

efficiently using a simple functional test of the type described above. The effectiveness of this procedure (Table 4.1) is demonstrated in Fig. 4.8.

4.5 Concluding remarks

The preceding sections have revealed a number of factors that are crucial to the process of thresholding. First, the need to avoid bias in threshold selection by arranging roughly equal populations in the dark and light regions of the intensity histogram; second, the need to work with the smallest possible subimage (or neighbourhood) size so that the intensity histogram has a well-defined valley despite variations in illumination; and third, the need for subimages to be sufficiently large so that statistics are good, permitting the valley to be located with precision.

A moment's thought will confirm that these conditions are not compatible and that a definite compromise will be needed in practical situations. In particular, it is generally not possible to find a neighbourhood size that can be applied everywhere in an image, on all occasions yielding roughly equal populations of dark and light pixels. Indeed, if the chosen size is small enough to span edges ideally, hence yielding unbiased local thresholds, it will be totally at a loss inside large objects. Attempting to avoid this situation by resorting to alternative methods of threshold calculation—as for the statistic of equation (4.1)—does not solve the problem since inherent in such methods is a built-in region size (in equation (4.1) the size is given by the summation limits). It is therefore not surprising that a number of workers have opted for variable-resolution and hierarchical techniques in an attempt to make thresholding more effective (Wu *et al.*, 1982; Wermser *et al.*, 1984; Kittler *et al.*, 1985).

At this stage we call to question the complications involved in such thresholding procedures—which become even worse when intensity distributions start to become multimodal. Note that the overall procedure is to find local intensity gradients in order to obtain accurate, unbiased estimates of thresholds, so that it then becomes possible to take a horizontal slice through a grey-scale image and hence, ultimately, find "vertical" (i.e. spatial) boundaries within the image. Why not use the gradients *directly* to estimate the boundary positions? Such an approach, for example, leads to no problems from large regions where intensity histograms are essentially unimodal, although it would be foolish to pretend that there are no other problems (see Chapters 5 and 7).

On the whole, the author takes the view that many approaches (region-growing, thresholding, edge detection, etc.), *taken to the limits of approximation*, will give equally good results. After all, they are all limited by the same physical effects—image noise, variability of lighting, the presence of shadows, etc. However, some methods are easier to coax into working well, or need minimal computation, or have other useful properties such as robustness. Thus thresholding can be a highly efficient means of aiding the interpretation of certain types of image but as soon as image complexity rises above a certain critical level, it suddenly becomes more effective and considerably less complicated to employ overt edge detection techniques. These are studied in depth in the next chapter. Meanwhile, we must not overlook the possibility of easing the thresholding process by going to some trouble to optimize the lighting system and to ensure that any worktable or conveyor is kept clean and white: this turns out to be a viable approach in a surprisingly large number of industrial applications.

The end result of thresholding is a set of silhouettes representing the shapes of objects: these constitute a "binarized" version of the original image. Many techniques exist for performing binary shape analysis and some of these are described in Chapter 6. Meanwhile, it should be noted that many features of the original scene—for example texture, grooves or other surface structure—will not be present in the binarized image. Although the use of multiple thresholds to generate a number of binarized versions of the original image can preserve relevant information present in the original image, this approach tends to be clumsy and impracticable, and sooner or later one is likely to be forced to return to the original grey-scale image for the required data.

4.6   Bibliographical and historical notes

Segmentation by thresholding started many years ago from simple beginnings, and in recent years has been refined into a set of mature procedures. In some ways there is evidence that these may have been pushed too far, or at least further than the basic concepts—which were after all intended to save computation—can justify. For example, Milgram's (1979) method of "convergent evidence" works by trying many possible threshold values and finding iteratively which of these is best at making thresholding agree with the results of edge detection—a method involving considerable computation. Pal *et al.* (1983) devised a technique based on fuzzy set theory, which analyses the consequences of applying all possible threshold levels and minimizes an "index of fuzziness": this technique also has the disadvantage of large computational cost. The grey-level transition matrix approach of Deravi and Pal (1983) involves constructing a matrix of adjacent pairs of image intensity levels and then seeking a threshold within this matrix. This achieves the capability for segmentation even in cases where the intensity distribution is unimodal but with the disadvantage of requiring large storage space.

The Chow and Kaneko method (1972) has been seen to involve considerable amounts of processing (though Greenhill and Davies, 1995, have sought to reduce it with the aid of neural networks). Nakagawa and Rosenfeld (1979) studied the method and developed it for cases of trimodal distributions but without improving computational load. For bimodal distributions, however, the approach of Kittler *et al.* (1985) should require far less processing because it does not need to find a best fit to pairs of Gaussian functions.

Several workers have investigated the use of relaxation methods for threshold selection. In particular, Bhanu and Faugeras (1982) devised a gradient relaxation scheme that is useful for unimodal distributions. However, relaxation methods as a class are well known to involve considerable amounts of computation even if they converge rapidly, since they involve several iterations of whole-image operations (see also Section 14.6). Indeed, in relation to the related region-growing approach, Fu and Mui (1981) stated explicitly that: "All the region extraction techniques process the pictures in an iterative manner and usually involve a great expenditure in computation time and memory". Finally, Wang and Haralick (1984) developed a recursive scheme for automatic multithreshold selection, which involves successive analysis of the histograms of relatively dark and relatively light edge pixels.

Fu and Mui (1981) provided a useful general survey on image segmentation: this has been updated by Haralick and Shapiro (1985). These papers review many topics that could not be covered in the present chapter for reasons of space and emphasis. A similar comment applies to Sahoo *et al.*'s (1988) survey of thresholding techniques.

As hinted in Section 4.4, thresholding (particularly local adaptive thresholding) has had many applications in optical character recognition. Amongst the earliest were the algorithms described by Bartz (1968) and Ullmann (1974): two highly effective algorithms have been described by White and Rohrer (1983).

During the 1980s the entropy approach to automatic thresholding evolved (e.g. Pun, 1981; Kapur *et al.*, 1985; Abutaleb, 1989; Pal and Pal, 1989): this approach proved highly effective, and its development has continued during the 1990s (e.g. Hannah *et al.*, 1995). Indeed, it may surprise the reader that midway through the 1990s thresholding and adaptive thresholding are still hot subjects (see for example Yang *et al.*, 1994; Ramesh *et al.*, 1995).

4.7 Problems

1. Using the methods of Section 4.3.3, model the intensity distribution obtained by finding all the edge pixels in an image and including also all pixels adjacent to these pixels. Show that while this gives a sharper valley than for the original intensity distribution, it is not as sharp as for pixels located by the Laplacian operator.
2. Because of the nature of the illumination, certain fairly flat objects are found to have a fine shadow around a portion of their boundary. Show that this could confuse an adaptive thresholder which bases its threshold value on the statistic of equation (4.1).

3. Consider whether it is more accurate to estimate a suitable threshold for a bimodal, dual-Gaussian distribution by (a) finding the position of the minimum, or (b) finding the mean of the two peak positions. What corrections could be made by taking account of the magnitudes of the peaks?

5

Locating Objects via Their Edges

5.1 Introduction

In Chapter 4 segmentation was tackled by the general approach of finding regions of uniformity in images — on the basis that the areas found in this way would have a fair likelihood of coinciding with the surfaces and facets of objects. The most computationally efficient means of following this approach was that of thresholding but this turns out for real images to be failure-prone or else quite difficult to implement satisfactorily. Indeed, to make it work well seems to require a multiresolution or hierarchical approach, coupled with sensitive measures for obtaining suitable local thresholds. Such measures have to take account of local intensity gradients as well as pixel intensities, and the possibility of proceeding more simply — by taking account of intensity gradients alone — was suggested.

In fact, edge detection has long been an alternative path to image segmentation and it is the method pursued in this chapter. Whichever way is inherently the better approach, edge detection has a considerable additional advantage in that it immediately reduces by a large factor (typically around 100) the redundancy inherent in the image data: this is useful because it immensely reduces both the space needed to store the information and the amount of processing subsequently required to analyse it.

Edge detection has gone through an evolution spanning more than 20 years. Two main methods of edge detection have been apparent over this period, the first of these being template matching (TM) and the second being the differential gradient (DG) approach. In either case the aim is to find where the intensity gradient magnitude g is sufficiently large to be taken as a reliable indicator of the edge of an object. Then g can be thresholded in a similar way to that in which intensity was thresholded in Chapter 4 (in fact, it is possible to look for local maxima of g instead of thresholding it but this possibility and its particular complexities are ignored until a later section). The TM and DG methods differ mainly in how they proceed to estimate g locally: however, there are also important differences in how they determine local edge orientation, which is an important variable in certain object detection schemes.

5.2 Basic theory of edge detection

Both DG and TM operators estimate local intensity gradients with the aid of suitable convolution masks. In the case of the DG type of operator, only two such masks are required—for the x and y directions. In the TM case, it is usual to employ up to 12 convolution masks capable of estimating local components of gradient in the different directions (Prewitt, 1970; Kirsch, 1971; Robinson, 1977; Abdou and Pratt, 1979).

In the TM approach, the local edge gradient magnitude (for short, the edge "magnitude") is approximated by taking the maximum of the responses for the component masks:

$$g = \max\ (g_i : i = 1 \text{ to } n) \tag{5.1}$$

where n is usually 8 or 12.

In the DG approach, the local edge magnitude may be computed vectorially using the nonlinear transformation:

$$g = (g_x^2 + g_y^2)^{1/2} \tag{5.2}$$

In order to save computational effort, it is common practice (Abdou and Pratt, 1979) to approximate this formula by one of the simpler forms:

$$g = |g_x| + |g_y| \tag{5.3}$$

or

$$g = \max\ (|g_x|, |g_y|) \tag{5.4}$$

which are, on average, equally accurate (Föglein, 1983).

In the TM approach, edge orientation is estimated simply as that of the mask giving rise to the largest value of gradient in equation (5.1). In the DG approach, it is estimated vectorially by the more complex equation:

$$\theta = \arctan\ (g_y/g_x) \qquad\qquad (5.5)$$

Clearly, DG equations (5.2) and (5.5) are more computationally intensive than TM equation (5.1), although they are also considerably more accurate. However, in some situations orientation information is not required; in addition, image contrast may vary widely, so there may apparently be little to be gained from thresholding a more accurate estimate of g. This may explain why so many workers have employed the TM instead of the DG approach. Since both approaches essentially involve estimation of local intensity gradients, it is not surprising that TM masks often turn out to be identical to DG masks—see Tables 5.1 and 5.2.

Table 5.1   Masks of well-known differential edge operators.

(a) Masks for the Roberts 2 × 2 operator

$$R_{x'} = \begin{bmatrix} 0 & 1 \\ -1 & 0 \end{bmatrix} \qquad\qquad R_{y'} = \begin{bmatrix} 1 & 0 \\ 0 & -1 \end{bmatrix}$$

(b) Masks for the Sobel 3 × 3 operator:

$$S_x = \begin{bmatrix} -1 & 0 & 1 \\ -2 & 0 & 2 \\ -1 & 0 & 1 \end{bmatrix} \qquad\qquad S_y = \begin{bmatrix} 1 & 2 & 1 \\ 0 & 0 & 0 \\ -1 & -2 & -1 \end{bmatrix}$$

(c) Masks for the Prewitt 3 × 3 "smoothed gradient" operator:

$$P_x = \begin{bmatrix} -1 & 0 & 1 \\ -1 & 0 & 1 \\ -1 & 0 & 1 \end{bmatrix} \qquad\qquad P_y = \begin{bmatrix} 1 & 1 & 1 \\ 0 & 0 & 0 \\ -1 & -1 & -1 \end{bmatrix}$$

In this table masks are presented in an intuitive format (viz. coefficients increasing in the positive x and y directions) by rotating the normal convolution format through 180°. This convention is employed throughout this chapter. The Roberts 2 × 2 operator masks (a) can be taken as being referred to axes x', y' at 45° to the usual x, y axes.

Table 5.2   Masks of well-known 3×3 template matching edge operators.

	0°	45°

(a) Prewitt masks:

$$\begin{bmatrix} -1 & 1 & 1 \\ -1 & -2 & 1 \\ -1 & 1 & 1 \end{bmatrix} \qquad \begin{bmatrix} 1 & 1 & 1 \\ -1 & -2 & 1 \\ -1 & -1 & 1 \end{bmatrix}$$

(b) Kirsch masks:

$$\begin{bmatrix} -3 & -3 & 5 \\ -3 & 0 & 5 \\ -3 & -3 & 5 \end{bmatrix} \qquad \begin{bmatrix} -3 & 5 & 5 \\ -3 & 0 & 5 \\ -3 & -3 & -3 \end{bmatrix}$$

(c) Robinson "3-level" masks:

$$\begin{bmatrix} -1 & 0 & 1 \\ -1 & 0 & 1 \\ -1 & 0 & 1 \end{bmatrix} \qquad \begin{bmatrix} 0 & 1 & 1 \\ -1 & 0 & 1 \\ -1 & -1 & 1 \end{bmatrix}$$

(d) Robinson "5-level" masks:

$$\begin{bmatrix} -1 & 0 & 1 \\ -2 & 0 & 2 \\ -1 & 0 & 1 \end{bmatrix} \qquad \begin{bmatrix} 0 & 1 & 2 \\ -1 & 0 & 1 \\ -2 & -1 & 0 \end{bmatrix}$$

The tables illustrates only two of the eight masks in each set: the remaining masks can in each case be generated by symmetry operations. For the 3-level and 5-level operators, four of the eight available masks are inverted versions of the other four (see text).

5.3   The template matching approach

Table 5.2 shows four sets of well-known TM masks for edge detection. These masks were originally (Prewitt, 1970; Kirsch, 1971; Robinson, 1977) introduced on an intuitive basis, starting in two cases from the DG masks shown in Table 5.1. In all cases the eight masks of each set are obtained from a given mask by permuting the mask coefficients cyclically. By symmetry, this is a good strategy for even permutations but symmetry alone does not justify it for odd permutations: the situation is explored in more detail below.

Note first that four of the "3-level" and four of the "5-level" masks can be generated from the other four of their set by sign inversion. This means that in either case only four convolutions need to be performed at each pixel neighbourhood, thereby saving computational effort. This is an obvious procedure if the basic idea of the TM approach is regarded as one of comparing intensity gradients in the eight directions. The two operators which do not employ this strategy were developed much earlier on some unknown intuitive basis.

Before proceeding, we note the rationale behind the Robinson "5-level" masks. These were intended (Robinson, 1977) to emphasize the weights of diagonal edges in order to compensate for the characteristics of the human eye, which tends to enhance vertical and horizontal lines in images. Normally, image analysis is concerned with computer interpretation of images and an isotropic response is required. Thus the "5-level" operator is a special-purpose one that need not be discussed further here.

These considerations show that the four template operators mentioned above have limited theoretical justification. It is therefore worth studying the situation in more depth and this is done in the next section.

5.4 Theory of 3 × 3 template operators

Before analysing the performance of TM operators, we note that they are likely to be used with a variety of types of edge, including in particular the "sudden step" edge, the "slanted step" edge, the "planar" edge, and various intermediate edge profiles (see Fig. 5.1). It is worth noting that the sudden step edge has very frequently been used as a paradigm with which to investigate and test the performance of edge template masks: this is highly convenient as its use eases much of the mathematical analysis, as will be seen below.

In what follows it is assumed that eight masks are to be used, for angles differing by 45°. In addition, four of the masks differ from the others only in sign, since this seems unlikely to result in any loss of performance. Symmetry requirements then lead to the following masks for 0° and 45°, respectively:

$$\begin{bmatrix} -A & 0 & A \\ -B & 0 & B \\ -A & 0 & A \end{bmatrix} \qquad \begin{bmatrix} 0 & C & D \\ -C & 0 & C \\ -D & -C & 0 \end{bmatrix}$$

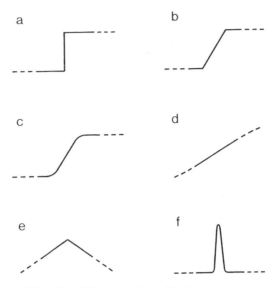

Fig. 5.1 Edge models: (a) sudden step edge; (b) slanted step edge; (c) smooth step edge; (d) planar edge; (e) roof edge; (f) line edge. The effective profiles of edge models are nonzero only within the stated neighbourhood. The slanted step and the smooth step are approximations to realistic edge profiles: the sudden step and the planar edge are extreme forms that are useful for comparisons (see text). The roof and line edge models are shown for completeness only and are not considered further in this chapter.

For a step edge, the computation devolves into a determination of the areas of various triangles and other shapes shown in Fig. 5.2. As a result the step edge responses are computed for the 0° mask as:

$$0° \text{ response} = 2A + B \tag{5.6}$$

$$45° \text{ response} = A + B \tag{5.7}$$

$$\alpha\text{-axis response} = (1 + X + Y - W)A + B$$
$$= 2(1 - W)A + B \tag{5.8}$$

and for the 45° mask as:

$$0° \text{ response} = C + D \tag{5.9}$$

$$45° \text{ response} = 2C + D \tag{5.10}$$

$$\alpha\text{-axis response} = (1 + T + U - V)C + D$$
$$= 2(1 - V)C + D \tag{5.11}$$

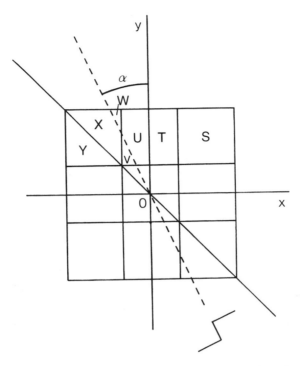

Fig. 5.2 Geometry for computation of step edge responses in a 3×3 neighbourhood: α, angle of the step edge; T, U, V, W, X, Y, areas of various triangles and other shapes; S, area of a single pixel (taken as unity).

Note that the α-axis response expressions given above are correct only for α in the range arctan $(1/3) \leqslant \alpha \leqslant$ arctan (1). Values for the areas V and W are given in terms of α by the expressions:

$$V = (1 - \tan \alpha)^2/8 \tan \alpha \qquad (5.12)$$

and

$$W = (3 \tan \alpha - 1)^2/8 \tan \alpha \qquad (5.13)$$

Applying these formulae to the 3-level and 5-level operators (see Table 5.2) immediately leads to the additional conditions A = C, B = D. Thus the operator responses for 0° and 45° step edges are exactly equal (it is easy to see that this is also true for the Prewitt and Kirsch operators). This was clearly one of the original design ideas in each case. It is also possible to determine the changeover angle α at which the 0° and 45° masks give equal responses. This occurs for

$$2(1 - W)A + B = 2(1 - V)C + D \tag{5.14}$$

which leads (for operators with A = C and B = D) to

$$W = V \tag{5.15}$$

Clearly, this means that $\alpha = \arctan(1/2)$, or 26.5°. Thus it has been shown that 26.5° is the changeover angle for *any* operator of the above type in which 45° masks are generated from 0° masks merely by permuting coefficients cyclically. A more complex calculation shows that 26.5° is the changeover angle for any set of masks in which the 0° mask has reflection symmetry in the x-axis and in which the 45° mask is obtained by permuting the coefficients cyclically. This explains the fact that all four masks in Table 5.2 have a changeover angle of 26.5°—i.e. in no case is the changeover angle equal to the ideal value of 22.5° (see for example, Table 5.3).

As noted earlier, equalization of the respective mask response in the 0° and 45° directions was essentially one of the main specifications underlying

Table 5.3 Angular variations for the Robinson 3-level TM operator.

Sudden step edge response			Planar edge response		
Edge angle	0° mask	45° mask	Edge angle	0° mask	45° mask
0.00	3.00	2.00	0.00	3.00	2.00
5.00	3.00	2.17	5.00	2.99	2.17
10.00	3.00	2.35	10.00	2.95	2.32
15.00	3.00	2.54	15.00	2.90	2.45
20.00	2.99	2.72	20.00	2.82	2.56
25.00	2.91	2.85	25.00	2.72	2.66
30.00	2.77	2.92	30.00	2.60	2.73
35.00	2.57	2.97	35.00	2.46	2.79
40.00	2.32	2.99	40.00	2.30	2.82
45.00	2.00	3.00	45.00	2.12	2.83

| Changeover angle = 26.5° | | | Changeover angle = 26.5° | | |

the masks of Table 5.2. However, the above discussion indicates that another relevant specification is that of making estimates of local edge orientation as accurate as possible: in practice, this means making the changeover angle equal to 22.5°.

To find how this affects the mask coefficients, we employ the strategy of ensuring that intensity gradients follow the rules of vector addition. If the pixel intensity values within a 3×3 neighbourhood are

$$
\begin{array}{ccc}
a & b & c \\
d & e & f \\
g & h & i
\end{array}
$$

then estimation of the 0°, 90° and 45° components of gradient by the earlier general masks gives:

$$g_0 = A(c + i - a - g) + B(f - d) \tag{5.16}$$

$$g_{90} = A(a + c - g - i) + B(b - h) \tag{5.17}$$

$$g_{45} = C(b + f - d - h) + D(c - g) \tag{5.18}$$

If vector addition is to be valid, then:

$$g_{45} = (g_0 + g_{90})/\sqrt{2} \tag{5.19}$$

Equating coefficients of a, b, c, d, e, f, g, h, i leads to the self-consistent pair of conditions:

$$C = B/\sqrt{2} \tag{5.20}$$

$$D = A\sqrt{2} \tag{5.21}$$

A further requirement is for the masks to give equal responses at 22.5°. This leads to the formula

$$B/A = \sqrt{2}\, \frac{9t^2 - (14 - 4\sqrt{2})t + 1}{t^2 - (10 - 4\sqrt{2})t + 1} \tag{5.22}$$

where $t = \tan 22.5°$, so that

$$B/A = (13\sqrt{2} - 4)/7 = 2.055 \tag{5.23}$$

This turns out to give a value for B/A in line with that for an optimized Sobel operator (see below).

The angular variations obtained with these masks are shown in Table 5.4. They are peculiar in giving optimal estimates of edge orientation for both step and planar edges and in giving relatively accurate estimates of edge magnitude for planar edges. However, their magnitude response for step edges is not particularly accurate. In this respect it is interesting to compare the results of Table 5.3 with those of Table 5.4.

Table 5.4   Angular variations for the "vector addition" operator.*

Sudden step edge response			Planar edge response		
Edge angle	0° mask	45° mask	Edge angle	0° mask	45° mask
0.00	4.06	2.87	0.00	4.06	2.87
5.00	4.06	3.12	5.00	4.04	3.11
10.00	4.06	3.38	10.00	3.99	3.32
15.00	4.06	3.65	15.00	3.92	3.51
20.00	4.05	3.92	20.00	3.81	3.67
25.00	3.97	4.10	25.00	3.68	3.81
30.00	3.82	4.21	30.00	3.51	3.92
35.00	3.62	4.27	35.00	3.32	3.99
40.00	3.37	4.31	40.00	3.11	4.04
45.00	3.06	4.32	45.00	2.87	4.06
Changeover angle = 22.5°			Changeover angle = 22.5°		

$$*0° \text{ mask:} \begin{bmatrix} -1.000 & 0.000 & 1.000 \\ -2.055 & 0.000 & 2.055 \\ -1.000 & 0.000 & 1.000 \end{bmatrix} \quad 45° \text{ mask:} \begin{bmatrix} 0.000 & 1.453 & 1.414 \\ -1.453 & 0.000 & 1.453 \\ -1.414 & -1.453 & 0.000 \end{bmatrix}$$

Finally, careful examination of equations (5.6)–(5.11) shows that coefficients B and D produce an isotropic contribution to the step edge response for the 0° and 45° masks over a wide range of angles. This suggests that for greatest angular sensitivity these coefficients should be kept small, although the situation is inevitably more complicated for a planar edge. On the other hand, for *least* angular sensitivity, and hence greater constancy and accuracy in the estimation of edge magnitude, we may set B = D and A = C = 0. As shown in Table 5.5, this strategy works well for step edges but is not good for planar edges. Note that in the masks of Table 5.5, only two of the coefficients are nonzero, so there is very little averaging effect which can help to suppress pixel intensity noise. These considerations show clearly that a number of competing factors are involved in the design of effective edge detection operators and that it will be impossible to meet all of them with a single set of masks: hence the desired application must be specified clearly in advance.

Table 5.5   Angular variations for a simple 3×3 TM operator.*

Sudden step edge response			Planar edge response		
Edge angle	0° mask	45° mask	Edge angle	0° mask	45° mask
0.00	1.00	1.00	0.00	1.00	1.00
5.00	1.00	1.00	5.00	1.00	1.08
10.00	1.00	1.00	10.00	0.98	1.16
15.00	1.00	1.00	15.00	0.97	1.22
20.00	1.00	1.00	20.00	0.94	1.28
25.00	1.00	1.00	25.00	0.91	1.33
30.00	1.00	1.00	30.00	0.87	1.37
35.00	1.00	1.00	35.00	0.82	1.39
40.00	1.00	1.00	40.00	0.77	1.41
45.00	1.00	1.00	45.00	0.71	1.41
Changeover angle = 22.5°			Changeover angle = 0°		

$$*0° \text{ mask:} \begin{bmatrix} 0 & 0 & 0 \\ -1 & 0 & 1 \\ 0 & 0 & 0 \end{bmatrix} \quad 45° \text{ mask:} \begin{bmatrix} 0 & 0 & 1 \\ 0 & 0 & 0 \\ -1 & 0 & 0 \end{bmatrix}$$

5.5   Summary—design constraints and conclusions

So far we have studied existing template masks for edge detection and examined the underlying theory. Detailed calculations have led to the design of TM operators having a variety of characteristics. At the same time, a number of design aims have become evident:

 (i)   the need to optimize accuracy in the estimation of edge magnitude;
 (ii)   the need to optimize accuracy in the estimation of edge orientation;
 (iii) the need to suppress noise during edge detection;
 (iv)   the need to optimize operators for different edge profiles.

These design aims have been found to conflict with each other. This means that tradeoffs between them exist and that compromises have to be made. In general, it is clear (Davies, 1986a) that:

(i)   different masks are needed for the accurate estimation of edge magnitude and orientation;
(ii) optimization of noise suppression imposes further conditions on TM masks;
(iii) obtaining sets of masks by permuting coefficients "cyclically" in a square neighbourhood is *ad hoc* and cannot be relied upon to produce useful results;
(iv) the vector addition approach to the design of template masks is optimal for planar edge profiles but also gives acceptable results for step edges;
(v)   there is some danger in tailoring masks specifically for step edge profiles, since they may then have poor responses for planar and other edge profiles.

Having obtained some insight into the process of designing TM masks for edge detection, we next move on to study the design of DG masks.

5.6   The design of differential gradient operators

This section studies the design of DG operators. These include the Roberts 2×2 pixel operator and the Sobel and Prewitt 3×3 pixel operators (Roberts, 1965; Prewitt, 1970; for the Sobel operator see Pringle, 1969; Duda and Hart, 1973, p. 271) (see Table 5.1). The Prewitt or "gradient smoothing" type of operator has been extended to larger pixel neighbourhoods by Prewitt (1970) and others (Brooks, 1978; Haralick, 1980) (see Table 5.6). In these instances the basic rationale is to model local edges by the best fitting plane over a convenient size of neighbourhood. Mathematically, this amounts to obtaining suitably weighted averages to estimate slope in the x and y directions. As pointed out by Haralick (1980), the use of equally weighted averages to measure slope in a given direction is incorrect: the proper weightings to use are given by the masks listed in Table 5.6. Thus the Roberts and Prewitt operators are apparently optimal, whereas the Sobel operator is not. More will be said about this below.

A full discussion of the edge detection problem involves consideration of the accuracy with which edge magnitude and orientation can be estimated when the local intensity pattern cannot be assumed to be planar. To analyse

Table 5.6 Masks for estimating components of gradient in square neighbourhoods.

	M_x	M_y

(a) 2 × 2 neighbourhood

$$\begin{bmatrix} -1 & 1 \\ -1 & 1 \end{bmatrix} \qquad\qquad \begin{bmatrix} 1 & 1 \\ -1 & -1 \end{bmatrix}$$

(b) 3 × 3 neighbourhood

$$\begin{bmatrix} -1 & 0 & 1 \\ -1 & 0 & 1 \\ -1 & 0 & 1 \end{bmatrix} \qquad\qquad \begin{bmatrix} 1 & 1 & 1 \\ 0 & 0 & 0 \\ -1 & -1 & -1 \end{bmatrix}$$

(c) 4 × 4 neighbourhood

$$\begin{bmatrix} -3 & -1 & 1 & 3 \\ -3 & -1 & 1 & 3 \\ -3 & -1 & 1 & 3 \\ -3 & -1 & 1 & 3 \end{bmatrix} \qquad\qquad \begin{bmatrix} 3 & 3 & 3 & 3 \\ 1 & 1 & 1 & 1 \\ -1 & -1 & -1 & -1 \\ -3 & -3 & -3 & -3 \end{bmatrix}$$

(d) 5 × 5 neighbourhood

$$\begin{bmatrix} -2 & -1 & 0 & 1 & 2 \\ -2 & -1 & 0 & 1 & 2 \\ -2 & -1 & 0 & 1 & 2 \\ -2 & -1 & 0 & 1 & 2 \\ -2 & -1 & 0 & 1 & 2 \end{bmatrix} \qquad\qquad \begin{bmatrix} 2 & 2 & 2 & 2 & 2 \\ 1 & 1 & 1 & 1 & 1 \\ 0 & 0 & 0 & 0 & 0 \\ -1 & -1 & -1 & -1 & -1 \\ -2 & -2 & -2 & -2 & -2 \end{bmatrix}$$

The above masks can be regarded as extended Prewitt masks. The 3 × 3 masks are Prewitt masks, included in this table for completeness. The 2 × 2 masks can be shown to be identical to Roberts operator masks rotated through 45°. In all cases weighting factors have been omitted in the interests of simplicity, as they are throughout this chapter.

the situation we again take the "worst case", i.e. that of a sudden step edge profile, which is as distinct as possible from the planar edge approximation.

There have been a number of analyses of the angular dependencies of edge detection operators for a step edge approximation. In particular, O'Gorman (1978) considered the variation of estimated versus actual angle resulting from a step edge observed within a square neighbourhood (see also

Brooks, 1978): note that the case considered was that of a *continuous* rather than a *discrete* lattice of pixels. This was found to lead to a smooth variation with angular error varying from zero at 0° and 45° to a maximum of 6.63° at 28.37° (where the estimated orientation was 21.74°), the variation for angles outside this range being replicated by symmetry. Abdou and Pratt (1979) obtained similar variations for the Sobel and Prewitt operators (in a *discrete* lattice), the respective maximum angular errors being 1.36° and 7.38° (Davies, 1984b). It seems that the Sobel operator has angular accuracy that is close to optimal because it is close to being a "truly circular" operator. This point is discussed in more detail below.

5.7   The concept of a circular operator

It was stated above that when step edge orientation is estimated in a square neighbourhood, an error of up to 6.6° can result. Such an error does not arise with a planar edge approximation, since fitting of a plane to a planar edge profile within a square window can be carried out exactly. Errors appear only when the edge profile differs from the ideal planar form, within the square neighbourhood—with the step edge probably being a "worst case".

One way to limit errors in the estimation of edge orientation might be to restrict observation of the edge to a circular neighbourhood. In the continuous case this is sufficient to reduce the error to zero for all orientations, since symmetry dictates that there is only one way of fitting a plane to a step edge within a circular neighbourhood, assuming that all planes pass through the same central point; the estimated orientation θ is then equal to the actual angle φ. A rigorous calculation along the lines indicated by Brooks (1976), which results in the following formula for a square neighbourhood (O'Gorman, 1978):

$$\tan \theta = 2 \tan \varphi / (3 - \tan^2 \varphi) \qquad 0° \leqslant \varphi \leqslant 45° \qquad (5.24)$$

leads to the formula

$$\tan \theta = \tan \varphi, \qquad \text{i.e. } \theta = \varphi \qquad (5.25)$$

for a circular neighbourhood (Davies, 1984b). Similarly, zero angular error results from fitting a plane to an edge of *any* profile within a circular neighbourhood, in the continuous approximation. Indeed, for an edge surface of arbitrary shape, the only problem is whether the mathematical best-fit plane coincides with one that is subjectively desirable (and, if not, a *fixed* angular correction will be required). Ignoring such cases, the basic problem is how to approximate a circular neighbourhood in a digitized image of small dimensions, containing typically 3×3 or 5×5 pixels.

A simple way to attempt this in a 3×3 window might be to approximate the neighbourhood by an octagon, giving the corner pixels a weight reflecting the active area of the octagon within them. A calculation of this type results in a relative corner weighting of 0.614—a value which already gives the Sobel operator significantly greater credibility. However, this approach is somewhat *ad hoc* and misleading. To proceed more systematically, we recall a fundamental principle stated by Haralick (1980):

> 'the fact that the slopes in two orthogonal directions determine the slope in any direction is well known in vector calculus. However, it seems not to be so well known in the image processing community.'

Essentially, appropriate estimates of slopes in two orthogonal directions permit the slope in any direction to be computed. For this principle to apply, appropriate estimates of the slopes have first to be made: if the components of slope are inappropriate they will not act as components of true vectors and the resulting estimates of edge orientation will be in error. This appears to be the main source of error with the Prewitt and other operators—it is not so much that the components of slope are in any instance incorrect, but rather that they are inappropriate for the purpose of vector computation since *they do not match one another adequately in the required way* (Davies, 1984b).

Following the arguments for the continuous case discussed earlier, slopes must be rigorously estimated within a circular neighbourhood. Then the operator design problem devolves into determining how best to simulate a circular neighbourhood on a discrete lattice so that errors are minimized. In order to carry this out, it is insufficient simply to weight the corner pixels after computing the masks, as implied by the octagon idea above. Instead, it is necessary to apply the circular (or octagonal) weighting while computing the masks, so that correlations between the gradient weighting and circular weighting factors are taken properly into account.

5.8 Detailed implementation of circular operators

In practice, the task of computing angular variations and error curves has to be tackled numerically, dividing each pixel in the neighbourhood into arrays of suitably small subpixels. Each subpixel is then assigned a gradient weighting (equal to the x or y displacement) and a neighbourhood weighting (equal to 1 for inside and 0 for outside a circle of radius r). Clearly, the angular accuracy of "circular" differential gradient edge detection operators must depend on the radius of the circular neighbourhood. In particular, poor accuracy would be expected for small values of r and reasonable accuracy for large values of r, as the discrete neighbourhood approaches a continuum.

The results of this study are presented in Fig. 5.3. The variations depicted represent (a) RMS angular errors and (b) maximum angular errors in the estimation of edge orientation. The structures on each variation are surprisingly smooth: they are so closely related and systematic that they can only represent statistics of the arrangement of pixels in neighbourhoods of various sizes. Details of these statistics are discussed in the next section.

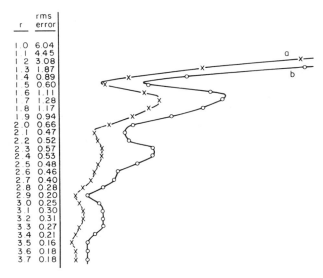

r	rms error
1.0	6.04
1.1	4.45
1.2	3.08
1.3	1.87
1.4	0.89
1.5	0.60
1.6	1.11
1.7	1.28
1.8	1.17
1.9	0.94
2.0	0.66
2.1	0.47
2.2	0.52
2.3	0.57
2.4	0.53
2.5	0.48
2.6	0.46
2.7	0.40
2.8	0.28
2.9	0.20
3.0	0.25
3.1	0.30
3.2	0.31
3.3	0.27
3.4	0.21
3.5	0.16
3.6	0.18
3.7	0.18

Fig. 5.3   Variations in angular error as a function of radius r: (a) RMS angular error; (b) maximum angular error.

Before proceeding with a detailed interpretation of the effects depicted in Fig. 5.3, three features will be apparent. First, as expected, there is a general trend to zero angular error as r tends to infinity: this is understandable, since the discrete neighbourhood is then tending towards a continuum. Second, there is a very marked periodic variation, with particularly good accuracy resulting where the circular operators best match the tessellation of the digital lattice. The third feature of interest is the fact that errors do not vanish for any finite value of r—clearly, the constraints of the problem do not permit more than the minimization of errors. Overall, these curves show that it is possible to generate a family of optimal operators (at the minima of the error curves), the first of which corresponds closely to an operator (the Sobel operator) that is known to be nearly optimal.

5.9 Structured bands of pixels in neighbourhoods of various sizes

In order to explain the variations shown in Fig. 5.3, a study was made (Davies, 1984b) of the numbers of pixels at various distances from a given pixel, in a square lattice. Since symmetry demands that in most cases 4 or 8 pixels lie at each valid distance from the central pixel (Table 5.7), the most useful measure of the relevant statistics seems to be the difference between consecutive pixel distances. These are listed out to quite a large distance (representing neighbourhoods of size of the order of 13×13) in Table 5.8. The histogram displayed in Table 5.8 indicates clearly that a few radial differences are substantially larger than the rest. It is possible to envisage these larger values as demarcating bands of pixels which are internally well packed and approximate continua as closely as is possible. Given such continua, circular neighbourhood approximations within them should be relatively accurate, whereas neighbourhoods that straddle bands would be expected to be poor approximations to continua. This substantially explains

Table 5.7 Numbers of pixels at different distances from central pixel.

Shell	Radius	No. of pixels	Shell	Radius	No. of pixels
a	0.000	1	q	5.657	4
b	1.000	4	r	5.831	8
c	1.414	4	s	6.000	4
d	2.000	4	t	6.083	8
e	2.236	8	u	6.325	8
f	2.828	4	v	6.403	8
g	3.000	4	w	6.708	8
h	3.162	8	x	7.000	4
i	3.606	8	y	7.071	12
j	4.000	4	z	7.211	8
k	4.123	8	A	7.280	8
l	4.243	4	B	7.616	8
m	4.472	8	C	7.810	8
n	5.000	12	D	8.000	4
o	5.099	8	E	8.062	16
p	5.385	8	F	8.246	8

Table 5.8   Differences in radii between shells of pixels.

Shells		Value	Histogram
			Difference between radii
a	b	1.00	*
b	c	0.41	*
c	d	0.59	*
d	e	0.24	*
e	f	0.59	*
f	g	0.17	*
g	h	0.16	*
h	i	0.44	*
i	j	0.39	*
j	k	0.12	*
k	l	0.12	*
l	m	0.23	*
m	n	0.53	*
n	o	0.10	*
o	p	0.29	*
p	q	0.27	*
q	r	0.17	*
r	s	0.17	*
s	t	0.08	*
t	u	0.24	*
u	v	0.08	*
v	w	0.31	*
w	x	0.29	*
x	y	0.07	*
y	z	0.14	*
z	A	0.07	*
A	B	0.34	*
B	C	0.19	*
C	D	0.19	*
D	E	0.06	*
E	F	0.18	*

the results of Fig. 5.3, and even suggests that these variations are not entirely specific to edge detection but reflect more fundamental limitations of digital image processing methodology. Thus the design of any image processing operator should take into account banding of pixels and look for solutions which suppress pixel position statistics as far as possible.

Table 5.9   Masks of "closed band" differential gradient edge operators.

(a) Band containing shells a–c (effective radius = 1.500)

$$\begin{bmatrix} -0.464 & 0.000 & 0.464 \\ -0.959 & 0.000 & 0.959 \\ -0.464 & 0.000 & 0.464 \end{bmatrix}$$

(b) Band containing shells a–e (effective radius = 2.121)

$$\begin{bmatrix} 0.000 & -0.294 & 0.000 & 0.294 & 0.000 \\ -0.582 & -1.000 & 0.000 & 1.000 & 0.582 \\ -1.085 & -1.000 & 0.000 & 1.000 & 1.085 \\ -0.582 & -1.000 & 0.000 & 1.000 & 0.582 \\ 0.000 & -0.294 & 0.000 & 0.294 & 0.000 \end{bmatrix}$$

(c) Band containing shells a–h (effective radius = 2.915)

$$\begin{bmatrix} 0.000 & 0.000 & -0.191 & 0.000 & 0.191 & 0.000 & 0.000 \\ 0.000 & -1.085 & -1.000 & 0.000 & 1.000 & 1.085 & 0.000 \\ -0.585 & -2.000 & -1.000 & 0.000 & 1.000 & 2.000 & 0.585 \\ -1.083 & -2.000 & -1.000 & 0.000 & 1.000 & 2.000 & 1.083 \\ -0.585 & -2.000 & -1.000 & 0.000 & 1.000 & 2.000 & 0.585 \\ 0.000 & -1.085 & -1.000 & 0.000 & 1.000 & 1.085 & 0.000 \\ 0.000 & 0.000 & -0.191 & 0.000 & 0.191 & 0.000 & 0.000 \end{bmatrix}$$

(d) Band containing shells a–i (effective radius = 3.500)

$$\begin{bmatrix} 0.000 & -0.646 & -0.815 & 0.000 & 0.815 & 0.646 & 0.000 \\ -0.963 & -1.997 & -1.000 & 0.000 & 1.000 & 1.997 & 0.963 \\ -2.458 & -2.000 & -1.000 & 0.000 & 1.000 & 2.000 & 2.458 \\ -2.962 & -2.000 & -1.000 & 0.000 & 1.000 & 2.000 & 2.962 \\ -2.458 & -2.000 & -1.000 & 0.000 & 1.000 & 2.000 & 2.458 \\ -0.963 & -1.997 & -1.000 & 0.000 & 1.000 & 1.997 & 0.963 \\ 0.000 & -0.646 & -0.815 & 0.000 & 0.815 & 0.646 & 0.000 \end{bmatrix}$$

In all cases only the x-mask is shown: the y-mask may be obtained by a trivial symmetry operation. Mask coefficients are accurate to ~0.003 but would in normal practical applications be rounded to 1- or 2-figure accuracy.

Table 5.10   Angular variations for the best operators tested.*

Actual angle (degrees)	Estimated angle (degrees)†						
	Prew	Sob	a–c	circ	a–e	a–h	a–i
0	0.00	0.00	0.00	0.00	0.00	0.00	0.00
5	3.32	4.97	5.05	5.14	5.42	5.22	5.01
10	6.67	9.95	10.11	10.30	10.81	10.28	9.94
15	10.13	15.00	15.24	15.52	15.83	14.81	14.74
20	13.69	19.99	20.29	20.64	20.07	19.73	19.90
25	17.72	24.42	24.73	25.10	24.62	25.00	25.34
30	22.62	28.86	29.14	29.48	29.89	30.02	30.16
35	28.69	33.64	33.86	34.13	35.43	34.86	34.91
40	35.94	38.87	39.00	39.15	40.30	39.71	40.01
45	45.00	45.00	45.00	45.00	45.00	45.00	45.00
RMS error	5.18	0.73	0.60	0.53	0.47	0.19	0.16

* *Key*: Prew, Prewitt; Sob, Sobel; a–c, theoretical optimum: closed band containing shells a–c; circ, actual optimum circular operator; a–e, theoretical optimum: closed band containing shells a–e; a–h, theoretical optimum: closed band containing shells a–h; a–i, theoretical optimum: closed band containing shells a–i.
† Values should be accurate to within 0.02° in each case.

The "closed band" operators predicted above are listed in Table 5.9: their angular variations appear in Table 5.10. It is seen that the Sobel operator, which is already the most accurate of the 3×3 edge gradient operators suggested previously, can be made some 30% more accurate by adjusting its coefficients to make it more circular. In addition, the closed bands idea indicates that the corner pixels of 5×5 or larger operators are best removed altogether: not only does this require less computation but also (e.g. compared with the extended Prewitt masks of Table 5.6) it actually improves performance!

It should be pointed out that many sets of mask coefficients within a 3×3 neighbourhood correspond to circular operators—i.e. some value of r can be found which generates the masks in the manner outlined above. However, the Sobel operator has been shown to be more ideal in that it also corresponds approximately to a closed band situation. Since it is then closer to the ideal analogue circular mask case, we may say that it is close to being a "truly circular" operator, thus justifying the statement at the end of Section 5.6. For neighbourhoods larger than 3×3, there are so many mask coefficients that arbitrarily designed operators will not in general be "circular" or (*a fortiori*) "truly circular".

Before leaving this topic, it should be noted that the optimal 3×3 masks obtained above numerically by consideration of circular operators are very close indeed to those obtained purely analytically in Section 5.4, for TM masks, following the rules of vector addition. In the latter case a value of 2.055 was obtained for the ratio of the two mask coefficients, whereas for circular operators the value $0.959/0.464 = 2.067 \pm 0.015$ is obtained. Clearly this is no accident and it is very satisfying that a coefficient that was formerly regarded as *ad hoc* (Kittler, 1983) is in fact optimizable and can be obtained in closed form (see Section 5.4).

5.10 The systematic design of differential edge operators

The family of "circular" differential gradient edge operators studied in Sections 5.6–5.9 incorporates only one design parameter—the radius r. Only a limited number of values of this parameter permit optimum accuracy for estimation of edge orientation to be attained.

It is worth considering what additional properties this one parameter can control and how it should be adjusted during operator design. Clearly, it affects the signal-to-noise ratio of the final edges and also the resolution (these affects are important in the design of any operator, for example in the design of an averaging filter). Signal-to-noise ratio varies linearly with the radius of the circular neighbourhood, since signal is proportional to area and Gaussian noise is proportional to $\sqrt{(\text{area})}$, i.e. to radius: this type of averaging process for improving signal-to-noise ratio is, of course, well known. Likewise, the accuracy of linear measurement is determined by the number of pixels over which averaging occurs and hence is proportional to operator radius. Resolution, in contrast, and the resulting "scale" of the edge image, varies inversely with radius, since relevant linear properties of the image are averaged over the active area of the neighbourhood. Note that the accuracy with which the positions of objects may be measured also depends on resolution; however, achievable resolution depends on available signal-to-noise ratio and, as has been seen, this itself varies with the radius of the neighbourhood. Here the final tradeoff depends on the scale of the noise one is attempting to suppress. Finally, computational load, and the associated cost of hardware for speeding up the processing, is generally at least in proportion to the number of pixels in the neighbourhood, and hence in proportion to r^2.

These facts explain various points that appear in the literature, such as the better fidelity of the Roberts' operator (see for example Bryant and Bouldin, 1979) and its higher noise levels (Abdou and Pratt, 1979). In addition, it is now seen that an exact tradeoff is operative, in that operator radius carries with it four immediate properties which have not always been seen to be intimately related: namely signal-to-noise ratio, resolution, accuracy and hardware/computational cost. Since none of these can be altered without the others being modified as well, one is always in an engineering situation of having to make a compromise to suit circumstances. On the other hand, the single design parameter concept is complicated by the discreteness of the digital image processing milieu; hence in practice, it is necessary to fall back on one or other of the "closed band" radii discussed above for the best combination of properties.

5.11 Problems with the above approach—some alternative schemes

Interesting though the above ideas may be, they have their own inherent problems. In particular, they take no account of the displacement E of the edge from the centre of the neighbourhood, or of the effects of noise in biasing the estimates of edge magnitude and orientation. Now it is possible to show that a Sobel operator gives *zero* error in the estimation of step edge orientation when $|\theta| \leqslant \arctan(1/3)$ and $|E| \leqslant (\cos\theta - 3\sin|\theta|)/2$. In fact, for a 3×3 operator of the form

$$\begin{bmatrix} -1 & 0 & 1 \\ -B & 0 & B \\ -1 & 0 & 1 \end{bmatrix} \qquad \begin{bmatrix} 1 & B & 1 \\ 0 & 0 & 0 \\ -1 & -B & -1 \end{bmatrix}$$

applied to the edge

a	$a + h(0.5 - E\sec\theta + \tan\theta)$	$a + h$
a	$a + h(0.5 - E\sec\theta)$	$a + h$
a	$a + h(0.5 - E\sec\theta - \tan\theta)$	$a + h$

Lyvers and Mitchell (1988) found that the estimated orientation is:

$$\varphi = \arctan[2B\tan\theta/(B+2)] \tag{5.26}$$

which immediately shows why the Sobel operator should give zero error for a specific range of θ and E. However, this is somewhat misleading, since considerable errors arise outside this region. Not only do they arise when $E = 0$, as assumed in the foregoing sections, but also they vary strongly with E. Indeed, the maximum errors for the Sobel and Prewitt operators rise to 2.90° and 7.43° respectively in this more general case (the corresponding RMS errors being 1.20° and 4.50°). Hence a full analysis should be performed to determine how to reduce the maximum and average errors. Lyvers and Mitchell (1988) carried out an empirical analysis and constructed a lookup table with which to correct the orientations estimated by the Sobel operator, the maximum error being reduced to 2.06°. Their approach was motivated by the "iterated Sobel" of Iannino and Shapiro (1979).

Another scheme that reduces the error is the moment-based operator of Reeves *et al.* (1983). This leads to Sobel-like 3×3 masks which are essentially identical to the 3×3 masks of Davies (1984b), both having $B = 2.067$ (for $A = 1$). However, the moment method can also be used to estimate the edge position E if additional masks are used to compute second-order moments of intensity. Hence it is possible to make a very significant improvement in performance by using a 2-D lookup table to estimate orientation: the result is that the maximum error is reduced from 2.83° to 0.135° for 3×3 masks, and from 0.996° to 0.0042° for 5×5 masks.

However, Lyvers and Mitchell (1988) found that much of this additional accuracy is lost in the presence of noise, and RMS standard deviations of edge orientation estimates are already around 0.5° for all 3×3 operators at 40 db signal-to-noise ratios. The reasons for this are quite simple. Each pixel intensity has a noise component which induces errors in its weighted mask components; the combined effects of these errors can be estimated assuming that they arise independently, so that their variances add (Davies, 1987k). Thus noise contributions to the x and y components of gradient can be computed. These provide estimates for the components of noise along and perpendicular to the edge gradient vector (Fig. 5.4): the edge orientation for a Sobel operator turns out to be affected by an amount $\sqrt{12}\sigma/4h$ radians, where σ is the standard deviation on the pixel intensity values and h is the edge contrast. This explains the angular errors given by Lyvers and Mitchell, if Pratt's (1978, p. 498) definition of signal-to-noise ratio (in db) is used:

$$S/N = 20 \log_{10} (h/\sigma) \qquad (5.27)$$

An alternative approach is that of Haralick (1984): this uses a "facet model" for edge detection and computes edge parameters from a polynomial surface fit at every pixel location. To estimate edge orientation accurately, a

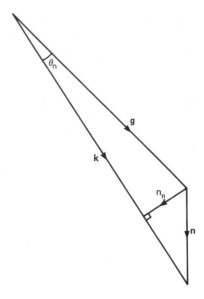

Fig. 5.4 Calculating angular errors arising from noise: **g**, intensity gradient vector; **n**, noise vector; **k**, resultant of intensity gradient and noise vector; n_n, normal component of noise; θ_n, noise-induced orientation error.

third-order model had to be used. This work was extended by Zuniga and Haralick (1987) using a special averaging technique called the integrated directional derivative (IDD). This involves integrating the directional derivative of the intensity over rectangular regions of length L and width W centred over each pixel and aligned in all possible directions θ: at any location, the edge magnitude is taken as the maximum value of the IDD, and the edge orientation as the direction which gives the maximum value. In practice equal values of L and W have been employed—1.8 for a 5×5 neighbourhood, 2.5 for 7×7. The theory permits new masks to be computed and, indeed, application of the operator then reduces merely to application of these masks with the usual formulae for edge direction and magnitude. Under zero-noise conditions, the 5×5 and 7×7 operators give worst bias values of 1.05° and 0.65°, respectively. These accuracies are not quite as good as for the moment-based detectors and certainly not as good as for the moment detectors with 2-D lookup tables. However, performance in the presence of noise (S/N ratio of 30 db or worse) is virtually identical. That it is possible to obtain such improvements in edge orientation accuracy is ultimately due to the availability of many adjustable mask coefficients, though clearly suitable strategies have to be found for deducing them.

A totally different approach was developed by Canny (1986). He used functional analysis to derive an optimal function for edge detection, starting with three optimization criteria—good detection, good localization, and only one response per edge under white noise conditions. The analysis is too technical to be discussed in detail here. However, the 1-D function found by Canny is accurately approximated by the derivative of a Gaussian: this is then combined with a Gaussian of identical σ in the perpendicular direction, truncated at 0.001 of its peak value, and split into suitable masks. Underlying this method is the idea of locating edges at local maxima of gradient magnitude of a Gaussian-smoothed image. In addition, the Canny implementation employs a hysteresis operation on edge magnitude in order to make edges reasonably connected. Finally, a multiple scale method is employed to analyse the output of the edge detector. More will be said about these points below. Lyvers and Mitchell (1988) tested the Canny operator and found it to be significantly less accurate for orientation estimation than the moment and IDD operators described above. In addition, it needed to be implemented using 180 masks and hence took enormous computation time. The latter investigation therefore seems to indicate that this operator is useful mainly for the principles it embodies concerning the underlying process of edge detection: however, the situation could change if a suitable fast hardware implementation became available (cf. the architecture mentioned by Ruff, 1987).

Other operators have been designed for detecting edges. Of particular interest is the Hueckel approach developed in the early 1970s: this involved expanding intensity variations over (ideally) rather large circular regions in terms of complete sets of Fourier-like orthonormal basis functions (Hueckel, 1971, 1973). This approach turns out to be quite computationally intensive and is not described in detail here: it suffices to note that the tests of Lyvers and Mitchell (1988) showed that it has certain problems of convergence, especially in the presence of noise, while it involves typically ten times the computation of other operators of comparable mask size.

An operator that has been of great historical importance is that of Marr and Hildreth (1980). The motivation for the design of this operator was the modelling of certain psychophysical processes in mammalian vision. The basic rationale is to find the Laplacian of the Gaussian-smoothed ($\nabla^2 G$) image and then to obtain a "raw primal sketch" as a set of zero-crossing lines. The Marr–Hildreth operator does not use any form of threshold since it merely assesses where the $\nabla^2 G$ image passes through zero. This feature is attractive, since working out threshold values is a difficult and unreliable task. However, the Gaussian smoothing procedure can be applied at a variety of scales, and in one sense the scale is a new parameter that substitutes for the threshold. In fact, a major feature of the Marr–Hildreth approach that has been very influential in later work (Witkin, 1983; Bergholm, 1986), is the fact that zero-crossings can be obtained at several scales, giving the potential for more powerful semantic processing: clearly, this necessitates finding means for combining all the information in a systematic and meaningful way. This may be carried out by bottom-up or top-down approaches, and there is much discussion in the literature about methods for carrying out these processes. However, it is worth remarking that in many (especially industrial inspection) applications, one is interested in working at a particular resolution and considerable savings in computation can then be made. It is also noteworthy that the Marr–Hildreth operator is reputed to require neighbourhoods of at least 35×35 for proper implementation (Brady, 1982). Nevertheless, other workers have implemented the operator in much smaller neighbourhoods, down to 5×5. Wiejak *et al.* (1985) showed how to implement the operator using linear smoothing operations to save computation. Lyvers and Mitchell (1988) reported orientation accuracies using the Marr–Hildreth operator that are not especially high ($2.47°$ for a 5×5 operator and $0.912°$ for a 7×7 operator, in the absence of noise).

It was noted above that those edge detection operators that are applied at different scales lead to different edge maps at different scales. In such cases it turns out that certain edges that are present at lower scales disappear at larger scales; in addition, edges that are present at both low and high scales appear shifted or merged at higher scales. Bergholm (1986) demonstrated the possible occurrence of elimination, shifting and merging, while Yuille and Poggio (1986) showed that edges that are present at high resolution should not disappear at some lower resolution. These aspects of edge location are now becoming well understood.

Finally, Overington and Greenway (1987) sound a cautionary note about the accuracy of the zero-crossing approach to edge detection. They found that it is significantly more accurate to estimate edge position and edge orientation by interpolating to local maximum values of edge gradient than by locating zero-crossings of the $\nabla^2 G$ image. The reason they gave for this is that accurate estimation of these zero-crossing positions relies on *third* derivatives of the image intensity function, which tend to be affected seriously by noise.

5.12   Concluding remarks

It will be clear from the above sections that the design of edge detection operators has by now been taken to quite an advanced stage, so that edges can be located to subpixel accuracy and orientated to fractions of a degree.

In addition, edge maps may be made at several scales and the results correlated as an aid to image interpretation. Unfortunately, some of the schemes that have been devised to achieve these things (and especially those outlined in the previous section) are fairly complex and tend to consume considerable computation. In many applications this complexity may not be justified because the application requirements are, or can reasonably be made, quite restricted. Furthermore, there is often the need to save computation for real-time implementation. For these reasons, the remainder of this book reverts to employing edges found by a single low-resolution detector. In practice this will normally be a Sobel operator, which provides a good balance between computational load and orientation accuracy. Indeed, virtually all the examples in Part 2 of the book have been implemented using a Sobel operator (in addition, the RMS error in estimating edge orientation using this operator was found by Lyvers and Mitchell to be 1.20°, and in later chapters this quantity will generally be taken as a round 1°). This does not in any way invalidate the latest methods, particularly those involving studies of edges at various scales: such methods come into their own in applications such as general scene analysis, where vision systems are required to cope with largely unconstrained image data.

This chapter has largely completed the task of segmentation of images at low level. The chapters that follow move on to consider the shapes of objects that have been found by the thresholding and edge detection schemes discussed in the last two chapters. In fact, Chapter 6 studies shapes by analysis of the regions over which objects extend, while Chapter 7 studies shapes by considering their boundary patterns.

5.13   Bibliographical and historical notes

As seen in the first few sections of this chapter, early attempts at edge detection tended to employ numbers of template masks that could locate edges at various orientations. Often these masks were *ad hoc* in nature, and after 1980 this approach finally gave way to the differential gradient approach that had already existed in various forms for a considerable period (see the influential paper by Haralick, 1980).

The Frei–Chen approach is of particular interest in that it takes a set of nine 3×3 masks forming a complete set within this size of neighbourhood—of which one tests for brightness, four test for edges and four test for lines (Frei and Chen, 1977). While interesting, the Frei–Chen edge masks do not correspond to those devised for optimal edge detection and it appears possible that edge detection in the set as a whole may have been compromised in some way in order to make a complete "toolbox". Indeed, Lacroix (1988) found errors in the Frei–Chen approach which may ultimately show how to rationalize the situation.

The Hueckel scheme (1971, 1973) provided an early alternative to these approaches, although it was rather computationally intensive and therefore perhaps not so widely used: to tackle this problem, a modified form of the scheme was developed by Mérõ and Vassy (1975) but it does not appear to have been used particularly widely.

Meanwhile, psychophysical work by Marr (1976), Wilson and Giese (1977) and others provided another line of development for edge detection. This led to the well-known paper by Marr and Hildreth (1980) which was highly influential in the following few years. This spurred others to think of alternative schemes, and the Canny (1986) operator emerged from this vigorous milieu. In fact, the Marr–Hildreth operator seems to have been the first to preprocess images in order to study them at different scales—a technique that has survived and expanded considerably (see for example, Witkin, 1983; Bergholm, 1986; Yuille and Poggio, 1986). The computational problems of the Marr–Hildreth operator have kept others thinking along more traditional lines, and the work by Reeves *et al.* (1983), Haralick (1984) and Zuniga and Haralick (1987) falls into this category. Lyvers and Mitchell (1988) reviewed many of these papers and made their own suggestions. Other studies (McIvor, 1988; Petrou and Kittler, 1988) show that the subject is by no means dead, although it is not improbable that diminishing returns are now occurring and that at this stage the main advances will be at a higher level: in this context the work of Sjöberg and Bergholm (1988), which finds rules for discerning shadow edges from object edges, is of great interest. Nevertheless, there are other nuances which are worth considering, such as the accuracy of edge orientation for partly saturated grey-scale images (Davies and Celano, 1993a); and the development of edge detectors which are suited to particular purposes, such as finding high accuracy subedges for use in conjunction with a Hough transform (Davies, 1992c).

5.14 Problems

1. Prove equations (5.22) and (5.23).
2. Check the results quoted in Section 5.11, giving the conditions under which the Sobel operator leads to zero error in the estimation of edge orientation. Proceed to prove equation (5.26).

6

Binary Shape Analysis

6.1 Introduction

Over the past 30 years 2-D shape analysis has provided the main means by which objects are recognized and located in digital images. Fundamentally, 2-D shape has been important because it uniquely characterizes many types of object, from keys to characters, from gaskets to spanners, and from fingerprints to chromosomes, while in addition it can be represented by patterns in simple binary images. Chapter 1 showed how the template matching approach leads to a combinatorial explosion even when fairly basic patterns are to be found, so preliminary analysis of images to find features constitutes a crucial stage in the process of efficient recognition and location. Thus the capability for binary shape analysis is a very basic requirement for practical visual recognition systems.

In fact, 30 years of progress have provided an enormous range of shape analysis techniques and a correspondingly large range of applications. Clearly, it will be impossible to cover the whole field within the confines of a single chapter—so completeness will not even be attempted (the alternative of a catalogue of algorithms and methods, all of which are covered only in brief outline, is eschewed). At one level, the main topics covered are examples with their own intrinsic interest and practical application, and at another level they introduce matters of fundamental principle. Recurring themes are the central importance of connectedness for binary images; the contrasts between local and global operations on images and between different representations of image data; the need to optimize accuracy and computational efficiency; and the compatibility of algorithms and hardware. The chapter starts with a discussion of how connectedness is measured in binary images.

6.2   Connectedness in binary images

This section begins with the assumption that objects have been segmented, by thresholding or other procedures, into sets of 1's in a background of 0's (see Chapters 2–4). At this stage it is important to realize that a second assumption is already being made implicitly—that it is easy to demarcate the boundaries between objects in binary images. However, in an image that is represented digitally in rectangular tessellation, a problem arises with the definition of connectedness. Consider a dumbell-shaped object[*]:

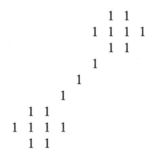

At its centre, this object has a segment of the form

$$0 \quad 1$$
$$1 \quad 0$$

which separates two regions of background. At this point diagonally adjacent 1's are regarded as being connected, whereas diagonally adjacent 0's are regarded as disconnected—a set of allocations which seems inconsistent. However, we can hardly accept a situation where a connected diagonal pair of 0's crosses a connected diagonal pair of 1's without causing a break in either case. Similarly, we cannot accept a situation in which a disconnected diagonal pair of 0's crosses a disconnected diagonal pair of 1's without there being a join in either case. Hence a symmetrical definition of connectedness is not possible and it is conventional to regard diagonal

---

[*] All unmarked image points are taken to have the binary value 0.

neighbours as connected only if they are foreground, i.e. the foreground is "8-connected" and the background is "4-connected". This convention is followed in the subsequent discussion.

6.3 Object labelling and counting

Now we have a consistent definition of connectedness, we can unambiguously demarcate all objects in binary images and should be able to devise algorithms for labelling them uniquely and counting them. Labelling may be achieved by scanning the image sequentially until a 1 is encountered on the first object; a note is then made of the scanning position, and a "propagation" routine is initiated to label the whole of the object with a 1: since the original image space is already in use, a separate image space has to be allocated for labelling. Next, the scan is resumed, ignoring all points already labelled, until another object is found; this is labelled with a 2 in the separate image space. This procedure is continued until the whole image has been scanned and all the objects have been labelled (Fig. 6.1). Implicit in this procedure is the possibility of propagating through a connected object. Suppose at this stage that no method is available for limiting the field of the propagation routine, so that it has to scan the whole image space. Then the propagation routine takes the form:

```
repeat
    for all points in image do
        if point is in object
            and next to a propagating region labelled N
        then label this point N also
until no further change;                                        (6.1)
```

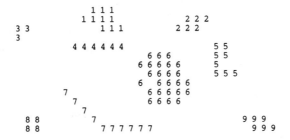

Fig. 6.1 A process in which all binary objects are labelled.

the kernel of the **repeat** loop being expressed more explicitly as:

```
PROP:
(* original image in A-space; labels inserted in P-space *)
[[ if (A0=1) & ((P1=N) or (P2=N) or (P3=N) or (P4=N) or
    (P5=N) or (P6=N) or (P7=N) or (P8=N)) then P0:=N ]];
```

$$(6.2)$$

Note the use here and below of the notation "&" for "**and**" to simplify tedious expressions (see Appendix A for a resumé of the special notations used in this book).

At this stage a fairly simple type of algorithm for object labelling is obtained, as shown in Table 6.1: the $[[+ \cdots +]]$ notation denotes a sequential scan over the image. Appendix A provides an important note on the limited validity of the above **repeat … until** finished construct for parallel algorithms.

It should be noticed that the above object counting and labelling routine requires a *minimum* of $2N + 1$ passes over the image space, and in practice the number will be closer to $NW/2$ where W is the average width of the objects: hence the algorithm is inherently rather inefficient. This prompts us to consider how the number of passes over the image could be reduced to save computation. One possibility would be to scan forwards through the image, propagating new labels through objects as they are discovered. While this would work mostly straightforwardly with convex objects, problems would be encountered with objects possessing concavities—e.g. "U" shapes—since different parts of the same object would end with different

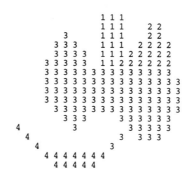

Fig. 6.2 Labelling U-shaped objects: a problem that arises in labelling certain types of object if a simple propagation algorithm is used. Some provision has to be made to accept "collisions" of labels although the confusion can be removed by a subsequent stage of processing.

Table 6.1   A simple algorithm for object labelling.

```
(* start with binary image containing objects in A-space *)
(* clear label space *)
[[ P0 := 0 ]];
(* start with no objects *)
N := 0;
(* look for objects using a sequential scan and propagate labels
through them *)
repeat (* search for an unlabelled object *)
  found := false;
  [[+ if(A0 = 1) & (P0 = 0) & not found then
        begin
          N := N + 1;
          P0 := N;
          found := true;
        end +]];
  if found then (* label the object just found *)
    repeat
      finished := true;
      [[ if (A0 = 1) & (P0 = 0) &
           ((P1 = N) or (P2 = N) or (P3 = N) or (P4 = N) or
           (P5 = N) or (P6 = N) or (P7 = N) or (P8 = N)) then
          begin
            P0 := N;
            finished := false;
          end ]];
    until finished;
until not found; (* i.e. no (more) objects *)
(* N is the number of objects found and labelled *)
```

labels, and also means would have to be devised for coping with "collisions" of labels (e.g. the largest local label could be propagated through the remainder of the object: see Fig. 6.2). Then inconsistencies could be resolved by a reverse scan through the image. However, this procedure will not resolve all problems that can arise, as in the case of more complex (e.g. spiral) objects. In such cases a general parallel propagation, repeatedly applied until no further labelling occurs, might be preferable—though as we have seen, such a process is inherently rather computationally intensive. However, it is implemented very conveniently on certain types of parallel SIMD processor (e.g. the CLIP4 computer: Fountain, 1987).

Ultimately, the least computationally intensive procedures for propagation involve a different approach: objects and parts of objects are labelled on a *single* sequential pass through the image, at the same time noting which labels coexist on objects. Then the labels are sorted separately, in a stage of abstract information processing, to determine how the initially rather *ad hoc* labels should be interpreted. Finally, the objects are relabelled appropriately in a second pass over the image (in fact, this latter pass is sometimes unnecessary, since the image data are merely labelled in an overcomplex manner and what is needed is simply a key to interpret them). The improved labelling algorithm now takes the form shown in Table 6.2. Clearly this algorithm with its single sequential scan is intrinsically far more efficient than the previous one, although the presence of particular dedicated hardware or a suitable

Table 6.2   The improved algorithm for object labelling.

```
( * clear label space * )
[[ P0 := 0 ]];
( * start with no objects * )
N := 0;
( * clear the table that is to hold the label coexistence data * )
for i := 1 to Nmax do
  for j := 1 to Nmax do
    coexist[i,j] := false;
( * label objects in a single sequential scan * )
[[+ if(A0 = 1) then
      if(P2 = 0) & (P3 = 0) & (P4 = 0) & (P5 = 0) then
        begin
          N := N + 1
          P0 = N;
        end
      else
        begin
          P0 := max(P2,P3,P4,P5);
          ( * now note which labels coexist in objects * )
          coexist[P0,P2] := true;
          coexist[P0,P3] := true;
          coexist[P0,P4] := true;
          coexist[P0,P5] := true;
        end +]];
analyse the coexist table and decide ideal labelling scheme;
relabel image if necessary;
```

SIMD processor might alter the situation and justify the use of alternative procedures.

It will be clear that minor amendments to the above algorithms permit the areas and perimeters of objects to be determined: thus objects may be labelled by their areas or perimeters instead of by numbers representing their order of appearance in the image. More important, the availability of propagation routines means that objects can be considered in turn in their entirety — if necessary by transferring them individually to separate image spaces or storage areas ready for unencumbered independent analysis. Evidently, if objects appear in individual binary spaces, maximum and minimum spatial coordinates are trivially measurable, centroids can readily be found and more detailed calculations of moments (see below) and other parameters can easily be undertaken.

6.4 Metric properties in digital images

Measurement of distance in a digital image is complicated by what is and what is not easy to calculate, given the types of neighbourhood operation that are common in image processing. However, it must be noted that rigorous measures of distance depend on the notion of a *metric*. A metric is defined as a measure $d(r_i, r_j)$ that fulfils three conditions:

(i) it must always be greater than or equal to zero, equality occurring only when points r_i and r_j are identical:

$$d(r_i, r_j) > 0, \quad i \neq j; \quad d(r_i, r_i) = 0 \tag{6.3}$$

(ii) it must be symmetrical:

$$d(r_i, r_j) = d(r_j, r_i) \tag{6.4}$$

(iii) it must obey the triangle rule:

$$d(r_i, r_j) + d(r_j, r_k) \geq d(r_i, r_k) \tag{6.5}$$

Clearly, distance can be estimated accurately in a digital image using the Euclidean metric:

$$d_\mathrm{E} = [(x_i - x_j)^2 + (y_i - y_j)^2]^{1/2} \tag{6.6}$$

However, it is interesting that the 8-connected and 4-connected definitions of connectedness given above also lead to measures that are metrics:

$$d_8 = \max \left(| x_i - x_j |, \; | y_i - y_j | \right) \tag{6.7}$$

and

$$d_4 = | x_i - x_j | + | y_i - y_j | \tag{6.8}$$

On the other hand, the latter two metrics are far from ideal in that they lead to constant distance loci which appear as squares and diamonds, rather than as circles. In addition, it is difficult to devise other more accurate measures which fulfil the requirements of a metric. Clearly, this means that if we restrict ourselves to local neighbourhood operations, we will not be able to obtain very high accuracy of measurement. Hence, if we require the maximum accuracy possible, it will be necessary to revert to using the Euclidean metric. For example, if we wish to draw an accurate ellipse, then we should employ some scheme such as that in Table 6.3 rather than attempting to employ neighbourhood operations.

Table 6.3   Routine for drawing an ellipse.

```
(* local integer coordinates are (x, y) *)
[[ A0 := 0 ]];
for i := 0 to 359 do
  begin
    theta := i * 2 * pi/360;
    (* now write coordinates to framestore registers *)
    x := round(a * sin(theta));
    y := round(b * cos(theta));
    A0 := 1;
  end;
```

Similarly, if we find two features in an image and wish to measure their distance apart, it is trivial to do this by "abstract" processing using the two pairs of coordinates but difficult to do it efficiently by operations taking place entirely within image space. Nevertheless, it is useful to see how much can be achieved in image space. This aim has given rise to such approximations as the following widely used scheme for rapid estimation of the perimeter of an object: treat horizontal and vertical edges normally (i.e. as unit steps), but treat diagonal edges as $\sqrt{2}$ times the step length. This rather limited approach (analysed more fully in Chapter 7) is acceptable for certain simple recognition tasks.

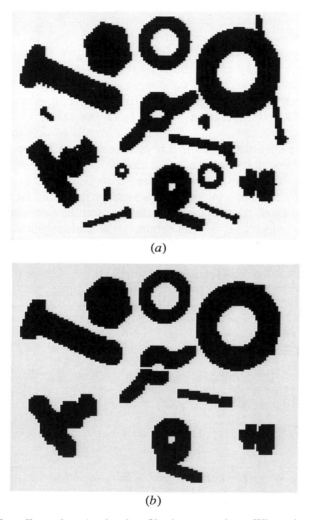

(a)

(b)

Fig. 6.3 The effect of a simple size filtering procedure. When size filtering is attempted by a set of N SHRINK and N EXPAND operations, larger objects are restored approximately to their original sizes but their shapes frequently become distorted or even fragmented. In this example, (b) shows the effect of applying two SHRINK and then two EXPAND operations to the image in (a).

6.5 Size filtering

Although the use of the d_4 and d_8 metrics is bound to lead to certain inaccuracies, it is useful to see what can be achieved with the use of local operations in binary images. This section studies how simple size filtering operations can be carried out, using merely local (3×3) operations. The basic idea is that small objects may be eliminated by applying a series of SHRINK operations. In fact, N SHRINK operations will eliminate an object (or those parts of an object) that are $2N$ or fewer pixels wide (i.e. across their narrower dimension). Of course this process shrinks *all* the objects in the image, but in principle a subsequent N EXPAND operations will restore the larger objects to their former size.

If complete elimination of small objects is required but perfect retention of larger objects, this will, however, not be achieved by the above procedure, since in many cases the larger objects will be distorted or fragmented by these operations (Fig. 6.3). To recover the larger objects in their *original* form, the proper approach is to use the shrunken versions as "seeds" from which to grow the originals via a propagation process. The algorithm of Table 6.4 is able to achieve this.

Table 6.4   Algorithm for recovering original forms of shrunken objects.

```
( * save original image * )
[[ C0 := A0 ]];
( * now SHRINK the original objects N times * )
for i :=1 to N do
  begin
    [[ sigma := A1 + A2 + A3 + A4 + A5 + A6 + A7 + A8;
      B0 := if sigma < 8 then 0 else A0 ]];
    [[ A0 := B0 ]];
  end;
( * next propagate the shrunken objects using the original image * )
repeat
  finished := true;
  [[ sigma := A1 + A2 + A3 + A4 + A5 + A6 + A7 + A8;
    if (A0 = 0) & (sigma > 0) & (C0 = 1) then
      begin
        A0 := 1;
        finished := false;
      end ]];
until finished;
```

Having seen how to remove whole (connected) objects that are every-where narrower than a specified width, it is possible to devise algorithms for removing any subset of objects that are characterized by a given range of widths: large objects may be filtered out by first removing lesser-sized objects and then performing a logical masking operation with the original image, while intermediate-sized objects may be filtered out by removing a larger subset and then restoring small objects that have previously been

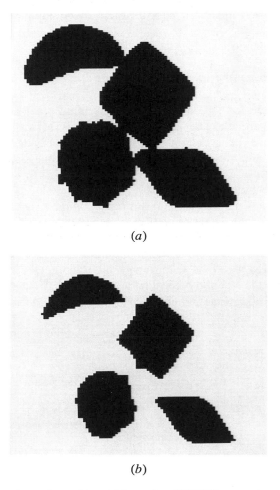

(a)

(b)

Fig. 6.4 Separation of touching objects by SHRINK operations. Here objects (chocolates) in (a) are shrunk (b) in order to separate them so that they may be counted reliably.

stored in a separate image space. Ultimately, all these schemes depend on the availability of the propagation technique, which in turn depends on the internal connectedness of individual objects.

Finally, note that EXPAND operations followed by SHRINK (or thinning—see below) operations may be useful for joining nearby objects, filling in holes, and so on. Numerous refinements and additions to these simple techniques are possible. A particularly interesting one is to separate the silhouettes of touching objects such as chocolates by a shrinking operation: this then permits them to be counted reliably (Fig. 6.4).

6.6   The convex hull and its computation

This section moves on to consider potentially more precise descriptions of shape, and in particular the *convex hull*: this is basically an analogue descriptor of shape, being defined (in two dimensions) as the shape enclosed by an elastic band placed around the object in question. The *convex deficiency* is defined as the shape that has to be added to a given shape to create the convex hull (Fig. 6.5).

The convex hull may be used to simplify complex shapes to provide a rapid indication of the extent of an object. A particular application might be locating the hole in a washer or nut, since the hole is that part of the convex deficiency that is *not* adjacent to the background. In addition, the convexity or otherwise of a figure is determined by finding whether the convex deficiency is (to the required precision) an empty set. A fuller description of the shape of an object may be obtained by means of *concavity trees*: here

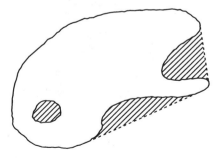

Fig. 6.5   Convex hull and convex deficiency. The convex hull is the shape enclosed on placing an elastic band around an object. The shaded portion is the convex deficiency that is added to the shape to create the convex hull.

the convex hull of an object is first obtained with its convex deficiencies, then the convex hulls and deficiencies of the convex deficiencies are found, then the convex hulls and deficiencies of these convex deficiencies—and so on until all the derived shapes are convex. Thus a tree is formed which can be used for systematic shape analysis and recognition (Fig. 6.6). We shall not dwell on this approach beyond noting its inherent utility and that at its core is the need for a reliable means of determining the convex hull of a shape.

A simple means of obtaining the convex hull (but not an accurate Euclidean one) is to repeatedly fill in the centre pixel of all neighbourhoods which exhibit a concavity, including each of the following:

$$\begin{matrix} 1 & 1 & 1 \\ 1 & 0 & 0 \\ 0 & 0 & 0 \end{matrix} \qquad \begin{matrix} 0 & 1 & 1 \\ 1 & 0 & 0 \\ 0 & 0 & 0 \end{matrix}$$

until no further change occurs. If, as in the second example, a corner pixel in the neighbourhood is 0 and is adjacent to two 1's, it makes no difference

(a)

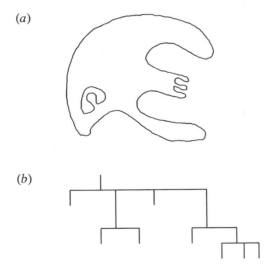

(b)

Fig. 6.6 A simple shape and its concavity tree. The shape in (a) has been analysed by repeated formation of convex hulls and convex deficiencies until all the constituent regions are convex (see text). The tree representing the entire process is shown in (b): at each node, the branch on the left is the convex hull and the branches on the right are convex deficiencies.

to the 8-connectedness condition whether the corner is 0 or 1. Hence we can pretend that it has been replaced by a 1. We then get a simple rule for determining whether to fill in the centre pixel: if there are four or more 1's around the boundary of the neighbourhood then there is a concavity which must be filled in. This can be expressed as follows:

```
BASICFILL:
[[ sigma := A1 + A2 + A3 + A4 + A5 + A6 + A7 + A8;
   B0 := if sigma > 3 then 1 else A0 ]];                    (6.9)
```

However, the corner pixels may not in fact be present. It is therefore necessary to correct the routine as follows[*]:

```
FILL:  [[ sigma :=A1 + A2 + A3 + A4 + A5 + A6 + A7 + A8
                 + (A1 & ~A2 & A3) + (A3 & ~A4 & A5)
                 + (A5 & ~A6 & A7) + (A7 & ~A8 & A1);
          B0 := if sigma > 3 then 1 else A0 ]];             (6.10)
```

Finally, a complex convex hull algorithm (Table 6.5) requires the FILL routine to be included in a loop, with a suitable flag to test for completion. Note that the algorithm is forced to perform a final pass in which nothing happens. Although this seems to waste computation, it is necessary as otherwise it will not be known that the process has run to completion. This

Table 6.5   The complete convex hull algorithm.

```
repeat
  finished := true;
  [[sigma := A1 + A2 + A3 + A4 + A5 + A6 + A7 + A8
       ·           + (A1 & ~A2 & A3) + (A3 & ~A4 & A5)
                   + (A5 & ~A6 & A7) + (A7 & ~A8 & A1);
     if (A0 = 0) & (sigma > 3) then
       begin
         B0 := 1;
         finished := false;
       end
     else B0 := A0 ]];
   [[ A0 := B0 ]];
until finished;
```

[*] For an explanation of this nonstandard Pascal notation, see Appendix A.

situation arises repeatedly in this sort of work, there being seven examples in this chapter alone.

The shapes obtained by the above algorithm are actually overconvex and approximate to octagonal (or degenerate octagonal) shapes such as the following:

```
    .  1  1  1  1  1  .  .  .  .  .  .  .  .  .  .
 1  1  1  1  1  1  1  1  1  1  .  .  .  .  1  1  1  1
 1  1  1  1  .  .  .  .  1  1  1  .  .  .  1  .  .  .
 1  1  1  .  .  .  .  .  .  1  1  1  1  1  .  .  .  .
 .  1  1  1  .  .  .  .  .  .  .  1  1  .  .  .  .
 1  1  1  1  .  1  .  .  .  .  .  .  .  .  .
    .  1  1  1  .  .  .  .  .  .  .  .  .
```

The method may be refined so that an approximately 16-sided polygon results (Hussain, 1988). This is achieved by forming the above "overconvex hull"; then forming an "underconvex hull" by finding the convex hull of the background with the constraint that it is permitted to eat into the convex deficiency but must not eat into the original object; then finding a median line (Fig. 6.7) between the over- and underconvex hulls by a suitable technique such as thinning (see Section 6.8)[*].

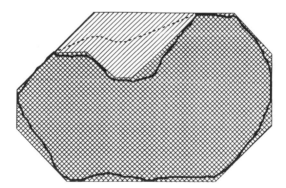

Fig. 6.7 Finding an approximate convex hull using over- and underconvex hulls. The shaded region is the overconvex hull of the original shape and the doubly-shaded region is the underconvex hull. A (potentially) 16-sided approximation to the ideal convex hull is found by thinning the singly-shaded region, as indicated by the dotted line.

---

[*] Although the effect of this algorithm is as described above, it is formulated in rather different terms which cannot be gone into here.

Although such techniques making use of solely local operations are possible, and are ideally implemented on certain types of parallel processor (Hussain, 1988), they are limited in accuracy and can only be improved with considerable ingenuity. Ultimately, for connected objects the alternative option of tracking around the perimeter of the object and deducing the position of the convex hull has to be adopted. In fact, the basic strategy for achieving this is simple—search for positions on the boundary that have common tangent lines. Note that changing from a body-orientated approach to a boundary-orientated approach can in principle save considerable computation. However, the details of this strategy are not trivial to implement, because of the effects of spatial quantization which complicate these inherently simple analogue ideas. Overall, it is worth noting that there is no escape from the necessity to take connectedness fully into account. Hence algorithms are required which can reliably track around object boundaries: this aspect of the problem is dealt with in a later section. Further details of convex hull algorithms based on this approach are not given here: the reader is referred to the literature for additional information (see for example, Rutovitz, 1975).

Before leaving the topic of convex hull determination, a possible complication should be noted: this arises when the convex hulls of the individual objects overlap (though not the original objects). In this case, the final convex hull is undefined: should it be the union of the separate convex hulls or the convex hull of their union? The local operator algorithm considered above will find the latter, whereas algorithms based on boundary tracking will find the former. In general, obtaining a convex hull from an original shape is not trivial, as tests must be incorporated to ensure that separate limbs are in all cases connected, and to ensure that the convex hulls of other objects are not inadvertently combined. Ultimately, such problems are probably best dealt with by transferring each object to its own storage area before starting convex hull computations (see also Section 6.3).

6.7 Distance functions and their uses

The distance function of an object is a very simple and useful concept in shape analysis. Essentially, each pixel in the object is numbered according to its distance from the background. As usual, background pixels are taken as 0's; then edge pixels are counted as 1's; object pixels next to 1's become 2's; next to those are the 3's; and so on throughout all the object pixels (Fig. 6.8).

```
                1 1 1
                1 2 1      1 1
        1       1 2 1      1 1
        1 1 1   1 2 1    1 1 1 1
        1 2 1 1 1 2 1 1 1 2 2 1
    1 1 2 2 1 1 2 2 2 2 2 2 1
    1 2 2 2 1 1 1 2 3 3 3 3 2 1
    1 2 3 2 2 2 2 3 4 4 3 2 1 1
    1 2 2 3 3 3 3 3 3 3 3 2 2 1
    1 1 2 2 2 2 2 2 2 3 3 2 2 1
      1 1 2 1 1 1 1 1 2 2 3 2 1 1
          1 1 1        1 1 2 2 2 1
  1           1        1 1 2 1 1
    1                  1 1 1
      1              1
  1 1 1 1 1 1 1
    1 1 1 1 1
```

Fig. 6.8 The distance function of a binary shape: the value at every pixel is the distance (in the d_8 metric) from the background.

The parallel algorithm of Table 6.6 finds the distance function of binary objects by propagation. Note that, as in the computation of convex hulls, this algorithm performs a final pass in which nothing happens; again, this is inevitable if it is to be certain that the process runs to completion. An alternative way of propagating the distance function is shown in Table 6.7. Note the complication in ensuring completion of the procedure; if it could be assumed that N passes are needed for objects of known maximum width,

Table 6.6 A parallel algorithm for propagating distance functions.

```
PPROP1:   (* Start with binary image containing objects in A-space *)
          begin
            [[ Q0 := A0*255]]
            N := 0;
            repeat
              finished := true;
              [[ if (Q0 = 255) (* in object & no answer yet *)
                 & ((Q1 = N) or (Q2 = N) or (Q3 = N) or (Q4 = N) or
                 (Q5 = N) or (Q6 = N) or (Q7 = N) or (Q8 = N))
                 (* next to an N *)
                 then
                   begin
                     Q0 := N + 1;
                     finished := false; (* some action has been taken *)
                   end ]];
              N:= N + 1;
            until finished;
          end;
```

Table 6.7   An alternative parallel algorithm for propagating distance functions.

```
PPROP2:  begin
              [[ Q0 := A0 * 255 ]];
              repeat
                finished := true;
                [[ minplusone := min(Q0 - 1,Q1,Q2,Q3,Q4,Q5,Q6,Q7,Q8) + 1;
                   if Q0 > minplusone then
                     begin
                       Q0 := minplusone;
                       finished := false; (* some action has been taken *)
                     end ]];
              until finished;
            end;
```

then the algorithm would be simply:

```
PPROP3:
for i := 1 to N do
  [[ Q0 := min(Q0 - 1, Q1, Q2, Q3, Q4, Q5, Q6, Q7, Q8) + 1 ]];
```
$$(6.11)$$

This routine operates entirely within a single image space and appears to be simpler and more elegant than PPROP1: however, note that in every pass over the image it writes a value at every location, which could under some circumstances require more computation (this is not true for PPROP1, which writes a value only at those object locations which are next to previously assigned values). For SIMD parallel processing machines, which have a processing element at every pixel location, this is no problem and PPROP3 may be more useful.

It is possible to perform the propagation of a distance function with far fewer operations if the process is expressed as two sequential operations, one being a normal forward raster scan and the other being a reverse raster scan:

```
SPROP1: begin
            [[ Q0 := A0 * 255 ]];
            [[+ minplusone := min(Q2, Q3, Q4, Q5) + 1;
                if Q0 > minplusone then Q0 := minplusone +]];
            [[- minplusone := min(Q6, Q7, Q8, Q1) + 1;
                if Q0>minplusone then Q0 := minplusone -]];
          end;
```
$$(6.12)$$

Note the notation being used to distinguish between forward and reverse raster scans (see Appendix A). A more succinct version of this algorithm is the following:

```
SPROP2:  begin
             [[    Q0  := A0 * 255 ]];
             [[+ Q0  := min(Q0 - 1, Q2, Q3, Q4, Q5) + 1 +]];
             [[- Q0  := min(Q0 - 1, Q6, Q7, Q8, Q1) + 1 -]];
         end;
```

$$(6.13)$$

An interesting application of distance functions is that of data compression. To achieve this, operations are carried out to locate those pixels which are local maxima of the distance function (Fig. 6.9), since storing these pixel values and positions permits the original image to be regenerated by a process of downwards propagation (see below). Note that although finding the local maxima of the distance function provides the basic information for data compression, the actual compression occurs only when the data are stored as a list of points rather than in the original picture format. In order to locate the local maxima, the following parallel routine may be employed:

```
LMAX1:  [[ maximum := max(Q1, Q2, Q3, Q4, Q5, Q6, Q7, Q8);
           B0 := if (Q0 > 0) & (Q0 >= maximum) then 1
                                                else 0 ]];  (6.14)
```

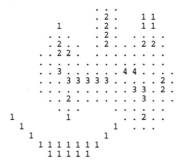

Fig. 6.9 Local maxima of the distance function of the shape shown in Fig. 6.8, the remainder of the shape being indicated by dots and the background being blank. Notice that the local maxima group themselves into clusters each containing points of equal distance function value, while clusters of different values are clearly separated.

Alternatively, the compressed data can be transferred to a single image space:

```
LMAX2:  [[ maximum := max(Q1, Q2, Q3, Q4, Q5, Q6, Q7, Q8);
           P0 := if (Q0 > 0) & (Q0 >= maximum) then Q0
                                                 else 0 ]];
```
$$(6.15)$$

Note that the local maxima that are retained for the purpose of data compression are not absolute maxima but are maximal in the sense of not being adjacent to larger values. If this were not so, insufficient numbers of points would be retained for completely regenerating the original object. As a result of this, it is found that the absolute maxima group themselves into clusters of connected points, each cluster having a common distance value and being separated from points of different distance values (Fig. 6.9). Thus the set of local maxima of an object is not a connected subset. This fact has an important bearing on skeleton formation (see below).

Having seen how data compression may be performed by finding local maxima of the distance function, it is relevant to consider a parallel downwards propagation algorithm (Table 6.8) for recovering the shapes of objects from an image into which the values of the local maxima have been inserted. Note again that if it can be assumed that at most N passes are needed to propagate through objects of known maximum width, then the algorithm becomes simply:

```
DPROP2:
  for i := 1 to N do
  [[ Q0 := max(Q0 + 1, Q1, Q2, Q3, Q4, Q5, Q6, Q7, Q8) - 1 ]];
```
$$(6.16)$$

Finally, it is interesting to note that while the local maxima of the distance function form a subset of points that may be used to reconstruct the object shape exactly, it is frequently possible to find a subset $\mathcal{M}$ of this subset that has this same property and thereby corresponds to a greater degree of data compression. This can be seen by appealing to Fig. 6.10. The reason that this is possible lies in the fact that although downwards propagation from $\mathcal{M}$ permits the object shape to be recovered immediately, it does not also provide a correct set of distance function values: if required these must be obtained by subsequent upwards propagation. It turns out that the process of finding a minimal set of points $\mathcal{M}$ involves quite a complex and computationally intensive algorithm, and is too specialized to describe in detail here (see Davies and Plummer, 1980).

Table 6.8   A parallel algorithm for recovering objects from local maxima of the
distance functions.

```
DPROP1:   ( * assume that input image is in Q-space, and that
            nonmaximum values have value 0, as at output of LMAX2
            routine * )
        repeat
          finished := true;
          [[ maxminusone := max(Q0 + 1,Q1,Q2,Q3,Q4,Q5,Q6,Q7,Q8) - 1;
            if Q0 <maxminusone then
              begin
                Q0 := maxminusone;
                finished := false; ( * some action has been taken * )
              end ]];
        until finished;
```

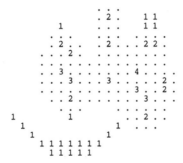

Fig. 6.10   Minimal subset of points for object reconstruction. The numbered pixels
may be used to reconstruct the entire shape shown in Fig. 6.8 and form a more
economical subset than the set of local maxima of Fig. 6.9.

6.8   Skeletons and thinning

Like the convex hull, the skeleton is a powerful analogue concept which
may be employed for the analysis and description of shapes in binary
images. A skeleton may be defined as a connected set of medial lines along
the limbs of a figure: for example, in the case of thick hand-drawn characters
the skeleton may be supposed to be the path actually travelled by the pen. In
fact, the basic idea of the skeleton is that of eliminating redundant information while retaining only the topological information concerning the shape

and structure of the object that can help with recognition. In the case of hand-drawn characters, the thickness of the limbs is taken to be irrelevant: it may be constant and therefore carry no useful information, or it may vary randomly and again be of no value for recognition (Fig. 1.2).

The definition presented above leads to the idea of finding the loci of the centres of maximal discs inserted within the object boundary. First, suppose the image space to be a continuum. Then the discs are circles and their centres form loci which may be modelled very conveniently when object boundaries are approximated by linear segments. In fact, sections of the loci fall into three categories:

(i)   they may be angle bisectors, i.e. lines which bisect corner angles and reach right up to the apexes of corners;

(ii) they may be lines that lie half-way between boundary lines;

(iii) they may be parabolas which are equidistant from lines and from the nearest points of other lines—namely, corners where two lines join.

Clearly, (i) and (ii) are special forms of a more general case.

These ideas lead to unique skeletons for objects with linear boundaries, and the concepts are easily generalizable to curved shapes. It turns out that this approach tends to give rather more detail than is commonly required, even the most obtuse corner having a skeleton line going into its apex (Fig. 6.11). Hence a thresholding scheme is often employed such that skeleton lines only reach into corners having a specified minimum degree of sharpness.

We now have to see how the skeleton concept will work in a digital lattice. Here we are presented with an immediate choice: which metric should we employ? If we select the Euclidean metric, we are liable to engender a considerable computational load (although many workers do choose this option). If we select the d_8 metric we will immediately lose accuracy but the computational requirements should be more modest (we do not here consider the d_4 metric, since we are dealing with foreground objects). In what follows we concentrate on the d_8 metric.

At this stage some thought shows that the application of maximal discs in order to locate skeleton lines amounts essentially to finding the positions of local maxima of the distance function. Unfortunately, as seen in the previous section, the set of local maxima does not form a connected graph within a given object: nor is it necessarily composed of thin lines, and indeed it may in places be 2 pixels wide. Thus problems arise in trying to use this approach to obtain a connected unit-width skeleton that can conveniently be used to represent the shape of the object. We shall return to

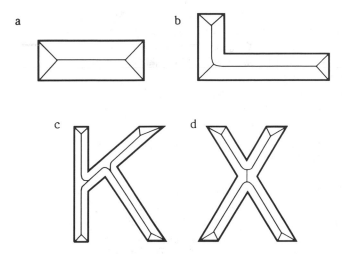

Fig. 6.11 Four shapes whose boundaries consist entirely of straight line segments. The idealized skeletons go right to the apex of each corner, however obtuse. In certain parts of shapes (b), (c) and (d), the skeleton segments are parts of parabolas rather than straight lines. As a result, the detailed shape of the skeleton (or the approximations produced by most algorithms operating in discrete images) is not exactly what might initially be expected or what would be preferred in certain applications.

this approach again below. Meanwhile, however, we pursue an alternative idea—that of thinning.

Thinning is perhaps the simplest approach to skeletonization. It may be defined as the process of systematically stripping away the outermost layers of a figure until only a connected unit-width skeleton remains (see for example, Fig. 6.12). A number of algorithms are available to implement this process, with varying degrees of accuracy, and we discuss below how a specified level of precision can be achieved and tested for. First, however, it is necessary to discuss the mechanism by which points on the boundary of a figure may validly be removed in thinning algorithms.

6.8.1 Crossing number

The exact mechanism for examining points to determine whether they can be removed in a thinning algorithm must now be considered. This may be decided by reference to the *crossing number* χ (chi) for the 8 pixels around the outside of a particular 3×3 neighbourhood. χ is defined as the total

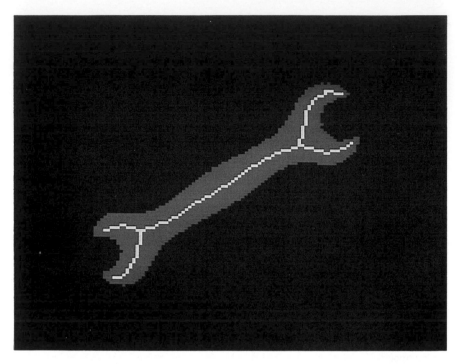

Fig. 6.12 Typical result of a thinning algorithm operating in a discrete lattice.

number of 0-to-1 and 1-to-0 transitions on going once round the outside of
the neighbourhood: this number is in fact twice the number of potential
connections joining the remainder of the object to the neighbourhood (Fig.
6.13). Unfortunately, the formula for χ is made more complex by the 8-
connectedness criterion. Basically, we would expect:

$$
\begin{aligned}
\texttt{badchi} := \ &(\texttt{A1} <> \texttt{A2}) + (\texttt{A2} <> \texttt{A3}) + (\texttt{A3} <> \texttt{A4}) \\
&+ (\texttt{A4} <> \texttt{A5}) + (\texttt{A5} <> \texttt{A6}) + (\texttt{A6} <> \texttt{A7}) \\
&+ (\texttt{A7} <> \texttt{A8}) + (\texttt{A8} <> \texttt{A1});
\end{aligned}
\tag{6.17}
$$

0 0 0	0 0 0	1 0 0	1 0 0	1 0 1	1 0 0	1 0 1
0 1 0	0 1 1	0 1 1	0 1 0	0 1 0	0 1 0	0 1 0
0 0 0	1 1 1	1 1 1	0 1 0	1 1 1	1 0 1	1 0 1
0	2	4	4	6	6	8

Fig. 6.13 Some examples of the crossing number values associated with given
pixel neighbourhood configurations (0, background; 1, foreground).

However, this is incorrect because of the 8-connectedness criterion. For example, in the case

$$
\begin{array}{ccc}
0 & 1 & 0 \\
0 & 1 & 1 \\
1 & 1 & 1
\end{array}
$$

the formula gives the value 4 for χ instead of 2. The reason is that the isolated 0 in the top right-hand corner does not prevent the adjacent 1's from being joined. It is therefore tempting to use the modified formula:

```
wrongchi := (A1 <> A3) + (A3 <> A5)
          + (A5 <> A7) + (A7 <> A1);        (6.18)
```

However, this too is wrong, since in the case

$$
\begin{array}{ccc}
0 & 0 & 1 \\
0 & 1 & 0 \\
1 & 1 & 1
\end{array}
$$

it gives the answer 2 instead of 4. It is therefore necessary to add four extra terms to deal with isolated 1's in the corners:

```
chi := (A1 <> A3) + (A3 <> A5) + (A5 <> A7) + (A7 <> A1)
     + 2 * ((~A1 & A2 & ~A3) + (~A3 & A4 & ~A5)
     + (~A5 & A6 & ~A7) + (~A7 & A8 & ~A1));        (6.19)
```

This (now correct) formula for crossing number gives values 0, 2, 4, 6 or 8 in different cases (Fig. 6.13). The rule for removing points during thinning is that points may only be removed if they are at those positions on the boundary of an object where χ is 2: when χ is greater than 2, the point *must* be retained, as it forms a vital connecting point between two parts of the object; in addition, when χ is 0 it must be retained since removing it would create a hole.

Finally, there is one more condition that must be fulfilled before a point can be removed during thinning—that the sum σ (sigma) of the eight pixel values around the outside of the 3×3 neighbourhood (see Chapter 2) must not be equal to 1. The reason for this is to preserve line ends, as in the following cases:

$$
\begin{array}{ccc}
0 & 0 & 0 \\
0 & 1 & 0 \\
0 & 1 & 0
\end{array}
\qquad
\begin{array}{ccc}
0 & 0 & 0 \\
0 & 1 & 0 \\
0 & 0 & 1
\end{array}
$$

Clearly, if line ends are eroded as thinning proceeds, the final skeleton will not represent the shape of an object (including the relative dimensions of its limbs) at all accurately (however, it is possible that we might sometimes wish to shrink an object while preserving connectedness, in which case this extra condition need not be implemented). Having covered these basics, we are now in a position to devise complete thinning algorithms.

6.8.2 Parallel and sequential implementations of thinning

Thinning is "essentially sequential" in that it is easiest to ensure that connectedness is maintained by arranging that only one point may be removed at a time. As indicated above, this is achieved by checking before removing a point that it has a crossing number of 2. Now imagine applying the "obvious" sequential algorithm of Table 6.9 to a binary image. Assuming a normal forward raster scan, the result of this process is to produce a highly distorted skeleton, consisting of lines along the right-hand and bottom edges of objects. It may now be seen that the $\chi = 2$ condition is

Table 6.9   An "obvious" sequential thinning algorithm.

```
STHIN:   repeat
             finished := true;
             [[+  sigma := A1 + A2 + A3 + A4 + A5 + A6 + A7 + A8;
                  chi := (A1 <> A3) + (A3 <> A5) + (A5 <> A7) + (A7 <> A1)
                       + 2 * ((~A1 & A2 & ~A3) + (~A3 & A4 & ~A5)
                          + (~A5 & A6 & ~A7) + (~A7 & A8 & ~A1));
                  if (A0 = 1) & (sigma <> 1) & (chi = 2) then
                     begin
                     A0 := 0;
                     finished := false;  (* some action has been taken *)
                     end  +]];
             until finished;
```

---

necessary but is not sufficient, since it says nothing about the order in which points are removed. To produce a skeleton that is unbiased, giving a set of truly medial lines, it is necessary to remove points as evenly as possible around the object boundary. A scheme that helps with this involves a novel processing sequence: mark edge points on the first pass over an image; on the second pass, strip points sequentially as in the above algorithm, *but only where they have already been marked*; then mark a new set of edge points; then perform another stripping pass; then repeat this marking and stripping sequence until no further change occurs. An early algorithm working on this principle is that of Beun (1973).

While the novel sequential thinning algorithm described above can be used to produce a reasonable skeleton, it would be far better if the stripping action could be performed symmetrically around the object, thereby removing any possible skeletal bias. In this respect a parallel algorithm should have a distinct advantage. However, parallel algorithms result in several points being removed at once: this means that lines 2 pixels wide will disappear (since masks operating in a 3×3 neighbourhood cannot "see" enough of the object to judge whether a point may validly be removed or not), and as a result shapes can become disconnected. The general principle for avoiding this problem is to strip points lying on different parts of the boundary in different passes, so that there is no risk of causing breaks. In fact, there are a very large number of ways of achieving this, by applying different masks and conditions to characterize different parts of the boundary. If boundaries were always convex the problem would no doubt be reduced; however, boundaries can be very convoluted and are subject to quantization noise, so the problem is a complex one. With so many potential solutions to the problem, we concentrate here on one that can conveniently be analysed and which gives acceptable results.

The method discussed is that of removing north, south, east and west points cyclically until thinning is complete. North points are defined as the following:

$$
\begin{array}{ccc}
\times & 0 & \times \\
\times & 1 & \times \\
\times & 1 & \times
\end{array}
$$

where $\times$ means either a 0 or a 1: south, east and west points are defined similarly. It is easy to show that all north points for which $\chi = 2$ and $\sigma \neq 1$ may be removed in parallel without any risk of causing a break in the skeleton—and similarly for south, east and west points. Thus a possible format for a parallel thinning algorithm in rectangular tessellation is the following:

```
repeat
   strip appropriate north points;
   strip appropriate south points;
   strip appropriate east points;
   strip appropriate west points;
until no further change;                              (6.20)
```

where the basic parallel routine for stripping "appropriate" north points is:

```
NSTRIP:
  [[ sigma := A1 + A2 + A3 + A4 + A5 + A6 + A7 + A8;
     chi:= (A1 <> A3) + (A3 <> A5) + (A5 <> A7)
                                          + (A7 <> A1)
          + 2 * ((~A1 & A2 & ~A3) + (~A3 & A4 & ~A5)
          + (~A5 & A6 & ~A7) + (~A7 & A8 & ~A1));
     BO := if (~A3 & A0 & A7) (* north point *)
              & (sigma <> 1) & (chi = 2)
           then 0 else A0 ]];                         (6.21)
```

(but extra code needs to be inserted to detect whether any changes have been made in a given pass over the image).

Algorithms of the above type can be highly effective, although their design tends to be rather intuitive and *ad hoc*. In a survey made by the author in 1981 (Davies and Plummer, 1981), a great many such algorithms exhibited problems. Ignoring cases where the algorithm design was insufficiently rigorous to maintain connectedness, four other problems were evident:

(i)   the problem of skeletal bias;
(ii) the problem of eliminating skeletal lines along certain limbs;
(iii) the problem of introducing "noise spurs";
(iv) the problem of slow speed of operation.

In fact, problems (ii) and (iii) are opposites in many ways: if an algorithm is designed to suppress noise spurs, it is liable to eliminate skeletal lines in some circumstances; contrariwise, if an algorithm is designed never to eliminate skeletal lines, it is unlikely to be able to suppress noise spurs. This situation arises since the masks and conditions for performing thinning are intuitive and *ad hoc*, and therefore have no basis for discriminating between valid and invalid skeletal lines: ultimately this is because it is difficult to build overt global models of reality into purely local operators. In a similar way, algorithms that proceed with caution, i.e. which do not remove object points in the fear of making an error or causing bias, tend to be slower in operation than they might otherwise be. Again, it is difficult to design algorithms that can make correct global decisions rapidly via intuitively designed local operators. Hence a totally different approach is needed if solving one of the above problems is not to cause difficulties with the others. Such an alternative approach is discussed in the next section.

6.8.3   Guided thinning

This section returns to the ideas of the early part of Section 6.8, where it was found that the local maxima of the distance function do not form an ideal skeleton because they appear in clusters and are not connected. In addition, the clusters are often 2 pixels wide. On the plus side, the clusters are accurately in the correct positions and should therefore not be subject to skeletal bias. Hence an ideal skeleton should result if (i) the clusters could be reconnected appropriately and (ii) the resulting structure could be reduced to unit width. In this respect it should be noted that a unit-width skeleton can only be perfectly unbiased where the object is an odd number of pixels wide: where the object is an even number of pixels wide, the unit-width skeleton is bound to be biased by $1/2$ pixel, and the above strategy is able to give a skeleton that is as unbiased as is in principle possible.

A simple means of reconnecting the clusters is to use them to guide a conventional thinning algorithm (see Section 6.8.2). As a first stage, thinning is allowed to proceed normally but with the proviso that no cluster points may be removed. This gives a connected graph which is in certain places 2 pixels wide. Then a routine is applied to strip the graph down to unit width. At this stage an unbiased skeleton (within $1/2$ pixel) should result. The main problem here is the presence of noise spurs. The opportunity now exists to eliminate these systematically by applying suitable global rules. A simple rule is that of eliminating lines on the skeletal graph that terminate in a local maximum of value (say) 1 (or, better, stripping them back to the next local maximum), since such a local maximum corresponds to fairly trivial detail on the boundary of the object. Thus the level of detail that is ignored can be programmed into the system (Davies and Plummer, 1981). The whole guided thinning process is shown in Fig. 6.14.

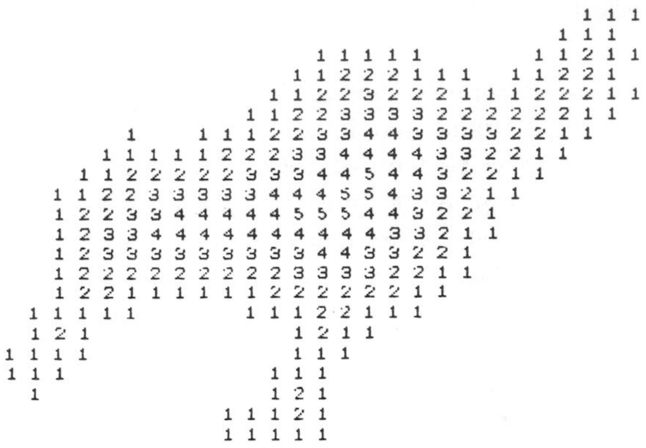

Fig. 6.14 (*continued*).

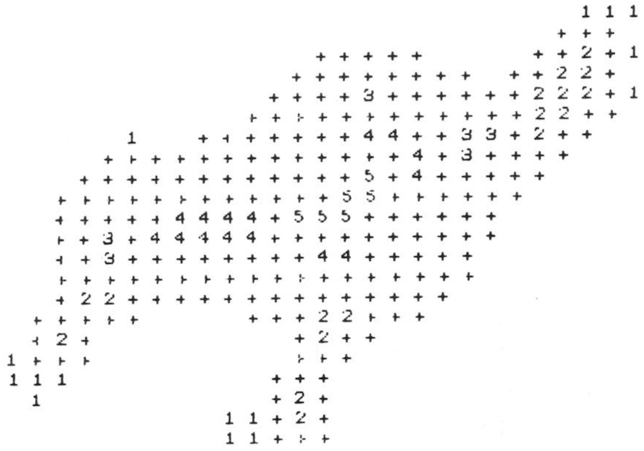

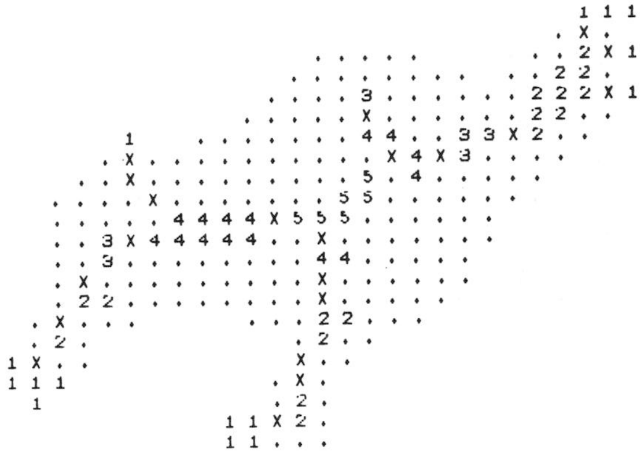

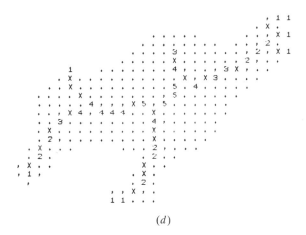

(*d*)

Fig. 6.14 Results of a guided thinning algorithm: (a) distance function on the original shape; (b) set of local maxima; (c) set of local maxima now connected by a simple thinning algorithm; (d) final thinned skeleton. The effect of removing noise spurs systematically, by cutting limbs terminating in a 1 back to the next local maximum, is easily discernible from the result in (d): the general shape of the object is not perturbed by this process.

6.8.4 A comment on the nature of the skeleton

At the beginning of Section 6.8, the case of character recognition was taken as an example and it was stated that the skeleton may be supposed to be the path travelled by the pen in drawing out the character. However, in one important respect this is not valid. The reason is seen both in the analogue reasoning and from the results of thinning algorithms (ultimately these reflect the same phenomena). Take the case of a letter K. The vertical limb on the left of the skeleton is most unlikely to turn out to be straight, as theoretically it will consist of two linear segments joined by two parabolic segments leading into the junction (Fig. 6.11). It will therefore be very difficult to devise a skeletonizing algorithm that will give a single straight line in such cases, for all possible orientations of the object. The most reliable means of achieving this would probably be via some high-level interpretation scheme that analyses the skeleton shape and deduces the ideal shape, e.g. by constrained least-squares line fitting.

6.8.5 Skeleton node analysis

Skeleton node analysis may be carried out very simply with the aid of the crossing number concept. Points in the middle of a skeletal line have a crossing number of 4; points at the end of a line have crossing number 2;

points at skeletal "T" junctions have a crossing number of 6; and points at skeletal "X" junctions have a crossing number of 8. However, there is a situation to beware of—places which look like a "+" junction:

$$
\begin{array}{ccc}
0 & 1 & 0 \\
1 & 1 & 1 \\
0 & 1 & 0
\end{array}
$$

In such places the crossing number is actually 0 (see formula), although the pattern is definitely that of a cross. At first the situation seems to be that there is insufficient resolution in a 3×3 neighbourhood to identify a "+" cross, the best option being to look for this particular pattern of 0's and 1's and use a more sophisticated construct than the 3×3 crossing number to check whether or not a cross is present. The problem is that of distinguishing between two situations such as:

$$
\begin{array}{ccccc}
0 & 0 & 0 & 0 & 0 \\
0 & 0 & 1 & 0 & 0 \\
0 & 1 & 1 & 1 & 0 \\
0 & 0 & 1 & 0 & 0 \\
0 & 0 & 0 & 0 & 0
\end{array}
\quad \text{and} \quad
\begin{array}{ccccc}
0 & 0 & 1 & 0 & 0 \\
1 & 0 & 1 & 0 & 0 \\
1 & 1 & 1 & 1 & 1 \\
0 & 0 & 1 & 0 & 0 \\
0 & 0 & 0 & 1 & 0
\end{array}
$$

However, further analysis shows that the first of these two cases would be thinned down to a dot (or a short bar), so that if a "+" node appears on the final skeleton (as in the second case) it actually signifies that a cross is present despite the contrary value of χ. Davies and Celano (1993b) have shown that the proper measure to use in such cases is the *modified* crossing number $\chi_{skel} = 2\sigma$, this crossing number being different from χ because it is required not to test whether points can be eliminated from the skeleton, but to ascertain the meaning of points that are at that stage *known* to lie on the final skeleton. Note that χ_{skel} can have values as high as 16 — it is not restricted to the range $0-8$!

Finally, note that sometimes insufficient resolution really is a problem, in that a cross with a shallow crossing angle appears as two "T" junctions:

$$
\begin{array}{cccccccc}
0 & 0 & 0 & 0 & 0 & 0 & 0 & 1 \\
1 & 1 & 1 & 0 & 0 & 1 & 1 & 0 \\
0 & 0 & 0 & 1 & 1 & 0 & 0 & 0 \\
0 & 1 & 1 & 0 & 0 & 1 & 1 & 1 \\
1 & 0 & 0 & 0 & 0 & 0 & 0 & 0
\end{array}
$$

Clearly, resolution makes it impossible to recognize an asterisk or more complex figure from its crossing number, within a 3×3 neighbourhood. Probably, the best solution is to label junctions tentatively, then to consider all the junction labels in the image, and to analyse whether a particular local combination of junctions should be reinterpreted—e.g. two "T" junctions may be deduced to form a cross. This is especially important in view of the distortions that appear on a skeleton in the region of a junction (see Section 6.8.4).

6.8.6 Application of skeletons for shape recognition

Shape analysis may be carried out simply and conveniently by analysis of skeleton shapes and dimensions. Clearly, study of the nodes of a skeleton (points for which there are other than two skeletal neighbours) can lead to the classification of simple shapes but not, for example, discrimination of all block capitals from each other. Many classification schemes exist which can complete the analysis, in terms of limb lengths and positions, and methods for achieving this are touched on in later chapters.

A similar situation exists for analysis of the shapes of chromosomes, which take the form of a cross or a "V". For small industrial components, more detailed shape analysis is called for; this can still be approached with the skeleton technique, by examination of distance function values along the lines of the skeleton. In general, shape analysis using the skeleton proceeds by examination in turn of nodes, limb lengths and orientations, and distance function values, until the required level of characterization is obtained.

The particular importance of the skeleton as an aid in the analysis of connected shapes is not only that it is invariant under translations and rotations, but also that it embodies what is for many purposes a highly convenient representation of the figure which (with the distance function values) essentially carries all the original information. If the original shape of an object can be deduced exactly from a representation, this is generally a good sign since it means that it is not merely an *ad hoc* descriptor of shape but that considerable reliance may be placed on it (compare other methods such as the circularity measure—see Section 6.9).

6.9 Some simple measures for shape recognition

There are many simple tests of shape that can be made to confirm the identity of products or to check for defects. These include measurements of product area and perimeter, length of maximum dimension, moments

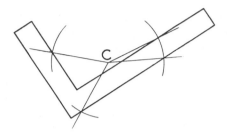

Fig. 6.15   Rapid product inspection by polar checking.

relative to the centroid, number and area of holes, area and dimensions of the convex hull and enclosing rectangle, number of sharp corners, number of intersections with a check circle and angles between intersections (Fig. 6.15), and numbers and types of skeleton nodes.

The list would not be complete without a mention of the widely used shape measure $C = \text{area}/(\text{perimeter})^2$. This quantity is often called "circularity" or "compactness", since it has a maximum value of $1/4\pi$ for a circle, decreases as shapes become more irregular, and approaches zero for long narrow objects: alternatively, its reciprocal is sometimes used, being called "complexity" since it increases in size as shapes become more complex. Note that both measures are dimensionless so that they are independent of size and are therefore sensitive only to the shape of an object. Other dimensionless measures of this type include rectangularity and aspect ratio.

All these measures have the property of characterizing a shape but not of describing it uniquely. Thus it is easy to see that there are in general many different shapes having the same value of circularity. Hence, these rather *ad hoc* measures are on the whole less valuable than approaches such as skeletonization (Section 6.8) or moments (Section 6.10) that can be used to represent and reproduce a shape to any required degree of precision. Nevertheless, it should be noted that rigorous checking of even one measured number to high precision often permits a machined part to be identified positively.

6.10   Shape description by moments

Moment approximations provide one of the more obvious means of describing 2-D shapes, and take the form of series expansions of the type:

$$M_{pq} = \sum_x \sum_y x^p y^q f(x, y) \qquad (6.22)$$

for a picture function $f(x, y)$; such a series may be curtailed when the approximation is sufficiently accurate. By referring axes to the centroid of the shape, moments can be constructed that are position-invariant: they can also be renormalized so that they are invariant under rotation and change of scale (Hu, 1962; see also Wong and Hall, 1978). The main value of using moment descriptors is that in certain applications the number of parameters may be made small without significant loss of precision—although the number required may not be clear without considerable experimentation. Moments can prove particularly valuable in describing shapes such as cams and other fairly round objects, although they have also been used in a variety of other applications including aeroplane silhouette recognition (Dudani *et al.*, 1977).

6.11 Boundary tracking procedures

The preceding sections have described methods of analysing shape on the basis of body representations of one sort or another—convex hulls, skeletons, moments, and so on. However, an important approach has so far been omitted—the use of boundary pattern analysis. This approach has the potential advantage of requiring considerably reduced computation, since the number of pixels to be examined is equal to the number of pixels on the boundary of any object rather than the much larger number of pixels within the boundary. Before proper use can be made of boundary pattern analysis techniques, means must be found for tracking systematically around the boundaries of all the objects in an image: in addition, care must be taken not to ignore any holes that are present or any objects within holes.

In one sense the problem has been analysed already, in that the object labelling algorithm of Section 6.3 systematically visits and propagates through all objects in the image. All that is required now is some means of tracking round object boundaries once they have been encountered. Quite clearly it will be useful to mark in a separate image space all points that have been tracked: alternatively, an object boundary image may be constructed and the tracking performed in this space, all tracked points being eliminated as they are passed.

It turns out that the latter procedure runs into the problem that objects having unit width in certain places may become disconnected. Hence we ignore this approach and adopt the previous one. There is still a problem when objects have unit-width sections, since these can cause badly designed tracking algorithms to choose a wrong path, going back around the previous

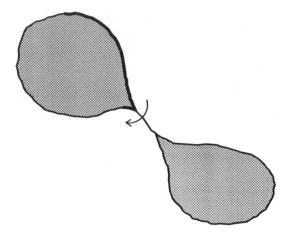

Fig. 6.16 A problem with an oversimple boundary tracking algorithm: the boundary tracking procedure takes a short-cut across a unit-width boundary segment instead of continuing and keeping to the left path at all times.

section instead of on to the next (Fig. 6.16). To avoid this circumstance it is best to adopt the following strategy:

(i) track round each boundary, keeping to the left path consistently;
(ii) stop the tracking procedure only when passing through the starting point in the original direction (or passing through the first two points in the same order).

Table 6.10   Basic procedure for tracking around a single object.

```
repeat
    (* find direction to move next *)
    start with current tracking direction;
    reverse it;
    repeat
      rotate tracking direction clockwise
    until the next 1 is met on outer pixels of 3×3 neighbourhood;
    record this as new current direction;
    move one pixel along this direction;
    increment boundary index;
    store current position in boundary list;
until (position = original position) & (direction = original
    direction)
```

Apart from necessary initialization at the start, a suitable tracking procedure is given in Table 6.10. Boundary tracking procedures such as this require careful coding; in addition, they have to operate rapidly despite the considerable amount of computation required for every boundary pixel (see the amount of code within the main **repeat** loop of the procedure of Table 6.10). For this reason it is normally best to implement the main direction-finding routine using lookup tables, or else using special hardware: details are, however, beyond the scope of this chapter.

Having seen how to track around the boundaries of objects in binary images, we are now in a position to embark on boundary pattern analysis. This is done in the following chapter.

6.12 Concluding remarks

This chapter has concentrated on rather traditional methods of performing image analysis—using image processing techniques. This has led naturally to area representations of objects, including for example moment and convex hull-based schemes, although the skeleton approach appeared as a rather special case in that it converts objects into graphical structures. An alternative schema is to represent shapes by their boundary patterns, after applying suitable tracking algorithms: this latter approach is considered in the following chapter. Meanwhile, it should be noted that connectedness has been an underlying theme in the present chapter, objects being separated from each other by regions of background, thereby permitting objects to be considered and characterized individually. Connectedness has been seen to involve rather more intricate computations than might have been expected, and this necessitates great care in the design of algorithms: this must partly explain why after so many years, new thinning algorithms are still being developed (e.g. Kwok, 1989; Choy *et al.*, 1995) (ultimately, these complexities arise because global properties of images are being computed by purely local means).

Although it will turn out that boundary pattern analysis is in certain ways more attractive than region pattern analysis, this comparison cannot be divorced from considerations of the hardware the algorithms have to run on. In this respect it should be noted that many of the algorithms of this chapter can be performed efficiently on SIMD processors, which have one processing element per pixel (see Chapter 24), whereas boundary pattern analysis will be seen to match better the capabilities of more conventional serial computers.

6.13 Bibliographical and historical notes

The development of shape analysis techniques has been particularly extensive: hence only a brief perusal of the history is attempted here. The all-important theory of connectedness and the related concept of adjacency in digital images was developed largely by Rosenfeld (see for example, Rosenfeld, 1970); for a review of this work, see Rosenfeld (1979). The connectedness concept led to the idea of a distance function in a digital picture (Rosenfeld and Pfaltz, 1966,1968), and the related skeleton concept (Pfaltz and Rosenfeld, 1967). However, the basic idea of a skeleton dates from the classic work by Blum (1967)—see also Blum and Nagel (1978). Important work on thinning has been carried out by Arcelli et al. (1975, 1981; Arcelli and di Baja, 1985) and parallels work by Davies and Plummer (1981). Note that the latter paper demonstrates a rigorous method for testing the results of any thinning algorithm, however generated, and in particular for detecting skeletal bias. More recently, Arcelli and Ramella (1995) have reconsidered the problem of skeletons in grey-scale images.

Returning momentarily to the distance function concept, which is of some importance for measurement applications, there have been important developments to generalize this concept and to make distance functions uniform and isotropic: see for example Huttenlocher et al. (1993).

Sklansky has carried out much work on convexity and convex hull determination (see for example, Sklansky, 1970; Sklansky et al., 1976), while Batchelor (1979) developed concavity trees for shape description. Early work by Moore (1968) using shrinking and expanding techniques for shape analysis has been developed considerably by Serra and his coworkers (see, for example, Serra, 1986), and Haralick et al. (1987) have generalized the underlying mathematical (morphological) concepts, including the case of grey-scale analysis: for more details, see Appendix B. Use of invariant moments for pattern recognition dates from the two seminal papers by Hu (1961, 1962) and has continued unabated ever since. The field of shape analysis has been reviewed a number of times by Pavlidis (1977, 1978), and he has also drawn attention to the importance of unambiguous ("asymptotic") shape representation schemes (Pavlidis, 1980).

Finally, the design of a modified crossing number χ_{skel} for the analysis of skeletal shape has not been considered until relatively recently: as pointed out in Section 6.8.5, χ_{skel} is different from χ since it judges the *remaining* (i.e. skeletal) points rather than points that *might* be eliminated from the skeleton (Davies and Celano, 1993b).

6.14 Problems

1. Write a routine for removing inconsistencies in labelling U-shaped objects by implementing a reverse scan through the image (see Section 6.3).
2. Write the full Pascal routine required to sort the lists of labels, to be inserted at the end of the algorithm of Table 6.2.
3. Devise an algorithm for determining the "underconvex" hull described in Section 6.6.
4. Show that, as stated in Section 6.8.2 for a parallel thinning algorithm, all north points may be removed in parallel without any risk of causing a break in the skeleton.
5. Show that the convex hull algorithm of Table 6.5 enlarges objects which may then coalesce. Find a simple means of locally inhibiting the algorithm to prevent objects from becoming connected in this way. Take care to consider whether a parallel or sequential process is to be used.

7

Boundary Pattern Analysis

7.1 Introduction

An earlier chapter showed how thresholding may be used to binarize grey-scale images and hence to present objects as 2-D shapes. However, this method of segmenting objects is successful only when considerable care is taken with lighting and when the object can be presented conveniently, e.g. as dark patches on a light background. In other cases, adaptive thresholding may help to achieve the same end: as an alternative, edge detection schemes can be applied, these generally being rather more resistant to problems of lighting. Nevertheless, thresholding of edge-enhanced images still gives certain problems: in particular, edges may peter out in some places and may be thick in others (Fig. 7.1). For many purposes, the output of an edge detector is ideally a connected unit-width line around the periphery of an object and steps will need to be taken to convert edges to this form.

Thinning algorithms can be used to reduce edges to unit thickness, whilst maintaining connectedness (Fig. 7.1(d)). Many algorithms have been developed for this purpose and the main problems here are: (i) slight bias and inaccuracy due to uneven stripping of pixels, especially in view of the fact that even the best algorithm can only produce a line that is locally within 1/2 pixel of the ideal position; and (ii) introduction of a certain number of noise spurs. The first of these problems can be minimized by using grey-scale edge thinning algorithms which act directly on the original grey-scale edge-enhanced image (e.g. Paler and Kittler, 1983). Noise spurs around object boundaries can be eliminated quite efficiently by removing lines that are shorter than (say) 3 pixels. Overall, the major problem to be

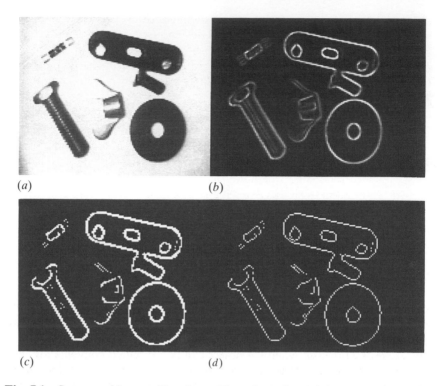

(a)   (b)

(c)   (d)

Fig. 7.1 Some problems with edges. The edge-enhanced image (b) from an original image (a) is thresholded as in (c). The edges so detected are found to peter out in some places and to be thick in other places. A thinning algorithm is able to reduce the edges to unit thickness (d) but *ad hoc* (i.e. not model-based) linking algorithms are liable to produce erroneous results (not shown).

dealt with is that of reconnecting boundaries that have become fragmented during thresholding.

A number of rather *ad hoc* schemes are available for relinking broken boundaries: for example, line ends may be extended along their existing directions—a very limited procedure since there are (at least for binary edges) only eight possible directions, and it is quite possible for the extended line ends not to meet. Another approach is to join line ends which are (i) sufficiently close together, and (ii) pointing in similar directions to each other and to the direction of the vector between the two ends. In fact, this approach can be made quite credible in principle but in practice it can lead to all sorts of problems as it is still *ad hoc* and not model driven. Hence adjacent lines which arise from genuine surface markings and from shadows may be arbitrarily linked together by such algorithms. In many situations it is therefore best if the process is model driven—for example by finding the best fit to some appropriate idealized boundary such as an ellipse. Yet another approach is that of relaxation labelling, which iteratively enhances the original image, progressively making decisions as to where the original grey levels reinforce each other. Thus edge linking is permitted only where evidence is available in the original image that this is permissible. A similar but computationally more efficient line of attack is the hysteresis thresholding method of Canny (1986): here intensity gradients above a certain upper threshold are taken to give definite indication of edge positions, whereas those above a second, lower threshold are taken to indicate edges only if they are adjacent to positions that have already been accepted as edges.

It may be thought that the Marr—Hildreth and related edge detectors do not run into these problems, because they give edge contours that are necessarily connected. However, the result of using methods that force connectedness is that sometimes (e.g. when edges are diffuse, or of low contrast, so that image noise is an important factor) parts of a contour will lack meaning; indeed, a contour may meander over such regions following noise rather than useful object boundaries. Furthermore, it is as well to note that the problem is not merely one of pulling low-level signals out of noise, but rather that sometimes there is no signal at all present that could be enhanced to a meaningful level. Reasons for this include lighting being such as to give zero contrast (as, for example, when a cube is lit in such a way that two faces have equal brightness) and occlusion. Lack of spatial resolution can also cause problems by merging together several lines on an object.

It is now assumed that all of these problems have been overcome by sufficient care with the lighting scheme, appropriate digitization and other means. It is also assumed that suitable thinning and linking algorithms have been applied so that all objects are outlined by connected unit-width boundary lines. At this stage, it should be possible (i) to *locate* the objects from the boundary image, (ii) to *identify* and *orientate* them accurately, and (iii) to *scrutinize* them for shape and size defects.

7.2 Boundary tracking procedures

Before objects can be matched from their boundary patterns, means must be found for tracking systematically around the boundaries of all the objects in an image: means have already been demonstrated for achieving this in the case of regions such as those that result from intensity thresholding routines

(Chapter 6). However, when a connected unit-width boundary exists, the problem of tracking is much simpler, since it is necessary only to move repeatedly to the next edge pixel storing the boundary information, for example, as a suitable pair of 1-D periodic coordinate arrays $x[i]$, $y[i]$: note that each edge pixel must be marked appropriately as we proceed around the boundary, to ensure (i) that we never reverse direction, (ii) that we know when we have been round the whole boundary once, and (iii) that we record which object boundaries have been encountered. As when tracking around regions, we must ensure that in each case we end by passing through the starting point in the same direction.

7.3   Template matching—a reminder

Before considering boundary pattern analysis, recall the object location problem discussed in Chapter 1. With 2-D images there is a significant matching problem because of the number of degrees of freedom involved; usually there will be two degrees of freedom for position and one for orientation. For an object of size 30×30 pixels in a 256×256 image, matching the position in principle requires some 256×256 trials, each one involving 30×30 operations. Since the orientation in general is also unknown, another factor of 360 is required to cope with varying orientation (i.e. one trial per degree of rotation), hence making a total of $256^2 \times 30^2 \times 360$, i.e. some 20 000 million operations. This is clearly an unacceptable situation, since even a 1 MIPS processor will take several hours to locate objects within the image. If there is more than one type of object present, or if the objects have various possible scales, the combinatorial explosion is even more serious.

7.4   Centroidal profiles

The matching problems outlined above make it attractive to attempt to locate objects in a smaller search space. In fact, this is possible to achieve very simply by matching the boundary of each object in a single dimension. Perhaps the most obvious such scheme uses an (r, θ) plot. Here the centroid of the object is first located[*]; then a polar coordinate system is set up

---

[*] Note that the position of the centroid is deducible directly from the list of boundary pixel coordinates—there is no need to start with a region-based description of the object for this purpose.

relative to this point and the object boundary is plotted as an (r, θ) graph—often called a "centroidal profile" (Fig. 7.2). Next, the 1-D graph so obtained is matched against the corresponding graph for an idealized object of the same type. Since the object in general has an arbitrary orientation, it is necessary to "slide" the idealized graph along that obtained from the image data until the best match is obtained. The match for each possible orientation α_j of the object is commonly tested by measuring the differences in radial distance between the boundary graph B and the template graph T for various values of θ and summing their squares to give a difference measure D_j for the quality of the fit:

$$D_j = \sum_i [r_B(\theta_i) - r_T(\theta_i + \alpha_j)]^2 \tag{7.1}$$

Alternatively, the absolute magnitudes of the differences are used:

$$D_j = \sum_i |r_B(\theta_i) - r_T(\theta_i + \alpha_j)| \tag{7.2}$$

a

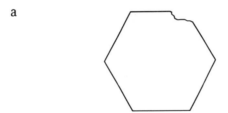

b

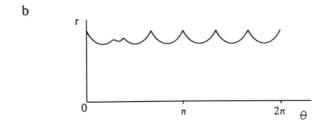

Fig. 7.2 Centroidal profiles for object recognition and scrutiny: (a) hexagonal nut shape in which one corner has been damaged; (b) centroidal profile, which permits both straightforward identification of the object and detailed scrutiny of its shape.

The latter measure has the advantage of being easier to compute and of being less biased by extreme or erroneous difference values. The factor of 360 referred to above still affects the amount of computation but at least the basic 2-D matching operation is reduced to 1-D. In fact, it is interesting that the scale of the problem is reduced in two ways, since both the idealized template and the image data are now 1-D. In the above example, the result is that the number of operations that are required to test each object drops to around 360^2 (which is only about 100 000), so there is a *very* substantial saving in computation.

The 1-D boundary pattern matching approach described above is able to identify objects and also to find their orientations. In fact, initial location of the centroid of the object also solves one other part of the problem as specified at the end of Section 7.1. At this stage it may be noted that the matching process also leads to the possibility of inspecting the object's shape as an inherent part of the process (Fig. 7.2). In principle, this combination of capabilities makes the centroidal profile technique quite powerful.

Finally, note that the method is able to cope with objects of identical shapes but different sizes. This is achieved by using the maximum value of r to normalize the profile, giving a variation (ρ, θ) where $\rho = r/r_{max}$.

7.5 Problems with the centroidal profile approach

In practice, there are several problems with the procedure outlined above. First, any major defect or occlusion of the object boundary can cause the centroid to be moved away from its true position, and the matching process will be largely spoiled (Fig. 7.3). Thus, instead of concluding that this is an object of type X with a specific part of its boundary damaged, the algorithm will most probably not recognize it at all. Such behaviour would be inadequate in many automated inspection applications, where positive identification and fault-finding are required, and the object would have to be rejected without a satisfactory diagnosis being made.

Second, the (r, θ) plot will be multivalued for a certain class of object (Fig. 7.4). This has the effect of making the matching process partly 2-D and leads to complication and excessive computation.

Third, the very variable spacing of the pixels when plotted in (r, θ) space is a source of complication. It leads to the requirement for considerable smoothing of the 1-D plots, especially in regions where the boundary comes close to the centroid—as for elongated objects such as spanners or screwdrivers (Fig. 7.5); however, in other places accuracy will be greater than necessary and the overall process will be wasteful. The problem arises

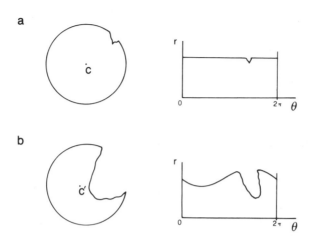

Fig. 7.3 Effect of gross defects on the centroidal profile: (a) a chipped circular object and its centroidal profile; (b) a severely damaged circular object and its centroidal profile: the latter is distorted not only by the break in the circle but also because of the shift in its centroid to C′.

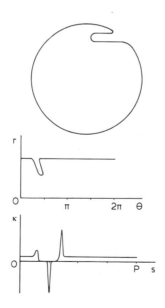

Fig. 7.4 Boundary pattern analysis via (r, θ) and (s, κ) plots.

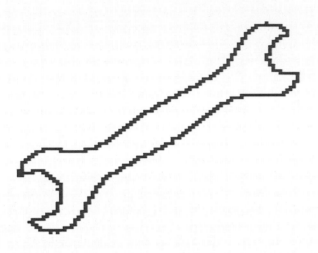

Fig. 7.5 A problem in obtaining a centroidal profile for elongated objects. This figure highlights the pixels around the boundary of an elongated object—a spanner—showing that it will be difficult to obtain an accurate centroidal profile for the region near the centroid.

because quantization should ideally be uniform along the θ-axis so that the two templates can be moved conveniently relative to one another to find the orientation of best match.

Finally, computation times can still be quite significant, so some time-saving procedure is required.

7.5.1 Some solutions

All four of the above problems can be tackled in one way or another, with varying degrees of success. The first problem, that of coping with occlusions and gross defects, is probably the most fundamental and the most resistant to satisfactory solution. For the method to work successfully, a stable reference point must be found within the object. The centroid is naturally a good candidate for this since the averaging inherent in its location tends to eliminate most forms of noise or minor defect: however, major distortions such as those arising from breakages or occlusions are bound to affect it adversely. The centroid of the boundary is no better, and may also be less successful at suppressing noise. Other possible candidates are the positions of prominent features such as corners, holes, centres of arcs and so on. In general, the smaller such a feature, the more likely it is to be missed altogether in the event of a breakage or occlusion, although the larger such a feature, the more likely it is to be affected by the defect. It turns out that circular arcs can be located accurately (at their centres) even if they are partly occluded (see Chapter 9), so these features are very useful for leading to suitable reference points. A set of symmetrically placed holes may sometimes be suitable, since even if one of them is obscured, one of the others is likely to be visible and can act as a reference point.

Clearly, such features can help the method to work adequately but their presence also calls into question the value of the 1-D boundary pattern matching procedure, since they make it likely that superior methods can be used for object recognition (see Parts 2 and 3). For the present we therefore accept (i) that severe complications arise when part of an object is missing or occluded, and (ii) that it may be possible to provide some degree of resistance to such problems by using a prominent feature as a reference point instead of the centroid. Indeed, the only significant *further* change that is required to cope with occlusions is that difference $(r_B - r_T)$ values of greater than (say) 3 pixels should be ignored, and the best match then becomes one for which the greatest number of values of θ give good agreement between B and T.

The second problem, of multivalued (r, θ) plots, is solved very simply by employing the heuristic of taking the smallest value of r for any given θ and then proceeding with matching as normal (here it is assumed that the boundaries of any holes present within the boundary of the object are dealt with separately, information about any object and its holes being collated at the end of the recognition process). This *ad hoc* procedure should in fact be acceptable when making a preliminary match of the object to its 1-D template, and may be discarded at a later stage when the orientation of the object is known accurately.

The third problem described above arises because of uneven spacing of the pixel boundaries along the θ dimension of the (r, θ) graph. To some extent this problem can be avoided by deciding in advance on the permissible values of θ and querying a list of boundary points to find which has the closest θ to each permissible value. Some local smoothing of the ordered set of boundary points can be undertaken but this is in principle unnecessary, since for a connected boundary, there will always be one pixel which is closest to a line from the centroid at a given value of θ.

The two-stage approach to matching hinted at above can also be used to help with the last of the problems mentioned above—the need to speed up the processing. First, a coarse match is obtained between the object and its 1-D template by taking θ in relatively large steps of (say) 5° and ignoring intermediate angles in both the image data and the template; then a better match is obtained by making fine adjustments to the orientations, obtaining a match to within 1°. In this way the coarse match is obtained perhaps 20 times faster than the previous full match, whereas the final fine match takes a relatively short time, since very few distinct orientations have to be tested.

This two-stage process can be optimized by making a few simple calculations. The coarse match is given by steps $\delta\theta$, so the computational load is proportional to $(360/\delta\theta)^2$, whereas the load for the fine match is proportional to $360\delta\theta$, giving a total load of:

$$\lambda = (360/\delta\theta)^2 + 360\delta\theta \qquad (7.3)$$

This should be compared with the original load of:

$$\lambda_0 = 360^2 \qquad (7.4)$$

Hence the load is reduced (and the algorithm speeded up) by the factor:

$$\eta = \lambda_0/\lambda = 1/[(1/\delta\theta)^2 + \delta\theta/360] \qquad (7.5)$$

This is a maximum for $d\eta/d\delta\theta = 0$, giving:

$$\delta\theta = (2 \times 360)^{1/3} \simeq 9° \qquad (7.6)$$

In practice this value of $\delta\theta$ is rather large and there is a risk that the coarse match will give such a poor fit that the object will not be identified: hence values of $\delta\theta$ in the range 2–5° are more usual (see for example, Berman *et al.*, 1985). Note that the optimum value of η is 26.8 and that this reduces only to 18.6 for $\delta\theta = 5°$, although it goes down to 3.9 for $\delta\theta = 2°$.

Another way of approaching the problem is to search the (r, θ) graph for some characteristic feature such as a sharp corner (this step constituting the coarse match), and then to perform a fine match around the object orientation so deduced. Clearly there are possibilities of error here, in situations where objects have several similar features—as in the case of a rectangle: however, the individual trials are relatively inexpensive and so it is worth invoking this procedure if the object possesses appropriate well-defined features. Note that it is possible to use the position of the maximum value, r_{max}, as an orientating feature but this is frequently inappropriate because a smooth maximum gives a relatively large angular error.

7.6 The (s, ψ) plot

It will be seen from the above considerations that boundary pattern analysis should usually be practicable except when problems from occlusions and gross defects can be expected. These latter problems do give motivation, however, for employing alternative methods if these can be found. In fact, the (s, ψ) graph has proved particularly popular since it is inherently better suited than the (r, θ) graph to situations where defects and occlusions may occur. In addition, it does not suffer from the multiple values encountered by the (r, θ) method.

The (s, ψ) graph does not require prior estimation of the centroid or some other reference point since it is computed directly from the boundary, in the form of a plot of the tangential orientation ψ as a function of boundary distance s. The method is not without its problems and, in particular, distance along the boundary needs to be measured accurately. The commonly used approach is to count horizontal and vertical steps as unit distance, and to take diagonal steps as distance $\sqrt{2}$: in fact this idea must be regarded as a rather *ad hoc* solution and the situation is discussed further in Section 7.10.

When considering application of the (s, ψ) graph for object recognition, it will immediately be noticed that the graph has a ψ value that increases by 2π for each circuit of the boundary, i.e. $\psi(s)$ is not periodic in s. The result is that the graph becomes essentially 2-D, i.e. the shape has to be matched by moving the ideal object template both along the s-axis and along the ψ-axis directions. Ideally, the template could be moved diagonally along the direction of the graph. However, noise and other deviations of the actual shape relative to the ideal shape mean that in practice the match must be at least partly 2-D, hence adding to the computational load.

One way of tackling this problem is to make a comparison with the shape of a circle of the same boundary length P. Thus, an $(s, \Delta\psi)$ graph is plotted which reflects the difference $\Delta\psi$ between the ψ expected for the shape and that expected for a circle of the same perimeter:

$$\Delta\psi = \psi - 2\pi s/P \qquad (7.7)$$

This expression helps to keep the graph 1-D, since $\Delta\psi$ automatically resets itself to its initial value after one circuit of the boundary (i.e. $\Delta\psi$ is periodic in s).

Next it should be noticed that the $\Delta\psi(s)$ variation depends on the starting position where $s = 0$ and that this is randomly sited on the boundary. It is useful to eliminate this dependence and this may be achieved by subtracting from $\Delta\psi$ its mean value μ. This gives the new variable:

$$\tilde{\psi} = \psi - 2\pi s/P - \mu \tag{7.8}$$

At this stage the graph is completely 1-D and is also periodic, being similar in these respects to an (r, θ) graph. Matching should now reduce to the straightforward task of sliding the template along the $\tilde{\psi}(s)$ graph until a good fit is achieved.

At this point, there should be no problems so long as (i) the scale of the object is known, and (ii) occlusions or other disturbances cannot occur. Suppose next that the scale is unknown: then the perimeter P may be used to normalize the value of s. If, however, occlusions *can* occur, then no reliance can be placed on P for normalizing s and hence the method cannot be guaranteed to work. This problem does not arise if the scale of the object is known, since a standard perimeter P_T can be assumed. However, the possibility of occlusion gives further problems which are discussed in the next section.

Another way in which the problem of nonperiodic $\psi(s)$ can be solved is by replacing ψ by its derivative $d\psi/ds$. Then the problem of constantly expanding ψ (which results in its increasing by 2π after each circuit of the boundary) is eliminated—the addition of 2π to ψ does not affect $d\psi/ds$ locally, since $d(\psi + 2\pi)/ds = d\psi/ds$. It should be noticed that $d\psi/ds$ is actually the local curvature function $\kappa(s)$ (see Fig. 7.4), so the resulting graph has a simple physical interpretation. Unfortunately, this version of the method has its own problems in that κ approaches infinity at any sharp corner. For industrial components, which frequently have sharp corners, this is a genuine practical difficulty and may be tackled by approximating adjacent gradients and ensuring that κ integrates to the correct value in the region of a corner (Hall, 1979).

Many workers take the (s, κ) graph idea further and expand $\kappa(s)$ as a Fourier series:

$$\kappa(s) = \sum_{n=-\infty}^{\infty} c_n \exp(2\pi i n s/P) \tag{7.9}$$

This results in the well-known Fourier descriptor method. In this method, shapes are analysed in terms of a series of Fourier descriptor components which are truncated to zero after a sufficient number of terms. Unfortunately, the amount of computation involved in this approach is considerable and there is a tendency to approximate curves with relatively few terms. In industrial applications where computations have to be performed in real time, this can give problems, so it is often more appropriate to match to the basic (s, κ) graph. In this way critical measurements between object features can be made with adequate accuracy in real time.

7.7   Tackling the problems of occlusion

Whatever means are used for tackling the problem of continuously increasing ψ, problems still arise when occlusions occur. However, the whole method is not immediately invalidated by missing sections of boundary as it is for the basic (r, θ) method. As noted above, the first effect of occlusions is that the perimeter of the object is altered, so P can no longer be used to indicate its scale. Hence the latter has to be known in advance: this is assumed in what follows. Another practical result of occlusions is that certain sections correspond correctly to parts of the object, whereas other parts correspond to parts of occluding objects; alternatively, they may correspond to unpredictable boundary segments where damage has occurred. Note that if the overall boundary is that of two overlapping objects, the observed perimeter P_B will be greater than the ideal perimeter P_T.

Segmenting the boundary between relevant and irrelevant sections is, *a priori*, a difficult task. However, a useful strategy is to start by making positive matches wherever possible and to ignore irrelevant sections—i.e. try to match as usual, ignoring any section of the boundary that is a bad fit. We can imagine achieving a match by sliding the template T along the boundary B. However, a problem arises since T is periodic in s and should not be cut off at the ends of the range $0 \leqslant s \leqslant P_T$. As a result, it is necessary to attempt to match over a length $2P_T$. At first sight it might be thought that the situation ought to be symmetrical between B and T. However, T is known in advance whereas B is partly unknown, there being the possibility of one or more breaks in the ideal boundary into which foreign boundary segments have been included; indeed, the positions of the breaks are unknown, so it is necessary to try matching the whole of T at all positions on B. Taking a length $2P_T$ in testing for a match effectively permits the required break to arise in T at any relevant position: see Fig. 7.6.

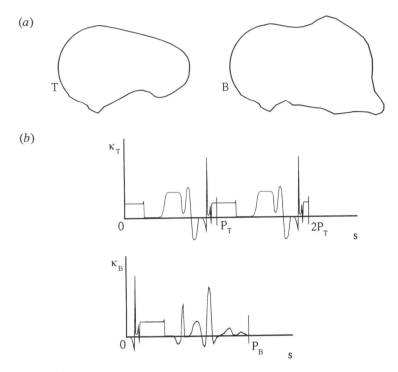

Fig. 7.6 Matching a template against a distorted boundary. When a boundary B is broken (or partly occluded) but continuous, it is necessary to attempt to match between B and a template T that is doubled to length $2P_T$, to allow for T being severed at any point: (a) the basic problem; (b) matching in (s, κ)-space.

When carrying out the match we basically use the difference measure:

$$D_{jk} = \sum_i [\psi_B(s_i) - (\psi_T(s_i + s_k) + \alpha_j)]^2 \tag{7.10}$$

where j and k are the match parameters for orientation and boundary displacement, respectively. Notice that the resulting D_{jk} is roughly proportional to the length L of the boundary over which the fit is reasonable. Unfortunately, this means that the measure D_{jk} appears to *improve* as L decreases; hence, when variable occlusions can occur, the best match must be taken as the one for which the greatest length L gives good agreement between B and T (this may be measured as the greatest number of values of s in the sum of equation (7.10) which give good agreement between B and

T, i.e. the sum over all i such that the difference in square brackets is numerically less than, say, 5°).

If the boundary is occluded in more than one place, then L is at most the largest single length of unoccluded boundary (not the total length of unoccluded boundary), since the separate segments will in general be "out of phase" with the template. This is a disadvantage when trying to obtain an accurate result, since extraneous matches add noise which degrades the fit that is obtainable—hence adding to the risk that the object will not be identified and reducing accuracy of registration. This suggests that it might be better to use only short sections of the boundary template for matching. Indeed, this strategy can be advantageous since speed is enhanced and registration accuracy can be retained or even improved by careful selection of salient features (note that nonsalient features such as smooth curved segments could have originated from many places on object boundaries and are not very helpful for identifying and accurately locating objects: hence it is reasonable to ignore them). In this version of the method we now have $P_T < P_B$ and it is necessary to match over a length P_T rather than $2P_T$, since T is no longer periodic (Fig. 7.7). Once various segments have been located, the boundary can be reassembled as closely as possible into its full form, and at that stage defects, occlusions and other distortions can be recognized overtly and recorded. Reassembly of the object boundaries can be performed by techniques such as the Hough transform and relational pattern matching techniques (see Chapters 10 and 14). Work of this type has been carried out by Turney *et al.* (1985), who found that the salient features should be short boundary segments where corners and other distinctive "kinks" occur.

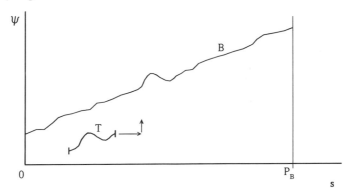

Fig. 7.7 Matching a short template to part of a boundary. A short template T, corresponding to part of an idealized boundary, is matched against the observed boundary B. Strictly speaking, matching in (s, ψ)-space is 2-D although there is very little uncertainty in the vertical (orientation) dimension.

Before leaving this topic, it is worth noting that $\tilde{\psi}$ can no longer be used when occlusions are present, since although the perimeter can be assumed to be known, the mean value of $\Delta\psi$ (equation (7.8)) cannot be deduced. Hence the matching task reverts to a 2-D search (though, as stated earlier, very little unrestrained search in the ψ direction need be made). However, in the case when small salient features are being sought, it is a reasonable working assumption that no occlusion occurs in any of the individual cases—a feature is either entirely present or entirely absent. Hence the average slope $\tilde{\psi}$ over T can validly be computed (Fig. 7.7) and this again reduces the search to 1-D (Turney *et al.*, 1985).

Overall, missing sections of object boundaries necessitate a fundamental rethink as to how boundary pattern analysis should be carried out. For quite small defects the (r, θ) method is sufficiently robust but in less trivial cases it is vital to use some form of the (s, ψ) approach, while for really gross occlusions it is not particularly useful to try to match for the full boundary: rather it is better to attempt to match small salient features. This sets the scene for the Hough transform and relational pattern matching techniques of later chapters (see Parts 2 and 3).

7.8   Chain code

This section briefly describes the chain code which was devised in 1961 by Freeman. This method recodes the boundary of a region into a sequence of octal digits, each one representing a boundary step in one of the eight basic directions relative to the current pixel position, denoted here by an asterisk:

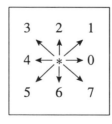

This code is important for presenting shape information in such a way as to save computer storage. In particular, note that the code ignores (and therefore does not store) the absolute position of the boundary pixels, so that only three bits are required for each boundary step. However, in order to regenerate a digital image, additional information is required, including the starting point of the chain and the length.

Chain code is useful in cutting down storage relative to the number of bits required to hold a list of the coordinates of all the boundary pixels (and *a fortiori* relative to the number of bits required to hold a whole binary image). In addition, chain code is able to provide data in a format that is suitable for certain recognition processes. However, chain coding may sometimes be too compressed a representation for convenient computer manipulation of the data: hence, in many cases direct (x, y) storage may be more suitable, or otherwise direct (s, ψ) coding. Note also that with modern computers, the advantage of reduced storage requirements is becoming less relevant than in the early days of pattern recognition.

7.9 The (r, s) plot

Before leaving the topic of 1-D plots for analysis and recognition of 2-D shapes, there is one further method that should be mentioned—the (r, s) plot. Freeman (1978) introduced this method as a natural follow on from the use of chain code (which necessarily involves the parameter s). Since quite a large family of possible 1-D plot representations exist, it is useful to demonstrate the particular advantages of any new one. In fact, the (r, s) plot is in many ways similar to the (r, θ) plot (particularly in using the centroid as a reference point) but it overcomes the difficulty of the (r, θ) plot sometimes being a multivalued function: for every value of s there is a unique value of r. However, against this advantage must be weighed the fact that it is not in general possible to recover the original shape from the 1-D plot, since an ambiguity is introduced whenever the tangent to the boundary passes through the centroid. To overcome this problem, a phase function $\Phi(s)$ (whose value is always either $+1$ or -1, the minus value corresponding to the curve turning back on itself) must be stored in addition to the function $r(s)$. Freeman made a slight variation on this scheme by relying on the fact that r is always positive, so that all the information can be stored in a single function which is the product of r and Φ. Note, however, that the derived function is no longer continuous, so minor changes in shape could lead to misleadingly large changes in cross-correlation coefficient during comparison with a template.

7.10 Accuracy of boundary length measures

Next we examine the accuracy of the idea expressed earlier, that adjacent pixels on an 8-connected curve should be regarded as separated by 1

pixel if the vector joining them is aligned along the major axes, and by
$\sqrt{2}$ pixels if the vector is in a diagonal orientation. It turns out that this
estimator in general overestimates the distance along the boundary. The
reason for this is quite simple to see by appealing to the following pair of
situations:

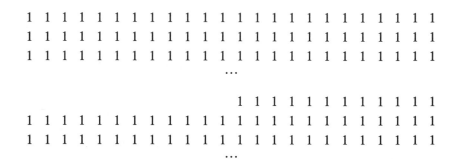

In either case we are considering only the top of the object. In the first
example, the boundary length along the top of the object is exactly that
given by the rule. However, in the second case the estimated length is
increased by amount $\sqrt{2} - 1$ because of the step. Now as the length of the
top of the object tends to large values, say p pixels, the actual length
approximates to p while the estimated length is $p + \sqrt{2} - 1$: thus a definite
error exists. Indeed, this error initially increases in importance as p
decreases, since the actual length of the top of the object (when there is one
step) is still

$$L = (1 + p^2)^{1/2} \simeq p \tag{7.11}$$

so the fractional error is

$$\xi \simeq (\sqrt{2} - 1)/p \tag{7.12}$$

which increases as p becomes smaller.

This result can be construed as meaning that the fractional error ξ in
estimating boundary length increases initially as the boundary orientation
ψ increases from zero. A similar effect occurs as the orientation
decreases from 45°. Thus the ξ variation possesses a maximum between
0° and 45°. This systematic overestimation of boundary length may be
eliminated by employing an improved model in which the length per
pixel is s_{m} along the major axes directions and s_{d} in diagonal directions.

A complete calculation (Kulpa, 1977; see also Dorst and Smeulders, 1987) shows that:

$$s_m = 0.948 \qquad (7.13)$$

and

$$s_d = 1.343 \qquad (7.14)$$

It is perhaps surprising that this solution corresponds to a ratio s_d/s_m that is still equal to $\sqrt{2}$, although the arguments given above make it obvious that s_m should be less than unity.

Unfortunately, an estimator that has just two free parameters can still permit quite large errors in estimating the perimeters of individual objects. To reduce this problem, it is necessary to perform more detailed modelling of the step pattern around the boundary (Koplowitz and Bruckstein, 1989), which seems certain to increase the computational load significantly.

It is important to underline that the basis of this work is to estimate the length of the original continuous boundary rather than that of the digitized boundary: furthermore, it must be noted that the digitization process loses information, so the best that can be done is to obtain a best estimate of the original boundary length. Thus employing the values 0.948, 1.343 given above, rather than the values 1, $\sqrt{2}$, reduces the estimated errors in boundary length measurement from 6.6% to 2.3%—but then only under certain assumptions about correlations between orientations at neighbouring boundary pixels (Dorst and Smeulders, 1987). For a more detailed insight into the situation, see Davies (1991c).

7.11 Concluding remarks

This chapter has been concerned with boundary pattern analysis. The boundary patterns were imagined to arise from edge detection operations which have been processed to make them connected and of unit width. However, if intensity thresholding methods were employed for segmenting images, boundary tracking procedures would also permit the boundary pattern analysis methods of this chapter to be used. Conversely, if edge detection operations led to the production of connected boundaries, these could be filled in by suitable algorithms (which are more tricky to devise than might at first be imagined!) (Ali and Burge, 1988) and converted to regions to which the binary shape analysis methods of Chapter 6 could be applied. Hence shapes are representable in region or boundary form: if they initially arise in one representation, they may be converted to the alternative representation. This means that boundary or regional means may be employed for shape analysis, as appropriate.

It might be thought that if shape information is available initially in one representation, the method of shape analysis that is employed should be one that is expressed in the same representation. However, this conclusion is dubious, since the routines required for conversion between representations (in either direction) require relatively modest amounts of computation. This means that it is important to cast the net quite widely in selecting the best available means for shape analysis and object recognition.

An important factor here is that a positive advantage is often gained by employing boundary pattern analysis, since computation should be inherently lower (in proportion to the numbers of pixels that are required to describe the shapes in the two representations). Another important determining factor which has been seen in the present chapter is that of occlusion. If occlusions are present, then many of the methods described in Chapter 6 operate incorrectly—as also happens for the basic centroidal profile method described early in this chapter. The (s, ψ) method then provides a good starting point. As has been seen, this is best applied to detect small salient boundary features, which can then be reassembled into whole objects by relational pattern matching techniques (see especially Chapter 14).

7.12 Bibliographical and historical notes

Many of the techniques described in this chapter have been known since the early days of image analysis. Boundary tracking has been known since 1961 when Freeman introduced his chain code. Indeed, Freeman is responsible for much subsequent work in this area (see for example, Freeman, 1974). Freeman (1978) introduced the notion of segmenting boundaries at "critical points" in order to facilitate matching: suitable critical points were corners (discontinuities in curvature), points of inflection, and curvature maxima. This work is clearly strongly related to that of Turney et al. (1985). Other workers have segmented boundaries into piecewise linear sections as a preliminary to more detailed pattern analysis (Dhome et al., 1983; Rives et al., 1985), since this procedure reduces information and hence saves computation (albeit at the expense of accuracy). Early work on Fourier boundary descriptors using the (r, θ) and (s, ψ) approaches was carried out by Rutovitz (1970), Barrow and Popplestone (1971) and Zahn and Roskies (1972); another notable paper in this area is that by Persoon and Fu (1977).

In an interesting development, Lin and Chellappa (1987) were able to classify partial (i.e. nonclosed) 2-D curves using Fourier descriptors.

At the beginning of the chapter it was noted that there are significant problems in obtaining a thin connected boundary for every object in an image. In the last few years, a new concept—that of active contour models (or "snakes")—has become popular and appears to solve many of these problems. It operates by growing an active contour which slithers around the search area until it ends in a minimum energy situation (Kass *et al.*, 1988). Unfortunately, space does not allow details of this fascinating topic to be explored further here (but see Chapter 17 for applications to vehicle location and some recent references).

Finally, it is worth remarking on the increased attention to accuracy evident over the past few years: this is seen for example in the length estimators for digitized boundaries discussed in Section 7.10 (see Kulpa, 1977; Dorst and Smeulders, 1987; Beckers and Smeulders, 1989; Koplowitz and Bruckstein, 1989; Davies, 1991c).

7.13 Problems

1. Devise a program for finding the centroid of a region, starting from an ordered list of the coordinates of its boundary points.
2. Devise a program for finding a thinned (8-connected) boundary of an object in a binary image.
3. Determine the saving in storage that results when the boundaries of objects that are held as lists of x and y coordinates are re-expressed in chain code; assume that the image size is 256×256 and that a typical boundary length is 256. How would the result vary (a) with boundary length, and (b) with image size?

Part 2
Intermediate-Level Processing

Intermediate-level image analysis is concerned with obtaining abstract information about images, starting with the images themselves: at this stage we are no longer interested in converting one image into another, as in the subject of image processing. In fact, intermediate-level processing was already being developed in parts of Chapter 7. However, what is special about Part 2 is that intermediate-level analysis is investigated for its own sake, in particular with the aid of transform methods that have been designed systematically for the purpose.

For the most part, the chapters in Part 2 result in abstract information on the positions and orientations of various image features: they do not aim to provide real-world data, a function that is left to Part 3 (thus Part 2 may indicate that a circle exists in one part of an image: it is left to Part 3 to interpret this as a wheel and to find any defects it may have).

8

Line Detection

8.1 Introduction

Straight edges are amongst the most common features of the modern world, arising in perhaps the majority of manufactured objects and components, not least in the very buildings in which we live. Yet it is arguable whether true straight lines ever arise in the natural state: possibly the only example of their appearance in virgin outdoor scenes is the horizon—although even this is clearly seen from space as a circular boundary! The surface of water is essentially planar, although it is important to realize that this is a deduction: the fact remains that straight lines seldom appear in completely natural scenes. Be all this as it may, it is clearly vital both in city pictures and in the factory to have effective means of detecting straight edges. This chapter studies available methods for locating these important features.

Historically, the Hough transform (HT) has been the main means of detecting straight edges, and over the quarter century or so since the method was originally invented (Hough, 1962) it has been developed and refined for this purpose. Hence this chapter concentrates on this particular technique; it also prepares the ground for applying the HT to the detection of circles, ellipses, corners, etc. in the next few chapters. We start by examining the original Hough scheme, even though it is now seen to be wasteful in computation, since important principles are involved.

8.2   Application of the Hough transform to line detection

The basic concept involved in locating lines by the HT is point–line duality. A point P can be defined either as a pair of coordinates or else in terms of the set of lines passing through it. The concept starts to make sense if we consider a set of collinear points P_i, then list the sets of lines passing through each of them, and finally note that there is just one line that is common to all these sets. Thus it is possible to find the line containing all the points P_i merely by eliminating those that are not multiple hits. Indeed, it is easy to see that if a number of noise points Q_j are intermingled with the signal points P_i, the method will be able to discriminate the collinear points from amongst the noise points at the same time as finding the line containing them, merely by searching for multiple hits. Thus the method is inherently robust against noise, as indeed it is in discriminating against currently unwanted signals such as circles.

In fact, the duality goes further. For just as a point can define (or be defined by) a set of lines, so a line can define (or be defined by) a set of points, as is obvious from the above argument. This makes the above approach to line detection a mathematically elegant one and it is perhaps surprising that the Hough detection scheme was first published as a patent (Hough, 1962) of an electronic apparatus for detecting the tracks of high-energy particles, rather than as a paper in a learned journal.

The form in which the method was originally applied involves parametrizing lines by the slope-intercept equation

$$y = mx + c \tag{8.1}$$

Every point on a straight edge is then plotted as a line in (m, c) space corresponding to all the (m, c) values consistent with its coordinates, and lines are detected in this space. The embarrassment of unlimited ranges of the (m, c) values (near-vertical lines require near-infinite values of these parameters) is overcome by using two sets of plots, the first corresponding to slopes of less than 1.0 and the second to slopes of 1.0 or more; in the latter case, equation (8.1) is replaced by the form:

$$x = \tilde{m}y + \tilde{c} \tag{8.2}$$

where

$$\tilde{m} = 1/m \tag{8.3}$$

The need for this rather wasteful device was removed by the Duda and Hart (1972) approach, which replaces the slope-intercept formulation with the so-called "normal" (θ, ρ) form for the straight line (see Fig. 8.1):

$$\rho = x \cos \theta + y \sin \theta \tag{8.4}$$

To apply the method using this form, the set of lines passing through each point P_i is represented as a set of sine curves in (θ, ρ) space: e.g. for point $P_1(x_1, y_1)$ the sine curve has equation:

$$\rho = x_1 \cos \theta + y_1 \sin \theta \tag{8.5}$$

Then multiple hits in (θ, ρ) space indicate, via their θ, ρ values, the presence of lines in the original image.

Each of the methods described above has the feature that it employs an "abstract" parameter space in which multiple hits are sought. Above we talked about "plotting" points in parameter space but in fact the means of looking for hits is to seek peaks which have been built by *accumulation* of

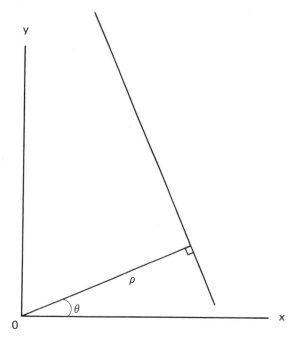

Fig. 8.1   Normal (θ, ρ) parametrization of a straight line.

data from various sources. Although it might be possible to search for hits by logical operations such as use of the AND function, the Hough method gains considerably by *accumulating evidence* for events by a *voting scheme*. It will be seen below that this is the source of the method's high degree of robustness.

Although the methods described above are mathematically elegant and are capable of detecting lines (or sets of collinear points—which may be completely isolated from each other) amid considerable interfering signals and noise, they are subject to severe computational problems. The reason for this is that every prominent point* in the original image gives rise to a great many votes in parameter space, so for a 256×256 image the (m, c) parametrization requires 256 votes to be accumulated, while the (θ, ρ) parametrization requires a similar number—say 360 if the θ quantization is to be fine enough to resolve $1°$ changes in line orientation.

A vital key to overcoming this problem was discovered by O'Gorman and Clowes (1976), who noted that points on lines are usually not isolated but instead are joined in fragments which are sufficiently long that their directions can be measured. Supposing that direction is known with high accuracy, it is then sufficient to accumulate just one vote in parameter space for every potential line point (in fact, if the local gradient is known with lesser accuracy then parameter space can be quantized more coarsely—say in $10°$ steps (O'Gorman and Clowes, 1976)—and again a single vote per point can be accumulated). Clearly, this method is much more modest in its computational requirements and it was soon adopted by other workers.

However, the computational load is still quite substantial: not only is a large 2-D storage area needed but this must be searched carefully for significant peaks—a tedious task if short line segments are being sought. Various means have been tried for cutting down computation further. Dudani and Luk (1978) tackled the problem by trying to separate out the θ and ρ estimations. They accumulated votes first in a 1-D parameter space for θ—i.e. a histogram of θ values (it must not be forgotten that such a histogram is itself a simple form of HT†). Having found suitable peaks in the θ histogram, they then built a ρ histogram for all the points that contributed votes to a given θ peak, and repeated this for all θ peaks. Thus two 1-D

---

* For the present purpose it does not matter in what way these points are prominent: they may in fact be edge points, dark specks, centres of holes, and so on. Later we shall consistently take them to be edge points.

† It is now common for any process to be called an HT if it involves accumulating votes in a parameter space, with the intention of searching for significant peaks to find properties of the original data.

spaces replace the original 2-D parameter space, with very significant savings in storage and load. Dudani and Luk applied their method to images of outdoor scenes containing houses: these images have the characteristics (i) that there are many parallel lines, (ii) that there are many lines that are naturally broken into smaller segments (e.g. parts of a roof line broken by windows), and (iii) that there are many short line segments. It seems that the method is particularly well adapted to such scenes, although it may not be so suitable for more general images. For example (as will also be seen with circle and ellipse detection), two-stage methods tend to be less accurate since the first stage is less selective: biased θ values may result from pairs of lines that would be well separated in a 2-D space. For the same reason, problems of noise tend to be more acute. In addition, any error in estimating θ values is propagated to the ρ determination stage, making values of ρ even less accurate (the latter problem may be reduced by searching the full (θ, ρ) space locally around the important θ values, although this procedure clearly increases computation again).

This section ends by noting that the more efficient methods outlined above require θ to be estimated for each line fragment. Before showing how this is carried out, it is worth noting that straight lines and straight edges are different and should ultimately be detected by different means. In fact, straight edges are probably more common than lines, true lines (such as telephone wires in outdoor scenes) being picked up by Laplacian-type operators and straight edges by edge detectors (such as the Sobel) described in Chapter 5. The two are different in that edges have a unique direction defined by that of the edge normal, whereas lines have 180° rotation symmetry and are not vectors. Hence application of the HT to the two cases will be subtly different. For concreteness, we here concentrate on straight *edge* detection, although minor modification to the techniques (changing the range of θ from 360° to 180°) permits this method to be adapted to lines.

To proceed with line detection, we first obtain the local components of intensity gradient, then deduce the magnitude g and threshold it to locate each edge pixel in the image. The gradient magnitude g may readily be computed from the components of intensity gradient:

$$g = (g_x^2 + g_y^2)^{1/2} \tag{8.6}$$

or using a suitable approximation (Chapter 5). θ may be estimated using the arctan function in conjunction with the local edge gradient components g_x, g_y:

$$\theta = \arctan (g_y/g_x) \tag{8.7}$$

However, since the arctan function has period π, an additional π may have to be added: this can be decided from the signs of g_x and g_y. Once θ is known, ρ can be found from equation (8.4).

8.3   The foot-of-normal method

The author has worked on an alternative means of saving computation (Davies, 1986c) which eliminates the use of trigonometric functions such as arctan by employing a different parametrization scheme. As noted earlier, the methods so far described all employ an abstract parameter space in which points bear no immediate relation to image space. In the author's scheme the parameter space is a second image space, which may be said to be congruent[*] to image space. In some ways this is highly convenient—for example, while setting up the algorithm—since working parameters and errors can be visualized and estimated more easily. Note also that it is common for there to be an additional space in the framestore which can be used for this purpose (see Chapter 2).

This type of parameter space is obtained in the following way. First, each edge fragment in the image is produced much as required previously so that ρ can be measured, but this time the foot of the normal from the origin is itself taken as a voting position in parameter space (Fig. 8.1). Clearly, the foot-of-normal position embodies all the information previously carried by the ρ and θ values, and the methods are mathematically essentially equivalent. However, the details differ, as will be seen.

The detection of straight edges starts with the analysis of (i) local pixel coordinates (x, y) and (ii) the corresponding local components of intensity gradient (g_x, g_y) for each edge pixel. Taking (x_0, y_0) as the foot of the normal from the origin to the relevant line (produced if necessary—see Fig. 8.2), it is found that

$$g_y/g_x = y_0/x_0 \tag{8.8}$$

$$(x - x_0)x_0 + (y - y_0)y_0 = 0 \tag{8.9}$$

These two equations are sufficient to compute the two coordinates (x_0, y_0) (this must be so since the point (x_0, y_0) is uniquely determined as the join of

---

[*] I.e. parameter space is like image space, *and* each point in parameter space holds information that is immediately relevant to the corresponding point in image space.

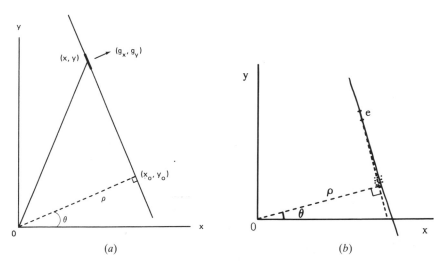

Fig. 8.2 Image space parametrization of a straight line: (a) parameters involved in the calculation (see text); (b) buildup of foot-of-normal positions in parameter space for a more practical situation, where the line is not exactly straight: e is a typical edge fragment leading to a single vote in parameter space.

the line to be detected and the normal through the origin). Solving for x_0 and y_0 gives

$$x_0 = v g_x \tag{8.10}$$

$$y_0 = v g_y \tag{8.11}$$

where

$$v = \frac{x g_x + y g_y}{g_x^2 + g_y^2} \tag{8.12}$$

Notice that these expressions involve only additions, multiplications and just one division.

For completeness, we also compute the distance of the line from the origin:

$$\rho = \frac{x g_x + y g_y}{(g_x^2 + g_y^2)^{1/2}} \tag{8.13}$$

and the vector from (x_0, y_0) to the current pixel (x, y): this has components

$$x - x_0 = wg_y \tag{8.14}$$

$$y - y_0 = -wg_x \tag{8.15}$$

where

$$w = \frac{xg_y - yg_x}{g_x^2 + g_y^2} \tag{8.16}$$

The magnitude of this vector is

$$s = \frac{|xg_y - yg_x|}{(g_x^2 + g_y^2)^{1/2}} \tag{8.17}$$

The last equation also turns out to be useful for image interpretation, as will be seen below. The next section considers the errors involved in estimating (x_0, y_0) from the raw data and shows how they may be minimized.

8.3.1 Error analysis

The most serious cause of error in estimating (x_0, y_0) is that involved in estimating edge orientation. Pixel location is generally accurate to within 1 pixel, and even when this is not quite valid, errors are largely averaged out during peak location in parameter space. In what follows we ignore errors in pixel location and concentrate on edge orientation errors. Figure 8.3 shows the geometry of the situation. A small angular error ε causes the estimate of (x_0, y_0) to shift to (x_1, y_1).

To a first approximation, the errors in estimating ρ and s are:

$$\delta\rho = \varepsilon s \tag{8.18}$$

$$\delta s = -\varepsilon\rho \tag{8.19}$$

Thus errors in both ρ and s due to angular error ε depend on distance from the origin. Clearly, these errors are minimized overall if the origin is chosen at the centre of the image.

The error ε arises from two main sources—intrinsic edge detector errors and noise. The intrinsic error of the Sobel operator is about 1°: thus image

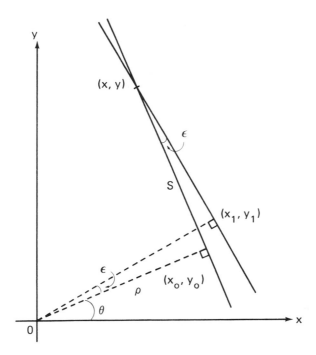

Fig. 8.3 Notation for analysis of errors with line parametrization.

noise will generally be the main cause of inaccuracy. In this respect, it should be noted that many mechanical components possess edges that are straight to a very high precision and that if each pixel of the image is quantized into at least 8 bits (256 grey levels), then aliasing[*] problems are minimized and the ground is laid for accurate line transforms.

Since errors in the foot-of-normal position are proportional to s and ρ, and these in turn are limited by the image size, errors may be reduced overall by dividing the image into a number of separate subimages each with an origin located at its centre. In principle, by this means errors may be reduced to arbitrarily small values. However, decomposing the picture into subimages may reduce the number of (useful) points constituting a peak in parameter space, thereby decreasing the accuracy with which that peak can

---

[*] Aliasing is the process whereby straight lines in digital images take a stepped form following sampling decisions that they pass through one pixel but not the next one. The process is often marked in graphics (binary) images and is less marked, though never absent, in grey-scale images. See also Chapter 23.

be located. Sometimes the number of useful points is limited by the available data, as when lines are short in length. In such cases, the subimage method helps by reducing the clutter of noise points in parameter space, thereby improving the detection of such small segments.

The subimage method should be used to increase accuracy and reduce noise when line segments are significantly shorter than the image size. The optimum subimage size depends on the lengths of the lines being detected and the lengths of the short line segments to be detected. If a long line is broken into two or more parts by analysis in subimages, this may not be too much of a disadvantage, since all the information in each line is still available at the end of the computation, although the data from the various subimage peaks will need to be correlated in order to get a complete picture. Indeed, it can be *advantageous* to break down long lines into shorter ones in this way, since this procedure gives crisp output data even if the line is slightly curved and slowly changes its orientation along its length. Thus the subimage approach permits an operator to be designed to locate lines of different lengths, or moderately curved lines of different shapes.

8.3.2 Quality of the resulting data

Although the foot-of-normal method is mathematically similar to the (θ, ρ) method, it is unable to determine line orientation to quite the same degree of accuracy. This is because it is mapped onto a discrete image-like parameter space so that the θ accuracy depends on the fractional accuracy in determining ρ—which in turn depends on the absolute magnitude of ρ. Hence for small ρ the orientation of a line that is predicted from the position of the peak in parameter space will be relatively inaccurate. Indeed, if ρ happens to be zero, then no information on the orientation of the line is available in parameter space (although, as has been seen, the *position* of the foot-of-normal is known accurately).

The difficulty can be overcome in two ways. The first is by referring to a different origin if ρ is too small. If for each edge point three noncollinear origins are available then it will always be possible to select a value of ρ which is greater than a lower bound. Errors are easy to compute if the subimage method is employed and the origins that are available are those of adjacent subimages: basically, the situation is that errors in estimating line orientation should not be worse than $2/S$ radians (for subimages of size S), although for large S they would be limited by those (around $1°$) introduced by the gradient operator.

The second and more rigorous way of obtaining accurate values of line orientation is by a two-stage method. In this case, of the original set of prominent points, the subset that contributed to a given peak in the first parameter space is isolated and made to contribute to a θ histogram, from which line orientation may be determined accurately.

Finally, it should be noted that it is often possible to ignore cases of small ρ or simply to acknowledge that cases of small ρ provide little line orientation information. While this is inelegant theoretically, the amount of redundancy of many images sometimes makes this practicable, thereby saving computation. For example, if a rectangular object is to be located, there is little chance that more than one side will have ρ close to zero (see Section 8.4). Hence the object can be located tentatively first of all from the peaks in parameter space for which ρ is large, and the interpretation confirmed with the aid of peaks for which ρ is small or zero. Appeal to Fig. 8.4 confirms that this can be a useful approach. Clearly, any outstanding instances of ambiguity can still be resolved by one of the methods outlined above.

8.3.3   Application of the foot-of-normal method

The methods outlined above have been tested on real images containing various mechanical components. For objects with straight edges varying between about 20 and 80 pixels in length, it was found convenient to work with subimages of size 64×64.

Typical results (in this case for 128×128 images with 64 grey levels) are shown in Fig. 8.4. Clearly, some of the objects in these pictures are grossly overdetermined by their straight edges, so low ρ values are not a major problem. For those peaks where $\rho > 10$, line orientation is estimated within approximately $2°$; as a result, these objects are located within 1 pixel and orientated within $1°$ by this technique, without the necessity for θ histograms. Notice that the curved edge in the object of Fig. 8.4(a) is orientated at each end in a way which could not easily have been achieved if the subimage method had not been used (see Section 8.3.2). Figures 8.4(b) and (c) contain some line segments that are not detected. This is due partly to their relatively low contrast, higher noise levels, above average fuzziness or short length. However, it is also due to the thresholds set on the initial edge detector and on the final peak detector: when these were set at lower values, additional lines were detected but other noise peaks also became prominent in parameter space, and each of these needed to be checked in detail to confirm the presence of the corresponding line in the image. This is one

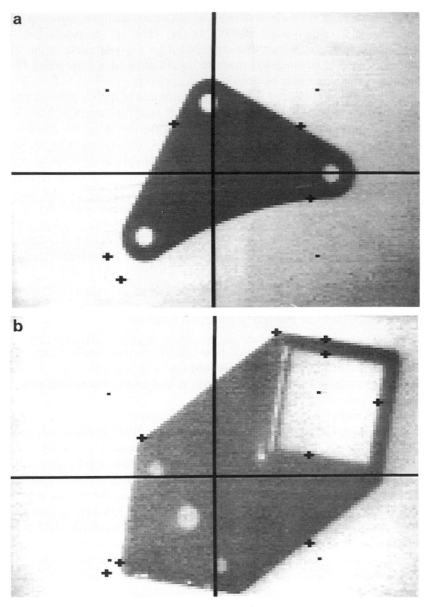

Fig. 8.4 (a) and (b) Results of image space parametrization of mechanical parts. The dot at the centre of each quadrant is the origin used for computing the image space transform. The crosses are the positions of peaks in parameter space which mark the individual straight edges (produced if necessary). For further explanation, see text. *Continued.*

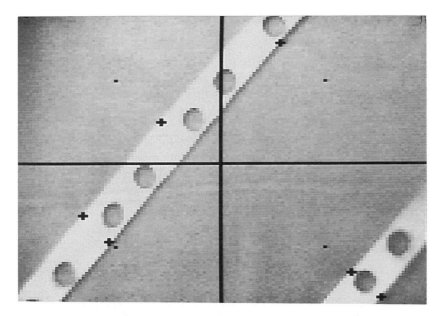

Fig. 8.4 (*continued*).

aspect of a general problem that arises right through the field of image analysis.

8.4 Longitudinal line localization

The preceding sections have provided a variety of means for locating lines in digital images and finding their orientations. However, these methods are insensitive to where along an infinite idealized line an observed segment appears. The reason for this is that the fit includes only two parameters. There is some advantage to be gained in this, in that partial occlusion of a line does not prevent its detection: indeed, if several segments of a line are visible, they can all contribute to the peak in parameter space, hence improving sensitivity. On the other hand, for full image interpretation, it is useful to have information about the longitudinal placement of line segments.

This is achieved by a further stage of processing (which may make line finding a two- or three-stage algorithm overall). The additional stage involves finding which points contributed to each peak in the main parameter space, and carrying out connectivity analysis in each case. Dudani and Luk (1978) called this process "*xy*-grouping". It is not vital that the line segments should be 4- or 8-connected—just that there should be sufficient points on them so that adjacent points are within a threshold distance apart, i.e. groups of points are merged if they are within the prespecified threshold distance (Dudani and Luk took this distance as 5 pixels). Finally, segments shorter than a certain minimum length can be ignored as too insignificant to help with image interpretation.

It should be noted that the distance s in the foot-of-normal method can be employed as a suitable grouping parameter, although an alternative is to use either the x or y projection, depending on the orientation of the line (i.e. use x if $|m| < 1$, and y otherwise).

8.5   Final line fitting

Dudani and Luk (1978) found it useful to include a least-squares fitting stage to complete the analysis of the straight edges present in an image. This is carried out by taking the x, y coordinates of points on the line as input data and minimizing the sum of squares of distances from the data points to the best-fit line. Their reason for adding such a stage is to eliminate local edge orientation accuracy as a source of error. This motivation is generally reasonable if the highest possible accuracy is required (e.g. obtaining line orientation to significantly better than 1°). However, many workers have criticized the use of the least-squares technique, since it tends to weight up the contribution of less accurate points—including those points which have nothing to do with the line in question but which arise from image "clutter". This criticism is probably justified, although the least-squares technique is convenient in yielding to mathematical analysis and in this case in giving explicit equations for θ and ρ in terms of the input data (Dudani and Luk, 1978). Dudani and Luk obtained the endpoints by reference to the best-fit line thus obtained.

8.6   Concluding remarks

This chapter has described a variety of techniques for finding lines and straight edges in digital images. These are all based on the HT, since the latter is used virtually universally for this purpose. However, the chapter has had the further function of indicating the importance of the HT throughout the field of image analysis. The reasons it is so important are that it permits certain types of global data to be extracted systematically from images and that it is able to ignore "local" problems, due for example to occlusions and noise. It is therefore an example of "intermediate-level" processing, which may be defined as the conversion of crucial features in the image to a more abstract form that is no longer restricted to the image representation.

The specific techniques covered have involved a particular parametrization of a straight line, and means of improving efficiency and accuracy. Speed is improved particularly by resorting to a two-stage line finding procedure—a method that is useful in other applications of the HT, as will be seen in later chapters. Accuracy tends to be cut down by such two-stage processing because of error propagation *and* because the first stage is liable to be subject to too many interfering signals. However, it is possible to take an approximate solution (in this case an approximation to a set of lines in an image) and to improve accuracy by including yet another stage of processing. Two examples of this were cited: one was adding a line orientation stage to the foot-of-normal method, and the other was the addition of a line fitting procedure to refine the output of an HT procedure (although there are problems with using the least-squares technique for this purpose). Again, later chapters illustrate these types of accuracy enhancing procedure applied in other tasks such as circle centre location.

Overall, use of the HT is central to image analysis and will be met many times in the remainder of the book. However, it is always important to bear in mind its limitations, the most obvious one being that it can involve considerable storage and computational effort. Further discussion of HT line finding schemes, including those aimed at saving computation, is found in Chapter 10.

8.7 Bibliographical and historical notes

The HT was developed in 1962 (Hough, 1962) with the aim of finding (straight) particle tracks in high-energy nuclear physics, and was brought into the mainstream image analysis literature much later by Rosenfeld (1969). Duda and Hart (1972) developed the method further and applied it to the detection of lines and curves in digital pictures. O'Gorman and Clowes (1976) soon developed a Hough-based scheme for finding lines efficiently, by making use of edge orientation information, at much the same time that Kimme *et al.* (1975) applied the same method (apparently independently) to the efficient location of circles. Many of the ideas for fast effective line finding described in this chapter arose in a paper by Dudani and Luk (1978). The author's foot-of-normal method (Davies, 1986c) was developed much later. During the 1990s, work in this area is still progressing—see for example Atiquzzaman and Akhtar's (1994) method for the efficient determination of lines together with their end coordinates and lengths. Other applications of the HT are covered in the following few chapters, while further methods for locating lines are dealt with in Chapter 10 on the generalized HT.

A few years ago there was a body of opinion that the HT is completely worked over and no longer a worthwhile topic for research. However, this view seems misguided and there appears to be continuing strong interest in the subject, as the following chapters show. After all, the computational difficulties of the method reflect a problem that is inescapable in image analysis as a whole, so development of methods must continue: in this context, note especially Sections 10.12 and 10.13.

Finally, some mention should be made of the Radon transform. This is formed by integrating the picture function $I(x, y)$ along infinitely thin straight strips of the image, with normal coordinate parameters (θ, ρ), and recording the results in a (θ, ρ) parameter space. The Radon transform is a generalization of the Hough transform for line detection (Deans, 1981). In fact, for straight lines the Radon transform reduces to the Duda and Hart (1972) form of the HT which, as remarked earlier, involves considerable computation: for this reason the Radon transform is not covered in depth in this book. The transforms of real lines have a characteristic "butterfly" shape (a pencil of segments passing through the corresponding peak) in parameter space. This phenomenon has been investigated by Leavers and Boyce (1987), who have devised special 3×3 convolution filters for sensitively detecting these peaks.

8.8 Problem

1. Write a Pascal procedure for implementing the arctan function of equation (8.7), taking account of the following practical requirements: (a) the need to add an extra π when necessary, (b) the need to avoid dividing by zero, and (c) the need to obtain an answer in the range 0 to 2π (the arctan function commonly yields values in the range $-\pi/2$ to $\pi/2$).

9

Circle Detection

9.1 Introduction

Location of round objects is important in many areas of image analysis but it is especially important in industrial applications such as automatic inspection and assembly. In the food industry alone, a very sizeable proportion of products are round—biscuits, cakes, pizzas, pies, jellies, oranges and so on (Davies, 1984c). In the automotive industry many circular components are used—washers, wheels, pistons, heads of bolts, etc., while round holes of various sizes appear in such items as casings and cylinder blocks. In addition, buttons and many other everyday objects are round. All these manufactured parts need to be checked, assembled, inspected and measured with high precision. Finally, objects can frequently be located by their holes, so finding round holes or features is part of a larger problem. Since round picture objects form a special category of their own, very efficient algorithms should be available for analysing digital images containing them. This chapter is addressed to some aspects of this problem.

An earlier chapter showed how objects may be found and identified using algorithms that involve boundary tracking (Table 9.1). Unfortunately, such algorithms are insufficiently robust in many applications: for example, they tend to be muddled by artefacts such as shadows and noise. This chapter shows that the Hough transform (HT) technique is able to overcome many of these problems. Indeed, it is found to be particularly good at dealing with all sorts of difficulties, including quite severe occlusions. It is important to notice that it achieves this not by adding robustness but by having robustness built in as an integral part of the technique.

Table 9.1   Procedure for finding objects using (r, θ) boundary graphs.

1. Locate edges within the image
2. Link broken edges
3. Thin thick edges
4. Track around object outlines
5. Generate a set of (r, θ) plots
6. Match (r, θ) plots to standard templates

This procedure is not sufficiently robust with many types of real data, e.g. in the presence of noise, distortions in product shape, etc.: in fact, it is quite common to find the tracking procedure veering off and tracking around shadows or other artefacts.

The application of the HT to circle detection is perhaps the most straightforward use of the technique. However, there are several enhancements and adaptations that can be applied in order to improve accuracy and speed of operation, and in addition to make the method work efficiently when detecting circles with a range of sizes. These modifications are studied below after covering the basic HT techniques.

9.2   Hough-based schemes for circular object detection

In the original HT method for finding circles (Duda and Hart, 1972), the intensity gradient is first estimated at all locations in the image and then thresholded to give the positions of significant edges. Then the positions of all possible centre locations—namely all points a distance R away from every edge pixel—are accumulated in parameter space, R being the anticipated circle radius. Parameter space can be a general storage area but when looking for circles it is convenient to make it congruent to image space: in that case possible circle centres are accumulated in a new plane of image space. Finally, parameter space is searched for peaks which correspond to the centres of circular objects. Since edges have nonzero width and noise will always interfere with the process of peak location, accurate centre location requires the use of suitable averaging procedures (Davies, 1984c; Brown, 1984).

This approach clearly requires a very large number of points to be accumulated in parameter space and so a revised form of the method has now become standard: in this approach, locally available edge orientation information at each edge pixel is used to enable the exact positions of circle centres to be estimated (Kimme et al., 1975). This is achieved by moving a distance R along the edge normal at each edge location. Thus the number of points accumulated is equal to the number of edge pixels in the image[*]: this represents a significant saving in computational load. For this to be possible, the edge detection operator that is employed must be highly accurate. Fortunately, the Sobel operator is able to estimate edge orientation to 1° and is very simple to apply (Chapter 5). Thus the revised form of the transform is viable in practice.

As was seen in Chapter 5, once the Sobel convolution masks have been applied, the local components of intensity gradient g_x and g_y are available, and the magnitude and orientation of the local intensity gradient vector can be computed using the formulae:

$$g = (g_x^2 + g_y^2)^{1/2} \tag{9.1}$$

and

$$\theta = \arctan\ (g_y/g_x) \tag{9.2}$$

However, use of the arctan operation is not necessary when estimating centre location coordinates (x_c, y_c) since the trigonometric functions can be made to cancel out:

$$x_c = x - R(g_x/g) \tag{9.3}$$

$$y_c = y - R(g_y/g) \tag{9.4}$$

the values of cos θ and sin θ being given by:

$$\cos\ \theta = g_x/g \tag{9.5}$$

$$\sin\ \theta = g_y/g \tag{9.6}$$

In addition, the usual edge thinning and edge linking operations—which normally require considerable amounts of processing—can be avoided if a little extra smoothing of the cluster of candidate centre points can be tolerated (Davies, 1984c) (Table 9.2). Thus this Hough-based approach can be a very efficient one for locating the centres of circular objects, virtually all the superfluous operations having been eliminated, leaving only edge

---

[*] We assume that objects are known to be *either* lighter *or* darker than the background, so that it is only necessary to move along the edge normal in one direction.

Table 9.2   A Hough-based procedure for locating circular objects.

1. Locate edges within the image
2. Link broken edges
3. Thin thick edges
4. For every edge pixel, find a candidate centre point
5. Locate all clusters of candidate centres
6. Average each cluster to find accurate centre locations

This procedure is particularly robust. It is largely unaffected by shadows, image noise, shape distortions and product defects. Note that stages 1–3 of the procedure are identical to stages 1–3 in Table 9.1. However, in the Hough-based method, computation can be saved, and accuracy actually increased, by omitting stages 2 and 3.

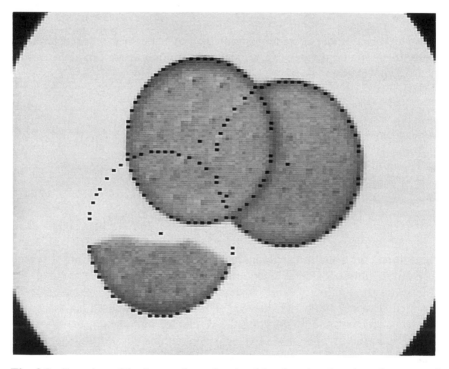

Fig. 9.1   Location of broken and overlapping biscuits, showing the robustness of the centre location technique. Accuracy is indicated by the black dots which are each within 1/2 pixel of the radial distance from the centre.

detection, location of candidate centre points and centre point averaging to be carried out. It is also relevant that the method is highly robust, in that if part of the boundary of an object is obscured or distorted, the object centre is still located accurately. The author has come across some remarkable instances of this (see for example, Figs 9.1 and 9.2). The reason for this useful property is clear from Fig. 9.3.

Despite the efficiency of the above technique, it still takes ~ 1 second to perform within a 128 × 128 pixel image on a conventional serial computer, over half of this time being due to the need to evaluate and threshold the intensity gradient over the whole image. Part of the reason for this is that the edge detector operates within a 3 × 3 neighbourhood and necessitates some 12 pixel accesses, four multiplications, eight additions, two subtractions and an operation for the evaluation of the square root of sum of squares (equation (9.1)). As seen in Chapter 5, the latter type of operation is

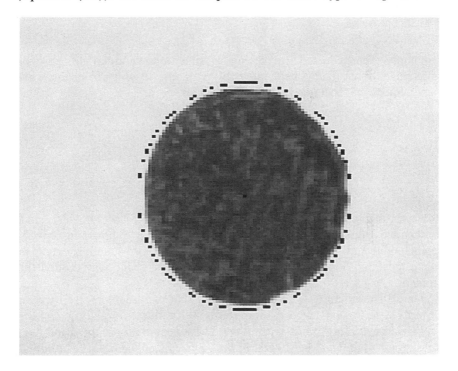

Fig. 9.2 Location of a biscuit with a distortion, showing a chocolate-coated biscuit with excess chocolate on one edge. Note that the computed centre has not been "pulled" sideways by the protuberances. For clarity, the black dots are marked 2 pixels outside the normal radial distance.

Fig. 9.3   Robustness of the Hough transform when locating the centre of a circular object. The circular part of the boundary gives candidate centre points that focus on the true centre, whereas the irregular broken boundary gives candidate centre points at random positions.

commonly approximated by taking a sum or maximum of absolute magnitudes of the component intensity gradients, in order to estimate the magnitude of the local intensity gradient vector:

$$g \simeq |g_x| + |g_y| \tag{9.7}$$

or

$$g \simeq \max(|g_x|, |g_y|) \tag{9.8}$$

However, this type of short-cut is not advisable in the present context, since an accurate value for the magnitude of this vector is required in order to compute with sufficient precision the position of the corresponding candidate centre location. However, lookup tables provide a viable way around this problem.

Overall, the dictates of accuracy imply that candidate centre location requires a nontrivial set of computational facilities including (preferably) use of floating-point arithmetic. There seems no escaping the need for special dedicated hardware if real-time speeds are to be attained. Nevertheless, substantial increases in speed are still possible by software means alone, as will be seen later in the chapter. Meanwhile we consider the problems that arise when images contain circles of many different radii, or for one reason or another radii are not known in advance.

9.3   The problem of unknown circle radius

There are a number of situations where circle radius is initially unknown. One such situation is where a number of circles of various sizes are being sought—as in the case of coins, or different types of washer. Another is where the circle size is variable—as for food products such as biscuits—so that some tolerance must be built into the system. In general, all circular objects have to be found and their radii measured. In such cases the standard technique is to accumulate candidate centre points simultaneously in a number of parameter planes in a suitably augmented parameter space, each plane corresponding to one possible radius value. The centres of the peaks detected in parameter space give not only the location of each circle in two dimensions but also its radius. Although this scheme is entirely viable in principle, there are several problems in practice:

   (i)   many more points have to be accumulated in parameter space;
  (ii)   parameter space requires much more storage;
 (iii)   significantly greater computational effort is involved in searching parameter space for peaks.

To some extent this is to be expected, since the augmented scheme enables more objects to be detected directly in the original image.

It is shown below that problems (ii) and (iii) may largely be eliminated. This is achieved by using just one parameter plane to store all the information for locating circles of different radii, i.e. accumulating not just one point per edge pixel but a whole line of points along the direction of the edge normal in this one plane. In practice, the line need not be extended indefinitely in either direction but only over the restricted range of radii over which circular objects or holes might be expected.

Even with this restriction, a large number of points are being accumulated in a single parameter plane, and it might be thought initially that this would lead to such a proliferation of points that almost any "blob" shape would lead to a peak in parameter space which might be interpreted as a circle centre. However, this is not so and significant peaks normally result only from genuine circles and some closely related shapes.

To understand the situation, consider how a sizeable peak can arise at a particular position in parameter space. This can happen only when a large number of radial vectors from this position meet the boundary of the object normally. In the absence of discontinuities in the boundary, a contiguous set of boundary points can only be normal to radius vectors if they lie on the arc of a circle (indeed, a circle could be *defined* as a locus of points which are normal to the radius vector and which form a thin connected line). If a limited number of discontinuities are permitted in the boundary it may be deduced that shapes like that of a fan[*] will also be detected using this scheme. Since it is in any case useful to have a scheme which can detect such shapes, the main problem is that there will be some ambiguity in interpretation—i.e. does peak P in parameter space arise from a circle or a fan? In practice it is quite straightforward to distinguish between such shapes with relatively little additional computation, the really important problem being to cut down the amount of computation needed to key into the initially unstructured image data. Indeed, it is often a good strategy to prescreen the image to eliminate most of it from further detailed consideration and then to analyse the remaining data with tools having much greater discrimination: this two-stage template matching procedure frequently leads to significant savings in computation (VanderBrug and Rosenfeld, 1977; Davies, 1988i).

A more significant problem arises because of errors in the measurement of local edge orientation. As stated earlier, edge detection operators such as the Sobel introduce an inherent inaccuracy of about 1°. Image noise typically adds a further 1° error, and for objects of radius 60 pixels the result is an overall uncertainty of about 2 pixels in the estimation of centre location. This makes the scheme slightly less specific in its capability for detecting objects. In particular, edge pixels can now be accepted when their orientations are not exactly normal to the radius vector, and an inaccurate position will be accumulated in parameter space as a possible centre point. Thus the method may occasionally accept cams or other objects with a spiral boundary as potential circles.

Overall, the scheme is likely to accept the following object shapes during its prescreening stage, and the need for a further discrimination stage will be highly application-dependent:

  (i)   circles of various sizes;
  (ii)   shapes such as fans which contain arcs of circles;
  (iii) spirals with almost normal radius vectors;
  (iv)   partly occluded or broken versions of these shapes.

If occlusions and breakages can be assumed to be small, then it will be safe to apply a threshold on peak size at say 70% of that expected from the

---

[*] In this chapter all references to "fan" mean a four-blade fan shape bounded by two concentric circles with eight equally spaced radial lines joining them.

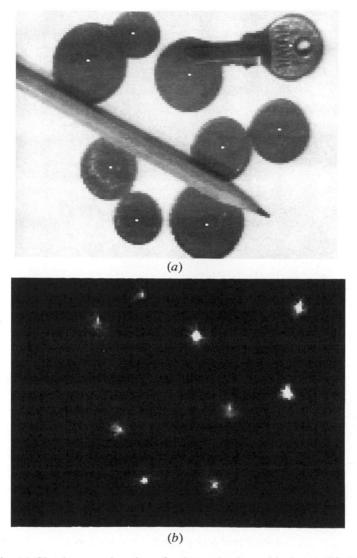

(a)

(b)

Fig. 9.4 (a) Simultaneous location of coins and a key with the modified Hough scheme: the various radii range from 10 to 17 pixels; (b) transform used to compute the centres indicated in (a). Detection efficiency is unimpaired by partial occlusions, shape distortions and glints. However, displacements of some centres are apparent; in particular, one coin (top left) has only two arcs showing: the fact that one of these is distorted, giving a lower curvature, leads to a displacement of the computed centre. The shape distortions are due to rather uneven illumination.

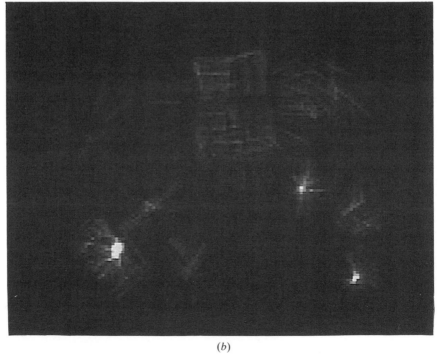

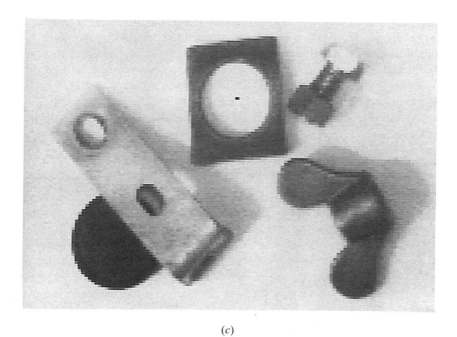

(c)

Fig. 9.5 (a) Accurate simultaneous detection of a lens cap and a wing nut when radii are assumed to range from 4 to 17 pixels; (b) response in parameter space that arises with such a range of radii: note the overlap of the transforms from the lens cap and the bracket; (c) hole detection in the image of (a) when radii are assumed to fall in the range −26 to −9 pixels (negative radii are used since holes are taken to be objects of negative contrast): clearly, in *this* image a smaller range of negative radii could have been employed.

smallest circle[*]: in that case it will be necessary to consider only whether the peaks arise from objects of types (i), (ii) or (iii). If, as usually happens, objects of types (ii) or (iii) are known *not* to be present, the method is immediately a very reliable detector of circles of arbitrary size. Sometimes specific rigorous tests for objects of types (ii) and (iii) will still be needed: they may readily be distinguished, for example, with the aid of corner detectors or from their radial intensity histograms (see Chapters 13 and 19).

Next, note that the information on radial distance has been lost by accumulating all votes in a single parameter plane. Hence a further stage of

---

[*] Note, however, that size-dependent effects can be eliminated by appropriate weighting of the votes cast at different radial distances.

analysis is needed to measure (or classify) object radius. This is so even if a stage of disambiguation is no longer required to confirm that the object is of the desired type. This extra stage of analysis normally involves negligible additional computation, because the search space has been narrowed down so severely by the initial circle location procedure (see Figs 9.4–9.6). The radial histogram technique already referred to can be used to measure the radius: in addition it can be used to perform a final object inspection function (see Chapter 19).

9.3.1 Experimental results

The method described above works much as expected, the main problems arising with small circular objects (of radii less than about 20 pixels) at low resolution (Davies, 1988b). Essentially, the problems are those of lack of discrimination in the precise shapes of small objects (see Figs 9.4–9.6), as anticipated above. More specifically, if a particular shape has angular deviations $\delta\psi$ relative to the ideal circular form, then the lateral displacement at the centre of the object is $r\delta\psi$. If this lateral displacement is 1 pixel or less, then the peak in parameter space is broadened only slightly and its height is substantially unchanged. Hence detection sensitivity is very similar to that for a circle. This model is broadly confirmed by the results of Fig. 9.6.

Theoretical analysis of the response can be carried out for any object by obtaining the shape of the evolute (envelope of the normals—see for example, Tuckey and Armistead, 1953) and determining the extent to which it spreads the peak in parameter space. However, performance is much as expected intuitively on the basis of the ideas already described, except that for small objects shape discrimination is relatively poor. As suggested earlier, this can frequently be turned to advantage in that the method becomes a circular feature detector for small radii (see Fig. 9.5, where a wing nut is located).

Finally, objects are detected reliably even when they are partly occluded. However, occlusions can result in the centres of small objects being "pulled" laterally (Fig. 9.4)—a situation that seldom occurs with the original Hough scheme. When this is a problem in practical situations, it may be reduced by (i) cutting down the range of values of r that are employed in any one parameter plane (and hence using more than one plane), and (ii) increasing resolution. Generally, subpixel accuracy of centre location cannot be expected when a single parameter plane is used to detect objects over a range of sizes exceeding a factor of about two: this is because of the amount

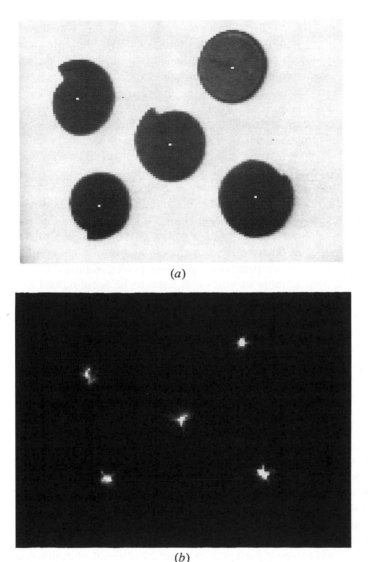

(a)

(b)

Fig. 9.6 (a) Simultaneous detection of four (approximately) equiangular spirals and a circle of similar size (16 pixels radius), for assumed range of radii 9–22; (b) transform of (a). In the spiral with the largest step, the step size is about 6 pixels and detection is possible only if some displacement of the centre is acceptable. In such cases the centre is effectively the point on the evolute of the shape in parameter space which happens to have the maximum value. Note that cams frequently have shapes similar to those of the spirals shown here.

of irrelevant "clutter" that appears in parameter space, which has the effect of reducing the signal-to-noise ratio (see Fig. 9.5). Similarly, attempting to find circular objects *and* holes simultaneously in the same parameter plane mitigates against good performance unless the *sum* of the ranges of radius values is restricted.

Thus there is a tradeoff between speed and accuracy with the approach outlined here. For high accuracy a relatively small range of values of r should be employed in any one parameter plane, whereas for high speed one parameter plane will often cover the detection of all sizes of circles with sufficient accuracy. There is also the possibility of two-stage recognition (essentially an extension of the approach already outlined in Section 9.3) in order to find circles first with moderate and then with optimal accuracy.

These results using the modified Hough scheme confirm that it is possible to locate objects of various radii within a parameter space of limited size. The scheme saves on storage requirements and on the amount of computation required to search parameter space for peaks, although it is not able to reduce the total number of votes that have to be accumulated in parameter space. A particular feature of the scheme is that it is essentially two-stage—the second stage being needed (when appropriate) for shape disambiguation and for size analysis. It should be noted that two-stage procedures are commonly used to make image analysis more efficient, as will be seen in other chapters of this book (see especially Chapter 11 on ellipse detection).

9.4   The problem of accurate centre location

The previous section analysed the problem of how to cope efficiently with images containing circles of unknown or variable size. This section examines how circle centres may be located with high (preferably subpixel) accuracy. This problem is relevant since high resolution is expensive both in electronic hardware and in the computation required to process the relatively large numbers of pixels that are involved. Hence it will be of advantage if high accuracy can be attained with images of low or moderate resolution.

There are several causes of inaccuracy in locating the centres of circles using the HT: (i) natural edge width may add errors to any estimates of the location of the centre (this applies whether or not a thinning algorithm is employed to thin the edge, as such algorithms themselves impose a degree of inaccuracy even when noise is unimportant); (ii) image noise may cause the radial position of the edge to become modified; (iii) image noise may cause variation in the estimates of local edge orientation; (iv) the object itself may be distorted, in which case the most accurate and robust estimate of circle centre is what is required (i.e. the algorithm should be able to ignore the distortion and take a useful average from the remainder of the boundary); (v) the object may appear distorted because of inadequacies in the method of illumination; and (vi) the Sobel or other edge orientation operator will have a significant inherent inaccuracy of its own (typically $1°$—see Chapter 5), which then leads to inaccuracy in the estimation of centre location.

Evidently, it is necessary to minimize the effects of all these potential sources of error. In applying the HT, it is usual to bear in mind that all possible centre locations that could have given rise to the currently observed edge pixel should be accumulated in parameter space. Thus it is clear that we should accumulate in parameter space not a single candidate centre point corresponding to the given edge pixel, but a point spread function (PSF) which may be approximated by a Gaussian error function: this will generally have different radial and transverse standard deviations. The radial standard deviation will commonly be 1 pixel—corresponding to the expected width of the edge—whereas the transverse standard deviation will frequently be at least 2 pixels, and for objects of moderate (60 pixels) radius the intrinsic orientation accuracy of the edge detection operator contributes at least 1 pixel to this standard deviation.

It would be convenient if the PSF arising from each edge pixel were isotropic, since errors could initially be ignored and then an isotropic error function could be convolved with the entire contents of parameter space, thus saving computation. However, this is not possible if the PSF is anisotropic, since the information on how to apply the local error function is lost during the process of accumulation.

When the centre need only be estimated to within 1 pixel, these considerations hardly matter. However, when it is desired to locate centre coordinates to within 0.1 pixel, each PSF will have to contain 100 points, all of which will have to be accumulated in a parameter space of much increased spatial resolution, or its equivalent[*], if the required accuracy is to be attained. However, a technique is available which can cut down the amount of computation from this sort of level, without significant loss of accuracy, as will be seen below.

---

[*] For example, some abstract list structure might be employed which effectively builds up to high resolution only where needed (see also the adaptive HT scheme described in Chapter 10).

9.4.1   Obtaining a method for reducing computational load

It is significant that the HT scheme outlined above for accurately locating circle centres is a completely parallel approach. Every edge pixel is examined independently, and only when this process is completed is a centre looked for. Since there is no obvious means of improving the underlying Hough scheme, a sequential approach appears to be required if computation is to be saved. Unfortunately, sequential methods are often less robust than parallel ones. For example, an edge tracking procedure can falter and go wandering off, following shadows and other artefacts round the image. In the current application, if a sequential scheme is to be successful in the sense of being fast, accurate *and* robust, it must be subtly designed and retain some of the features of the original parallel approach. The situation is now explored in more detail.

First, note that most of the inaccuracy in calculating the position of the centre arises from transverse rather than radial errors. Sometimes transverse errors can fragment the peak that appears in the region of the centre, thereby necessitating a considerable amount of computation to locate the real peak accurately. Thus it is reasonable to concentrate on eliminating transverse errors.

This immediately leads to the following strategy: find a point D in the region of the centre and use it to obtain a better approximation A to the centre by moving from the current edge pixel P a distance equal to the expected radius r in the direction of D (Fig. 9.7). Then repeat the process one edge pixel at a time until all edge pixels have been taken into account. Although intuition indicates that this strategy will lead to a good final approximation to the centre, the situation is actually quite complex. In fact, there are two important questions:

(i)   Is the process guaranteed to converge?
(ii) How accurate is the final result?

It turns out that there is a small range of boundary pixels for which the method described above gives a location A that is *worse* than the initial approximation D (see Fig. 9.8). However, this does not mean that our intuition is totally wrong, and theory confirms the usefulness of the approach when d is small ($d \ll r$). Figure 9.9 shows a set of graphs which map out the loci of points forming second approximations to the centre positions for initial approximations at various distances d from the centre. In each graph the locus is that obtained on moving around the periphery of the

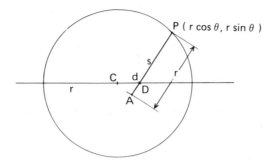

Fig. 9.7 Arrangement for obtaining improved centre approximation.

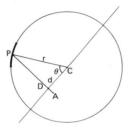

Fig. 9.8 Boundary points which lead to a worse approximation.

circle. Generally, for small d it is seen that the second approximation is on average a significantly better approximation than the initial approximation. Indeed, theory shows that for small d, the mean result will be an improvement by a factor close to 1.6 (the limiting value as $d \rightarrow 0$ is $\pi/2$: Davies, 1988e). In order to prove this result, it is first necessary to show that as d approaches zero, the shape of the locus becomes a circle whose centre is half-way between the original approximation D and the centre C, and whose radius is $d/2$ (Fig. 9.9).

At first, an improvement factor of only ~ 1.6 seems rather insignificant—until it is remembered that it should be obtainable for *every* circle boundary pixel. Since there are typically 200–400 such pixels (e.g. for $r \approx 50$), the overall improvement can in principle be virtually infinite (limitations on the actual accuracy are discussed below). However, to maximize the gain in accuracy, the order in which the algorithm processes boundary pixels should clearly be randomized (if this were not done, two adjacent pixels might often appear consecutively, the second giving negligible additional improvement in centre location).

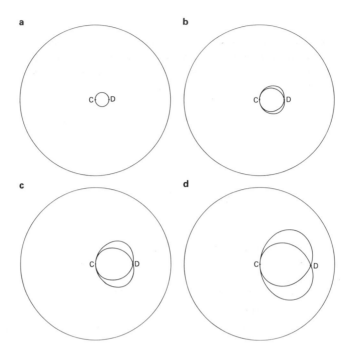

Fig. 9.9 Loci obtained for various starting approximations: (a) $d = 0.17r$; (b) $d = 0.32r$; (c) $d = 0.50r$; (d) $d = 0.68r$.

9.4.2   Improvements on the basic scheme

In practice, convergence occurs initially much as theory suggests. However, after a time no further improvement occurs, the system settling down to a situation where it is randomly wandering around the ideal centre, sometimes getting better and sometimes getting worse. The reason for this is that the particular approximation obtained at this stage depends more on the noise on individual edge pixels than on the underlying convergence properties discussed above.

To improve the situation further, the noise on the edge pixels must be suppressed by some sort of averaging procedure. Since the positions of the successive approximations produced by this method after a certain stage are essentially noise limited, the best final approximation will result from averaging all of these by accumulating all the intermediate x and y values and finding the means. Further, just as the best final approximation is obtained by averaging, so the best intermediate approximations are obtained by a similar averaging process—in which the x and y values so far accumulated are divided by the current number. Thus at this stage of the algorithm it is not necessary to use the previously calculated position as the starting point for the next approximation, but some position derived from it. In fact, if the kth position that is accumulated is (u_k, v_k), then the kth best estimate is given by:

$$x_k = (1/k)\sum_{i=1}^{k} u_i \qquad (9.9)$$

$$x_k = (1/k)\sum_{i=1}^{k} v_i \qquad (9.10)$$

In practice, typically 400 boundary pixels are available, each of which gives an independent measurement of position. On balance we would expect each of the component (x and y) inaccuracies to be averaged and thereby reduced by a factor of $\sqrt{200}$ (Davies, 1987h). Hence the linear accuracy should improve from around 1 pixel to around 0.1 pixels.

We now summarize the various stages of calculation in the algorithm. In fact, there are three:

(i)   find the position of the centre to within 2 pixels by the usual Hough technique;

(ii) use the procedure as originally outlined until it settles down with an accuracy of 0.5 pixels—i.e. until it is limited by radial errors rather than basic convergence problems;

(iii) use the averaging version of the procedure (equations (9.9) and (9.10)) to obtain the final result, with an accuracy of 0.1 pixels.

In practice it is possible to eliminate stage (ii), stage (i) then being used to get within 1 pixel of the centre and stage (iii) to get down to 0.1 pixels of the centre. The important point here is that to retain robustness and accuracy most of the work is done by parallel processing procedures (stage (i)) or procedures that emulate parallel processing (stage (iii)).

To confirm that stage (iii) emulates parallel processing, note that if a set of random points near the centre were chosen as initial approximations, then a parallel procedure could be used to accumulate a new set of better approximations to the centre, by using radial information from each of the edge pixels. However, the actual stage (iii) is better than this as it uses the current best estimate of the position of the centre as the starting position for each approximation: this is possible only if the calculation is being done sequentially rather than in parallel. Such a method is only an improvement if it is known that the initial approximation is good enough and if edge pixels that disagree badly with the currently known position of the centre can be discarded, as being due for example to a nearby object. Clearly a mechanism has to be provided for this eventuality within the algorithm. A useful strategy is to eliminate from consideration those edge pixels which would on their own point to centre locations further than 2 pixels from the approximation indicated by stage (i).

9.4.3 Discussion

The description given above may seem to imply that instead of using a successive approximation approach we could continue to use the same initial approximation for all boundary pixels, and then average the results at the end. However, this would not be valid, since a fixed bias of $0.64d$ in an unknown direction would then result and there would be no way of removing it. Thus the only real options are a set of random starting approximations or the sequence of approximations outlined above.

One thing that would upset the algorithm would be the presence of gross occlusions in the object, making less than about 30% of it visible, so that the centre position might be known accurately in one direction but would be uncertain in a perpendicular direction. Unfortunately, such uncertainties would not be apparent with this method, since the absence of information might well lead to little transverse motion of the apparent centre, which would then retain the same incorrect position.

9.4.4 Practical details

The basis of the algorithm for improving on an approximation D (x_i, y_i) to a circle of centre C is that of moving a distance r from a general boundary pixel P (x, y) in the direction of D. Thus

$$x_{i+1} = x - (r/s)(x - x_i) \qquad (9.11)$$

$$y_{i+1} = y - (r/s)(y - y_i) \qquad (9.12)$$

where

$$s = [(x - x_i)^2 + (y - y_i)^2]^{1/2} \qquad (9.13)$$

Clearly this is a computationally efficient process, since it does not require the evaluation of any trigonometric functions—the most complex operations involved being the calculation of various ratios and the evaluation of a square root.

It appears generally to be adequate to randomize boundary pixels (see Section 9.3) by treating them cyclically in the order they appear in a normal raster scan of the image, and then processing every nth one[*] until all have been used (Davies, 1988e). The number n will normally be in the range 50–150 (or for general circles with N boundary pixels, take $n \simeq N/4$). One advantage to be gained from randomizing boundary pixels is that processing can be curtailed at any stage when sufficient accuracy has been attained, thereby saving computation. This technique is particularly valuable in the case of food products such as biscuits, where accuracy is generally not as much of a problem as speed of processing.

The algorithm is straightforward to set up, parameters being adjusted on a heuristic basis rather than adhering to any specific theory. This does not seem to lead to any real problems or to compromise accuracy. The results of centre finding seem to be generally within 0.1–0.2 pixels (see below) for accurately made objects where such measurements can meaningfully be made. This rarely applies to food products such as biscuits, which are commonly found to be circular to a tolerance of only 5–10%.

When the algorithm was tested using an accurately made dark metal disc, accuracy was found to be in the region of 0.1 pixels (Davies, 1988e). However, some difficulty arose because of uncertainty in the *actual* position of the centre. Indeed, a slight skewness in the original image was found to contribute to this type of error—possibly about 0.05 pixels in this case. The reason for this appeared to lie in a slight nonuniformity in the lighting of the metal disc. It became clear that it will generally be difficult to eliminate such effects and that 0.1 pixel accuracy is close to the limit of measurement with such images. However, other tests using simulated (accurate) circles showed that the method has an intrinsic accuracy of better than 0.1 pixels.

Other improvements to the technique involve weighting down the earlier less accurate approximations in obtaining the average position of the centre. This can be achieved by the heuristic of weighting approximations in proportion to their number in the sequence: this strategy seems capable of finding centre locations within ~0.05 pixels—at least for cases where radii

---

[*] The final working algorithm is slightly more complex: if the nth pixel has been used, take the *next* one that has *not* been used. This simple strategy is sufficiently random and also overcomes problems of n sometimes being a factor of N.

are around 45 pixels and where there are some 400 edge pixels providing information about the position of each centre. Although accuracy is naturally data-dependent, the approach fulfils the specification originally set out—of locating the centres of circular objects to 0.1 pixels or better, with minimal computation. As a byproduct of the computation, the radius r can be obtained with exceptionally high accuracy—generally within 0.05 pixels (Davies, 1988e).

9.5   Overcoming the speed problem

The previous section studied how the accuracy of the HT circle detection scheme could be improved substantially, with modest additional computational cost. This section examines how circle detection may be carried out with significant improvement in speed. There is a definite tradeoff between speed and accuracy, so that greater speed inevitably implies lower accuracy. However, at least the method to be adopted permits graceful degradation to take place.

There appear to be only a few possible means of overcoming the speed problem described in Section 9.2. One is to sample the image data; another is to attempt to use a more rudimentary type of edge detection operator; and a third is to try to invent a totally new strategy for centre location.

Sampling the image data tends to mean looking at (say) every third or fourth line, to obtain a speed increase by a factor of three or so. However, improvements beyond this are unlikely, as we would then be looking at much less of the boundary of the object and the method would tend to lose its robustness. An overall improvement in speed by more than an order of magnitude appears to be unattainable.

Using a more rudimentary edge detection operator would necessarily give much lower accuracy. For example, the Roberts cross operator (Roberts, 1965) could be used but then the angular accuracy would drop by a factor of five or so (Abdou and Pratt, 1979), and in addition the effects of noise would increase markedly so that the operator would achieve nowhere near its intrinsic accuracy. With the Roberts cross operator, noise in any *one* of the four pixels in the neighbourhood will make estimation of orientation totally erroneous, thereby invalidating the whole strategy of the centre finding algorithm.

In this scenario the author attempted (Davies, 1987l) to find a new strategy that would have some chance of maintaining the accuracy and robustness of the original approach, while improving speed by at least an order of magnitude. In particular, it would be useful to aim for improvement by a factor of three by sampling, and by another factor of three by employing a very small neighbourhood. By opting for a 2-element neighbourhood, the capability of estimating edge orientation is lost: however, this approach still permits an alternative strategy based on bisecting horizontal and vertical chords of a circle. Clearly, this requires two passes over the image to determine the centre position fully. A similar amount of computation is needed as for a single application of a 4-element neighbourhood edge detector, so it should be possible to gain the desired factor of three in speed relative to a Sobel operator via the reduced neighbourhood size. Finally, the overall technique involves much less computation, the divisions, multiplications and square root calculations being replaced by 2-element averaging operations, thereby giving a further gain in speed. Thus the gain obtained by using this approach could be appreciably over an order of magnitude.

9.5.1 More detailed estimates of speed

This section presents formulae from which it is possible to estimate the gain in speed that would result by applying the above strategy for centre location. First, the amount of computation involved in the original Hough-based approach is modelled by:

$$T_0 = N^2 s + S t_0 \tag{9.14}$$

where T_0 is the total time taken to run the algorithm on an $N \times N$ pixel image; s is the time taken per pixel to compute the Sobel operator and to threshold the intensity gradient g; S is the number of edge pixels that are detected; and t_0 is the time taken to compute the position of a candidate centre point.

Next, the amount of computation in the basic chord bisection approach is modelled as:

$$T = 2(N^2 q + Q t) \tag{9.15}$$

where T is the total time taken to run the algorithm; q is the time taken per pixel to compute a 2-element x or y edge detection operator and to perform thresholding; Q is the number of edge pixels that are detected in one or other scan direction; and t is the time taken to compute the position of a candidate centre point coordinate (either x_c or y_c). In equation (9.15), the factor of 2 results because scanning occurs in both horizontal and vertical directions. If the image data are now sampled by examining only a proportion a of all possible horizontal and vertical scan lines, the amount of computation becomes:

$$T = 2\alpha(N^2q + Qt)$$ (9.16)

The gain in speed from using the chord bisection scheme with sampling is therefore:

$$G = \frac{N^2s + St_0}{2\alpha(N^2q + Qt)}$$ (9.17)

Typical values of relevant parameters for (say) a biscuit of radius 32 pixels in a 128 × 128 pixel image are:

$$N^2 = 16\,384$$
$$S \simeq Q \simeq 200$$
$$t_0/s \simeq 1$$
$$s/q \simeq 12/2 = 6$$
$$t_0/t \simeq 5$$
$$\alpha \simeq 1/3$$

so that

$$G \simeq \frac{16\,384 + 200}{(2/3)(16\,384/6 + 200/5)}$$
$$\simeq \frac{16\,584 \times 1.5}{2731 + 40}$$
$$\simeq 8.98$$

This is clearly a substantial gain, and we now consider more carefully how it could arise. If we assume $N \gg S, Q$, we get:

$$G \simeq s/2\alpha q$$ (9.18)

This is obviously the product of the sampling factor $1/\alpha$ and the gain from applying an edge detection operator twice in a smaller neighbourhood (i.e. applied both horizontally and vertically). In the above example $1/\alpha = 3$ and

$s/2q = 3$, this figure resulting from the ratio of the numbers of pixels involved. This would give an ideal gain of 9, so the actual gain in this case is not much changed by the terms in t_0 and t.

It is important that the strategy also involves reducing the amount of computation in determining the centre from the edge points that are found. However, in all cases the overriding factor involved in obtaining an improvement in speed is the sampling factor $1/\alpha$. If this factor could be improved further, then significant additional gains in speed could be obtained—in principle without limit, but in practice the situation is governed by how robust the algorithm really is.

9.5.2   Robustness

Robustness can be considered relative to two factors. The first is the amount of noise in the image; the second is the amount of signal distortion that can be tolerated. Fortunately, both the original HT and the chord bisection approach lead to peak finding situations, and if there is any distortion of the object shape, then points are thrown into relatively random locations in parameter space and consequently do not have a significant direct impact on the accuracy of peak location. However, they do have an indirect impact in that the signal-to-noise ratio is reduced, so that accuracy is impaired. In fact, if a fraction β, of the original signal is removed, leaving a fraction $\gamma = 1 - \beta$, either due to such distortions or occlusions or else by the deliberate sampling procedures already outlined, then the number of independent measurements of the centre location drops to a fraction γ of the optimum. This means that the accuracy of estimation of the centre location drops to a fraction $\sqrt{\gamma}$ of the optimum. Since noise affects the optimum accuracy directly, we have in principle shown the result of both major factors governing robustness.

What is important here is that the effect of sampling is substantially the same as that of signal distortion, so that the more distortion that must be tolerated, the higher the value α has to have. This principle applies both to the original Hough approach and to the chord bisection algorithm. However, the latter does its peak finding in a different way—via 1-D rather than 2-D averaging processes. As a result, it turns out to be somewhat more robust than the standard HT in its capability for accepting a high degree of sampling.

This gain in capability for accepting sampling carries with it a set of related disadvantages: highly textured objects may not easily be located by the method; high noise levels may muddle it via the production of too many false edges; and (since it operates by specific x and y scanning mechanisms) there must be a sufficiently small amount of occlusion and other gross distortion that a significant number of scans (both horizontally and vertically) pass through the object. Ultimately this means that the method does not tolerate more than about one-quarter of the circumference of the object being absent. Thus it cannot in general be used to handle images such as the one shown in Fig. 9.1. These factors are self-evident

(a)   (b)

(c)

Fig. 9.10 Successful object location using the chord bisection algorithm for the same initial image, using successive step sizes of 2, 4 and 8 pixels. The black dots show the positions of the horizontal and vertical chord bisectors, and the white dot shows the position found for the centre.

and further discussion of this point is curtailed, although the method does have the advantage that the degree of robustness that exists is clearly discernible so that its viability can easily be confirmed during the initial setting up phase.

9.5.3 Experimental results

Tests (Davies, 19871) with the image in Fig. 9.10 showed that gains in speed of more than 25 can be obtained, with values of α down to less than 0.1 (i.e. every 10th horizontal and vertical line scanned). To obtain the same accuracy and reliability with smaller objects (Fig. 9.11), a larger value of α was needed: this can be viewed as keeping the number of samples per object roughly constant. The maximum gain in speed was then by a factor of about 10. The results for broken circular products (Figs 9.12–9.14) are self-explanatory; they indicate the limits to which the method can be taken and confirm that it is straightforward to set the

Fig. 9.11 Location of small circular objects (coins): each object was checked after location to test whether its size fell within a given range. The illustration shows the centres of two objects of radius 27 pixels which were confirmed in this way.

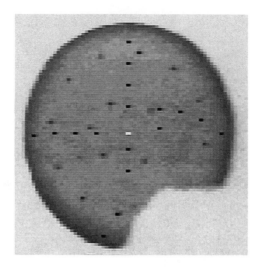

Fig. 9.12 Successful location of a broken object using the chord bisection algorithm: only about one-quarter of the ideal boundary is missing.

Fig. 9.13 Location of a round object amid parts of another round object: the algorithm has no difficulty in finding the slightly defective object but none of the parts of the other round object proves sufficiently complete to permit detection.

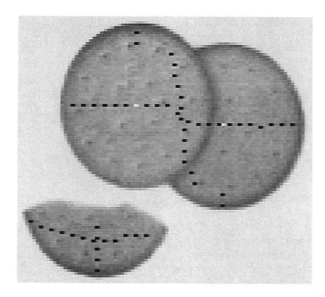

Fig. 9.14   A test on overlapping and broken biscuits: the overlapping objects are successfully located, albeit with some difficulty but there is no chance of finding the centre of the broken biscuit since over one-half of the ideal boundary is missing.

algorithm up by eye. An outline of the complete algorithm is given in Table 9.3.

Figure 9.15 shows the effect of adjusting the threshold in the 2-element edge detector. The result of setting it too low is seen in Fig. 9.15(a). Here the surface texture of the object has triggered the edge detector, and the chord midpoints give rise to a series of false estimates of the centre coordinates. Figure 9.15(b) shows the result of setting the threshold at too high a level, so that the number of estimates of centre coordinates is reduced and sensitivity suffers. The images in Fig. 9.10 were obtained with the threshold adjusted intuitively but its value is nevertheless clearly close to the optimum. However, a more rigorous approach can be taken by optimizing a suitable criterion function. There are two obvious functions: (i) the number of accurate centre predictions n; and (ii) the speed–sensitivity product. The latter can be written in the form $\sqrt{n}/T$ where T is the execution time (the reason for the square root being as in Section 9.4.2). The two methods of optimization make little difference in the example of Fig. 9.15. However, had there been a strong regular texture on the object, the situation would have been rather different.

Table 9.3　Outline of the fast centre-finding algorithm.

```
y := 0;
repeat
  scan horizontal line y looking for start and end of each object;
  calculate midpoints of horizontal object segments;
  accumulate midpoints in 1-D parameter space (x space);
  (* note that the same space, x space, is used for all lines y *)
  y := y + d ;
until y > ymax;

x := 0;
repeat
  scan vertical line x looking for start and end of each object;
  calculate midpoints of vertical object segments;
  accumulate midpoints in 1-D parameter space (y space);
  (* note that the same space, y space, is used for all lines x *)
  x := x + d;
until x > xmax;

find peaks in x space;
find peaks in y space;
test all possible object centres arising from these peaks;
(* the last step is necessary only if ∃ > 1 peak in each space *)
(* d is the horizontal and vertical step-size (= 1/alpha) * )
```

9.5.4 Summary

The centre location procedure described above is normally over an order of magnitude faster than the standard Hough-based approach: typically, execution times are reduced from about 2.5 to as low as 0.1 seconds for a 128×128 image. In a number of industrial applications this could mean that it will operate without special-purpose hardware accelerators. Robustness is good both in noise suppression capability and in the ability to negotiate partial occlusions and breakages of the object being located. In fact, the method tolerates at least one-quarter of the circumference of an object being absent for one reason or another: while this is not as good as for the original HT approach, it is adequate for many real applications. In addition, it is entirely clear on applying the method to real image data whether the latter would be likely to muddle the algorithm—since a visual indication of performance is available as an integral part of the method.

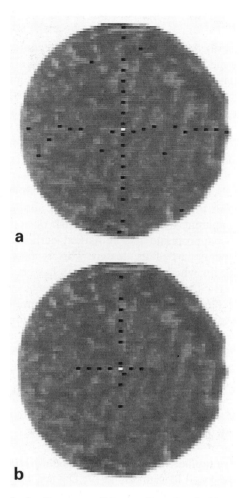

Fig. 9.15 Effect of misadjustment of the gradient threshold: (a) effect of setting the threshold too low, so that surface texture muddles the algorithm; (b) loss of sensitivity on setting the threshold too high.

Robustness is built into the algorithm in such a way as to emulate the standard Hough-based approach. Indeed, it is possible to interpret the method as being itself a Hough-based approach, since the centre coordinates are each accumulated in their own 1-D "parameter spaces". These considerations give further insight into the robustness of the standard Hough technique.

9.6   Concluding remarks

This chapter has described techniques for circle detection, starting with the HT approach. While the Hough technique is found to be effective and highly robust against occlusions, noise and other artefacts, it incurs considerable storage and computation—especially if it is required to locate circles of unknown radius or if high accuracy is required. Techniques have been described whereby the latter two problems can be tackled much more efficiently; in addition, a method has been described for markedly reducing the computational load involved in circle detection. The general circle location scheme is a type of HT but although the other two methods are related to the HT, they are distinct methods which draw on the HT for inspiration. As a result they are rather less robust, although the degree of robustness is quite apparent and there is little risk of applying them when their use would be inappropriate. As remarked earlier, the HT achieves robustness as an integral part of its design. This is also true of the other techniques described, for which considerable care has been taken to include robustness as an intrinsic part of the design rather than as an afterthought.

As in the case of line detection, a trend running through the design of circle detection schemes is the deliberate splitting of algorithms into two or more stages. This is useful for keying into the important and relevant parts of an image prior to finely discriminating one type of object or feature from another, or prior to measuring dimensions or other characteristics accurately. Indeed, the concept can be taken further, in that all the algorithms discussed in this chapter have improved their efficiency by searching first for edge features in the image. The concept of two-stage template matching is therefore deep-seated in the methodology of the subject and is developed further in later chapters—notably those on ellipse and corner detection. Although two-stage template matching is a standard means of increasing efficiency (VanderBrug and Rosenfeld, 1977; Davies, 1988i), it is not at all obvious that it is always possible to increase efficiency by this means. It appears to be in the nature of the subject that ingenuity is needed to discover ways of achieving this.

9.7   Bibliographical and historical notes

The Hough transform was developed in 1962 and first applied to circle detection by Duda and Hart (1972). However, the now standard HT technique, which makes use of edge orientation information to reduce computation, only emerged three years later (Kimme et al., 1975). The author's work on circle detection for automated inspection required real-time implementation and also high accuracy. This spurred the development of the three main techniques described in Sections 9.3–9.5 of this chapter (Davies, 1987l, 1988b, e). In addition, the author has considered the effect of noise on edge orientation computations, showing in particular their effect in reducing the accuracy of centre location (Davies, 1987k).

Yuen *et al.* (1989) reviewed various existing methods for circle detection using the HT. In general their results confirmed the efficiency of the method of Section 9.3 for unknown circle radius, although they found that the two-stage process involved can sometimes lead to slight loss of robustness. It appears that this problem can be reduced in some cases by using a modified version of the algorithm of Gerig and Klein (1986). Note that the Gerig and Klein approach is itself a two-stage procedure: it is discussed in detail in the next chapter. More recently, Pan *et al.* (1995) have increased the speed of computation of the HT by prior grouping of edge pixels into arcs, for an underground pipe inspection application.

The two-stage template matching technique and related approaches for increasing search efficiency in digital images were known by 1977 (Nagel and Rosenfeld, 1972; Rosenfeld and VanderBrug, 1977; VanderBrug and Rosenfeld, 1977), and have undergone further development since then— especially in relation to particular applications such as those described in this chapter (Davies, 1988i).

9.8 Problem

1. Prove the result of Section 9.4.1, that as D approaches C and d approaches zero (Fig. 9.9), the shape of the locus becomes a circle on DC as diameter.

10

The Hough Transform and Its Nature

'An open research issue is to find a parameterization that yields more nearly radially symmetric psf's for lines (or show why it cannot exist).'

(Brown, 1983)

10.1 Introduction

In the previous few chapters it has been seen that the Hough transform (HT) is of great importance for the detection of features such as lines and circles, and for finding relevant image parameters (cf. the dynamic thresholding problem of Chapter 4). This makes it worthwhile to see the extent to which the method can be generalized so that it can detect arbitrary shapes. The work of Merlin and Farber (1975) and Ballard (1981) was crucial historically and led to the development of the generalized Hough transform (GHT). The GHT is studied in this chapter, showing first how it is implemented and then examining how it is optimized and adapted to particular types of image data. This requires us to go back to first principles, taking spatial matched filtering as a starting point.

Having developed the relevant theory, it is applied to the important case of line detection, showing in particular how detection sensitivity is optimized. Finally, the computational problems of the GHT are examined and fast HT techniques are considered.

10.2   The generalized Hough transform

This section shows how the standard Hough technique is generalized so that it can detect arbitrary shapes. In principle it is trivial to achieve this. First, we need to select a localization point L within a template of the idealized shape. Then, we need to arrange so that, instead of moving from an edge point a *fixed* distance R directly along the local edge normal to arrive at the centre, as for circles, we move an appropriate *variable* distance R in a variable direction φ so as to arrive at L: R and φ are now functions of the local edge normal directions θ (Fig. 10.1). Under these circumstances votes will peak at the preselected object localization point L. The functions $R(\theta)$ and $\varphi(\theta)$ can be stored analytically in the computer algorithm, or for completely arbitrary shapes they may be stored as a lookup table. In either case the scheme is beautifully simple in principle but two complications arise in practice. The first arises because some shapes have features such as concavities and holes, so that several values of R and φ are required for certain values of θ (Fig. 10.2). The second arises because we are going from an isotropic shape (a circle) to an anisotropic shape which may be in a completely arbitrary orientation.

To cope with the first of these complications, the lookup table (usually called the "R-table") must contain a list of the positions $\mathbf{r}$, relative to L, of all points on the boundary of the object for each possible value of edge orientation θ (or a similar effect must be achieved analytically): then, on encountering an edge fragment in the image whose orientation is θ, estimates of the position of L may be obtained by moving a distance (or distances) $\mathbf{R} = -\mathbf{r}$ from the given edge fragment. Clearly, if the R-table has multivalued entries (i.e. several values of $\mathbf{r}$ for certain values of θ), only one of these entries (for given θ) can give a correct estimate of the position

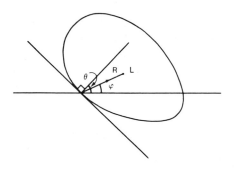

Fig. 10.1   Computation of the generalized Hough transform.

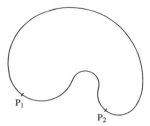

Fig. 10.2   A shape exhibiting a concavity: certain values of θ correspond to several points on the boundary and hence require several values of R and φ—as for points P_1 and P_2.

of L. However, at least the method is guaranteed to give optimum sensitivity, since all relevant edge fragments contribute to the peak at L in parameter space. This property of optimal sensitivity reflects the fact that the GHT is a form of spatial matched filter: this property is analysed in more detail below.

The second complication arises because any shape other than a circle is anisotropic. Since in most applications (including industrial applications such as automated assembly) object orientations are initially unknown, the algorithm has to obtain its own information on object orientation. This means adding an extra dimension in parameter space (Ballard, 1981). Then each edge point contributes a vote in each plane in parameter space at a position given by that expected for an object of given shape and given orientation. Finally, the whole of parameter space is searched for peaks, the highest points indicating both the locations of objects and their orientations. Clearly, if object size is also a parameter, the problem becomes far worse and this complication is ignored here (although the method of Section 9.3 is clearly relevant).

The changes made in proceeding to the GHT leave it just as robust as the HT circle detector described previously. This gives an incentive to improve the GHT so as to limit the computational problems in practical situations. In particular, the size of the parameter space must be cut down drastically both to save storage and to curtail the associated search task. Considerable ingenuity has been devoted to devising alternative schemes and adaptations to achieve this. Important special cases are those of ellipse detection and polygon detection, and in each of these definite advances have been made: ellipse detection is covered in Chapter 11 and polygon detection in Chapter 13. Here we proceed with some more basic studies of the GHT.

10.3   Setting up the generalized Hough transform—some relevant questions

The next few sections explore the theory underpinning the GHT, with the aim of clarifying how to optimize it systematically for specific circumstances. It is relevant to ask what is happening when a GHT is being computed. Although the HT has been shown to be equivalent to template matching (Stockman and Agrawala, 1977) and also to spatial matched filtering (Sklansky, 1978), further clarification is required. In particular, there are three problems (Davies, 1987c) that need to be addressed, as follows:

(i) *The parameter space weighting problem.* In introducing the GHT, Ballard mentioned the possibility of weighting points in parameter space according to the magnitudes of the intensity gradients at the various edge pixels. But when should gradient weighting be used in preference to uniform weighting?

(ii) *The threshold selection problem.* When using the GHT to locate an object, edge pixels are detected and used to compute candidate positions for the localization point L (see Section 10.2): to achieve this it is necessary to threshold the edge gradient magnitude. How should the threshold be chosen?

(iii) *The sensitivity problem.* Optimum sensitivity in detecting objects does not automatically provide optimum sensitivity in locating objects, and *vice versa.* How should the GHT be optimized for these two criteria?

To understand the situation and solve these problems it is necessary to go back to first principles. A start is made on this in the next section.

10.4   Spatial matched filtering in images

To discuss the questions posed in Section 10.3 it is necessary to analyse the process of spatial matched filtering: in principle, this is the ideal method of detecting objects, since it is well known (Rosie, 1966) that a filter that is matched to a given signal detects it with optimum signal-to-noise ratio under white noise[*] conditions (North, 1943; Turin, 1960). (For a more recent discussion of this topic, see Davies, 1993a.)

---

[*] White noise is noise that has equal power at all frequencies: in image science white noise is understood to mean equal power at all *spatial* frequencies. The significance of this is that noise at different pixels is completely uncorrelated but is subject to the same grey-scale probability distribution—i.e. it has potentially the same range of amplitudes at all pixels.

Mathematically, using a matched filter is identical to correlation with a signal (or "template") of the same shape as the one to be detected (Rosie, 1966). Here "shape" is a general term meaning the amplitude of the signal as a function of time or spatial location.

When applying correlation in image analysis, changes in background illumination cause large changes in signal from one image to another and from one part of an image to another. The varying background level prevents straightforward peak detection in convolution space. The matched filter optimizes signal-to-noise ratio only in the presence of white noise: whereas this is likely to be a good approximation in the case of radar signals, this is not generally true in the case of images. For ideal detection, the signal should be passed through a "noise-whitening filter" (Turin, 1960), which in the case of objects in images is usually some form of high-pass filter: this must be applied prior to correlation analysis. However, this is likely to be a computationally expensive operation.

If we are to make correlation work with near optimal sensitivity but without introducing a lot of computation, other techniques must be employed. In the template matching context, the following possibilities suggest themselves:

(i)   adjust templates so that they have a mean value of zero, to suppress the effects of varying levels of illumination in first order;
(ii) break up templates into a number of smaller templates each having a zero mean: then as the sizes of subtemplates approach zero, the effects of varying levels of illumination will tend to zero in second order;
(iii) apply a threshold to the signals arising from each of the subtemplates, so as to suppress those that are less than the expected variation in signal level.

If these possibilities fail, only two further strategies appear to be available:

(iv) Readjust the lighting system—an important option in industrial inspection applications, although it may give little improvement when a number of objects can cast shadows or reflect light over each other;
(v)   use a more "intelligent" (e.g. context sensitive) object detection algorithm, although this will almost certainly be computationally intensive.

10.5 From spatial matched filters to generalized Hough transforms

To proceed, we note that items (i) to (iii) listed above essentially amount to a specification of the GHT. First, breaking up the templates into small subtemplates each having a zero mean and then thresholding them is analogous, and in many cases identical, to a process of edge detection (see for example the templates used in the Sobel and similar operators). Next, locating objects by peak detection in parameter space clearly corresponds to the process of reconstructing whole template information from the subtemplate (edge location) data. What is important here is that these ideas reveal how the GHT is related to the spatial matched filter. Basically, *the GHT can be described as a spatial matched filter which has been modified, with the effect of including integral noise whitening, by breaking down the main template into small zero-mean templates and thresholding the individual responses before object detection.*

Small templates do not permit edge orientation to be estimated as accurately as large ones. Although the Sobel edge detector is in principle accurate to about 1° (see Chapter 5), there is a deleterious effect if the edge of the object is fuzzy. In such a case it is not possible to make the subtemplates very small, and an intermediate size should be chosen which gives a suitable compromise between accuracy and sensitivity.

Employing zero-mean templates results in the absolute signal level being reduced to zero and only local relative signal levels being retained. Thus the GHT is not a true spatial matched filter: in particular, it suppresses the signal from the bulk of the object, retaining only that near its boundary. As a result the GHT is highly sensitive to object position but is not optimized for object detection.

Thresholding of subtemplate responses has much the same effect as employing zero-mean templates, although it may remove a small proportion

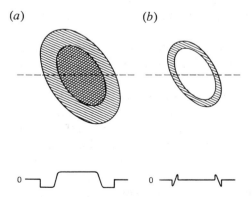

Fig. 10.3 The idea of a perimeter template: both the original spatial matched filter template (a) and the corresponding "perimeter template" (b) have a zero mean (see text). The lower illustrations show the cross-sections along the dotted lines.

of the signal giving positional information. This makes the GHT even less like an ideal spatial matched filter and further reduces the sensitivity of object detection. The thresholding process is particularly important in the present context since it provides a means of saving computational effort *without losing significant positional information.* On its own this characteristic of the GHT would correspond to a type of perimeter template around the outside of an object (see Fig. 10.3). This must not be taken as excluding all of the interior of the object, since any high-contrast edges within the object will facilitate location.

10.6 Gradient weighting versus uniform weighting

The first problem described in Section 10.1 was that of how best to weight plots in parameter space in relation to the respective edge gradient magnitudes. To find an answer to this problem, it should now only be necessary to go back to the spatial matched filter case to find the ideal solution, and then to determine the corresponding solution for the GHT in the light of the discussion in Section 10.4. First, note that the responses to the subtemplates (or to the perimeter template) are proportional to edge gradient magnitude. With a spatial matched filter, signals are detected optimally by templates of the same shape. Each contribution to the spatial matched filter response is then proportional to the local magnitude of the signal and to that of the template. In view of the correspondence between (i) using a spatial matched filter to locate objects by looking for peaks in convolution space and (ii) using a GHT to locate objects by looking for peaks in parameter space, we should use weights proportional to the gradients of the edge points *and* proportional to the *a priori* edge gradients.

There are two ways in which the choice of weighting is important. First, the use of uniform weighting implies that all edge pixels whose gradient magnitudes are above threshold will effectively have them reduced to the threshold value, so that the signal will be curtailed: this can mean that the signal-to-noise ratio of high-contrast objects will be reduced significantly. Second, the widths of edges of high-contrast objects will be broadened in a crude way by uniform weighting (see Fig. 10.4) but under gradient weighting this broadening will be controlled, giving a roughly Gaussian edge profile: thus the peak in parameter space will be narrower and more rounded, and the object reference point L can be located more easily and with greater accuracy. This effect is visible in Fig. 10.5, which also shows clearly the relatively increased noise level that results from uniform weighting.

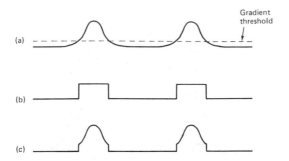

Fig. 10.4 Effective gradient magnitude as a function of position within a section across an object of moderate contrast, thresholded at a fairly low level: (a) gradient magnitude for original image data and gradient thresholding level; (b) uniform weighting: the effective widths of edges are broadened rather crudely, adding significantly to the difficulty of locating the peak in parameter space; (c) gradient weighting: the position of the peak in parameter space can be estimated in a manner that is basically limited by the shape of the gradient profile for the original image data.

Note also that low gradient magnitudes correspond to edges of poorly known location, while high values correspond to sharply defined edges: thus the accuracy of the information relevant to object location is proportional to the magnitude of the gradient at each of the edge pixels, and appropriate weighting should therefore be used.

10.6.1 Calculation of sensitivity and computational load

The aim of this subsection is to underline the above ideas by working out formulae for sensitivity and computational load. It is assumed that p objects of size around $n \times n$ are being sought in an image of size $N \times N$.

Correlation requires $N^2 n^2$ operations to compute the convolutions for all possible positions of the object in the image. Using the perimeter template the number of basic operations is reduced to $\sim N^2 n$, corresponding to the reduced number of pixels in the template. The GHT requires $\sim N^2$ operations to locate the edge pixels, plus a further pn operation to plot the points in parameter space.

The situation for sensitivity is rather different. With correlation the results for n^2 pixels are summed, giving a signal proportional to n^2, although the noise (assumed to be independent at every pixel) is proportional to n: this is because of the well-known result that the noise powers of various

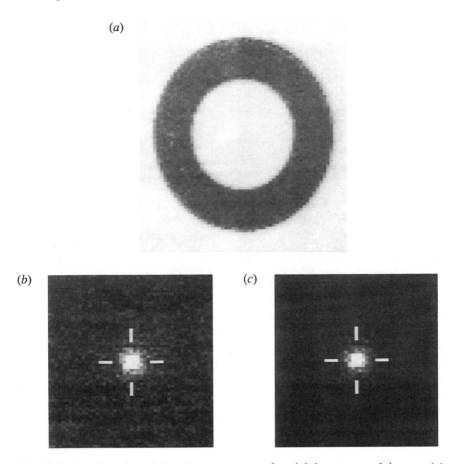

Fig. 10.5 Results of applying the two types of weighting to a real image: (a) original image; (b) results in parameter space for uniform weighting; (c) results for gradient weighting. The peaks (which arise from the outer edges of the washer) are normalized to the same levels in the two cases: the increased level of noise in (b) is readily apparent. In this example the gradient threshold is set at a low level (around 10% of the maximum level) so that low-contrast objects can also be detected.

independent noise components are additive (Rosie, 1966). Overall this results in the signal-to-noise ratio being proportional to n. The perimeter template possesses only $\sim n$ pixels, and here the overall result is that the signal-to-noise ratio is proportional to $\sqrt{n}$. The situation for the GHT is inherently identical to that for the perimeter template method, so long as plots in parameter space are weighted proportional to edge gradient g

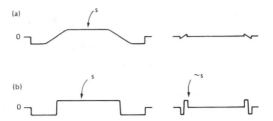

Fig. 10.6 Effect of edge gradient on perimeter template signal: (a) low edge gradient: signal is proportional to gradient; (b) high edge gradient: signal saturates at value of s.

multiplied by *a priori* edge gradient G. It is now necessary to compute the constant of proportionality α. Take s as the average signal, equal to the intensity (assumed to be roughly uniform) over the body of the object, and S as the magnitude of a full matched filter template. In the same units, g (and G) is the magnitude of the signal within the perimeter template (assuming the unit of length equals 1 pixel). Then $\alpha = 1/sS$. This means that the perimeter template method and the GHT method lose sensitivity in two ways—first because they look at less of the available signal, and second because they look where the signal is low. For a high value of gradient magnitude, which occurs for a step edge (where most of the variation in intensity takes place within the range of 1 pixel), the values of g and G saturate out, so that they are nearly equal to s and S (see Fig. 10.6). Under these conditions the perimeter template method and the GHT have sensitivities that depend only on the value of n.

Table 10.1   Formulae for computational load and sensitivity[*].

	Template matching	Perimeter template matching	Generalized Hough transform
Number of operations	$O(N^2n^2)$	$O(N^2n)$	$O(N^2) + O(pn)$
Sensitivity	$O(n)$	$O(\sqrt{n}gG/sS)$	$O(\sqrt{n}gG/sS)$
Maximum sensitivity[†]	$O(n)$	$O(\sqrt{n})$	$O(\sqrt{n})$

[*] This table gives formulae for computational load and sensitivity when p objects of size $n \times n$ are sought in an image of size $N \times N$. The intensity of the image within the whole object template is taken as s and the value for the ideal template is taken as S: corresponding values for intensity gradient within the perimeter template are g and G.
[†] Maximum sensitivity refers to the case of a step edge, for which $g \approx s$ and $G \approx S$ (see Fig. 10.6).

Table 10.1 summarizes the situation discussed above. The oft-quoted statement that the computational load of the GHT is proportional to the number of perimeter pixels, rather than to the much greater number of pixels within the body of an object, is only an approximation. In addition, this saving is not obtained without cost: in particular, the sensitivity (signal-to-noise ratio) is reduced (at best) as the square root of object area/perimeter (note that area and perimeter are measured in the same units, so it is valid to find their ratio).

Finally, the absolute sensitivity for the GHT varies as gG. As contrast changes so that $g \rightarrow g'$, we see that $gG \rightarrow g'G$: i.e. sensitivity changes by a factor of g'/g. Hence theory predicts that sensitivity is proportional to contrast. Although this result might have been anticipated, we now see that it is valid only under conditions of gradient weighting.

10.7 Summary

The above sections examined the GHT and found a number of factors involved in optimizing it, as follows.

(i)   Each point in parameter space should be weighted in proportion to the intensity gradient at the edge pixel giving rise to it, *and* in proportion to the *a priori* gradient, if sensitivity is to be optimized, particularly for objects of moderate to high contrast.

(ii) The ultimate reason for using the GHT is to save computation. The main means by which this is achieved is by ignoring pixels having low magnitudes of intensity gradient. If the threshold of gradient magnitude is set too high, fewer objects are in general detected; if it is set too low, computational savings are diminished. Suitable means are required for setting the threshold but little reduction in computation is possible if the highest sensitivity in a low-contrast image is to be retained.

(iii) The GHT is inherently optimized for the location of objects in an image but is not optimized for the detection of objects. This means that it may miss low-contrast objects which are detectable by other methods that take the whole area of an object into account. However, this consideration is often unimportant in applications where signal-to-noise ratio is less of a problem than finding objects quickly in an uncluttered environment.

Overall, it is clear that the GHT is a spatial matched filter only in a particular sense, and as a result it has suboptimal sensitivity. The main advantage of the technique is that it is highly efficient, overall computational load in principle being proportional to the relatively few pixels on the perimeters of objects rather than to the much greater numbers of pixels within them. In addition, by concentrating on the boundaries of objects, the GHT retains its power to locate objects accurately. It is thus important to distinguish clearly between sensitivity in *detecting* objects and sensitivity in *locating* them.

10.8 Applying the generalized Hough transform to line detection

This section considers the effect of applying the GHT to line detection, this being relevant because it should be known how sensitivity may be optimized. Straight lines fall into the category of features such as concavities which give multivalued entries in the R-table of the GHT—i.e. certain values of θ give rise to a whole set of values of $\mathbf{r}$. Here we regard a line as a complete object and aim to detect it by a localization point at its midpoint M. Taking the line as having length S and direction given by unit vector $\mathbf{u}$ along its length, then the transform of a general edge point E at $\mathbf{e}$ on the line is the set of points:

$$P = \{\mathbf{e} + s\mathbf{u}; \qquad -S/2 \leqslant s \leqslant S/2\} \qquad (10.1)$$

or, more accurately (since the GHT is constructed so as to accumulate evidence for the existence of objects at particular locations), the transform function $T(\mathbf{x})$ is unity at these points and zero elsewhere:

$$T(\mathbf{x}) = \begin{cases} 1; & \mathbf{x} \in P \\ 0; & \mathbf{x} \notin P \end{cases} \qquad (10.2)$$

Thus each edge pixel gives rise to a *line* of votes in parameter space equal in length to the whole length of the line in the original shape. It is profitable to explore briefly why lines should have such an odd PSF, which contrasts strongly with the point PSF for circles. The reason is that a straight line is very accurately locatable normal to its length, but very inaccurately locatable along its length (for an infinite line, there is *zero* accuracy of location along the length, and complete precision for location normal to the length). Hence lines cannot have a radially symmetrical PSF in a parameter space that is congruent to image space (this solves a research issue raised by Brown (1983)—see quotation at the head of the chapter).

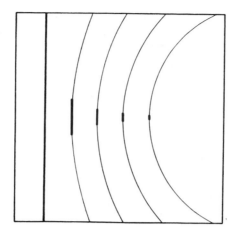

Fig. 10.7 Longitudinal accuracy as a function of curvature: localization points are selected on the lines to which they refer, for clarity of presentation. The sizes of the localization points indicate the relative accuracy in the longitudinal and normal directions.

However, circles and lines do not form totally distinct situations but are two extremes of a more general situation. This is indicated in Fig. 10.7, where it is seen that progressively lowering the curvature for objects within the (limited) field of view leads to progressive lowering of the accuracy with which the object may be located in the direction along the visible part of the perimeter, but to no loss in accuracy in the normal direction.

Unfortunately, although the method of computing the transforms of edge points described above ensures optimum sensitivity and helps to localize the line longitudinally, it also increases computation. An alternative approach which might be expected to overcome this problem is discussed below.

10.9 An instructive example

We now consider how a composite shape such as a semicircle should be detected. In fact, this is very straightforward if the GHT is employed; it is especially convenient if the centre of the circular region is chosen as the localization point L. First, normals on the circular region all pass through L, so the point transforms are found with little computation for this particular set of edge pixels. Second, the points on the linear region of the boundary each transform into a line of votes, all passing through L (see equation (10.2)). Hence *every* point on the boundary of the semicircle contributes a vote at L in the same parameter space.

Thus the transform is readily assembled from the separate parts of the boundary, and sensitivity is automatically optimized. In addition, both the transform of the linear segment and that of the circular part of the boundary peak at L, so the whole transform is markedly peaked at L. However, as has been seen, computation can be excessive since each point on the linear segment contributes many votes in parameter space. Hence it is worth considering whether other solutions to the semicircle problem could be found which would require less computation. One obvious way would be to detect the straight and curved portions of the semicircle separately and then to combine the two sets of data. It is well known how to use HT methods to locate straight edges in images (Chapter 8). However, applying this more complex multistage process has two disadvantages: first, the usual HT line parametrization schemes are not guaranteed to give optimal detection sensitivity; and second, they encode only the lateral position of the line and not its longitudinal position. This means that when entering the straight line data in parameter space in order to compute the final GHT, it is necessary to take account of the uncertainty in the longitudinal position of the line. Since there is complete uncertainty in longitudinal position, all that can be done is to record a uniform probability of the localization point of the line being along one axis. This is clearly a worse situation than that for the (pure) GHT parametrization scheme.

Another disadvantage of the above scheme is that if we have a barely detectable signal from an object such as a semicircle, the straight edge section taken on its own may not be detected: hence it will not be possible to transfer the result of straight edge detection into parameter space as required. At high signal-to-noise ratios this problem does not really arise but at low signal-to-noise ratios it can arguably be overcome by deferring nonlinear processes of detection, and adding the straight and circular edge results into the same parameter space prior to object detection. However, the computational load is high and the method is in general impracticable, since an entire parameter space has to be recoded and then entered into the GHT parameter space.

10.10   Tradeoffs to reduce computational load

A possible means of saving computation involves modifying line parametrization to include fewer points in the transform represented by equations (10.1) and (10.2), i.e. S can be reduced to a smaller value S'. Clearly this will reduce sensitivity, although it seems that an optimum value of S' will have to be found on a practical basis. Ultimately, the basic principle is that S' must remain large enough to iron out the effects of individual noisy edge pixels, so that the accuracy of peak location is not compromised.

Figure 10.8 shows the result of applying GHT line parametrization in the case of an image containing a semicircle. In this case computation has been saved by using only one-half the optimum number of points per straight edge pixel ($S' = S/2$), since this gives a reasonable compromise between sensitivity and computational effort (Davies, 1987d).

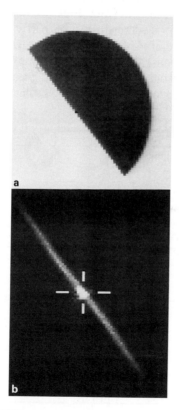

Fig. 10.8 Effect of the GHT on a real semicircle: (a) original image; (b) transform. In this case, computation is saved by using half the optimum number of plots per straight edge pixel (see text).

10.11   The effects of occlusions for objects with straight edges

The last few sections have shown that the GHT gives optimum sensitivity of line detection only at the expense of transforming each edge point into a whole line of votes in parameter space. However, there is a resulting advantage, in that the GHT formulation includes a degree of intrinsic longitudinal localization of the line. We now consider this aspect of the situation more carefully.

While studying line transforms obtained using the GHT line parametrization scheme, it is relevant to consider what happens in situations where objects may be partly occluded. It will already be clear that sensitivity for line detection degrades gracefully, as for circle detection. But what happens to the line localization properties? In fact, the results are rather surprising and gratifying, as will be seen from Figs 10.9–10.12 (Davies, 1989d). These figures show that the transforms of partly occluded lines have a distinct peak at the centre of the line in all cases where both ends are visible, and a flat-topped response otherwise, indicating the exact degree of uncertainty in the position of the line centre (see Figs 10.9 and 10.10). Naturally, this effect is spoiled if the line PSF that is used is of the wrong

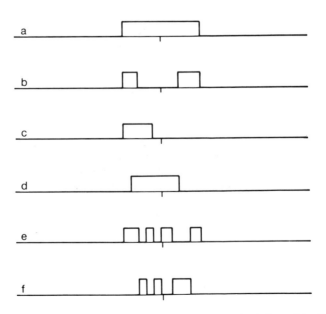

Fig. 10.9   Various types of occlusion of a line: (a) original line; (b)–(f) line with various types of occlusion (see text).

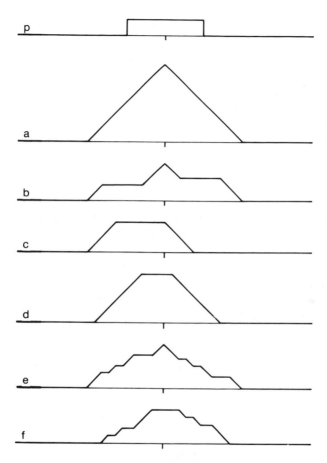

Fig. 10.10 Transforms of a straight line and of the respective occluded variants shown in Fig. 10.9 (p, PSF). It is easy to see why response (a) takes the form shown. The result of correlating a signal with its matched filter is the autocorrelation function of the signal. Applying these results to a short straight line (Fig. 10.9(a)) gives a matched filter with the same shape (p), and the resulting profile in convolution space is the autocorrelation function (a).

length, and then an uncertainty in the centre position again arises (Fig. 10.11).

Figure 10.12 shows a rectangular component being located accurately despite gross occlusions in each of its sides. The fact that L is independently the highest point in either branch of the transform at L is ultimately because two ends in each *pair* of parallel sides of the rectangle are visible.

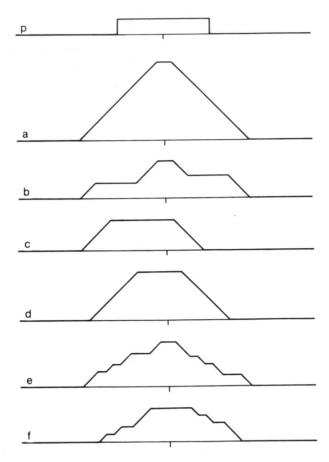

Fig. 10.11 Transforms of a straight line and of the respective occluded variants shown in Fig. 10.9, obtained by applying a PSF that is slightly too long. The results should be compared with those in Fig. 10.10 (p, PSF).

Thus the GHT parametrization of the line is valuable not only in giving optimum sensitivity in the detection of objects possessing straight edges, it also gives maximum available localization of these straight edges in the presence of severe occlusions. In addition, if the ends of the lines are visible, then the centres of the lines are locatable precisely—a property that arises automatically *as an integral part of the procedure*, no specific line-end detectors being required. Thus the GHT with the specified line parametrization scheme can be considered as a process which automatically takes possible occlusions into account in a remarkably "intelligent" way. This

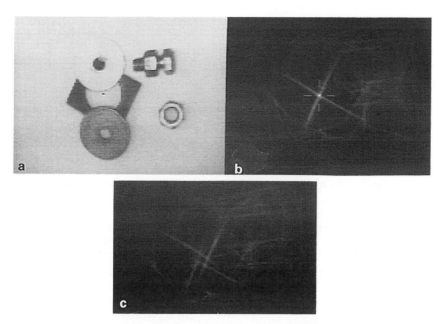

Fig. 10.12 Location of a severely occluded rectangular object: (a) original image; (b) transform assuming that the short sides are orientated between 0 and $\pi/2$; (c) transform assuming that the short sides are orientated between $\pi/2$ and π. Note that the rectangular object is precisely located by the peak in (b), despite gross occlusions over much of its boundary. The fact that the peak in (b) is larger than those in (c) indicates the orientation of the rectangular object within $\pi/2$ (Davies, 1989c) (for further details, see Chapter 13). Notice that occlusions *could* become so great that only one peak was present in either parameter plane *and* the remaining longer side was occluded down to the length of the shorter sides: clearly, there would then be insufficient information to indicate which range of orientations of the rectangle was correct, and an inherent ambiguity would remain.

section ends by summarizing, in Table 10.2, the differences between the normal and the GHT parametrizations of the straight line.

10.12   Fast implementations of the Hough transform

The foregoing sections have made it abundantly clear that the GHT has a severe weakness in that it demands considerable computation. The problem arises particularly in respect of the number of planes needed in parameter space to accommodate transforms for different object orientations and sizes;

Table 10.2   Comparison between the two parametrizations of a straight line.

Normal parametrization	GHT parametrization
Localized relative to origin of coordinates	Localized relative to the line: independent of any origin
Basic localization point is foot-of-normal from origin	Basic localization point is midpoint of line
Provides no direct information on line's longitudinal position	Provides definite information on line's longitudinal position
Takes no account of length of line: this information not needed	Takes account of length of line: must be given this information, for optimal sensitivity
Not directly compatible with GHT	Directly compatible with GHT
Not guaranteed to give maximum sensitivity	Gives maximum sensitivity even when line is combined with other boundary segments

when line detection by the GHT is involved, an additional problem arises in that a whole line of votes is required for each edge pixel. Quite clearly, significant improvements in speed are needed before the GHT can achieve its potential in practical instances of arbitrary shapes. This section considers some important developments in this area.

The source of the computational problems of the HT is the huge size the parameter space can take. Typically, a single plane parameter space has much the same size as an image plane (this will normally be so in those instances where parameter space is congruent to image space), but when many planes are required to cope with various object orientations and sizes the number of planes is likely to be multiplied by a factor of around 300 for each extra dimension. Hence the total storage area will then involve some 10 000 million accumulator cells. Clearly, reducing the resolution might just make it possible to bring this down to 100 million cells, although accuracy will start to suffer. Further, when the HT is stored in such a large space, the search problem involved in locating significant peaks becomes formidable.

Fortunately, data are not stored at all uniformly in parameter space and this provides a key to solving both the storage and the subsequent search problem. Indeed, the fact that parameter space is to be searched for the most prominent peaks—in general, the highest and the sharpest ones—means that the process of detection can start during accumulation. Furthermore, initial accumulation can be carried out at relatively low resolution, and then resolution can be increased locally as necessary in order to provide both the required accuracy and the separation of nearby peaks. In this context it is natural to employ a binary splitting procedure, i.e. to repeatedly double the resolution locally in each of the dimensions of parameter space (e.g. the x dimension, the y dimension, the orientation dimension and the size dimension, where these exist) until resolution is sufficient: a tree structure may conveniently be used to keep track of the situation.

Such a method (the "fast" HT) was developed by Li et al. (1985; Li and Lavin, 1986). Illingworth and Kittler (1987) found this method to be insufficiently flexible in dealing with real data and produced a revised version (the "adaptive" HT) which permits each dimension in parameter space to change its resolution locally in tune with whatever the data demand, rather than insisting on some previously devised rigid structure. In addition, they employed a 9×9 accumulator array at each resolution rather than the theoretically most efficient 2×2 array, since this was found to permit better judgements to be made on the nature of the local data. This approach seemed to work well with fairly clean images, but doubts were later cast on its effectiveness with complex images (Illingworth and Kittler, 1988): the most serious problem to be overcome here is that, at coarse resolutions, extended patterns of votes from several objects can overlap, giving rise to spurious peaks in parameter space. Since all of these peaks have to be checked at all relevant resolutions, the whole process can consume more computation than it saves. Clearly, optimization in multiresolution peak-finding schemes is complex[*] and data-dependent, and so discussion is curtailed here. The reader is referred to the original research papers (Li et al., 1985; Li and Lavin, 1986; Illingworth and Kittler, 1987) for implementation details.

---

[*] Ultimately, the problem of system optimization for analysis of complex images is a difficult one, since in conditions of low signal-to-noise ratio even the eye may find it difficult to interpret an image and may "lock on" to an interpretation that is incorrect. It is worth noting that in general image interpretation work there are many variables to be optimized— sensitivity, efficiency/speed, storage, accuracy, robustness, etc.—and it is seldom valid to consider any of these individually. Often tradeoffs between just two such variables can be examined and optimized but in real situations multivariable tradeoffs should be considered. This is a very complex task and it is one of the purposes of this book to show clearly the serious nature of these types of optimization problem, although at the same time it can only guide the reader through a limited number of basic optimization processes.

An alternative scheme is the hierarchical HT (Princen *et al.*, 1989a); so far it has been applied only to line detection. The scheme can most easily be envisaged by considering the foot-of-normal method of line detection described in Section 8.3. Rather small subimages of just 16×16 pixels are taken first of all, and the foot-of-normal positions determined. Then each foot-of-normal is tagged with its orientation and an identical HT procedure is instigated to generate the foot-of-normal positions for line segments in 32×32 subimages, this procedure being repeated as many times as necessary until the whole image is spanned at once. The paper by Princen *et al.* discusses the basic procedure in detail and also elaborates necessary schemes for systematically grouping separate line segments into full-length lines in a hierarchical context (in this way the work extends the techniques described in Section 8.4). An interesting detail is that successful operation of the method requires subimages with 50% overlap to be employed at each level. The overall scheme appears to be as accurate and reliable as the basic HT method but does appear to be faster and to require significantly less storage.

10.13   The approach of Gerig and Klein

The Gerig and Klein approach was first demonstrated in the context of circle detection but was only mentioned in passing in Chapter 9. This is because it is an important *approach* that has much wider application than merely to circle detection. The motivation for it has already been noted in the previous section—namely, the problem of extended patterns of votes from several objects giving rise to spurious peaks in parameter space.

Ultimately, the reason for the extended pattern of votes is that each edge point in the original image can give rise to a very large number of votes in parameter space: the tidy case of detection of circles of known radius is somewhat unusual, as will be seen particularly in the chapters on ellipse detection and 3-D image interpretation. Hence, in general *most* of the votes in parameter space are in the end unwanted and serve only to confuse. Ideally, we would like a scheme in which each edge point gives rise only to the single vote corresponding to the localization point of the particular boundary on which it is situated. While this ideal is not initially realizable, it can be engineered by the "back-projection" technique of Gerig and Klein (1986). Here all peaks and other positions in parameter space to which a given edge point contributes are examined, and a new parameter space is built in which only the vote at the strongest of these peaks is retained (there is the greatest probability, but no certainty, that it belongs to the *largest* such peak). This second parameter space thus contains no extraneous clutter and weak peaks are hence found much more easily: this gives objects with highly fragmented or occluded boundaries much more chance of being detected. Overall, the method avoids many of the problems associated with setting arbitrary thresholds on peak height—in principle no thresholds are required in this approach.

The scheme can be applied to any HT detector that throws multiple votes for each edge point. Thus it appears to be widely applicable and is capable of improving robustness and reliability at an intrinsic expense of approximately doubling computational effort (however, set against this is the relative ease with which peaks can be located—a factor which is highly data-dependent). Note that the method is another example in which a two-stage process is used for effective recognition.

Other interesting features of the Gerig and Klein method must be omitted here for reasons of space, although it should be pointed out that, rather oddly, the published scheme ignores edge orientation information as a means of reducing computation.

10.14 Concluding remarks

The Hough transform was introduced in Chapter 8 as a line detection scheme and then used in Chapter 9 for detecting circles. In those chapters it appeared as a rather cunning method for aiding object detection; although it was seen to offer various advantages, particularly in its robustness in the face of noise and occlusion, there appeared to be no real significance in its rather novel voting scheme. The present chapter has shown that, far from being a trick method, the HT is much more general an approach than originally supposed: indeed, it embodies the properties of the spatial matched filter and is therefore capable of close-to-optimal sensitivity for object detection. However, this does not prevent its implementation from entailing considerable computational load, and significant effort and ingenuity has been devoted to overcoming this problem, both in general and in specific cases. The general case is tackled by the schemes discussed in the previous two sections, while two specific cases—those of ellipse and polygon detection—are considered in detail in the following chapters. It is important not to underestimate the value of specific solutions, both because such shapes as lines, circles, ellipses and polygons cover a large proportion of (or approximations to) manufactured objects, and because methods for coping with specific cases have a habit (as for the original HT!) of becoming more general as other workers see possibilities for developing the underlying techniques. For further discussion and critique of the whole HT approach, see Chapter 25.

10.15 Bibliographical and historical notes

Although the HT was introduced as early as 1962, a number of developments—including especially those of Merlin and Farber (1975) and Kimme *et al.* (1975)—were required before the GHT was developed in its current standard form (Ballard, 1981). By that time the HT was already known to be formally equivalent to template matching (Stockman and Agrawala, 1977) and to spatial matched filtering (Sklansky, 1978). However, the questions posed in Section 10.3 were only answered much later (Davies, 1987c), the necessary analysis being reproduced in Sections 10.4–10.7. Sections 10.8–10.11 reflect the author's work (Davies, 1987d,j, 1989d) on line detection by the GHT, which was aimed particularly at optimizing sensitivity of line detection, although deeper issues of tradeoffs between sensitivity, speed and accuracy are also involved.

By 1985, the computational load of the HT became the critical factor preventing its more general use—particularly as the method could by that time be used for most types of arbitrary shape detection, with well-attested sensitivity and considerable robustness. Preliminary work in this area had been carried out by Brown (1984), with the emphasis on hardware implementations of the HT. Li *et al.* (1985; Li and Lavin, 1986) showed the possibility of much faster peak location by using parameter spaces that are not uniformly quantized. This work was developed further by Illingworth and Kittler (1987), although there are clear signs that the issue is not yet closed (see for example, Illingworth and Kittler, 1988; Princen *et al.*, 1989a,b; Davies, 1992h). The future will undoubtedly bring many further developments: unfortunately, the solutions may be rather complex and difficult (or at least tedious) to program because of the contextual richness of real images. However, an important development has been the randomized Hough transform (RHT), pioneered by Xu and Oja (1993) amongst others: it involves casting votes until specific peaks in parameter space become evident, thereby saving unnecessary computation.

Accurate peak location remains an important aspect of the HT approach. Properly, this is the domain of robust statistics which handles the elimination of outliers (of which huge numbers arise from the clutter of background points in digital images – see Appendix D). However, Davies (1992g) has shown a computationally efficient means of accurately locating HT peaks; he has also found why in many cases peaks may appear narrower than *a priori* considerations would indicate (Davies, 1992b). Kiryati and Bruckstein (1991) have tackled aliasing effects which can arise with the Hough transform, and which have the effect of cutting down accuracy.

Over time, the GHT approach has been broadened by geometric hashing, structural indexing, and other approaches (e.g. Gavrila and Groen, 1992; Califano and Mohan, 1994). At the same time, a probabilistic approach to the subject has been developed (Stephens, 1991) which puts it on a firmer footing (though rigorous implementation would in most cases demand excessive computational load). Finally, Grimson and Huttenlocher (1990) warn against the blithe use of the GHT for complex object recognition tasks, because of the false peaks that can appear in such cases, though their paper is probably overpessimistic. For further review of the subject, see Leavers (1992, 1993).

11

Ellipse Detection

11.1 Introduction

It was seen in the previous chapter that very many manufactured goods are circular or else possess distinctive circular features: this makes it imperative to have an efficient set of algorithms for detecting circular shapes in digital images. However, in many situations such objects are viewed obliquely and so efficient ellipse detectors are also required. Furthermore, some objects are actually elliptical and this also gives motivation for finding effective ellipse detection algorithms.

This chapter describes some of the algorithms that have been developed for this purpose. We concentrate on parallel algorithms, since sequential methods have already been covered in Chapter 7. In addition, parallel algorithms such as the Hough transform tend to be more robust: in fact, all the algorithms described here are based on one or other form of the Hough transform. Although Duda and Hart (1972) considered curve detection, the first method that made use of edge orientation information to reduce the amount of computation was that of Tsuji and Matsumoto (1978). This ingenious method is described in the next section.

11.2 The diameter bisection method

The diameter bisection method of Tsuji and Matsumoto (1978) is very simple in concept. First, a list is compiled of all the edge points in the image. Then, the list is sorted to find those that are antiparallel, so that they could lie at opposite ends of ellipse diameters; next, the positions of the centre points of the connecting lines for all such pairs are taken as voting positions in parameter space (Fig. 11.1). As for circle location, the parameter space that is used for this purpose is congruent to image space. Finally, the positions of significant peaks in parameter space are located to identify possible ellipse centres.

Naturally, in an image containing many ellipses and other shapes, there will be very many pairs of antiparallel edge points and for most of these the centre points of the connecting lines will lead to nonuseful votes in parameter space. Clearly, such clutter leads to wasted computation. However, it is a principle of the Hough transform that votes must be accumulated in parameter space at all points which *could in principle* lead to correct object centre location: it is left to the peak finder to find the voting positions that are most likely to correspond to object centres.

Not only does clutter lead to wasted computation but the method itself is computationally expensive. This is because it examines all *pairs* of edge points, and there are many more such pairs than there are edge points (m edge points would lead to $\binom{m}{2} \approx m^2/2$ pairs of edge points). Indeed, since there are likely to be at least 1000 edge points in a typical 256×256 image, the computational problems can be formidable. The situation is examined more carefully in Section 11.6.

It should be noted that the basic method is not particularly discriminating about ellipses. It picks out many symmetrical shapes—any indeed that possess 180° rotation symmetry, including rectangles, ellipses, circles, or

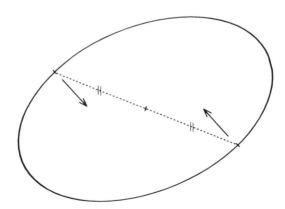

Fig. 11.1   Principle of the diameter bisection method. A pair of points is located for which the edge orientations are antiparallel. If such a pair of points lies on an ellipse, the midpoint of the line joining the points will be situated at the centre of the ellipse.

"ovals" such as ellipses with compressed ends. In addition, the basic scheme sometimes gives rise to a number of false identifications even in an image in which only ellipses are present (Fig. 11.2). However, Tsuji and Matsumoto (1978) also proposed a technique by which true ellipses can be distinguished. The basis of the technique is the property of an ellipse that the lengths of perpendicular semidiameters OP, OQ obey the relation:

$$1/OP^2 + 1/OQ^2 = 1/R^2 = \text{constant} \qquad (11.1)$$

To proceed, the set of edge points that contribute to a given peak in parameter space is used to construct a histogram of R values (the latter being obtained from equation (11.1)). If a significant peak is found in this histogram, then there is clear evidence of an ellipse at the specified location in the image. If two or more such peaks are found, then there is evidence of a corresponding number of concentric ellipses in the image. If, however, no such peaks are found, then a rectangle, oval or other symmetrical shape may be present and each of these would need its own identifying test.

It will be clear from the above description that the method relies on there being an appreciable number of pairs of edge points on an ellipse lying at opposite ends of diameters: hence there are strict limits on the amount of the boundary that must be visible (Fig. 11.3). Finally, it should not go unnoticed that the method wastes the signal available from unmatched edge points.

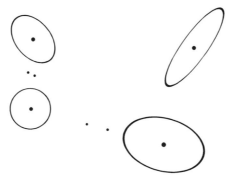

Fig. 11.2   Result of using the basic diameter bisection method. The large dots show true ellipse centres found by the method, while the smaller dots show positions at which false alarms commonly occur. Such false alarms are eliminated by applying the test described in the text.

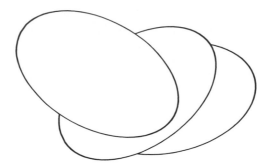

Fig. 11.3 Limitations of the diameter bisection method: of the three ellipses shown, only the lowest one cannot be located by the diameter bisection method.

These considerations have led to a search for further methods of ellipse detection.

11.3   The chord–tangent method

The chord–tangent method was devised by Yuen *et al.* (1988) and makes use of another simple geometric property of the ellipse. Again pairs of edge points are taken in turn, and for each point of the pair, tangents to the ellipse are constructed and found to cross at T; the midpoint of the connecting line is found at M; then the equation of the line TM is found and all points that lie on the portion MK of this line are accumulated in parameter space (Fig. 11.4) (clearly, T and the centre of the ellipse lie on the opposite sides of M). Finally, peak location proceeds as before.

The proof that this method is correct is trivial. Symmetry ensures that the method works for circles, and projective properties then ensure that it also works for ellipses. Under projection, straight lines project into straight lines, midpoints into midpoints, tangents into tangents, and circles into ellipses; in addition, it is always possible to find a viewpoint such that a circle can be projected into a given ellipse.

Unfortunately, this method suffers from significantly increased computation, since so many points have to be accumulated in parameter space. This is obviously the price to be paid for greater applicability. However, computation can be minimized in at least three ways: (i) cutting down the lengths of the lines of votes accumulated in parameter space by taking account of the expected sizes and spacings of ellipses; (ii) not pairing edge points initially if they are too close together or too far apart; and (iii)

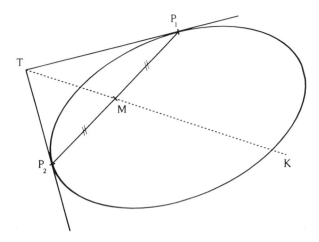

Fig. 11.4 Principle of the chord–tangent method. The tangents at P_1 and P_2 meet at T and the midpoint of P_1P_2 is M. The centre C of the ellipse lies on the line TM produced. Notice that M lies between C and T. Hence the transform for points P_1 and P_2 need only include the portion MK of this line.

eliminating edge points once they have been identified as belonging to a particular ellipse (clearly, the latter two techniques are also applicable to the diameter bisection method).

Section 11.5 explores in more detail whether computation can be saved by reverting to the GHT approach. Meanwhile, it is necessary to consider how ellipse detection can be augmented by suitable procedures for finding ellipse orientation, size and shape.

11.4 Finding the remaining ellipse parameters

Although it has proved possible to locate ellipses in digital images by simple geometric constructions based on well-known properties of the ellipse, a more formal approach is required to determine other ellipse parameters. Accordingly, we write the equation of an ellipse in the form:

$$Ax^2 + 2Hxy + By^2 + 2Gx + 2Fy + C = 0 \qquad (11.2)$$

an ellipse being distinguished from a hyperbola by the additional condition:

$$AB > H^2 \qquad (11.3)$$

This condition guarantees that A can never be zero and that the ellipse equation may without loss of generality be rewritten with $A = 1$; in any case, only ratios between the parameters have physical significance. This leaves five parameters, which can be related to the position of the ellipse, its orientation, and its size and shape (or eccentricity).

Having located the centre of the ellipse, we may select a new origin of coordinates at its centre (x_c, y_c); the equation then takes the form:

$$x'^2 + 2Hx'y' + By'^2 + C' = 0 \tag{11.4}$$

where

$$x' = x - x_c; \qquad y' = y - y_c \tag{11.5}$$

It now remains to fit to equation (11.4) the edge points that gave evidence for the ellipse centre under consideration. The problem will in general be vastly overdetermined. Hence an obvious approach is the method of least squares. Unfortunately, this technique tends to be very sensitive to outlier points and is therefore liable to be inaccurate. An alternative is to employ some form of Hough transformation. Since Hough parameter spaces require storage that increases exponentially with the number of unknown parameters, it is tempting to try to separate the determination of some of the parameters from that of others. For example, we can follow Tsukune and Goto (1983) by differentiating equation (11.4):

$$x' + By' \, dy'/dx' + H(y' + x' \, dy'/dx') = 0 \tag{11.6}$$

Then dy'/dx' can be determined from the local edge orientation at (x', y') and a set of points accumulated in the new (H, B) parameter space. When a peak is eventually located in (H, B) space, the relevant data (a subset of a subset of the original set of edge points) can again be used with equation (11.4) to obtain a histogram of C' values, from which the final parameter for the ellipse can be obtained.

The following formulae must be used to determine the orientation and semi-axes values of an ellipse in terms of H, B and C':

$$\theta = (1/2) \arctan [2H/(1 - B)] \tag{11.7}$$

$$a^2 = \frac{-2C'}{(B + 1) - [(B - 1)^2 + 4H^2]^{1/2}} \tag{11.8}$$

$$b^2 = \frac{-2C'}{(B + 1) + [(B - 1)^2 + 4H^2]^{1/2}} \qquad (11.9)$$

Mathematically, θ is the angle of rotation that diagonalizes the second-order terms in equation (11.4); having performed this diagonalization, the ellipse is then essentially in the standard form $\bar{x}^2/a^2 + \bar{y}^2/b^2 = 1$, so a and b are determined.

It should be noted that the above method finds the five ellipse parameters in *three* stages: first the positional coordinates are obtained, then the orientation, and finally the size and eccentricity*. This three-stage calculation involves less computation but compounds any errors—in addition, edge orientation errors, though low, become a limiting factor. For this reason, Yuen *et al.* (1988) tackled the problem by speeding up the HT procedure itself rather than by avoiding a direct assault on equation (11.4): i.e. they aimed at a fast implementation of a thoroughgoing second stage which finds all the parameters of equation (11.4) in one 3-D parameter space. Their fast adaptive HT procedure has already been described in Chapter 10 and is not covered again here.

At this stage it is clear that reasonably optimal means are available for finding the orientation and semiaxes of an ellipse once its position is known: the weak point in the process appears to be that of finding the ellipse initially. Indeed, the two approaches for achieving this that have been described above are particularly computationally intensive, mainly because they examine all pairs of edge points; hence it seems worth considering again the GHT approach, which locates objects by taking edge points singly, to see whether any gains can be achieved by this means.

11.5 Reducing computational load for the generalized Hough transform method

The complications described in Chapter 10 when the GHT is used to detect anisotropic objects clearly apply to ellipse detection: the various possible orientations of the object lead to the need to employ a large number of planes in parameter space. So far this chapter has avoided considering the GHT for these reasons. However, it will be seen below that by accumulating

---

*Strictly, the eccentricity is $e = (1 - b^2/a^2)^{1/2}$ but in most cases we are more interested in the ratio of semiminor to semimajor axes, b/a.

the votes for all possible orientations in a *single* plane in parameter space, significant savings in computation can frequently be made. Basically, the idea is largely to eliminate the enormous storage requirements of the GHT by using only one instead of 360 planes in parameter space, while at the same time reducing by a large factor the computation involved in the final search for peaks. Such a scheme may have concomitant disadvantages such as the production of spurious peaks, and this aspect will have to be examined carefully.

To achieve these aims, it is necessary to analyse the shape of the point spread function (PSF) to be accumulated for each edge pixel, in the case of an ellipse of unknown orientation. We start by taking a general edge fragment at a position defined by ellipse parameter ψ and deducing the bearing of the centre of the ellipse relative to the local edge normal (Fig. 11.5). Working first in an ellipse-based axes system, for an ellipse with semimajor and semiminor axes a and b, respectively, it is clear that:

$$x = a \cos \psi \tag{11.10}$$

$$y = b \sin \psi \tag{11.11}$$

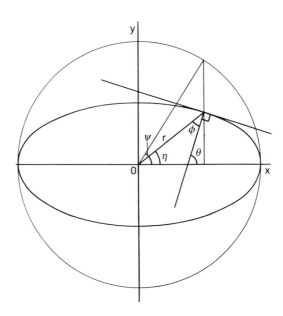

Fig. 11.5   Geometry of an ellipse and its edge normal.

Hence

$$dx/d\psi = -a \sin \psi \qquad (11.12)$$

$$dy/d\psi = b \cos \psi \qquad (11.13)$$

giving

$$dy/dx = -(b/a) \cot \psi \qquad (11.14)$$

Hence the orientation of the edge normal is given by:

$$\tan \theta = (a/b) \tan \psi \qquad (11.15)$$

At this point we wish to deduce the bearing φ of the centre of the ellipse relative to the local edge normal. From Fig. 11.5:

$$\varphi = \theta - \eta \qquad (11.16)$$

where

$$\tan \eta = y/x = (b/a) \tan \psi \qquad (11.17)$$

and

$$\tan \varphi = \tan (\theta - \eta)$$
$$= \frac{\tan \theta - \tan \eta}{1 + \tan \theta \tan \eta} \qquad (11.18)$$

Substituting for $\tan \theta$ and $\tan \eta$, and then rearranging, gives:

$$\tan \varphi = \frac{(a^2 - b^2)}{2ab} \sin 2\psi \qquad (11.19)$$

In addition:

$$r^2 = a^2 \cos^2 \psi + b^2 \sin^2 \psi \qquad (11.20)$$

To obtain the PSF for an ellipse of unknown orientation, we now simplify matters by taking the current edge fragment to be at the origin and orientated with its normal along the u-axis (Fig. 11.6). The PSF is then the locus of all possible positions of the centre of the ellipse. To find its form it is merely required to eliminate ψ between equations (11.19) and (11.20). This is facilitated by re-expressing r^2 in double angles (the significance of double angles lies in the 180° rotation symmetry of an ellipse):

$$r^2 = \frac{a^2 + b^2}{2} + \frac{a^2 - b^2}{2} \cos 2\psi \qquad (11.21)$$

After some manipulation the locus is obtained as:

$$r^4 - r^2(a^2 + b^2) + a^2 b^2 \sec^2 \varphi = 0 \qquad (11.22)$$

which can, in the edge-based coordinate system, also be expressed in the form:

$$v^2 = (a^2 + b^2) - u^2 - a^2 b^2 / u^2 \qquad (11.23)$$

Unfortunately this is not a circle, and is not even symmetrical about axes through its centre. The shape of the PSF for various values of a and b is shown in Fig. 11.7. For *very* highly eccentric ellipses, the PSF is approximated by the arcs of two circles, one of radius a and the other of radius a^2/b. It will be seen that in the other extreme, of ellipses of low eccentricity, the PSF is approximated by an ellipse. In general, however, it is easy to see

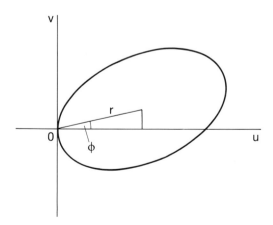

Fig. 11.6 Geometry for finding the PSF for ellipse detection by forming the locus of the centres of ellipses touching a given edge fragment.

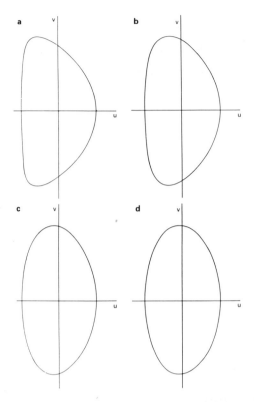

Fig. 11.7 Typical PSF shapes for detection of ellipses with various eccentricities: (a) ellipse with $a/b = 21.0$, $c/d = 1.1$; (b) ellipse with $a/b = 5.0$, $c/d = 1.5$; (c) ellipse with $a/b = 2.0$, $c/d = 3.0$; (d) ellipse with $a/b = 1.4$, $c/d = 6.0$. Notice how the PSF shape approaches a small ellipse of aspect ratio 2.00 as eccentricity tends to zero. The semimajor and semiminor axes of the PSF are $2d$ and d, respectively.

that, since the original ellipse possesses regions along the principal axes where the radii of curvature are approximately constant with values a^2/b and b^2/a, the PSF also possesses regions at either end of one diameter where the radii of curvature are $(a^2/b) - b$ and $a - (b^2/a)$: this makes it clear that the shape is not symmetrical, and also that it should become approximately symmetrical when $a \simeq b$.

Defining two new parameters:

$$c = (a + b)/2 \tag{11.24}$$

$$d = (a - b)/2 \tag{11.25}$$

and then taking an approximation of small d, the locus is obtained in the form:

$$(u + c + d^2/2c)^2/d^2 + v^2/4d^2 = 1 \qquad (11.26)$$

Neglecting the term $d^2/2c$ which provides a small correction to u, the PSF is itself approximately an ellipse of semimajor axis $2d$ and semiminor axis d. Making use of this fact simplifies implementation but it is practicable only for low-eccentricity ellipses where $d \lesssim 0.1c$, so that $a \lesssim 1.2b$. However, in practice, where ellipses are small and the PSF is only a few pixels across, it is a reasonable approximation to insist only that $a < 2b$.

Implementation is simplified since a universal lookup table (ULUT) for ellipse detection can be compiled, which is independent of the size and eccentricity of the ellipse to be detected so long as the eccentricity is not excessive. Since ellipses are common features in industrial and other applications—arising both from elliptical objects and from oblique views of circles—this factor should be an important consideration in many applications. Thus a single ULUT is compiled and stored and it then needs only to be scaled and positioned to produce the PSF in a given instance of ellipse detection.

11.5.1 Practical details

Having constructed a ULUT for ellipse detection, the detection algorithm has to scale it, position it and rotate it so that points can be accumulated in parameter space. *A priori*, it would be imagined that a considerable amount of trigonometric computation is involved in this process. However, it is possible to avoid calculating angles directly (e.g. using the arctan function) by always working with sines and cosines; this is rendered possible partly because such edge orientation operators as the Sobel give two components (g_x, g_y) for the intensity gradient vector (this has now been seen to happen in several line and circle detection schemes—see Chapters 8 and 9). Hence a very considerable amount of computation can be saved. This is clearly demonstrated by the simplicity of the final code for computing a given contribution to the PSF (Table 11.1).

Figure 11.8 shows the result of testing the above scheme on an image of some O-rings lying on a slope of arbitrary direction. The O-rings are found accurately and with a fair degree of robustness, i.e. despite overlapping and partial occlusion (up to 40% in one case). In several cases, incidental transforms from points on the inner edges of the O-rings overlap other

Table 11.1   Algorithm for ellipse detection.

```
for all pixels in image do (* scan over image *)
  begin
    findgx; (* apply edge operator *)
    findgy;
    g := magnitude[gx, gy]; (* lookup magnitude *)
    if g > gthreshold then (* if edge point then *)
      for i := 1 to tablesize do (* run over PSF *)
        begin
          dx := c + d * xtable[i]; (* scale and position *)
          dy := d * ytable[i];
          xc := xedge - round((dx * gx - dy * gy)/g);
          yc := yedge - round((dx * gy + dy * gx)/g);
          (* rotate and position *)
          vote[xc, yc] := vote[xc, yc] + 1; (* accumulate vote *)
        end;
  end; (* end image scan *)
```

In this section of code, it is arbitrarily assumed (i) that the magnitude of the vector (g_x, g_y) is available from a lookup table, and (ii) that all calculations are carried out in floating-point form until the final "rounding" stage where voting positions are being estimated to the nearest pixel.

transforms from points on the outer edges, although only the latter are actually employed usefully here for peak finding. Hence the scheme is able to overcome problems resulting from additional clutter in parameter space.

Figure 11.8 also shows the arrangement of points in parameter space that results from applying the PSF to every edge point on the boundary of an ellipse: the pattern is somewhat clearer in Fig. 11.9. In either case it is seen to contain a high degree of structure. For an ideal transform, there would be no structure apart from the main peak. Looking at the situation in another way, in deducing the locus of points forming the PSF we found all points that could possibly be at the centre of an ellipse when we had no knowledge about its orientation: all points on the PSF *not* falling on the peak at the centre of the ellipse would hopefully be randomly (and fairly uniformly) distributed nearby. However, for a shape such as an ellipse—which has a particular structure and symmetry of its own—this is not the situation. In fact, a full computation of the positions of votes in parameter space yields very many candidate positions, and to find strong structure in the transform it is necessary to look for *envelopes* to the system of PSFs. This is a sufficiently complex task in this case to be left to experiment, but one or two observations on the results are useful and relevant. First, where there is a lot

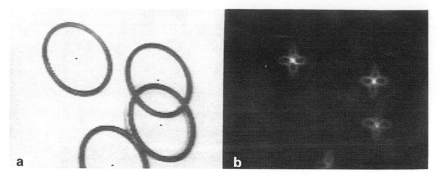

Fig. 11.8   Applying the PSF to detection of tilted circles: (a) off-camera 128×128 image of a set of circular O-rings on a 45° slope of arbitrary direction; (b) transform in parameter space: notice the peculiar shape of the ellipse transform, which is close to a "four-leaf clover" pattern. Part (a) also shows the positions of the centres of the O-rings as located from (b): accuracy is limited by the presence of noise, shadows, clutter and available resolution, to an overall standard deviation of about 0.6 pixels.

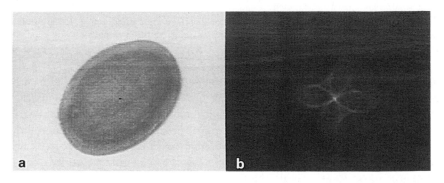

Fig. 11.9   Applying the PSF to detection of elliptical objects: (a) off-camera 128×128 image of an elliptical bar of soap of arbitrary orientation; (b) transform in parameter space: in this case the clover-leaf pattern is better resolved. Accuracy of location is limited partly by distortions in the shape of the object but the peak location procedure results in an overall standard deviation of the order of 0.5 pixels.

of local symmetry in the starting shape (the ellipse)—and specifically where the curvature on the boundary is approximately constant—this leads automatically to strong envelope lines in the transform, as indicated in Fig. 11.10, where the on-axes regions of the original ellipse have essentially been moved radially (and are hence also compressed laterally) until they pass through its centre. These lines seem practically to form a "four-leaved

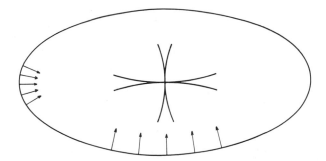

Fig. 11.10 Formation of the ellipse transform by moving the regions of the ellipse boundary lying roughly along the principal axes radially inwards until they pass through the centre of the ellipse. The reason for this peculiar construction lies in the fact that the prominent lines in the transform are the envelopes of all possible PSFs: the necessary coherence is provided by the nearly constant radii of curvature roughly along the directions of the principal axes.

clover" pattern, although the loops become more diffuse on moving away from the centre, in accordance with the envelope idea expressed above.

The important problem here is whether the structure in the transform is misleading, i.e. whether it gives rise to sharply focused peaks at positions other than the main peak. In fact this appears not to happen, the main and most intense structure being the cross pattern at the centre. This is clearly strongest and most focused (both practically and theoretically) near the centre while, in addition, it goes up to double the height quite sharply as the two branches cross at the very centre. The overall conclusion is that the additional structure does no particular harm, although it gives rise to interesting complications if two ellipses overlap so that their centres are quite close. Hence it is more difficult to use the method described here to disentangle two such ellipses than it would be to disentangle two circular objects with a similar degree of overlap. The way around this problem is to check individual high points in the transform by interrogating the original image—see for example, Bolles and Cain (1982), Davies (1987g) and Chapters 12 and 14.

It is interesting that more help would be obtained in interpreting the transform in such cases if the PSFs were regarded not as mere loci of the centres of ellipses of different orientations touching a given edge fragment, but as loci with proper weights, unit weight corresponding to unit boundary distance on the original ellipse. This is too complex a problem to be pursued here, but it may be noted that curvature information immediately leads to the prediction of larger weights for the part of the PSF corresponding to the two regions of the ellipse along the minor axes.

Finally, note that accuracy will be lost if the PSF is not guaranteed to place a vote at the centre of the ellipse for every edge pixel. This means that the PSF should be a connected chain of pixels: hence the size of the LUT must vary with the size of the PSF—or, if a large ULUT is used, a small PSF should be able to skip over a proportion of the entries. In the examples of Figs 11.8 and 11.9, the PSF contained 50 and 100 votes, respectively.

11.6 Comparing the various methods

This section briefly compares the computational loads for the methods of ellipse detection discussed above. To make fair comparisons, we concentrate on ellipse location *per se* and ignore any additional procedures concerned with (i) finding other ellipse parameters, (ii) distinguishing ellipses from other shapes, or (iii) separating concentric ellipses. We start by examining the GHT method and the diameter bisection method.

First, suppose that an $N \times N$ pixel image contains p ellipses of identical size given by the parameters a, b, c, d defined in Section 11.3. By ignoring noise and general background clutter, we shall be favouring the diameter bisection method, as will be seen below. Next, the discussion is simplified by supposing that the computational load resides mainly in the calculation of the positions at which votes should be accumulated in parameter space—the effort involved in locating edge pixels and in locating peaks in parameter space is much smaller.

Under these circumstances the load for the GHT method may be approximated by the product of the number of edge pixels and the number of points per edge pixel that have to be accumulated in parameter space, the latter being equal to the number of points on the PSF. Hence the load is proportional to:

$$L_A \simeq p \times 2\pi c \times 2\pi (2d + d)/2 = 6\pi^2 pcd$$

$$\simeq 60 pcd \qquad\qquad (11.27)$$

where the ellipse has been taken to have relatively low eccentricity so that the PSF itself approximates to an ellipse of semiaxes $2d$ and d.

For the diameter bisection method the actual voting is a minor part of the algorithm—as indeed it is in the GHT method (see the snippet of code listed in Table 11.1). In either case most of the computational load concerns edge orientation calculations or comparisons. Assuming that these calculations and comparisons involve similar inherent effort, it is fair to assess the load for the diameter bisection method as:

$$L_B \simeq \left(\frac{p \times 2\pi c}{2} \right) \simeq (2\pi pc)^2/2$$

$$\simeq 20p^2c^2 \tag{11.28}$$

Hence

$$L_B/L_A \simeq pc/3d \tag{11.29}$$

When a is close to b, as for a circle, $L_A \rightarrow 0$ and then the diameter bisection method becomes a very poor option. However, in many cases it is found that a is close to $2b$, so that c is close to $3d$. The ratio of the loads then becomes:

$$L_B/L_A \simeq p \tag{11.30}$$

It is possible that p will be as low as 1 in some cases: however, such cases are likely to be rare and are offset by applications where there is significant background image clutter and noise, or where all p ellipses have other edge detail giving irrelevant signals that can be considered as a type of self-induced clutter (see the O-ring example of Fig. 11.8). The overall image "clutter factor" would normally be expected to be at least two.

It is also possible that some of the pairs of edge points in the diameter bisection method can be excluded before they are considered, e.g. by giving every edge point a range of interaction related to the size of the ellipses. This would tend to reduce the computational load by a factor of the order of (but not as small as) p. However, the computational overhead required for this would not be negligible.

Overall, there seems no doubt that the GHT method will be significantly faster than the diameter bisection method in most real applications, the diameter bisection method being at a definite disadvantage when image clutter and noise is strong. The only situation where the diameter bisection method might win appears to be where an image contains a minimal number of highly eccentric ellipses. By comparison, the chord–tangent method always requires more computation than the diameter bisection method, since not only does it examine every pair of edge points but also it generates a *line* of votes in parameter space for each pair.

Against these computational limitations must be noted the different characteristics of the methods. First, the diameter bisection method is not particularly discriminating, in that it locates many symmetrical shapes, as remarked earlier. The chord–tangent method is selective for ellipses but is not selective about their size or eccentricity. The GHT method is selective about all of these factors. These types of discriminability, or lack of it, can turn out to be advantageous or disadvantageous, depending on the application: hence we do no more here than draw attention to these different properties. It is also relevant that the diameter bisection method is rather less robust than the other methods: this is so since if one edge point of an antiparallel pair is not detected, then the other point of the pair cannot contribute to detection of the ellipse—a factor that does not apply for the other two methods since they take all edge information into account.

11.7 Concluding remarks

This chapter has examined methods for ellipse detection in digital images. The simplest such scheme is the diameter bisection method, which is unfortunately not very robust because many pairs of diameter ends need to be visible before sufficient signal can be built up. One alternative is the chord–tangent method; this is highly robust yet at the penalty of considerable additional computation. Another alternative is the GHT, but when applied to anisotropic shapes this method requires that one plane in parameter space be used for each discriminable orientation of the shape. However, if a single plane parameter space is used for this purpose, the GHT approach frequently requires less computational effort than the other two methods. This is because they are based on using pairs of edge points to deduce candidate ellipse centres, whereas the GHT takes edge points one at a time. However, the three methods differ markedly in their shape discriminability characteristics and this should be noted in real applications.

In particular, the diameter bisection method locates ellipses of all eccentricities, size and orientations, and also other symmetrical shapes; the chord–tangent method locates ellipses of all eccentricities, sizes and orientations but discriminates against other shapes; and the GHT locates ellipses of all orientations but discriminates against ellipses of the wrong eccentricity and size and against other shapes. However, the GHT can be adapted to detect many other shapes, whereas the diameter bisection method is limited and the chord–tangent method is specific to ellipses.

With the GHT, the avoidance of many planes in parameter space is one aspect of a more general problem. For example, we might need to use many planes in parameter space to allow for variation in object size. It is also possible that we might know the size and orientation of an ellipse but not its eccentricity, many planes being needed to cope with this situation. In Chapter 9 it was shown that circles of many sizes can be sought simultaneously in a single plane in parameter space, by a method related to that described here (namely using a PSF consisting of a radial line in place of a single point): it is clear that this method will carry over to ellipses and other shapes (i.e. replace the $R(\theta)$ in the basic transform in turn by $\alpha R(\theta)$, for all α in a given range, leaving $\varphi(\theta)$ unchanged). Similarly, it is evident from the derivation in Section 11.5 that the GHT approach carries over to other shapes when orientation is unknown. Finally, a similar extension can be made when eccentricity is unknown but size and orientation are known. Thus we hypothesize generally that *parameter space can be compressed by one degree of freedom from the ideal* in order to save computation. We also note that *it is not possible to compress parameter space by more than one degree of freedom from the ideal*: if we tried to do this, virtually any shape would be able to trigger the detector and there would be a great many false alarms. As an example, if we try to detect ellipses of unknown orientation and size in a single plane in parameter space, the PSF becomes an extended blob which is unable to discriminate between a very great variety of shapes (in general, the PSF becomes an *area* which is the result of convolving the two basic PSF *curves* in parameter space: indeed, it is not possible to find two such curves that do not produce an area when convolved). These observations appear both to extend and ultimately to limit the capabilities of the GHT approach for object location, and are by no means limited to ellipse detection.

11.8 Bibliographical and historical notes

This chapter is based on relatively few papers, all of which were written within the last 20 years: first, that of Tsuji and Matsumoto (1978), whose insight led to a totally new ellipse detector; second, that of Tsukune and Goto (1983), which aimed at more accurate determination of ellipse parameters; third, that of Yuen *et al.* (1988), whose detector made good certain shortcomings of the previous two methods; and fourth, that of the author (Davies, 1989a), which developed the GHT idea of Ballard (1981) in order to save computation. The contrasts between these methods are many and intricate, as the chapter has shown. In particular, the idea of saving dimensionality in the implementation of the GHT appears also in a general circle detector (Davies, 1988b). Finally, the necessity for a multistage approach to determination of ellipse parameters now seems proven, although somewhat surprisingly the optimum number of such stages appears to be just two.

The fact that as late as 1989 basic new ellipse detection schemes were being developed says something about the science of image analysis: even today the toolbox of algorithms is incomplete, and the science of how to choose between items in the toolbox, or how, *systematically*, to develop new items for the toolbox, is immature. Further, although all the parameters for specification of such a toolbox *may* be known, knowledge about the possible tradeoffs between them is still at a rather primitive stage. Nevertheless, there are moves to greater degrees of robustness with real data by explicit inclusion of errors and error propagation (Ellis *et al.*, 1992). In addition, increased attention has been given to the verification stage of the Hough approach (Ser and Siu, 1995).

Finally, some work has been carried out on the detection of superellipses, which are intermediate in shape between ellipses and rectangles, though the technique used (Rosin and West, 1995) was that of segmentation trees rather than Hough transforms (nonspecific detection of superellipses can of course be achieved by the diameter bisection method—see Section 11.2).

11.9 Problems

1. Derive equations (11.7)–(11.9) by the method suggested in the text.
2. Determine which of the methods described in this chapter will detect (a) hyperbolas, (b) curves of the form $Ax^3 + By^3 = 1$, (c) curves of the form $Ax^4 + Bx + Cy^4 = 1$.
3. Prove equation (11.1) for an ellipse. *Hint*: Write the coordinates of P and Q in suitable parametric forms, and then use the fact that $OP \perp OQ$ to eliminate one of the parameters from the left-hand side of the equation.

12

Hole Detection

12.1 Introduction

The last few chapters have been particularly concerned with the direct location of objects from their shapes. Where objects have simple shapes this is a viable approach but in more complex situations they may have to be located from their features. In the case of larger objects and assemblies these features may be components of the very shapes discussed so far. However, there are two important types of feature that have not so far been covered: these are small holes and corners. Corners are dealt with in the following chapter. This chapter considers the location of small holes: not only can these be used to locate objects and to identify them positively, but also they are important in the task of accurately measuring object dimensions. A later chapter considers carefully how the presence of an object may be deduced once its hole features (or a subset of them) have been found.

We start by considering the template matching approach, which seems *a priori* the most obvious means of searching an image for small holes. Later sections examine alternative means of carrying out this task. It should be noted that although the chapter is ostensibly about hole location, it is also about *methods* of searching for such features: in this sense, holes are used as convenient vehicles which allow the relevant principles to be aired.

12.2 The template matching approach

When considering features such as small holes it is convenient to have a mental model with which to work. We can start by imagining a round, relatively dark region some 1–2 pixels in diameter surrounded by a lighter region which is part of some object. The obvious means of detecting such a feature is by template matching. This involves applying a Laplacian type of operator, and in a 3×3 neighbourhood this may take the form of one of the following convolution masks:

$$\begin{bmatrix} -1 & -1 & -1 \\ -1 & 8 & -1 \\ -1 & -1 & -1 \end{bmatrix} \qquad \begin{bmatrix} -2 & -3 & -2 \\ -3 & 20 & -3 \\ -2 & -3 & -2 \end{bmatrix}$$

Having applied the convolution, the resulting image is thresholded to indicate regions where holes might be present: note that if dark holes are regions of negative contrast which are to be located after applying one of the above masks, positions of high *negative* intensity must be sought. It should be noted in passing that the coefficients in each of the above masks sum to zero so that they are insensitive to varying levels of background illumination.

If the holes in question are larger than 1–2 pixels in diameter, the above masks will prove inadequate. Indeed, they will indicate the edges of the holes rather than their centres, and they will also indicate the edges of most objects of similar contrast; in any case, the edges are located with low efficiency since the signal originates from only one-half of the mask. Clearly, larger masks are required if larger holes are to be detected efficiently at their centres. For very large holes, the masks that are required are so large that the method becomes impracticable. For example, holes 15 pixels in diameter require masks of size around 22×22 pixels; remembering that these have to be applied at all positions in an image, and assuming that the latter has size 256×256 pixels, over 30 million operations are involved. Hence a better method must be found for locating larger holes. Similar comments clearly apply if any other type of feature is being sought: indeed, in many cases the problem is then worse, since most such features are not circular, and templates of many different orientations have to be applied in order to locate them effectively.

This problem has been tackled by many workers over the past decade or so. Indeed, several powerful general techniques have been developed for this purpose. They include "ordered search" (Nagel and Rosenfeld, 1972), and "coarse–fine" (Rosenfeld and VanderBrug, 1977) and "two-stage" (VanderBrug and Rosenfeld, 1977) template matching; there has also been considerable interest in the basic (Stockman and Agrawala, 1977) and generalized (Ballard, 1981) HT approaches because of their higher inherent efficiency. The next few sections study the lateral histogram technique, which offers significant advantages in certain applications—particularly for corner and (as described below) round object detection. A later section appraises the situation, taking account of the odd intensity profiles of some holes, and the fact that large holes may be located by the circle detection procedures discussed in Chapter 9.

12.3 The lateral histogram technique

The lateral histogram technique involves projecting an image on two or more axes by summing pixel intensities (see Fig. 12.1) and using the resulting histograms to identify objects in the image. It has been used previously (Pavlidis, 1968; Ogawa and Taniguchi, 1979) as an aid to pattern recognition in binary images, and has recently been applied to the problem of corner detection in grey-scale images (Wu and Rosenfeld, 1983). Here we consider its application to the detection of small round objects and holes, again in grey-scale images.

The basic technique is inherently attractive, in that it requires only about $2N^2$ pixel operations in an $N \times N$ image. However, it turns out that if several similar objects are present in an image, the histograms can be interpreted in a number of ways. Making the method work in practical situations therefore involves finding efficient procedures for resolving ambiguities: it is thus necessary to study the additional computational load these procedures impose. This is done in Sections 12.4–12.6 below.

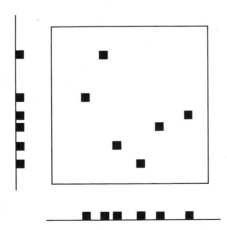

Fig. 12.1 The basis of the lateral intensity histogram technique. It is clear that, taken on their own, lateral histograms do not predict object locations uniquely.

12.4   The removal of ambiguities in the lateral histogram technique

To resolve the ambiguities that arise with the lateral histogram technique, an obvious strategy is to list all candidate object sites (which arise at the intersections of horizontal and vertical lines corresponding to "hits" in the lateral histograms) and to examine them in turn to check rigorously for the presence of an object (Wu and Rosenfeld, 1983). Basically, the lateral histograms are then used to provide rapid screening of the picture data: all relevant locations are checked subsequently (e.g. by normal template matching) to determine their significance. As a result of this procedure, any artefacts due to random combinations of signals are in the end eliminated. The computational implications of this procedure are assessed below. Meanwhile, it is worth noting that if the initial level of ambiguity in a particular sequence of images turns out to be high, then it is often possible to rectify the situation by computing the lateral histograms of *subimages*. An alternative strategy is studied in Section 12.6.

12.4.1   Computational implications of the need to check for ambiguities

This section shows how many basic computations are required in order to complete the process of object detection and identification. For simplicity, it is assumed that images of size $N \times N$ pixels contain p identical objects of size approximately $n \times n$.

First, the number of operations required to obtain each of the lateral intensity histograms is of the order of N^2; the number of operations required to perform template matching in each of the two histograms is Nn; and the number of operations required to confirm the presence of an object at each possible object site is around n^2. Thus the total number of operations for the complete algorithm is:

$$R_{\text{lat}} = 2aN^2 + 2bNn + cp^2n^2 \tag{12.1}$$

For convenience we shall take $a \simeq b \simeq c$, since this should be a reasonable approximation and sufficiently accurate to give a useful indication of the efficiency of the method.

A typical practical case is that of locating holes of diameter 5 pixels in an image of 256×256 pixels. If there are 20 holes and a template of size 7×7

pixels is required to locate a hole of diameter 5 pixels, then the total number of operations is:

$$R_{lat} = a(2 \times 256^2 + 2 \times 256 \times 7 + 20^2 \times 7^2) \tag{12.2}$$

$$= 2 \times 256 \times 301.3a = 256^2 \times 2.35a$$

This result should be compared with the result for straight template matching, which is:

$$R_{tem} = cN^2n^2 \tag{12.3}$$

Since we have already decided to make the approximation $c \approx a$, in the given example it is found that:

$$R_{tem} = 256^2 \times 7^2a = 256^2 \times 49a \tag{12.4}$$

Here the speedup factor is more than 20, so the saving in computational effort is substantial. It is interesting that the checking and 1-D matching procedures reduce the overall saving achieved by the lateral histogram approach very little. The gain in this instance is effectively $n^2/2$. It becomes even more substantial for larger objects (e.g. around 30 for 20 holes of diameter 7 pixels, 50 for 10 holes of diameter 9 pixels, and 400 for 3 holes of diameter 30 pixels).

It is now useful to examine the relative sizes of the terms in equation (12.1). First, the second term is invariably smaller than the first and can essentially be ignored. However, the third term can become so large that it can dominate over the first: in such a situation it is even possible that the lateral histogram method would require more computational effort than straightforward template matching. Suppose that the computational load of the lateral histogram technique increases to equal that of pure template matching. Then

$$2N^2 + 2Nn + p^2n^2 = N^2n^2 \tag{12.5}$$

Hence

$$2/n^2 + 2/Nn + p^2/N^2 = 1 \tag{12.6}$$

Since $N \gg 1$ and $n^2 \gg 1$, it follows that $p \approx N$. This shows that the lateral histogram technique does not save computation, and using it is not justifiable, if the number of objects approaches N. The danger sign is perhaps when the third term is just equal to the first. This occurs when

$$p = \sqrt{2N/n} \tag{12.7}$$

This value of p can be interpreted as the point at which object projections start to overlap and interpretation consequently becomes more difficult. It therefore seems that use of the method should be restricted to cases when $p \leqslant N/n$, so that (i) interpretation is straightforward, and (ii) a significant saving in computational effort can result. These ideas give a precise indication of when it is profitable to use the subimage method mentioned in Section 12.4: an image should be broken into subimages whenever the image data make $N < np$. This is discussed in more detail below.

   Finally, the lateral histogram method incorporates feedback by virtue of its two-stage mode of operation. This permits it to adapt optimally to variations in the image data, so that it is both rapid and accurate in its task of object location. However, image variability needs to be monitored in order to confirm that operation can take place within the limits indicated above.

12.4.2   Further detail of the subimage method

If an image is broken down into subimages, then the number of objects in each subimage should vary approximately as the square of subimage size. Suppose that the size of a subimage is $\tilde{N} \times \tilde{N}$ and that the number of objects it contains is p. Defining

$$a = N/\tilde{N} \tag{12.8}$$

we expect

$$\tilde{p} = p/a^2 \tag{12.9}$$

The equation corresponding to equation (12.1) is then:

$$\tilde{R}_{\text{lat}} = a(2\tilde{N}^2 + 2\tilde{N}n + \tilde{p}^2 n^2) \tag{12.10}$$

Multiplying by a^2 to obtain a new equation for the whole image, gives:

$$R'_{\text{lat}} = a(2N^2 + 2aNn + p^2 n^2/a^2) \tag{12.11}$$

Thus the second term is increased while the third is reduced substantially.

Differentiating equation (12.11) with respect to α gives:

$$\frac{dR'_{\text{lat}}}{d\alpha} = a(2Nn - 2p^2n^2/\alpha^3) \qquad (12.12)$$

This is positive at $\alpha = 1$ for p in the range $p < \sqrt{(N/n)}$. Hence if p is small (typically $p < 4$), the subimage method incurs additional computation and should not be used. However, the minimum of R_{lat} as α varies is shallow and there is little to be gained from using the method until $p > N/n$, as indicated in Section 12.4.1. Under these circumstances, α should be adjusted so that:

$$\tilde{N} \simeq n\bar{p} \qquad (12.13)$$

Then it is found that:

$$\alpha \simeq np/N \qquad (12.14)$$

and hence:

$$\tilde{N} \simeq N^2/np \qquad (12.15)$$

Substituting for α in equation (12.11) now gives

$$R'_{\text{lat}} \simeq a(2N^2 + 2pn^2 + N^2) \qquad (12.16)$$

The significance of this result is that the third term is limited by the subimage method to one-half the first term and is no longer able to exceed it by a large factor. The second term remains relatively unimportant since it varies as p instead of p^2. This means that there is rarely likely to be any penalty from using the subimage method. However, there is a practical limitation, which is now considered.

The practical limitation involves the problem of "edge effects" between subimages. If an object falls on the border between two subimages, then it may escape detection. To overcome this problem it is necessary to permit subimages to overlap—a factor that will plainly reduce efficiency. The amount by which subimages will have to overlap is equal to the diameter of the objects being detected, since when an object starts to move out of one subimage it must already be included totally in an adjacent subimage. This means that N must be replaced by $\tilde{N} + n$ in equation (12.10) (N must now be defined by equation (12.8) and is no longer the subimage size). As a consequence, the additional terms $4N\alpha n + 4\alpha^2 n^2$ appear in equation (12.11). These dominate over the first term if $\alpha \gtrsim N/2n$, i.e. $n \gtrsim \tilde{N}/2$.

However, there is one way of reducing this problem and getting back substantially to the level indicated by equation (12.16), namely, by hierarchical addition of rows and columns of pixel intensities while computing the lateral histograms. For example, sub-subimages of some fraction of the size of the subimages may be computed first: these may then be added in appropriate groups. If the size of the subimages is reasonably large $(\tilde{N} + n \gg 1)$, the final set of summations consumes relatively little computational effort and equation (12.16) is a realistic approximation.

This subsection has shown that when $p > N/n$, the amount of computational effort needed to locate objects can be kept within reasonable bounds (of order $3N^2$) by use of the subimage method. These bounds are remarkably close to those (of order $2N^2$) which apply when $p \ll N/n$. Bounds of size N^2 would of course be ideal. It is not expected that significantly different results would apply if the coefficients a, b, c were not exactly equal as assumed above.

12.5   Application of the lateral histogram technique for object location

We now consider what types of image data can be analysed conveniently with the aid of lateral histograms. An important feature of the approach is that it is not limited to use in binary images, since all the procedures that are employed (e.g. template matching) are equally applicable to grey-scale images. Another feature of the approach is that objects are identified from their 1-D templates. This means that they may be difficult to locate in the lateral histograms if these are very orientation-dependent. Thus the method is well adapted to the location of round objects. In fact, a great many products, from foodstuffs to machined parts, are almost exactly circular (Davies, 1984c), so the approach is frequently valuable. In addition, very many components contain accurately machined circular holes and these can be used to detect the products in which they appear (Chapter 14). It is significant that the method is able to locate objects whose intensity profiles do not possess well-defined edges (see Fig. 12.2): in such situations the HT method cannot easily be applied, since its use requires rapid location and orientation of edges.

Thus the lateral histogram approach is valuable for the efficient location of (i) small round objects or holes, and (ii) objects that are fairly large, round and somewhat fuzzy or low in contrast. If the number of objects is not excessive, they should be located rapidly and reliably. The author has tested

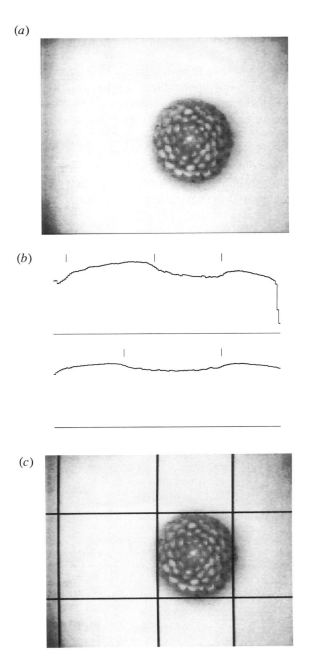

Fig. 12.2 (a) Image of a decorative candle which appears as a fairly diffuse object with a fuzzy edge; (b) lateral histograms; (c) edges of the object as detected by differentiating the histograms and locating peaks: the false peak resulting from strongly varying backgrounds is easily eliminated.

the method on a variety of data from foodstuffs to mechanical parts, and has confirmed its value in these applications for images of up to 128×128 pixels. For images of these sizes the subimage method did not have to be used but it seems clear that for images of 512×512 or larger, the subimage method will be required.

Figure 12.3 shows the histograms obtained for images of metal bars containing holes (see Fig. 12.4). All the holes were located by the method, despite their high variability and noise content and despite the rather rudimentary templates used to analyse the histograms (see Fig. 12.5). One or

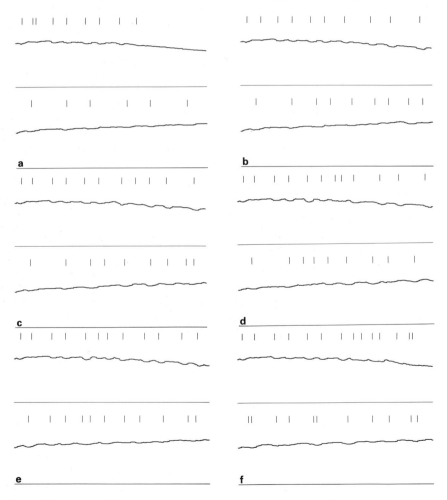

Fig. 12.3 Lateral histograms for images of metal bars containing holes for original versions of images in Fig. 12.4.

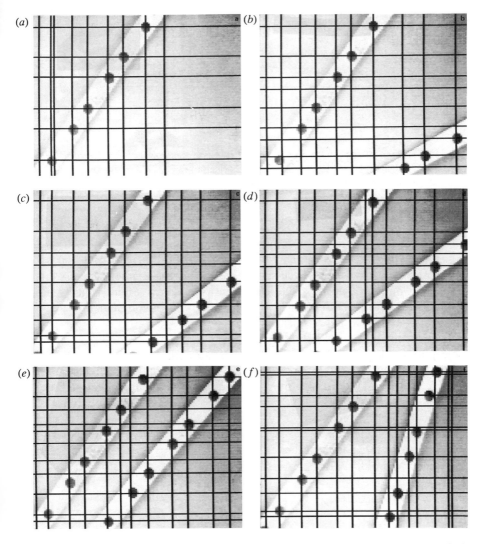

Fig. 12.4 Result of applying lateral histograms shown in Fig. 12.3 to find candidate hole positions.

two comments should be made about the initial application of lateral histograms to find candidate hole locations (see Fig. 12.4). First, occasional false positives arose from noise or strong corners (Fig. 12.4(a–c,f)). Second, on some occasions lateral juxtaposition of holes caused "pulling" and merging of estimated candidate positions (Fig. 12.4(b–e)). Third, in two cases holes were located in only one histogram (Fig. 12.4(e,f)): more will be said about this in Section 12.5.1 below. These deficiencies are easily corrected in the final template matching procedure.

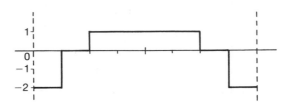

Fig. 12.5   Form of the 1-D templates used to locate the hole profiles in the lateral histograms of Fig. 12.3. Greater sophistication was not necessary in this instance.

12.5.1   Limitations of the approach

When applying lateral histograms for object detection in industrial images, the most serious drawback is that some objects may mask others. Ideally this problem would not arise, since the intensity patterns of all objects are incorporated additively in the histograms, and any suspicion of an object at a particular location would be checked in detail.

   In practice the most likely masking error arises when the edges of an object are masked by other objects on both sides (if only one of an object's edges is so masked, it will almost certainly be detected). This effect is seen in Fig. 12.4(e) and 12.4(f), although in the latter case one of the masking objects is a strong corner in the image. There is quite a low probability of this occurring. The probability may be estimated by considering the chance of throwing p balls into N boxes, arranged in line, so that three balls fall into *any* set of three adjacent boxes. The probability of missing an object in this way is approximately:

$$P = \left[N \binom{p}{3} (1/N)^3 \right]^2 \tag{12.17}$$

a function that varies extremely rapidly with p. This probability equals 0.01 when

$$\binom{p}{3} (1/N)^2 = 0.1$$

i.e.

$$p(p-1)(p-2) = 0.6N^2$$

Hence

$$p \simeq 0.6^{1/3} N^{2/3} \qquad (12.18)$$

For $N = 128$ this results in $p \simeq 21$.

It is instructive to estimate the effect of noise by including an uncertainty v of the order of 2 pixels in object edge positions. The result of this is effectively to reduce resolution, so that N has to be replaced by N/v in equation (12.17). Then P is 0.01 when

$$p \simeq 0.6^{1/3} (N/v)^{2/3} \qquad (12.19)$$

For $N = 128$ it is now found that $p \simeq 13$. These values are generally in accordance with observations for industrial images (Davies, 1987g). Basically, the above calculation reveals that there is an upper limit on the density of objects that can be located reliably by this method.

On some occasions the predictions of the above statistical model do not match what is found in practice. This is because of correlations between holes on machined parts which permit alignments to occur more often than predicted by chance alone. There appear to be two ways around this difficulty: one is to accept that some holes or other features will be missed and to use point pattern matching techniques (Chapter 14) to locate and identify complex objects; the other is to use a sequential recognition technique in which those features which are positively identified are subtracted from the lateral histograms, thereby giving an increased chance of locating the remaining features.

As already seen from Fig. 12.4, false positives can also arise with the lateral histogram technique: however, since the method involves use of feedback, false positives are rigorously checked in every case and the final incidence of false predictions can be no worse than for normal template matching. Clearly, the number of false positives is strongly related to the amount of noise present in the image: further discussion of this is curtailed, as the effect is so data-dependent.

12.6 A strategy based on applying the histograms in turn

This section studies an alternative strategy that has been used for locating peaks in Hough parameter space (IBM, 1986). Here the first histogram is computed and analysed before calculating the second histogram. In addition, the second histogram is not calculated in full—instead, a number of subhistograms are calculated in strips of the image where objects are known (from the first histogram) to lie. Clearly, this strategy has some similarity to the subimage method, although it is aimed at saving computational effort by calculating and analysing only those parts of the second histogram that are useful for finding objects. However, it is not obvious that the method saves computation when a large number of objects are present, since the strips analysed by the second set of histograms may overlap and as a result even more computation could be required than in the original method of Section 12.4.

We now estimate the number of operations required in this alternative ("strip") strategy. As before, obtaining the first histogram requires some N^2 operations, and analysing it requires Nn operations. Calculating the set of subhistograms requires Nn operations per object, and analysing them requires a further Nn operations per object. Thus the total number of operations is:

$$R_{str} = aN^2 + bNn + apNn + bpNn$$
$$= a(N^2 + Nn + 2pNn) \tag{12.20}$$

where we have again taken $b \approx a$.

We now wish to find how R_{str} compares with R_{lat}. Writing

$$S = R_{lat} - R_{str} = a(N^2 + Nn - 2pNn + p^2n^2) \tag{12.21}$$

we try finding where $S = 0$. This gives a quadratic in p which has no real roots. However, for $p = 0$:

$$R_{lat} = a(2N^2 + 2Nn) \tag{12.22}$$

and

$$R_{str} = a(N^2 + Nn) \tag{12.23}$$

Hence the strip strategy always requires less computation.

To find when the methods are closest, we differentiate S:

$$\frac{dS}{dp} = a(2pn^2 - 2Nn) \tag{12.24}$$

This is zero for $p = N/n$, and then:

$$R_{lat} = a(3N^2 + 2Nn) \tag{12.25}$$
$$R_{str} = a(3N^2 + Nn) \tag{12.26}$$

so the amounts of computation are similar. This is highlighted by substituting the values used in Section 12.4.1, which give:

$$R_{str} = a(256^2 + 256 \times 7 + 2 \times 20 \times 256 \times 7)$$
$$= 256 \times 543a = 256^2 \times 2.12a \tag{12.27}$$

which is only marginally smaller than indicated by equation (12.2).

The strip strategy is now compared with the subimage method. Applying equation (12.16), the difference is found as:

$$S' = R'_{lat} - R_{str} = a(2N^2 + 2pn^2 - Nn - 2pNn) \tag{12.28}$$

In this case the equation $S' = 0$ has one real root:

$$p = (N/n)(N - n/2)/(N - n) \simeq N/n \tag{12.29}$$

Thus the strip strategy is optimal for $p \lesssim N/n$, and the subimage method is optimal for $p \gtrsim N/n$. This means that the strip strategy gives the better solution for the example quoted earlier. However, as indicated above, there is very little gain in this case relative to the original strategy. In addition, it is to be expected that the original strategy will be slightly more robust, since there is less chance of an object being masked by noise if two complete lateral histograms are analysed.

12.7 Appraisal of the hole detection problem

So far, much of this chapter has been given over to a study of the lateral histogram technique. This is justifiable since it is a highly efficient means of locating holes, although this is not its only application. We return here to a more careful appraisal of methods for hole detection.

At this stage an obvious question is whether the usual methods for circular object detection can be applied to hole detection. In principle this is undoubtedly practicable, in many cases the only change arising because holes have negative contrast. This means that if the HT technique is used, votes will have to be accumulated on moving a distance $-R$ along the direction of the local edge normal.

However, small holes are rarely found to have an intensity pattern which matches that of the model introduced in Section 12.2. Figure 12.6 shows some of the possibilities. The basic problem is that it is virtually impossible

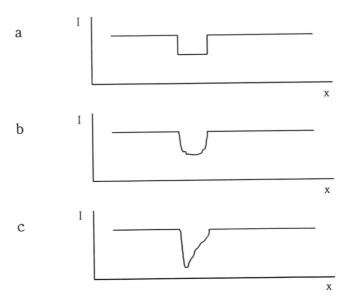

Fig. 12.6 (a) Cross-sectional intensity pattern of an idealized hole; (b) a more commonly found case; (c) the effect of the hole being lit obliquely, a significant part of the hole being in shadow.

to make lighting totally diffuse: on the contrary, it usually has significant directionality, so there will be a region of shadow on one side of the hole. Assuming that the holes are infinitely deep, but are narrow, the region of shadow may well stretch half-way across the width of each hole. If the holes are little more than shallow depressions in the surface, there may be no shadows but luminous intensity will be graded across each hole. Finally, complications arise because holes are often only a few pixels across, so the arrangements of pixel intensities are unlikely to be smoothly varying; in any case, noise can have a significant effect on the appearance of a hole.

To summarize, there are four factors which can affect the form of a hole in a digitized image: (i) isotropic variations in intensity over the hole region, (ii) anisotropic variations in intensity, (iii) digitization effects (both spatial and grey-scale), and (iv) noise. Since it is unlikely to be possible to make a perfect working model of a hole, even in respect of (i), methods for detecting holes need to include a significant amount of averaging to overcome these effects. For example, when an HT circle detector is applied in such cases, for diameters of 10 pixels or less, edge orientation accuracy and edge position accuracy are frequently poor. In addition, the boundary has very few edge points, so little averaging can occur; hence peaks in parameter space are ill defined and the accuracy of centre location is poor (as a percentage measurement it is very poor, while absolutely it may be little better than ± 1 pixel). In contrast, the lateral histogram method incorporates significant averaging and usually arrives at reasonable estimates of hole position. The remaining problem is that factor (ii) can result in a systematic bias in estimates of hole position, although it is possible that this can be measured and corrected for.

Finally, it should be remarked that the HT approach requires use of an edge detector. If a Sobel operator is employed, the total amount of computation involved will be at least $2 \times 8 \times N^2 = 16N^2$, many times that required by the lateral histogram method.

Despite these arguments, there must be some point at which a hole becomes so large that it is better considered as a circle. This point depends particularly on image resolution but it also varies with the lighting scheme and with the detailed characteristics (shape, smoothness, etc.) of the hole, as indicated above. Hence the optimum solution is highly data-dependent. Suffice it to say that if holes are smaller than about 16 pixels in diameter they are probably best detected by the lateral histogram method, and if larger than this they are probably best treated as normal circular objects and detected by an HT method.

Finally, we consider briefly the algorithm of Kelley and Gouin (1984). This works by finding the midpoints of horizontal and vertical chords of dark areas in an image and locating holes at the centre of the resulting cross. While the method apparently worked adequately on Kelley and Gouin's data for holes around 12 pixels in diameter, it is liable to fail on less ideal images (e.g. lower contrast or where there are interfering features) since ultimately it relies on the presence of the four edge pixels at the ends of the horizontal and vertical diameters of the holes. Overall, there seems to be no reason to abandon the conclusion that an effective hole-finder needs to incorporate significant averaging, both for general robustness and for accuracy of centre location.

12.8 Concluding remarks

This chapter has aimed to show some of the problems involved in detecting small holes. For all but the smallest holes, template matching is likely to be too computationally intensive and the lateral histogram technique offers a useful solution: it is able to make significant savings in computation since it reduces the dimensionality of the basic search task from two dimensions to one. However, quite large holes are generally best treated as circular objects and detected by an HT technique.

Yet again problems of lighting have arisen, and with them the task of problem specification. Without an accurate problem specification, diagnosis is liable to be inaccurate and algorithm design suboptimal. However, this chapter has been able to show the parameters that are important in the task of hole detection.

12.9 Bibliographical and historical notes

Hole detection is inherently a search problem that would normally involve template matching (correlation). Where template masks are large, some form of speedup is vital and basic methods for achieving this were devised in the 1970s (Nagel and Rosenfeld, 1972; Rosenfeld and VanderBrug, 1977; VanderBrug and Rosenfeld, 1977). At much the same time it was discovered (Stockman and Agrawala, 1977) that the HT technique (Hough, 1962; Ballard, 1981) is a form of template matching and offers significant savings in computation if edge orientation information is taken into account (Kimme et al., 1975). Lateral intensity histograms were developed for character recognition and later used for corner detection (Pavlidis, 1968; Ogawa and Taniguchi, 1979; Wu and Rosenfeld, 1983). Subsequently the author used them for hole detection (Davies, 1985a, 1987g) and Sections 12.3–12.6 embody this work: see also Davies and Barker (1990). Perhaps oddly, the literature gives far greater attention to corner detection than to hole detection (see Chapter 13).

12.10 Problems

1. Calculate the bias in the estimated position of a hole if it is illuminated from light at an angle 30° to the vertical and has depth three times its width.
2. Find a formula giving the accuracy of a Hough-based hole detector as a function of hole diameter. What other information would be needed to choose between this and other hole detection schemes?
3. Estimate how the basic accuracy of a lateral histogram hole detector varies with the size of the image being used. What limits the accuracy of hole detection with the lateral histogram technique? Is there any possibility of improving it?

13

Polygon and Corner Detection

13.1 Introduction

This chapter is concerned with the efficient detection of polygons and corners. Polygons are ubiquitous amongst industrial objects and components, and it is of vital importance to have suitable means for detecting them. For example, many manufactured objects are square, rectangular or hexagonal, and many more have components which possess these shapes. Clearly, polygons can be detected by the two-stage process of first finding straight lines and then deducing the presence of the shapes. However, in this chapter we first focus on the *direct* detection of polygonal shapes. There are two reasons for this: first, the algorithms for direct detection are often particularly simple; and second, direct detection is inherently more sensitive (in a signal-to-noise sense), so there is less chance of objects being missed in a jumble of other shapes. The basic method used is the GHT, and it will be seen how GHT storage requirements and search time can be minimized by exploiting symmetry and other means.

After exploring the possibilities for direct detection of polygons, we move on to more indirect detection, by concentrating on corner features. While this approach to object detection may be somewhat less sensitive, it is more general, in that it permits objects with curved boundaries to be detected if

Portions of this chapter are reprinted, with permission, from *Proceedings of the 8th International Conference on Pattern Recognition, Paris (27–31 October 1986)*, pp. 495–497 (Davies, 1986d).

the latter are interspersed with corners. It also has the advantage of permitting shape distortions to be taken into account if suitable algorithms to infer the presence of objects can be devised (see Chapters 14–16).

13.2 The generalized Hough transform

It was seen in an earlier chapter that the GHT is a form of spatial matched filter and is therefore ideal in that it gives optimal detection sensitivity. Unfortunately, when the precise orientation of an object is unknown, votes have to be accumulated in a number of planes in parameter space, each plane corresponding to one possible orientation of the object—a process involving considerable computation. In the previous chapter, it was shown that GHT storage and computational effort can be reduced, in the case of ellipse detection, by employing a single plane in parameter space. This approach appeared to be a useful one which should be worth trying in other cases, including that of polygon detection. To proceed, we first examine straight line detection by the GHT.

13.2.1 Straight edge detection

We start by taking a straight edge as an object in its own right, and using the GHT to locate it; in this case the PSF of a given edge pixel is simply a line of points of the same orientation and of length equal to that of the (idealized) straight edge (see Chapter 10): the transform is the autocorrelation function of the shape, and in this case has a simple triangular profile (Fig. 10.10). The peak of this profile is the position of the localization point (in this case the midpoint M of the line). Thus the straight edge can be fully localized in two dimensions, and this is achieved while retaining the full sensitivity of a spatial matched filter.

We now need to see how best to incorporate this line parametrization scheme into a full object detection algorithm. Basically, the solution is simple—by the addition of a suitable vector to move the localization point from M (the midpoint of the line) to the general point L (Chapter 10).

Thus the GHT approach is a natural one when searching images for polygonal shapes such as squares, rectangles, hexagons and so on, since whole objects can be detected directly in one parameter space: however, note that use of the HT (θ, ρ) parametrization would require identification of lines in one parameter space followed by further analysis to identify whole objects. Furthermore, in situations of low signal-to-noise ratio it is possible that some such objects would not be visible using a feature-based scheme when they *would* be found by a whole object matched filter or its GHT equivalent (Chapter 10).

13.3 Application to the detection of regular polygons

When applied to an object such as a square whose boundary consists of straight edges, the parametrization scheme outlined above takes the following form: construct a transform in parameter space by moving inwards from every boundary pixel by a distance equal to half the side S of the square and accumulate a line of votes of length S at each location, parallel to the local edge orientation; then the pixel having the highest accumulation of votes is the centre of the square. This technique was used to produce Fig. 13.1.

It is clear that the GHT can be used in this way to detect all types of regular polygon, since in all such cases the centre of the object can be chosen as the localization point L: this means that to determine the positions at which lines of votes will be accumulated in parameter space, it is necessary merely to specify a displacement by the same distance along each edge normal. Taking a narrow rectangle as the lowest-order regular polygon (i.e. having two sides), and the circle as the highest-order polygon with an infinite number of short sides, it is apparent from Fig. 13.2 how the transform changes with the number of sides.

In the above scheme, notice that votes can be accumulated in parameter space using locally available edge orientation information and that the method is not muddled by changes in object orientation. This is true only for

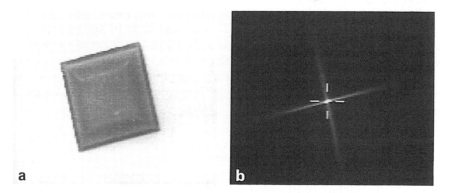

Fig. 13.1 (a) Original off-camera grey-scale image of a square; (b) derived transform.

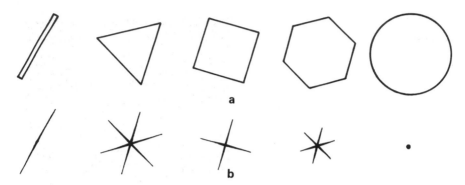

Fig. 13.2 (a) Set of polygons; (b) associated transforms. The widths of the lines in (b) represent transform strength: ideally, the transform lines would have zero width.

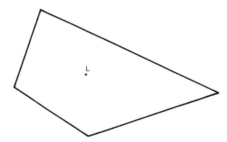

Fig. 13.3 Geometry of a typical irregular polygon: the localization point L is at different distances from each of the sides of the polygon.

regular polygons, where the sides are *equidistant* from the localization point L. For more general shapes, including the quadrilateral in Fig. 13.3, the standard method of applying the GHT is to have a set of voting planes in parameter space, each corresponding to a possible orientation of the object. To find general shapes and orientate them within 1°, 360 parameter planes appear to be required: this would incur a serious storage and search problem. We show next how this problem may be minimized in the case of irregular polygons.

13.4 The case of an arbitrary triangle

The case of an arbitrary triangle can be tackled by making use of the well-known property of the triangle that it possesses a unique inscribed circle. Thus a localization point can be defined which is equidistant from the three

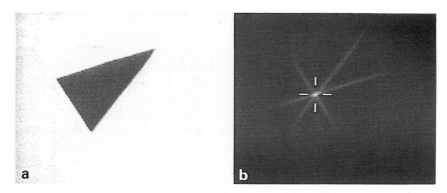

Fig. 13.4 (a) Off-camera image of a triangle; (b) transform in a single parameter plane, making use of a localization point L which is equidistant from the three sides. An optimized transform requires four parameter planes (see Fig. 13.7). © *1986 IEEE.*

sides. This means that the transform of any triangle can be found by moving into the triangle a specified distance from any side and accumulating the results in a single parameter plane (see Fig. 13.4). Clearly, such a parametrization can cope with random orientations of a triangle of given size and shape. However, this result is not wholly optimal since the separate transforms for each of the three sides do not peak at exactly the same point (except for an equilateral triangle): neither should the lengths of the edge pixel transforms be the same. Each of these factors results in some loss of sensitivity. Thus this solution optimizes the situation for storage and search speed but violates the condition for optimal sensitivity. If desired, this tradeoff could be reversed by reverting to the use of a large number of storage planes in parameter space, as described in Section 13.3. A better approach will be described in Section 13.6. Finally, note that the method is applicable only to a very restricted class of irregular polygons.

13.5 The case of an arbitrary rectangle

The case of a rectangle is characterized by greater symmetry than for an arbitrary quadrilateral. In this case the number of parameter planes required to cope with variations in orientation may be reduced from a large number (e.g. 360—see Section 13.3) to just two. This is achieved in the following way. Suppose first that the major axis of the rectangle (of sides $A, B: A > B$) is known to be orientated between 0° and 90°. Then we need to accumulate votes by moving a distance $B/2$ from edge pixels oriented between 0° and

90° and accumulating a line of votes of length A in parameter space, and by moving a distance $A/2$ from edge pixels orientated between 90° and 180° and accumulating a line of votes of length B in parameter space. This is very similar to the situation for a square (see Section 13.3). However, suppose that the major axis was instead known to be orientated between 90° and 180°. Then we would need to change the method of transform suitably by interchanging the values of A and B. Since we do not know which supposition is true, we have to try both methods of transform and for this we need two planes in parameter space. If a strong peak is obtained in the first plane, clearly the first supposition is true and the rectangle major axis is orientated between 0° and 90°—and *vice versa* if the strong peak is in the second plane in parameter space. In either case, the position of the strong peak indicates the position of the centre of the rectangle (see Fig. 13.5).

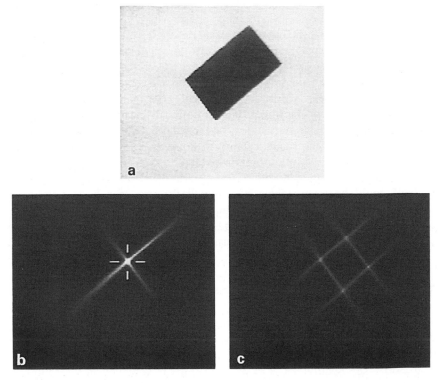

Fig. 13.5   (a) Off-camera image of a rectangle; (b), (c) optimized transforms in two parameter planes. The first plane (b) contains the highest peak, thereby indicating the position of the rectangle and its orientation within 90°.   © *1986 IEEE*.

In summary, the largest peak in either plane indicates (i) which assumption about the orientation of the rectangle was justified and (ii) the precise location of the rectangle. Apart from its inherent value, this example is important in giving a clue of how to develop the parametrization as symmetry is gradually reduced.

13.6   Lower bounds on the numbers of parameter planes

This section generalizes the solution of the previous section. First, note that two planes in parameter space are needed for optimal rectangle detection, and that each of these planes may be taken to correspond to a range of orientations of the major axis.

For arbitrarily orientated polygons of known size and shape it is clear that votes must be accumulated in a number of planes in parameter space. Suppose that the polygon has N sides and that we are using M parameter planes to locate it. Since the complete range of possible orientations of the object is 2π, each of the M planes will have to be taken to represent a range $2\pi/M$ of object orientations. On detecting an edge pixel in the image, it is not clear from its orientation which of the N sides of the object it originated from, and it is necessary to insert a line of votes at appropriate positions in N of the parameter planes. When every edge pixel has been processed in this way, the plane with the highest peak again indicates both the location of the object and its approximate orientation. Already it is clear that we must ensure that $M > N$, or else false peaks will start to appear in at least one of the planes, implying some ambiguity in the location of the object. Having obtained some feel for the problem, we now study the situation in more detail in order to determine the precise restriction on the number of planes.

Take $\{\varphi_i: i = 1, \ldots, N\}$ as the set of angles of the edge normals of the polygon in a standard orientation α. Next, suppose an edge of edge normal orientation θ is found in the image. If this is taken to arise from side i of the polygon, we are assuming that the polygon has orientation $\alpha + \theta - \varphi_i$, and we accumulate a line of votes in parameter plane $K_i = (\alpha + \theta - \varphi_i)/(2\pi/M)$ (here we force an integer result in the range 0 to $M - 1$). Clearly it is necessary to accumulate N lines of votes in the set of N parameter planes $\{K_i: i = 1, \ldots, N\}$, as stated above. However, additional false peaks will start to appear in parameter space if an edge point on any one side of the polygon gives rise to more than one line of votes in the same parameter plane. To ensure that this does not occur it is necessary to ensure that $K_i \neq K_j$ for different i, j. Thus we must have

$$|(\varphi_j - \varphi_i)/(2\pi/M)| \geq 1 \qquad (13.1)$$

This result has the simple interpretation that the minimum exterior angle of the polygon ψ_{min} must be greater than $2\pi/M$, the exterior angle being (by definition) the angle through which one side of the polygon has to be rotated before it coincides with an adjacent side (Fig. 13.6). Thus the minimum exterior angle describes the maximum amount of rotation before an unnecessary ambiguity is generated about the orientation of the object. If no unnecessary ambiguity is to occur, the number of planes M must obey the condition

$$M \geqslant 2\pi/\psi_{min} \tag{13.2}$$

We are now in a position to re-assess all the results so far obtained with polygon detection schemes. Consider regular polygons first: if there are N sides, then the common exterior angle is $\psi = 2\pi/N$, so the number of planes required is inherently

$$M = 2\pi/\psi = 2\pi/(2\pi/N) = N \tag{13.3}$$

However, symmetry dictates that all N planes contain the same information, so only one plane is required.

Consider next the rectangle case: here the exterior angle is $\pi/2$ and the inherent number of planes is four. However, symmetry again plays a part, and just two planes are sufficient to contain all the relevant information, as has already been demonstrated.

In the triangle case, for a nonequilateral triangle the minimum exterior angle is typically greater than $\pi/2$ but is never as large as $2\pi/3$. Hence at least four planes are needed to locate a general triangle in an arbitrary orientation (see Fig. 13.7). Note, however, that for a special choice of localization point fewer planes are required—as shown in Section 13.4.

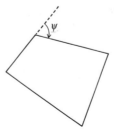

Fig. 13.6   The exterior angle of a polygon, ψ: the angle through which one side has to be rotated before it coincides with an adjacent side.

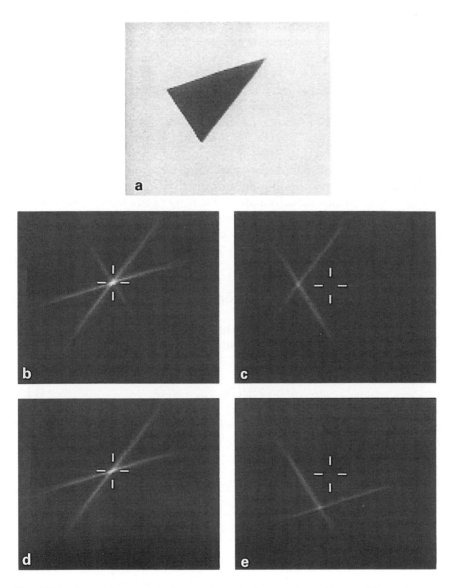

Fig. 13.7 An arbitrary triangle and its transform in four planes: (a) localization point L selected to optimize sensitivity; (b)–(e) four planes required in parameter space.

Next, for a polygon that is *almost* regular, the minimum exterior angle is *just* greater than $2\pi/N$, so the number of planes required is $N + 1$. In this case it is at first surprising that as symmetry increases, the requirement jumps from $N + 1$ planes not to N planes but straight to one plane.

Now consider a case where there is significant symmetry but the polygon cannot be considered to be fully regular. Figure 13.8(a) exhibits a semiregular hexagon for which the set of identity rotations forms a group of order 3. Here the minimum exterior angle is the same as for a regular hexagon. Thus the basic rule predicts that we need a total of six planes in parameter space. However, symmetry shows that only two of these contain different information. Generalizing this result, the number of planes actually required is N/G rather than N, G being the order of the group of identity rotations of the object. To prove this more general result note that we need to consider only object orientations in the range 0–$2\pi/G$ instead of 0–2π. Hence condition (13.2) now becomes

$$M \geqslant (2\pi/G)/\psi_{\min} = (2\pi/G)/(2\pi/N) = N/G \qquad (13.4)$$

Suppose instead that the hexagon had 3-fold rotation symmetry but had two sets of exterior angles with slightly different values (see Fig. 13.8(b)). Then

$$M \geqslant (2\pi/G)/\psi_{\min} > N/G \qquad (13.5)$$

and at best we will have $M = (N/G) + 1$, which is 3 for the hexagon shape of Fig. 13.8(b).

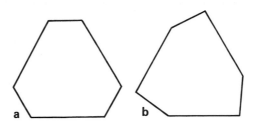

a   b

Fig. 13.8   (a) Semiregular hexagon with three-fold rotation symmetry and having all its exterior angles equal: two planes are required in parameter space in this case; (b) a similar hexagon with two slightly different exterior angles: here three planes are needed in parameter space.

13.7   An extension of the triangle result

The reader may already have noticed that the triangle result can be extended to optimize the case of the arbitrary quadrilateral if L is chosen appropriately. Essentially, what is required is to take the vertex with the smallest exterior angle and choose a localization point which lies on the corresponding internal

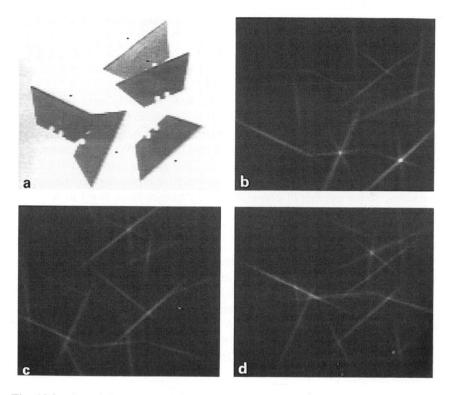

Fig. 13.9 Set of lino-cutter blades (a) which have been located from just three planes (b)–(d) in parameter space: careful choice of L has excluded the smallest two exterior angles, leaving two exterior angles of 121°. Hence seven planes would have had to be used to optimize sensitivity (in this case giving an increase in sensitivity of about 10%). In this example a simple peak height thresholding procedure was able to find all the objects without giving false alarms. However, if the degree of occlusion had been any greater, additional procedures would have been required to identify positively some of the objects. Once the objects have been located, the most difficult task involved in keying into the image has been completed: object orientation can then be achieved by a variety of methods requiring relatively little computation.

angle bisector. Then the sides adjacent to this vertex will be the same distance from L and ignoring this particular exterior angle will not be a limitation, since the ambiguity thereby introduced will already have been taken care of! Indeed, this operation can be repeated, finding the next smallest exterior angle and ensuring that L again lies on the corresponding internal angle bisector. Since a point is completely defined once it is known to lie on two intersecting lines, there is no further freedom to ignore exterior angles, and the third smallest exterior angle now determines the number of planes required in parameter space.

For an arbitrary quadrilateral it may be deduced that we require a minimum of three planes in parameter space if the relevant two (nonexcluded) exterior angles are greater than 120° (see for example, Fig. 13.9). In other cases many planes may still be needed. The only such case worth considering here is that of a polygon of N sides whose exterior angles are all approximately equal once the two smallest have been excluded. If the smallest remaining exterior angle is just less than $2\pi(N - 2)$, the number of planes needed is $N - 1$. As confirmation of this result, note that eliminating two angles from consideration is for this purpose equivalent to eliminating two sides, so the number of planes required changes from $N + 1$ to $N - 1$, as already shown.

An interesting example of this result is provided by the semicircle. Here the minimum exterior angle is zero, so an infinite number of planes in parameter space are in principle required. However, all cases of this particular exterior angle can be eliminated simultaneously by placing L on the angle bisectors: we choose L to lie at the centre of the circle of which the semicircular arc is part. Thus the minimum remaining exterior angle is $\pi/2$ and only four planes are required in parameter space. The main peak in parameter space is then as shown in Fig. 10.8. In this case there is no additional rotational symmetry to modify the result.

13.8 Discussion

At this stage it is relevant to investigate in more detail what happens as the number of planes in parameter space is reduced. Consider first what happens if only one plane is employed. Then each edge pixel that is located in the image could in principle have resulted from any of the N sides of the polygon. To ensure that sensitivity is optimized each such pixel must be permitted to place a vote at the position of L in parameter space: this means assuming in turn that the edge pixel is on each of the N sides of the polygon, then moving along the edge normal an appropriate distance and placing a line of votes at that position. As seen from Fig. 13.10, this results in a proliferation of lines in parameter space and gives rise to a great many peaks where these lines cross each other. Fortunately, the lines are quite well localized and so the situation is much better than for the normal HT case of line finding, where potential polygon centres would occur for all possible crossovers out to infinity.

In fact, the highest peak arising from one body is that at L, since all sides of the object "focus" on this point. However, it is often difficult to select the main peak in this way since a great many other peaks arise, all of which have to be located and checked. In the presence of possible occlusions and object defects (such as are looked for in industrial inspection applications) it will not be clear from peak intensities alone which is the main peak corresponding to L. Thus it is worth trying to reduce the total number of peaks to be investigated in parameter space. In essence, what has been demonstrated so far is that it is not necessary to go to the extreme of including planes for all possible object orientations: we merely need to have

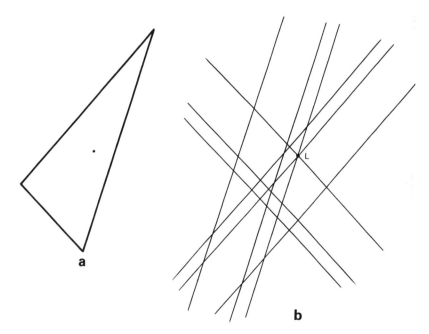

Fig. 13.10 A single plane (b) in parameter space used to accumulate all the information about the location of the triangle (a). There is a proliferation of lines of votes and a large number of false peaks arise—including some which, through obvious coincidences, could be of similar height to the main peak.

a sufficient number so that the number of peaks is reduced significantly (indeed, if planes were included for all possible object orientations, most of them would be empty and storage would be wasted).

To illustrate this consider the case of an arbitrary triangle, where L is at a general location, not at the centre of the inscribed circle. In this case, there are three possible distances from L to each of the sides, and thus each of the three sides gives rise to three parallel lines in parameter space (Fig. 13.10). The total number of intersections to be checked is then basically

$$\binom{3}{2} 3^2 = 27,$$

of which three should be coincident at L. In general, for an N-sided polygon the number is

$$\binom{N}{2} N^2 = N(N - 1)N^2/2,$$

which is of the order of $N^4/2$ for $N \gg 1$. Clearly, with such large numbers of peaks to be checked it is preferable to use more than one plane in parameter space.

In the case of an arbitrary triangle, the number of peaks in the four separate planes of parameter space by the present approach is six (of which three appear coincident at L). For a general polygon, this number is highly data-dependent. However, if M is $N + 1$, as will happen for an almost regular polygon, there are N planes in parameter space each containing $N - 1$ lines and having

$$\binom{N - 1}{2}$$

peaks, plus one plane containing N lines and having

$$\binom{N}{2}$$

peaks which are coincident at L. Counting the peaks coincident at L as a single peak, the following formulae are obtained respectively for the total numbers of peaks in one plane and in $N+1$ planes, for almost regular N-sided polygons:

$$P_1 = N^3(N-1)/2 - N(N-1)/2 + 1 \qquad (13.6)$$

$$P_{N+1} = N(N-1)(N-2)/2 + 1 \qquad (13.7)$$

Hence P_1 is $O(N^4)$ while P_{N+1} is $O(N^3)$. Table 13.1 shows that the scheme described above gives a very worthwhile reduction in the numbers of peaks in parameter space.

In the earlier sections of this chapter the view was taken that the number of planes in parameter space should be minimized without introducing unnecessary ambiguity. This section has refined this view of the situation by reinterpreting the need to avoid ambiguity in terms of reducing the number of false peaks in parameter space. Unfortunately, for objects which have small values of ψ_{min}, it may well be impracticable to employ the numbers of planes that theory indicates as necessary for this purpose. It is then necessary to find the best tradeoff between the number of planes and the number of peaks. Intuition suggests that for practical polygons a useful compromise will be achieved in which M is still $\sim N$ (say $M < 2N$). In such cases, by comparison with the results of equations (13.1) and (13.2), one might still expect of the order of N^3 peaks in parameter space.

Finally, it is also worth mentioning that one advantage of using a number of parameter planes is that some information on the orientation of the object is obtained as an integral part of the process of locating it. This may well influence the choice of a suitable value for M.

Table 13.1 Number of planes in parameter space for different polygons.

N	P_1	P_{N+1}
3	25	4
4	91	13
5	241	31
6	526	61
7	1009	106
8	1765	169

In this table, P_1 is the number of peaks in a parameter space consisting of a single plane, and P_{N+1} is the number of peaks in a space containing $N+1$ planes, when the polygon has N sides and is almost regular.

13.9   Determining orientation

Having located a given polygon, it is usually necessary to determine its orientation. One way of proceeding is to find first all the edge points that contribute to a given peak in parameter space, and to construct a histogram of their edge orientations. The orientation η of the polygon can be found if the positions of the peaks in this histogram are now determined and matched against the set $\{\varphi_i: i = 1, \ldots, N\}$ of the angles of the edge normals for the idealized polygon in standard orientation α (Section 13.6).

However, this method is wasteful in computation, since it does not make use of the fact that the main peak in parameter space has been found to lie in a *particular* plane K. Hence we can take the set of (averaged) polygon side orientations θ from the histogram already constructed, identify which sides had which orientations, deduce from each θ a value for the polygon orientation $\eta = \alpha + \theta - \varphi_i$ (see Section 13.6), and enter this value into a histogram of estimated η values: the peak in this histogram will eventually give an accurate value for the polygon orientation η.

It may be noted that there are strong similarities, but also significant differences, between the procedure described above and that described in Chapter 11 for the determination of ellipse parameters. In particular, it should be noted that both procedures are two-stage ones, in which it is computationally more efficient to find object location before object orientation.

13.10   Why corner detection?

The above analysis has shown the possibilities for direct detection of polygons, but it has also revealed the limitations of the approach: these arise specifically with more complex polygons—such as those having concavities or many sides. In such cases, it may be easier to revert to using line or corner detection schemes coupled with the abstract pattern matching approaches discussed in Chapter 14.

It has been noted in previous chapters that objects are generally located most efficiently from their features. Prominent features have been seen to include straight lines, arcs, holes and corners. Corners are particularly important since they may be used to orientate objects and to provide measures of their dimensions—a knowledge of orientation being vital on some occasions, for example if a robot is to find the best way of picking up the object, and measurement being necessary in most inspection applications. Hence accurate, efficient corner detectors are of great relevance in machine vision. In the remainder of the chapter we shall focus on corner detection procedures.

In fact, workers in this field have devoted considerable ingenuity to devising corner detectors, sometimes with surprising results (see, for example, the median-based detector described below). We start by considering what is perhaps the most obvious detection scheme— that of template matching.

13.11 Template matching

Following our experience with template matching methods for edge detection (Chapter 5), it would appear to be straightforward to devise suitable templates for corner detection. These would have the general appearance of corners, and in a 3×3 neighbourhood would take forms such as the following:

$$
\begin{bmatrix}
-4 & 5 & 5 \\
-4 & 5 & 5 \\
-4 & -4 & -4
\end{bmatrix}
\qquad
\begin{bmatrix}
5 & 5 & 5 \\
-4 & 5 & -4 \\
-4 & -4 & -4
\end{bmatrix}
$$

the complete set of eight templates being generated by successive 90° rotations of the first two shown. An alternative set of templates was suggested by Bretschi (1981). As for edge detection templates, the mask coefficients are made to sum to zero so that corner detection is insensitive to absolute changes in light intensity. Ideally, this set of templates should be able to locate all corners and to estimate their orientation to within 22.5°.

Unfortunately, corners vary very much in a number of their characteristics, including in particular their degree of pointedness[*], internal angle and the intensity gradient at the boundary. Hence it is quite difficult to design optimal corner detectors. In addition, corners are generally insufficiently pointed for good results to be obtained with the 3×3 template masks shown above. Another problem is that in larger neighbourhoods, not only do the masks become larger but also more of them are needed to obtain optimal corner responses, and it rapidly becomes clear that the template matching approach is likely to involve excessive computation for practical corner detection. The alternative is to approach the problem analytically, somehow deducing the ideal response for a corner at any arbitrary orientation, and thereby bypassing

---

[*] The term "pointedness" is used as the opposite to "bluntness", the term "sharpness" being reserved for the total angle η through which the boundary turns in the corner region, i.e. π minus the internal angle.

the problem of calculating very many individual responses to find which one gives the maximum signal. The methods described in the remainder of this chapter embody this alternative philosophy.

13.12   Second-order derivative schemes

Second-order differential operator approaches have been used widely for corner detection and mimic the first-order operators used for edge detection. Indeed, the relationship lies deeper than this. By definition, corners in grey-scale images occur in regions of rapidly changing intensity levels. By this token they are detected by the same operators that detect edges in images. However, corner pixels are much rarer[*] than edge pixels—by one definition, they arise where two relatively straight-edged fragments intersect. Thus it is useful to have operators that detect corners *directly*, i.e. *without unnecessarily locating edges*. To achieve this sort of discriminability it is clearly necessary to consider local variations in image intensity up to at least second order. Hence the local intensity variation is expanded as follows:

$$I(x, y) = I(0, 0) + I_x x + I_y y + I_{xx} x^2/2 + I_{xy} xy + I_{yy} y^2/2 + \cdots \quad (13.8)$$

where the suffices indicate partial differentiation with respect to x and y and the expansion is performed about the origin $x_0 (0, 0)$. The symmetrical matrix of second derivatives is:

$$\mathcal{I}_{(2)} = \begin{bmatrix} I_{xx} & I_{xy} \\ I_{yx} & I_{yy} \end{bmatrix} \qquad \text{where} \qquad I_{xy} = I_{yx} \qquad (13.9)$$

This gives information on the local curvature at x_0. In fact, a suitable rotation of the coordinate system transforms $\mathcal{I}_{(2)}$ into diagonal form:

$$\tilde{\mathcal{I}}_{(2)} = \begin{bmatrix} I_{\tilde{x}\tilde{x}} & 0 \\ 0 & I_{\tilde{y}\tilde{y}} \end{bmatrix} = \begin{bmatrix} \kappa_1 & 0 \\ 0 & \kappa_2 \end{bmatrix} \qquad (13.10)$$

where appropriate derivatives have been reinterpreted as principal curvatures at x_0.

---

[*] We might imagine a 256×256 image of 64K pixels, of which 1000 ($\sim 2\%$) lie on edges and a mere 30 ($\sim 0.05\%$) are situated at corner points.

We are particularly interested in rotationally invariant operators and it is significant that the trace and determinant of a matrix such as $\mathcal{I}_{(2)}$ are invariant under rotation. Thus we obtain the Beaudet (1978) operators:

$$\text{Laplacian} = I_{xx} + I_{yy} = \kappa_1 + \kappa_2 \tag{13.11}$$

and

$$\text{Hessian} = \det(\mathcal{I}_{(2)}) = I_{xx}I_{yy} - I_{xy}^2 = \kappa_1\kappa_2 \tag{13.12}$$

It is well known that the Laplacian operator gives significant responses along lines and edges and hence is not particularly suitable as a corner detector. On the other hand, Beaudet's "DET" operator does not respond to lines and edges but gives significant signals near corners: it should therefore form a useful corner detector. However, DET responds with one sign on one side of a corner and with the opposite sign on the other side of the corner: at the point of real interest—on the corner—it gives a null response. Hence rather more complicated analysis is required to deduce the presence and exact position of each corner (Dreschler and Nagel, 1981; Nagel, 1983). The problem is clarified by Fig. 13.11. Here the dotted line shows the path of maximum horizontal curvature for various intensity values up the slope. The DET operator gives maximum response at positions P and Q on this line, and the parts of the line between P and Q must be explored to find the "ideal" corner point C where DET is zero.

Perhaps to avoid rather complicated procedures of this sort, Kitchen and Rosenfeld (1982) examined a variety of strategies for locating corners, starting from the consideration of local variation in the directions of edges. They found a highly effective operator which estimates the projection of the local rate of change of gradient direction vector along the horizontal edge tangent direction, and showed that it is mathematically identical to calculating the horizontal curvature κ of the intensity function I. To obtain a realistic indication of the strength of a corner they multiplied κ by the magnitude of the local intensity gradient g:

$$\begin{aligned}
C = \kappa g &= \kappa (I_x^2 + I_y^2)^{1/2} \\
&= \frac{I_{xx}I_y^2 - 2I_{xy}I_xI_y + I_{yy}I_x^2}{I_x^2 + I_y^2}
\end{aligned} \tag{13.13}$$

Finally, they used the heuristic of nonmaximum suppression along the edge normal direction to localize the corner positions further.

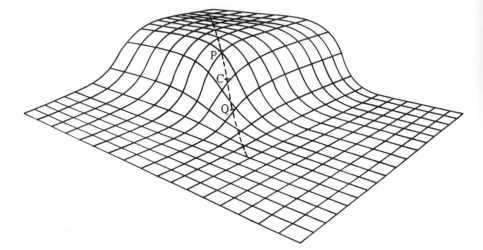

Fig. 13.11 Sketch of an idealized corner, taken to give a smoothly varying intensity function. The dotted line shows the path of maximum horizontal curvature for various intensity values up the slope. The DET operator gives maximum responses at P and Q, and it is required to find the ideal corner position C where DET gives a null response.

In 1983 Nagel was able to show that the Kitchen and Rosenfeld (KR) corner detector using nonmaximum suppression is mathematically virtually identical to the Dreschler and Nagel (DN) corner detector. A year later, Shah and Jain (1984) studied the Zuniga and Haralick (ZH) corner detector (1983) based on a bicubic polynomial model of the intensity function: they showed that this is essentially equivalent to the KR corner detector. However, the ZH corner detector operates rather differently in that it thresholds the intensity gradient and then works with the subset of edge points in the image, only at that stage applying the curvature function as a corner strength criterion. By making edge detection explicit in the operator, the ZH detector eliminates a number of false corners that would otherwise be induced by noise.

The inherent near-equivalence of these three corner detectors need not be overly surprising, since in the end the different methods would be expected to reflect the same underlying physical phenomena (Davies, 1988d). However, it is gratifying that the ultimate result of these rather mathematical formulations is interpretable by something as easy to visualize as horizontal curvature multiplied by intensity gradient.

13.13 A median-based corner detector

An entirely different strategy for detecting corners was developed by Paler
et al. (1984). It adopts an initially surprising and rather nonmathematical
approach based on the properties of the median filter. The technique
involves applying a median filter to the input image, and then forming
another image that is the difference between the input and the filtered
images. This difference image contains a set of signals which are interpreted
as local measures of corner strength.

Clearly, it seems risky to apply such a technique since its origins suggest
that, far from giving a correct indication of corners, it may instead unearth all
the noise in the original image and present this as a set of "corner" signals.
Fortunately, analysis shows that these worries may not be too serious. First, in
the absence of noise, strong signals are not expected in areas of background;
nor are they expected near straight edges, since median filters do not shift or
modify such edges significantly (see Chapter 3). However, if a window is
moved gradually from a background region until its central pixel is just over a
convex object corner, there is no change in the output of the median filter:
hence there is a strong difference signal indicating a corner.

Paler *et al.* (1984) analysed the operator in some depth and concluded that
the signal strength obtained from it is proportional to (i) the local contrast,
and (ii) the "sharpness" of the corner. The definition of sharpness they used
was that of Wang *et al.* (1983), meaning the angle η through which the
boundary turns. Since it is assumed here that the boundary turns through a
significant angle (perhaps the whole angle η) within the filter neighbourhood, the difference from the second-order intensity variation approach is a
major one. Indeed, it is an implicit assumption in the latter approach that
first- and second-order coefficients describe the local intensity characteristics
reasonably rigorously, the intensity function being inherently continuous
and differentiable. Thus the second-order methods may give unpredictable
results with pointed corners where directions change within the range of a
few pixels. Although there is some truth in this, it is worth looking at the
similarities between the two approaches to corner detection before considering the differences. We proceed with this in the next subsection.

13.13.1 Analysing the operation of the median detector

This subsection considers the performance of the median corner detector
under conditions where the grey-scale intensity varies by only a small

amount within the median filter neighbourhood region. This permits the performance of the corner detector to be related to low-order derivatives of the intensity variation, so that comparisons can be made with the second-order corner detectors mentioned earlier.

To proceed we assume a continuous analogue image and a median filter operating in an idealized circular neighbourhood. For simplicity, since we are attempting to relate signal strengths and differential coefficients, noise is ignored. Next, recall (Chapter 3) that for an intensity function that increases monotonically with distance in some arbitrary direction $\tilde{x}$ but which does not vary in the perpendicular direction $\tilde{y}$, the median within the circular window is equal to the value at the centre of the neighbourhood. This means that the median corner detector gives zero signal if the horizontal curvature is locally zero.

If there is a small horizontal curvature κ, the situation can be modelled by envisaging a set of constant-intensity contours of roughly circular shape and approximately equal curvature, within the circular window which will be taken to have radius a (Fig. 13.12). Consider the contour having the median intensity value. The centre of this contour does not pass through the centre of the window but is displaced to one side along the negative $\tilde{x}$-axis. Furthermore, the signal obtained from the corner detector depends on this displacement. If the displacement is D, it is easy to see that the corner signal is $Dg_{\tilde{x}}$ since $g_{\tilde{x}}$ allows the intensity change over the distance D to be estimated (Fig. 13.12). The remaining problem is to relate D to the horizontal curvature κ. A formula giving this relation has already been obtained in Chapter 3. The required result is:

$$D = \kappa a^2 / 6 \tag{13.14}$$

so the corner signal is

$$C = Dg_{\tilde{x}} = \kappa g_{\tilde{x}} a^2 / 6 \tag{13.15}$$

Note that C has the dimensions of intensity (contrast), and that the equation may be re-expressed in the form:

$$C = (g_{\tilde{x}} a) \times (2a\kappa)/12 \tag{13.16}$$

so that, as in the formulation of Paler et al. (1984), corner strength is closely related to corner contrast and corner sharpness.

To summarize, the signal from the median-based corner detector is proportional to horizontal curvature and to intensity gradient. Thus this

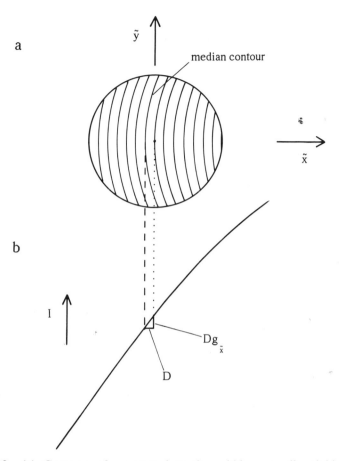

Fig. 13.12 (a) Contours of constant intensity within a small neighbourhood: ideally, these are parallel, circular and of approximately equal curvature (the contour of median intensity does not pass through the centre of the neighbourhood); (b) cross-section of intensity variation, indicating how the displacement D of the median contour leads to an estimate of corner strength.

corner detector gives an identical response to the three second-order intensity variation detectors discussed in Section 13.12, the closest practically being the KR detector. However, this comparison is valid only when second-order variations in intensity give a complete description of the situation. Clearly the situation might be significantly different where corners are so pointed that they turn through a large proportion of their total angle within the median neighbourhood. In addition, the effects of noise might be expected to be rather different in the two cases, as the median filter is particularly good at suppressing impulse noise. Meanwhile, for small horizontal curvatures, there ought to be no difference in the positions at which median and second-order derivative methods locate corners, and accuracy of localization should be identical in the two cases.

13.13.2   Practical results

Experimental tests with the median approach to corner detection have shown that it is a highly effective procedure (Paler *et al.*, 1984; Davies, 1988d). Corners are detected reliably and signal strength is indeed roughly proportional both to local image contrast and to corner sharpness (see Fig. 13.13).

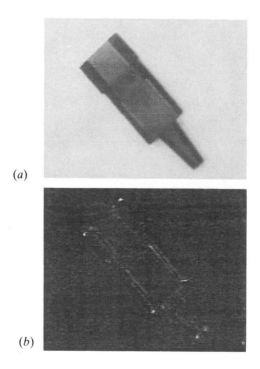

(a)

(b)

Fig. 13.13   (a) Original off-camera 128 × 128 6-bit grey-scale image; (b) result of applying the median-based corner detector in a 5 × 5 neighbourhood. Note that corner signal strength is roughly proportional both to corner contrast and to corner sharpness.

Noise is more apparent for 3×3 implementations and this makes it better to use 5×5 or larger neighbourhoods to give good corner discrimination. However, the fact that median operations are slow in large neighbourhoods, and that background noise is still evident even in 5×5 neighbourhoods, means that the basic median-based approach gives poor performance by comparison with the second-order methods. However, both of these disadvantages are virtually eliminated by using a "skimming" procedure, in which edge points are first located by thresholding the edge gradient, and the edge points are then examined with the median detector to locate the corner points (Davies, 1988d). With this improved method, performance is found to be generally superior to that for (say) the KR method in that corner signals are better localized and accuracy is enhanced. Indeed, the second-order methods appear to give rather fuzzy and blurred signals which contrast with the sharp signals obtained with the improved median approach (Fig. 13.14).

At this stage the reason for the more blurred corner signals obtained using the second-order operators is not clear. Basically, there is no valid rationale for applying second-order operators to pointed corners, since higher derivatives of the intensity function will become important and will at least in principle interfere with their operation. However, it is evident that the second-order methods will probably give strong corner signals when the tip of a pointed corner appears anywhere in their neighbourhood, so there is likely to be a minimum blur region of radius a for any corner signal. This appears to explain the observed results adequately. It is worth noting that the sharpness of signals obtained by the KR method may be improved by nonmaximum suppression (Kitchen and Rosenfeld, 1982; Nagel, 1983). However, this technique can also be applied to the output of median-based corner detectors: hence, the fact remains that the median-based method gives inherently better localized signals than the second-order methods (although for low-curvature corners the signals have similar localization).

Overall, the inherent deficiencies of the median-based corner detector can be overcome by incorporating a skimming procedure, and then the method becomes superior to the second-order approaches in giving better localization of corner signals. The underlying reason for the difference in localization properties appears to be that the median-based signal is ultimately sensitive only to the particular few pixels whose intensities fall near the median contour within the window, whereas the second-order operators use typical convolution masks which are in general sensitive to the intensity values of all the pixels within the window. Thus the KR operator tends to give a strong signal when the tip of a pointed corner is present anywhere in the window.

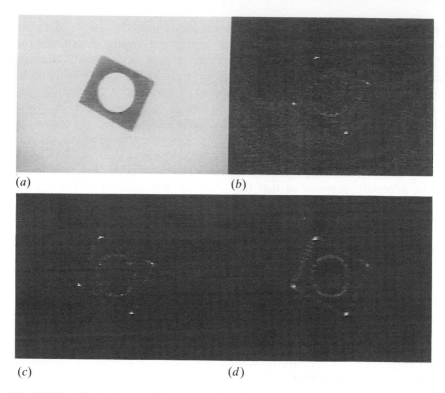

(a)   *(b)*

(c)   *(d)*

Fig. 13.14   Comparison of the median and KR corner detectors: (a) original
128×128 grey-scale image; (b) result of applying a median detector; (c) result of
including a suitable gradient threshold; (d) result of applying a KR detector. The
considerable amount of background noise is saturated out in (a) but is evident from
(b). To give a fair comparison between the median and KR detectors, 5×5
neighbourhoods are employed in each case, and nonmaximum suppression
operations are not applied: the same gradient threshold is used in (c) and (d).

13.14   The Hough transform approach to corner detection

In certain applications it can happen that corners rarely appear in the
idealized pointed form suggested earlier. Indeed, in food applications
corners are commonly chipped, crumbly or rounded, and are bounded by
sides that are by no means straight lines (Fig. 13.15). Even mechanical
components may suffer chipping or breakage. Hence it is useful to have
some means of detecting these types of nonideal corner. The generalized HT
provides a suitable technique (Davies, 1986b, 1988a).

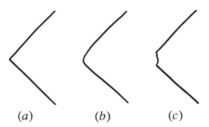

(a) (b) (c)

Fig. 13.15 Types of corner: (a) pointed; (b) rounded; (c) chipped. Corners of type (a) are normal with metal components, those of type (b) are usual with biscuits and other food products, whereas those of type (c) are common with food products but rarer with metal parts.

The method used is a variation on that described in Section 13.3 for locating objects such as squares. Instead of accumulating a line of candidate centre points parallel to each side after moving laterally inwards a distance equal to half the length of a side, only a small lateral displacement is required when locating corners. The lateral displacement that is chosen must be sufficient to offset any corner bluntness, so that the points that are located by the transform lines are consistently positioned relative to the idealized corner point (Fig. 13.16). In addition, the displacement must be sufficiently large that enough points on the sides of the object contribute to the corner transforms to ensure adequate sensitivity.

In fact, there are two relevant parameters for the transform: D, the lateral displacement, and T, the length of the transform line arising from each edge point. These parameters must be related to the following parameters for the object being detected: B, the width of the corner bluntness region (Fig. 13.16), and S, the length of the object side (for convenience the object is assumed to be a square). E, the portion of a side that is to contribute to a corner peak, is simply related to T:

$$T = E \qquad\qquad (13.17)$$

Since we wish to make corner peaks appear as close to idealized corners as possible, we may also write:

$$D = B + E/2 \qquad\qquad (13.18)$$

These definitions now allow the sensitivity of corner location to be optimized. Note first that the error in the measurement of each corner position is proportional to $1/\sqrt{E}$, since our measurement of the relevant part

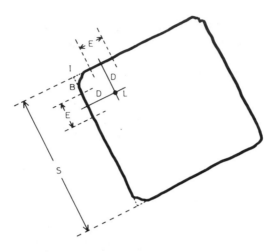

Fig. 13.16 Geometry for locating blunt corners: B, region of bluntness; D, lateral displacement employed in edge pixel transform; E, distance over which position of corner is estimated; I, idealized corner point; L, localization point of a typical corner; S, length of a typical side. The degree of corner bluntness shown here is common amongst food products such as fishcakes.

of each side is effectively being averaged over E independent signals. Next, note that the purpose of locating corners is ultimately to measure (i) the orientation of the object and (ii) its linear dimensions. Hence the fractional accuracy is proportional to:

$$A = (S - 2D)\sqrt{E} \tag{13.19}$$

Eliminating D now gives:

$$A = (S - 2B - E)\sqrt{E} \tag{13.20}$$

Differentiating, we find:

$$dA/dE = (S - 2B - 3E)/2\sqrt{E} \tag{13.21}$$

Accuracy can now be optimized by setting $dA/dE = 0$, so that:

$$E = S/3 - 2B/3 \tag{13.22}$$

making

$$D = S/6 + 2B/3 \tag{13.23}$$

Hence there is a clear optimum in accuracy, and as $B \longrightarrow 0$ it occurs for $D \simeq S/6$, whereas for rather blunt or crumbly types of object (typified by biscuits) $B \simeq S/8$ and then $D \simeq S/4$.

Note now that if optimal object detection and location had been the main aim, this would have been achieved by setting $E = S - 2B$ and detecting all the corners together at the centre of the square. It is also possible to see that by optimizing object orientation, object detection is carried out with reduced sensitivity, since only a length $2E$ of the perimeter contributes to each signal, instead of a length $4S - 8B$.

The calculations presented above are important for a number of reasons: (i) they clarify the point that tasks such as corner location are carried out not in isolation but as parts of larger tasks which are optimizable; (ii) they show

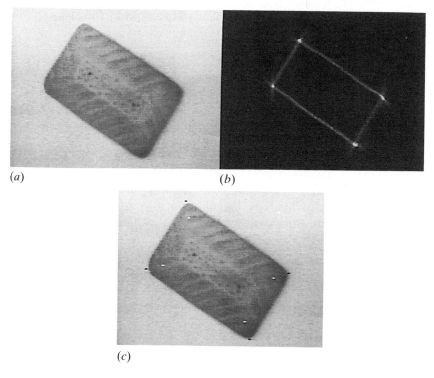

(a)   (b)

(c)

Fig. 13.17 Example of the Hough transform approach: (a) original image of a biscuit (128 × 128 pixels, 64 grey levels); (b) transform with lateral displacement around 22% of the shorter side; (c) image with transform peaks located (white dots) and idealized corner positions deduced (black dots). The lateral displacement employed here is close to the optimum for this type of object.

that the ultimate purpose of corner detection may be either to locate objects accurately, or to orientate and measure them accurately; and (iii) they indicate that there is an optimum region of interest E for any operator that is to locate corners. This last factor is of relevance when selecting the optimum size of neighbourhood for the other corner detectors described in this chapter.

Although the calculations above emphasized sensitivity and optimization, the underlying principle of the HT method is to find corners by interpolation rather than by extrapolation, as the former is generally a more accurate procedure. The fact that the corners are located away from the idealized corner positions is to some extent a disadvantage (although it is difficult to see how a corner can be located directly if it is missing). However, once the corner peaks have been found, for known objects the idealized corner positions may be deduced (Fig. 13.17)—although it is as well to note that this step is unlikely to be necessary if the real purpose of corner location is to orientate the object and to measure its dimensions.

13.15   The lateral histogram approach to corner detection

Before leaving the topic of corner detection it is worth considering again the lateral histogram method. Note that although this was introduced in Chapter 12 as a means of hole detection, the method was in fact used for corner detection before being used for hole detection (Wu and Rosenfeld, 1983; Davies, 1987g). As stated earlier, it is a highly efficient procedure, which can examine an image in some $3N^2$ operations: it should therefore be particularly valuable for detecting moderately blunt corners which would otherwise require the use of large neighbourhood corner detectors.

The analysis of Chapter 12 indicated that the lateral histogram approach is useful only when isotropic features are being sought in images. Clearly, this applies to round objects and small holes but is invalid for corners. However, more detailed analysis shows that corners can generally be detected by this means. Figure 13.18 shows the basic reason for this. The lateral histograms (obtained by summing the pixel intensities horizontally in one case and vertically in the other) give wedge-shaped response patterns in one of three forms. In all three cases the relevant parts of the intensity profiles are bounded by two straight lines, and as a result a strong signal will be obtained on applying a 1-D Laplacian. Ignoring signs, the resulting signals will be peaked for x and y values equal to those of the various corners in the image. Hence sets of candidate corner points can readily be obtained and these can then be interrogated to find the true corner points. The amount of

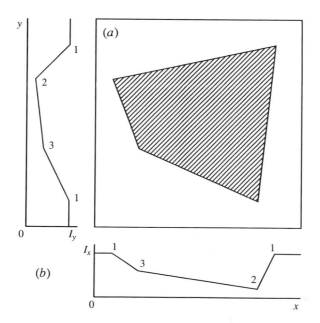

Fig. 13.18 Lateral histograms of a simple polygonal object. Examination of polygons such as that in (a) shows that lateral histograms for regions containing corners in general appear in one of the three forms shown in (b), or in upside down or laterally inverted versions of these forms. Note that there are exceptional cases when a side bounding a corner is parallel to the x- or y-axis.

computation involved is similar to that calculated in Chapter 12 for hole detection.

One problem that arises with this method is that those corners for which an adjacent side is aligned along the x or y direction will give a peculiar response to the 1-D Laplacian operator, with peaks on either side of the relevant corner coordinates. However, this situation can be detected by applying a 1-D edge detector to each of the lateral histograms: where this gives a strong response, adjacent Laplacian responses can be ignored.

The method is frequently highly effective although, as noted in the case of holes, some features can be masked by others. Furthermore, there is the possibility that holes can mask corners and *vice versa*. In any case, once the density of features becomes large, masking effects can become a major problem. Although such effects can be cut down by the subimage method, it is possible that in extreme cases the method will be inapplicable. In addition, noise is sometimes a particular problem. Overall, the main practical problem with this method is the fact that its response is anisotropic, so that different thresholds are needed to detect corners at different orientations—and clearly the correct threshold cannot be known until the corner has been located!

13.16 Corner orientation

This chapter has so far considered the problem of corner detection as relating merely to corner location. However, corners differ from holes in that they are not isotropic, and hence are able to provide orientation information. Such information can be used by procedures that collate the information from various features in order to deduce the presence and positions of objects containing them. In Chapter 14 it will be seen that orientation information is valuable in immediately eliminating a large number of possible interpretations of an image, and hence of quickly narrowing down the search problem and saving computation.

Clearly, when corners are not particularly pointed, or are detected within rather small neighbourhoods, the accuracy of orientation will be somewhat restricted. However, orientation errors will seldom be worse than 90°, and will often be less than 20°. Although these accuracies are far worse than those (around 1°) for edge orientation (see Chapter 5), nonetheless they provide valuable constraints on possible interpretations of an image.

Here we consider only simple means of estimating corner orientation. Basically, once a corner has been located accurately, it is a rather trivial matter to estimate the orientation of the intensity gradient at that location. The main problem that arises is that a slight error in locating the corner gives a bias in estimating its orientation towards that of one or other of the bounding sides—and for a sharp pointed corner this can mean an error approaching 90°. The simplest solution to this problem is to find the mean orientation over a small region surrounding the estimated corner position: then without excessive computation the orientation accuracy will usually be within the 20° limit suggested above.

13.17 Concluding remarks

This chapter has reviewed the use of the GHT for the direct detection of polygons, with the particular aim of minimizing storage and computation requirements. In some cases only one parameter plane is needed to detect a polygonal object: commonly, a number of parameter planes of the order of the number of sides of the polygon is required, but this still represents a significant saving compared with the numbers needed to detect general shapes. It has been demonstrated how optimal savings can be achieved by taking into account the symmetry possessed by many polygon shapes such as squares and rectangles.

The analysis has shown the possibilities for direct detection of polygons, but it has also revealed the limitations of this approach: these arise specifically with more complex polygons—such as those having concavities or many sides. In such cases, it may be easier to revert to using line or corner detection schemes coupled with the abstract pattern matching approaches discussed in Chapter 14. Indeed, the latter approach offers the advantage of greater resistance to shape distortions if suitable algorithms to infer the presence of objects can be found.

Hence it was also appropriate for the chapter to examine corner detection schemes. Apart from the obvious template matching procedure, which is strictly limited in its application, and its derivative, the lateral histogram method, three main approaches have been described. The first was the second-order derivative approach, which includes the KR, DN and ZH methods—all of which embody the same basic schema; the second was the median-based method, which turned out to be equivalent to the second-order methods—but only in situations where corners have smoothly varying intensity functions; and the third was a Hough-based scheme, which is particularly useful when corners are liable to be blunt, chipped or damaged. The analysis of this last scheme revealed interesting general points about the specification and design of corner detection algorithms.

Although the median-based approach has obvious inherent disadvantages —its slowness and susceptibility to noise—these can largely be overcome by a skimming (two-stage template matching) procedure. Then the method becomes essentially equivalent to the second-order derivative methods for low-curvature intensity functions, as stated above. However, for pointed corners, higher-order derivatives of the intensity function are bound to be important, and as a result the second-order methods are based on an unjustifiable rationale and indeed are found to blur corner signals, hence reducing accuracy of corner localization. This problem does not arise for the median-based method, which seems to offer rather superior performance in practical situations. Clearly, however, it should not be used when image noise is excessive, as then the skimming procedure will either (with a high threshold) remove a proportion of the corners with the noise, or else (with a low threshold) not provide adequate speedup to counteract the computational load of the median filtering routine.

13.18 Bibliographical and historical notes

The initial sections of this chapter arose from the author's own work which was aimed at finding optimally sensitive means of locating shapes, starting with spatial matched filters. Early work in this area included that of Sklansky (1978), Ballard (1981), Brown (1983) and the author (Davies, 1987c) which compared the HT approach with spatial matched filters: these general investigations led to more detailed studies of how straight edges are optimally detected (Davies, 1987d,j), and thence to polygon detection (Davies, 1986d, 1989c). Although optimal sensitivity is not the only criterion for effectiveness of object detection, nonetheless it is of great importance, since it governs both effectiveness *per se* (especially in the presence of "clutter" and occlusion) and the accuracy with which objects may be located.

The emphasis on spatial matched filters seemed to lead inevitably to use of the GHT, with its accompanying computational problems. It should be plain from the present chapter that although these problems are formidable, systematic analysis can make a significant impact on them.

Corner detection is a subject which has been developing for well over a decade. The scene was set for the development of parallel corner detection algorithms by Beaudet's (1978) work on rotationally invariant image operators. This was soon followed by Dreschler and Nagel's (1981) sophisticated second-order corner detector: the motivation for this research was to map the motion of cars in traffic scenes, corners providing the key to unambiguous interpretation of image sequences. One year later, Kitchen and Rosenfeld (1982) had completed their study of corner detectors based mainly on edge orientation, and had developed the second-order KR method described in Section 13.12. The years 1983 and 1984 saw the development of the lateral histogram detector, the second-order ZH detector and the median-based detector (Wu and Rosenfeld, 1983; Zuniga and Haralick, 1983; Paler *et al.*, 1984). Subsequently, the author's work on the detection of blunt corners and on analysing and improving the median-based detector appeared (Davies, 1986b, 1988a,d, 1992a): these papers form the basis of Sections 13.13 and 13.14. Meanwhile, other methods had been developed, such as the Plessey algorithm (Noble, 1988): the latter is based on analysis of local first-order derivatives and does not seem to be directly equivalent to the existing second-order methods; space does not permit a detailed appraisal of this work here. The much later algorithm of Seeger and Seeger (1994) is also of interest because it appears to work well on real (outdoor) scenes in real time without requiring the adjustment of any parameters.

The considerable recent interest in corner detection underlines the point made in Section 13.10, that good point feature detectors are vital to efficient image interpretation, as will be seen in detail in subsequent chapters.

13.19 Problems

1. How would a quadrant of a circle be detected optimally? How many planes would be required in parameter space?
2. Find formulae for the numbers of planes required in parameter space for optimal detection of (a) a parallelogram and (b) a rhombus.
3. Find how far the signal-to-noise ratio is below the optimum for an isosceles triangle of sides 1, 5 and 5 cm which is detected in a single plane parameter space.
4. By examining suitable binary images of corners, show that the median corner detector gives a maximal response within the corner boundary rather than half-way down the edge outside the corner. Show how the situation is modified for grey-scale images. How will this affect the value of the gradient noise-skimming threshold to be used in the improved median detector?
5. Sketch the response patterns for a $[-1 \ 2 \ -1]$ type of 1-D Laplacian operator applied to corner detection in lateral histograms for an image containing a triangle, (a) if the triangle is randomly oriented, and (b) if one or more sides are parallel to the x- or y-axis. In case (b), how is the situation resolved?
6. Prove equation (13.6), starting with the following formula for curvature:

$$\kappa = (d^2y/dx^2)/[1 + (dy/dx)^2]^{3/2}$$

Hint: First express dy/dx in terms of the components of intensity gradient, remembering that the intensity gradient vector (I_x, I_y) is oriented along the edge normal; then replace the x, y variation by I_x, I_y variation in the formula for κ.

Part 3
Application-Level Processing

Part 3 covers a variety of topics related to the interpretation and application of vision in the real world. These topics include deduction of the presence and interrelations of objects in 2-D and 3-D scenes, examination of objects for inspection and assembly, consideration of the lighting and hardware systems required to acquire the images and perform the necessary processing in real time, and further detail of recognition processes. In fact, the chosen topics are so variegated, and interact with each other at such a variety of different levels, that there is no really natural ordering. Nevertheless, the reader should not have too much difficulty in finding relevant information. Finally, it should be noted that Appendix C is strongly linked with the work of Part 3, and specifically Chapters 15–17.

14

Abstract Pattern Matching Techniques

14.1 Introduction

In the foregoing chapters it has been seen how objects having quite simple shapes may be located in digital images via the Hough transform. For more complex shapes this approach tends to require excessive computation: in general, the way this problem may be overcome is to locate objects from their features. Suitable salient features include small holes, corners, straight, circular or elliptical segments, and indeed any readily localizable subpatterns: earlier chapters have shown how such features may be located. However, at some stage it becomes necessary to find methods for collating the information from the various features, in order to recognize and locate the objects containing them. This task is studied in the present chapter.

It is perhaps easiest to envisage the feature collation problem when the features themselves are unstructured points carrying no directional information—nor indeed any attributes other than their x, y coordinates in the image. Then the object recognition task is often called "point pattern matching"[*]. The features that are closest to unstructured points are small holes, such as the "docker" holes in many types of biscuit. Corners can also be considered as points if their other attributes—including sharpness, orientation, etc.—are ignored. In what follows we start with point features and then see how the attributes of more complex types of feature can be included in recognition schemes.

---

[*] Note that this term is sometimes used not just for object recognition but for initial matching of two stereo views of the same scene.

Overall, it is worth bearing in mind that it is highly efficient to use small high-contrast features for object detection, since the computation involved in searching an image decreases with the size of the template used. As will be clear from the preceding chapters, the main disadvantage resulting from such an approach to object detection is the loss in sensitivity (in a signal-to-noise sense) due to the greatly impoverished information content of the point feature image. However, even the task of identifying objects from a rather small number of point features is far from trivial and frequently involves considerable computation, as will be seen below. We start by studying a graph-theoretic approach to point pattern matching which involves the "maximal clique" concept.

14.2   A graph-theoretic approach to object location

This section considers a commonly occurring situation which involves considerable constraints—objects appearing on a horizontal worktable or conveyor at a known distance from the camera. It is also assumed (i) that objects are flat or can appear in only a restricted number of stances in three dimensions, (ii) that objects are viewed from directly overhead, and (iii) that perspective distortions are small. In such situations the objects may in principle be identified and located from very few point features. Since such features are taken to have no structure of their own, it will be impossible to locate an object uniquely from a single feature, although positive identification and location would be possible using two features if these were distinguishable and if their distance apart were known. For truly indistinguishable point features, an ambiguity remains for all objects not possessing 180° rotation symmetry. Hence at least three point features are in general required to locate and identify objects at known range. Clearly, noise and other artefacts such as occlusions modify this conclusion. In fact, when matching a template of the points in an idealized object with the points present in a real image, we may find:

(i)   a great many feature points may be present because of multiple instances of the chosen type of object in the image;

(ii) additional points may be present because of noise or clutter from irrelevant objects and structure in the background;

(iii) certain points that should be present are missing because of noise or occlusion, or because of defects in the object being sought.

These problems mean that we should in general be attempting to match a subset of the points in the idealized template to various subsets of the points in the image. If the point sets are considered to constitute *graphs* with the point features as *nodes*, the task devolves into the mathematical problem of subgraph-subgraph isomorphism, i.e. finding which subgraphs in the image graph are isomorphic[*] to subgraphs of the idealized template graph. Of course, there may be a large number of matches involving rather few points: these would arise from sets of features that *happen* (see for example item (ii) above) to lie at valid distances apart in the original image. The most significant matches will involve a fair number of features and will lead to correct object identification and location. Clearly, a point feature matching scheme will be most successful if it finds the most likely interpretation by searching for solutions with the greatest internal consistency—i.e. with the greatest number of point matches per object.

Unfortunately, the scheme of things presented above is still too simplistic in many applications as it is insufficiently robust against distortions. In particular, optical (e.g. perspective) distortions may arise, or the objects themselves may be distorted, or by resting partly on other objects they may not be quite in the assumed stance: hence distances between features may not be exactly as expected. These factors mean that some tolerance has to be accepted in the distances between pairs of features and it is common to employ a threshold such that interfeature distances have to agree within this tolerance before matches are accepted as potentially valid. Clearly, distortions lay more strain on the point matching technique and make it all the more necessary to seek solutions with the greatest possible internal consistency. Thus as many features as possible should be taken into account in locating and identifying objects. The maximal clique approach is intended to achieve this.

As a start, as many features as possible are identified in the original image and these are numbered in some convenient order such as the order of appearance in a normal TV raster scan. The numbers then have to be matched against the letters corresponding to the features on the idealized object. A systematic way of achieving this is by constructing a *match graph* (or *association graph*) in which the nodes represent feature assignments, and arcs joining nodes represent pairwise compatibilities between assignments. To find the best match it is then necessary to find regions of the match graph where the cross-linkages are maximized. To achieve this, *cliques* are sought within the match graph. A clique is a *complete subgraph*—i.e. one for which all pairs of nodes are connected by arcs. However, the previous arguments indicate that if one clique is completely

---

[*] i.e. of the same basic shape and structure.

included within another clique, it is likely that the larger clique represents a
better match—and indeed *maximal cliques* can be taken as leading to the
most reliable matches between the observed image and the object model.

Figure 14.1(a) illustrates the situation for a general triangle: for simplicity, the figure takes the observed image to contain only one triangle and
assumes that lengths match exactly and that no occlusions occur. The match
graph in this example is shown in Fig. 14.1(b): there are nine possible
feature assignments, six valid compatibilities and four maximal cliques,
only the largest corresponding to an exact match.

Figure 14.2(a) shows the situation for the less trivial case of a quadrilateral, the match graph being shown in Fig. 14.2(b). In this case there are 16
possible feature assignments, 12 valid compatibilities and seven maximal
cliques. If occlusion of a feature occurs, this will (taken on its own) reduce
the number of possible feature assignments and also the number of valid
compatibilities: in addition, the number of maximal cliques and the size of

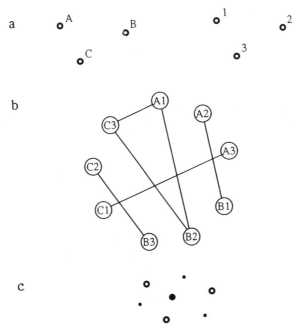

Fig. 14.1 A simple matching problem—a general triangle: (a) basic labelling of
model (*left*) and image (*right*); (b) match graph; (c) placement of votes in
parameter space. In (b) the maximal cliques are: (1) A1, B2, C3; (2) A2, B1; (3)
B3, C2; and (4) C1, A3. In (c) the following notation is used: O, positions of
observed features; ●, positions of votes, ●, position of main voting peak.

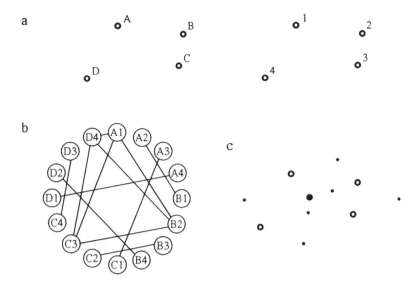

Fig. 14.2 Another matching problem—a general quadrilateral: (a) basic labelling of model (*left*) and image (*right*); (b) match graph; (c) placement of votes in parameter space (notation as in Fig. 14.1).

the largest maximal clique will be reduced. On the other hand, noise or clutter can add erroneous features. If the latter are at arbitrary distances from existing features, then the number of possible feature assignments will be increased but there will not be any more compatibilities in the match graph, so the latter will have only trivial additional complexity. However, if the extra features appear at *allowed* distances from existing features, this will introduce extra compatibilities into the match graph and make it more tedious to analyse. In the case shown in Fig. 14.3, both types of complication—an occlusion and an additional feature—arise: there are now eight pairwise assignments and six maximal cliques, rather fewer overall than in the original case of Fig. 14.2. However, the important factor is that the largest maximal clique still indicates the most likely interpretation of the image and that the technique is inherently highly robust.

When using methods such as the maximal clique approach which involve repetitive operations, it is useful to look for means of saving computation. It turns out that when the objects being sought possess some symmetry, economies can be made. Consider the case of a parallelogram (Fig. 14.4). Here the match graph has 20 valid compatibilities and there are ten maximal cliques. Of these, the largest two have equal numbers of nodes and *both*

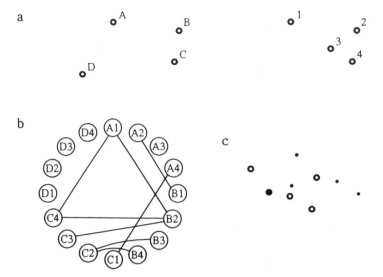

Fig. 14.3 Matching when one feature is occluded and another is added: (a) basic labelling of model (*left*) and image (*right*); (b) match graph; (c) placement of votes in parameter space (notation as in Fig. 14.1).

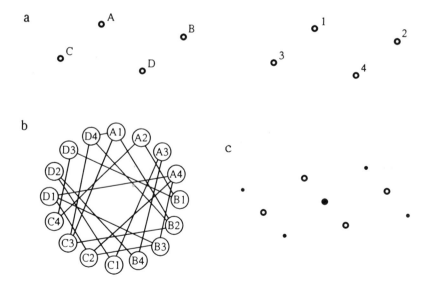

Fig. 14.4 Matching a figure possessing some symmetry: (a) basic labelling of model (*left*) and image (*right*); (b) match graph; (c) placement of votes in parameter space (notation as in Fig. 14.1).

identify the parallelogram within a symmetry operation. This means that the maximal clique approach is doing more computation than absolutely necessary: this can be avoided by producing a new "symmetry-reduced" match graph after relabelling the model template in accordance with the symmetry operations (see Fig. 14.5). This gives a much smaller match graph with half the number of pairwise compatibilities and half the number of maximal cliques. In particular, there is only one nontrivial maximal clique: note, however, that its size is not reduced by the application of symmetry.

14.2.1 A practical example—locating cream biscuits

Figure 14.6(a) shows one of a pair of cream biscuits which are to be located from their "docker" holes—this strategy being advantageous since it has the potential for highly accurate product location prior to detailed inspection (in this case the purpose is to locate the biscuits accurately from the holes, and then to check the alignment of the biscuit wafers and detect any excess cream around the sides of the product). The holes found by a simple template matching routine are indicated in Fig. 14.6(b): the template used is rather small and, as a result, the routine is fairly fast but fails to locate all

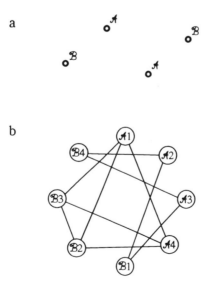

Fig. 14.5 Using a symmetry-reduced match graph: (a) relabelled model template; (b) symmetry-reduced match graph.

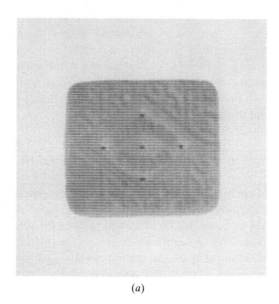

(a)

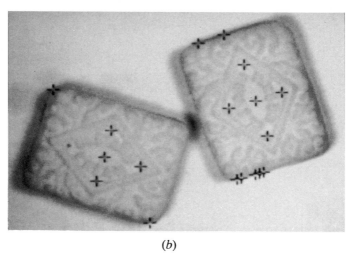

(b)

Fig. 14.6 (a) A typical cream sandwich biscuit; (b) a pair of cream sandwich biscuits with crosses indicating the result of applying a simple hole detection routine; (c) the two biscuits reliably located by the GHT from the hole data in (b): the isolated small crosses indicate the positions of single votes.

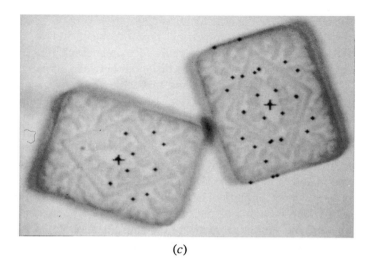

(c)

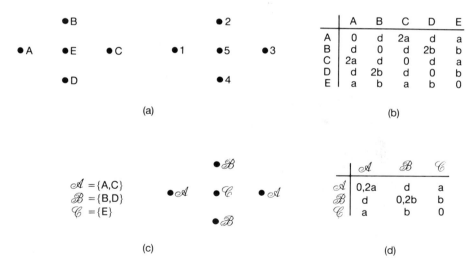

	A	B	C	D	E
A	0	d	2a	d	a
B	d	0	d	2b	b
C	2a	d	0	d	a
D	d	2b	d	0	b
E	a	b	a	b	0

(a) (b)

𝒜 = {A,C}
ℬ = {B,D}
𝒞 = {E}

	𝒜	ℬ	𝒞
𝒜	0,2a	d	a
ℬ	d	0,2b	b
𝒞	a	b	0

(c) (d)

Fig. 14.7 Interfeature distances for holes on cream biscuits: (a) basic labelling of model (*left*) and image (*right*); (b) allowed distance values; (c) revised labelling of model using symmetric set notation; (d) allowed distance values. The cases of zero interfeature distance in the final table can be ignored as they do not lead to useful matches.

holes; in addition, it can give false alarms. Hence an "intelligent" algorithm must be used to analyse the hole location data.

Clearly this is a highly symmetrical type of object and so it should be beneficial to employ the symmetry-reduced match graph described above. To proceed it is helpful to tabulate the distances between all pairs of holes in the object model (Fig. 14.7(b)). Then this table can be regrouped to take account of symmetry operations (Fig. 14.7(d)). This will help when we come to build the match graph for a particular image. Analysis of the data in the above example shows that there are two nontrivial maximal cliques, each corresponding correctly to one of the two biscuits in the image. Note, however, that the reduced match graph does not give a *complete* interpretation of the image: it locates the two objects but it does not confirm uniquely which hole is which. In particular, for a given starting hole of type $\mathcal{A}$, it is not known which is which of the two holes of type $\mathcal{B}$. This can be ascertained by applying simple geometry to the coordinates in order to determine (say) which hole of type $\mathcal{B}$ is reached by moving around the centre hole E in a clockwise sense.

14.3 Possibilities for saving computation

In these examples the checking of which subgraphs are maximal cliques is a simple problem. However, in real matching tasks it can quickly become unmanageable (the reader is encouraged to draw the match graph for an image containing two objects of seven points!).

Table 14.1 shows what is perhaps the most obvious type of algorithm for finding maximal cliques. It operates by examining in turn all cliques of a given number of nodes and finding what cliques can be constructed from them by adding additional nodes (bearing in mind that any additional nodes must be compatible with all existing nodes in the clique). This permits all cliques in the match graph to be identified. However, an additional step is needed to eliminate (or relabel) all cliques that are included as subgraphs of a new larger clique before it is known which cliques are maximal.

In view of the evident importance of finding maximal cliques, many algorithms have been devised for the purpose. It is probable that the best of these is now close to the fastest possible speed of operation. Unfortunately, the optimum execution time is known to be bounded not by a polynomial in M (for a match graph containing maximal cliques of up to M nodes) but by a much faster varying function. Specifically, the task of finding maximal cliques is akin to the well-known travelling salesman problem and is known to be "NP-complete", implying that it runs in exponential time (see

Table 14.1 A simple maximal clique algorithm.

```
set clique size to 2;
( * this is the size already included by the match graph * )
while newcliques = true ( * new cliques still being found * ) do
  begin
    increment clique size;
    set newcliques = false;
    for all cliques of previous size do
      begin
        set all cliques of previous size to status maxclique;
        for all possible extra nodes do
        if extra node is joined to all existing nodes in clique
          then
            begin
              store as a clique of current size;
              set newcliques = true;
            end;
      end;
    ( * the larger cliques have now been found * )
    for all cliques of current size do
      for all cliques of previous size do
        if all nodes of smaller clique are included
          in current clique
            then set smaller clique to status not maxclique;
    ( * the subcliques have now been relabelled * )
  end;
```

Section 14.4.1). For a single-processor machine, run times may be a few seconds for values of M up to about 6, and of the order of minutes for values of M up to about 12: for M greater than 15 some *hours* of processing may be required. In practical situations there are several ways of tackling this problem:

(i) use the symmetry-reduced match graph wherever possible;
(ii) choose the fastest available maximal clique algorithm;
(iii) write critical loops of the maximal clique algorithm in machine code;
(iv) build special hardware or multiprocessor systems to implement the algorithm;
(v) use the LFF method (see below: this means searching for cliques of small M and then working with an alternative method);

(vi) use an alternative sequential strategy, which may however not be guaranteed to find all the objects in the image;

(vii) use the GHT approach (see Section 14.4).

Of these methods, the first should be used wherever applicable. Methods (ii)–(iv) amount to improving the implementation and are subject to diminishing returns: it should be noted that the execution time varies so rapidly with M that even the best software implementations are unlikely to give a practical increase in M of more than 2 (i.e. $M \rightarrow M + 2$). Likewise, dedicated hardware implementations may only give increases in M of the order of 4–6. Method (v) is a "short-cut" approach which proves highly effective in practice. The idea is to search for specific subsets of the features of an object, and then to hypothesize that the object exists and go back to the original image to check that it is actually present. Bolles and Cain (1982) devised this method when looking for hinges in quite complex images. In principle, the method has the disadvantage that the particular subset of an object that is chosen as a cue may be missing because of occlusion or some other artefact. Hence it may be necessary to look for several such cues on each object. This is an example of further deviation from the matched filter paradigm, which reduces detection sensitivity yet again. The method is called the local-feature-focus (LFF) method because objects are sought by cues or local foci.

The maximal clique approach is a type of exhaustive search procedure and is effectively a parallel algorithm. This has the effect of making it highly robust but is also the source of its slow speed. An alternative is to perform some sort of sequential search for objects, stopping when sufficient confidence is attained in the interpretation or partial interpretation of the image. For example, the search process may be terminated when a match has been obtained for a certain minimum number of features on a given number of objects. Such an approach may be useful in some applications and will generally be considerably faster than the full maximal clique procedure when M is greater than about 6. An analysis of several tree-search algorithms for subgraph isomorphism was carried out by Ullmann (1976): the paper tests algorithms using artificially generated data and it is not clear how they relate to real images. The success or otherwise of all nonexhaustive search algorithms must, however, depend critically on the particular types of image data being analysed: hence it is difficult to give further general guidance on this matter (but see Section 14.7 below for additional comments on search procedures).

The final method listed above is based on the GHT. In many ways this provides an ideal solution to the problem since it presents an exhaustive search technique that is essentially equivalent to the maximal clique approach, while not falling into the NP-complete category. This may seem contradictory, since any approach to a well-defined mathematical problem should be subject to the mathematical constraints known to be involved in its solution. However, although the abstract maximal clique problem is known to be NP-complete, the *subset* of maximal clique problems that arises from 2-D image-based data may well be solved with less computation by other means, and in particular by a 2-D technique. This special circumstance does appear to be valid but unfortunately it offers no possibility of solving general NP-complete problems by reference to the specific solutions found using the GHT approach! The GHT approach is described in the following section.

14.4 Using the generalized Hough transform for feature collation

This section describes how the GHT can be used as an alternative to the maximal clique approach, to collate information from point features in order to find objects. Initially, we consider situations where objects have no symmetries—as for the cases of Figs 14.1–14.3.

To apply the GHT, we first list all features and then accumulate votes in parameter space at every possible position of a localization point L consistent with each *pair* of features (Fig. 14.8). This strategy is particularly suitable in the present context, as it corresponds to the pairwise assignments used in the maximal clique method. To proceed it is necessary merely to use the interfeature distance as a lookup parameter in the GHT *R*-table. For indistinguishable point features this means that there must be two entries for the position of L for each value of the interfeature distance. Note that we have assumed that no symmetries exist and that all pairs of features have

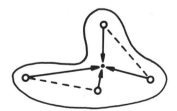

Fig. 14.8 Method for locating L from pairs of feature positions: each pair of feature points gives two possible voting positions in parameter space, when objects have no symmetries. When symmetries are present, certain pairs of features may give rise to up to four voting positions: this is confirmed on careful examination of Fig. 14.6(c).

different interfeature distances. If this were not so, then more than two vectors would have to be stored in the R-table per interfeature distance value.

To illustrate the procedure it is applied first to the triangle example of Fig. 14.1. Figure 14.1(c) shows the positions at which votes are accumulated in parameter space. There are four peaks with heights of 3, 1, 1, 1, it being clear that, in the absence of complicating occlusions and defects, the object is locatable at the peak of maximum size. Next, the method is applied to the general quadrilateral example of Fig. 14.2: this leads to seven peaks in parameter space, whose sizes are 6, 1, 1, 1, 1, 1, 1 (Fig. 14.2(c)).

Close examination of Figures 14.1–14.3 indicates that every peak in parameter space corresponds to a maximal clique in the match graph. Indeed, there is a one-to-one relation between the two. In the uncomplicated situation being examined here, this is bound to be so for any general arrangement of features within an object, since every pairwise compatibility between features corresponds to two potential object locations, one correct and one that can be correct only from the point of view of that pair of features. Hence the correct locations all add to give a large maximal clique and a large peak in parameter space, whereas the incorrect ones give maximal cliques each containing two wrong assignments and each corresponding to a false peak of size 1 in parameter space. This situation still applies even when occlusions occur or additional features are present (see Fig. 14.3). The situation is slightly more complicated when symmetries are present, the two methods each deviating in a different way: space does not permit the matter to be explored in depth here but the solution for the case of Fig. 14.4(a) is presented in Fig. 14.4(c). Overall, it seems simplest to assume that there is still a one-to-one relationship between the solutions from the two approaches.

Finally, consider again the example of Section 14.2.1 (Fig. 14.6(a)), this time obtaining a solution by the GHT. Figure 14.6(c) shows the positions of candidate object centres as found by the GHT. The small isolated crosses indicate the positions of single votes, and those very close to the two large crosses lead to voting peaks of weights 10 and 6 at these respective positions. Hence object location is both accurate and robust, as required.

14.4.1 Computational load

This subsection compares the computational requirements of the maximal clique and GHT approaches to object location. For simplicity, imagine an image that contains just one wholly visible example of the object being sought. Also, suppose that the object possesses n features and that we are trying to recognize it by seeking all possible pairwise compatibilities, whatever their distance apart (as for all examples in Section 14.2).

For an object possessing n features, the match graph contains n^2 nodes (i.e. possible assignments), and there are

$$\binom{n^2}{2} = n^2(n^2 - 1)/2$$

possible pairwise compatibilities to be checked in building the graph. The amount of computation at this stage of the analysis is $O(n^4)$. To this must be added the cost of finding the maximal cliques. Since the problem is NP-complete, the load rises at a rate which is faster than polynomial, and probably exponential in n^2 (Gibbons, 1985).

Now consider the cost of getting the GHT to find objects via pairwise compatibilities. As has been seen, the total height of all the peaks in parameter space is in general equal to the number of pairwise compatibilities in the match graph. Hence the computational load is of the same order, $O(n^4)$. Next comes the problem of locating all the peaks in parameter space. In this case parameter space is congruent to image space. Hence for an $N \times N$ image only N^2 points have to be visited in parameter space and the computational load is $O(N^2)$. Note, however, that an alternative strategy is available in which a running record is kept of the relatively small numbers of voting positions in parameter space. The computational load for this strategy will be $O(n^4)$: although of a higher *order*, this often represents less computation in practice.

The reader may have noticed that the basic GHT scheme as outlined so far is able to locate objects from their features but does not determine their orientations. However, orientations can be computed by running the algorithm a second time and finding all the assignments that contribute to each peak. Alternatively, the second pass can aim to find a different localization point within each object. In either case the overall task should be completed in little over twice the time, i.e. still in $O(n^4 + N^2)$ time.

Although the GHT at first appears to solve the maximal clique problem in polynomial time, what it actually achieves is to solve a real-space template matching problem in polynomial time; as remarked earlier, it does not solve an *abstract* graph-theoretic problem in polynomial time. The overall lesson is that the graph theory representation is not well matched to real space, not that real space can be used to solve abstract NP-complete problems in polynomial time.

14.5   Generalizing the maximal clique and other approaches

This section considers how the graph matching concept can be generalized to cover alternative types of feature and also various attributes of features. The earlier discussion was restricted to point features and in particular to small holes which were supposed to be isotropic. Corners were also taken as point features by ignoring attributes other than position coordinates. Both holes and corners seem to be ideal, in that they give maximum localization and hence maximum accuracy for object location. Straight lines and straight edges at first appear to be rather less well suited to the task. However, more careful thought shows that this is not so. One possibility is to use straight lines to deduce the positions of corners which can then be used as point features, although this approach is not as powerful as might be hoped because of the abundance of irrelevant line crossings that are thrown up (in this context, note that the maximal clique method is inherently capable of sorting the true corners from the false alarms). A more elegant solution is simply to determine the angles between pairs of lines, referring to a lookup table for each type of object to determine whether each pair of lines should be marked as compatible in the match graph. Once the match graph has been built, an optimal match can be found as before (although ambiguities of scale will arise which will have to be resolved by further processing). These possibilities significantly generalize the ideas of the foregoing sections.

Other types of feature generally have more than two specifying parameters, one of which may be contrast and the other size. This applies for most holes and circular objects, although for the smallest (i.e. barely resolvable) holes it is sometimes most practicable to take the central dip in intensity as the measured parameter. For straight lines the relevant size parameter is the length (we here count the line ends, if visible, as points which will already have been taken into account). Corners may have a number of attributes, including contrast, sharpness and orientation—none of which is likely to be known to high accuracy. Finally, more complex shapes such as ellipses have orientation, size and eccentricity, and again contrast may be a usable attribute (generally, the latter is a less reliable measure because of possible variation in the background).

One other attribute is worthy of mention and that is colour. Many features have a measurable colour and clearly this may be used to aid recognition.

With so much information, it is pertinent to ask how it may be used to help locate objects. For convenience this is discussed in relation to the maximal clique method. In fact, the answer is very simple. When compatibilities are being considered and the arcs are being drawn in the match graph, *any* available information may be taken into account in deciding whether a pair of features in the image matches a pair of features in the object model. In Section 14.2 the discussion was simplified by taking interfeature distances as the only relevant measurements. However, it is quite acceptable to describe the features in the object model more fully and to insist that they all match within prespecified tolerances. For example, holes and corners may be permitted to lead to a match only if the former are of the correct size, the latter are of the correct sharpness and orientation, and the distances between these features are also appropriate. All relevant information has to be held in suitable lookup tables. In general, the gains easily outweigh the losses, since a considerable number of potential interpretations will be eliminated—hence making the match graph significantly simpler and reducing, in many cases by a large factor, the amount of computation that is required to find the maximal cliques. Note, however, that there is a limit to this, since in the absence of occlusions and erroneous tolerances on the additional attributes, the number of nodes in the match graph cannot be less than the square of the number of features in an object, and the number of nodes in a maximal clique will be unchanged.

Thus extra feature attributes are of very great value in cutting down computation: they are also useful in making interpretation less ambiguous. This latter property is "obvious" but not always realizable. In particular, extra attributes help in this way only if (i) some of the features on an object are missing, through occlusion or for other reasons such as breakage, or (ii) if the distance tolerances are so large as to make it unclear which features in the image match with those in the model.

Suppose next that the distance attributes become very imprecise, either because of shape distortions or else because of unforeseen rotations in 3-D. It is worth enquiring how far we can proceed under these circumstances. In fact, in the limit of low distance accuracy, we may only be able to employ an "adjacent to" descriptor. This parallels the situation in general scene analysis, where use is made of a number of relational descriptors such as "on top of", "to the left of", and so on. Such possibilities are considered in the next section.

14.6   Relational descriptors

The previous section showed how additional attributes could be incorporated into the maximal clique formalism so that the effects of diminishing accuracy of distance measurement could be accommodated. This section considers what happens when the accuracy of distance measurement drops to zero and we are left just with relational attributes such as "adjacent to", "near", "inside", "underneath", "on top of", and so on. We start by taking adjacency as the basic relational attribute. To illustrate the approach, imagine a simple outdoor scene where various rules apply to segmented regions. These rules will be of the type "sky may be adjacent to forest", "forest may be adjacent to field", and so on. Note that "adjacent to" is not transitive, i.e. if P is adjacent to Q, and Q adjacent to R, this does not imply that P is adjacent to R (in fact, it is quite likely that P will not be adjacent to R, as Q may well separate the two regions completely!).

The rules for a particular type of scene may be summarized as in Table 14.2. Now consider the scene shown in Fig. 14.9. Applying the rules for adjacency from Table 14.2 will be seen to permit four different solutions, in which regions 1–3 are respectively:

   (i)   sky, forest, field (the correct interpretation);
  (ii)   sky, forest, sky;
 (iii)   field, forest, sky; or
 (iv)   field, forest, field.

Evidently there are too few constraints. Possible constraints are the following: *sky is above field and forest, sky is blue, field is green*, and so on. Of these, the first is a binary relation like adjacency, while the other two are unary constraints. It is easy to see that two such constraints are required to resolve completely the ambiguity in this particular example.

Paradoxically, adding further regions can make the situation inherently less ambiguous. This is because other regions are less likely to be adjacent to all the original regions, and in addition may act in such a way as to label

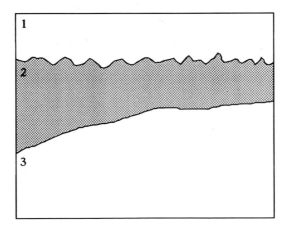

Fig. 14.9   Regions in a simple scene: 1, sky; 2, forest; 3, field.

Table 14.2 Adjacency table for simple scene.

	Sky	Forest	Field
Sky	—	✓	—
Forest	✓	—	✓
Field	—	✓	—

This table is relevant to the scene in Fig. 14.9. Ticks indicate that regions must be adjacent: dashes indicate that regions must not be adjacent (i.e. adjacency is not optional).

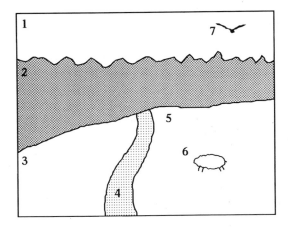

Fig. 14.10 Regions in a more complex scene: 1, sky; 2, forest; 3, field; 4, road; 5, field; 6, sheep; 7, bird.

them uniquely. This is seen in Fig. 14.10 which is interpreted by reference to Table 14.3, although the keys to unique interpretation are the notions that *any small white objects must be sheep*, and *any small dark objects must be birds*. On analysing the available data the whole scene can now be interpreted unambiguously. As expected, the maximal clique technique can be applied successfully, although to achieve this assignments must be taken to be compatible not only when image regions are adjacent and their interpretations are marked as adjacent, but also when image regions are nonadjacent and their interpretations are marked as nonadjacent (the reason for this is easily seen by drawing an analogy between adjacency and distance, adjacent and nonadjacent corresponding respectively to zero and a significant distance apart). Note that in this scene analysis type of situation, it is *very* tedious to apply the maximal clique method manually because there tend to

Table 14.3   Adjacency table for more complex scene.

	Sky	Forest	Field	Road	Sheep	Bird
Sky	—	✓	—	—	—	✓
Forest	✓	—	✓	✓	—	—
Field	—	✓	—	✓	✓	—
Road	—	✓	✓	—	—	—
Sheep	—	—	✓	—	—	—
Bird	✓	—	—	—	—	—

This table is relevant to the scene in Fig. 14.10. Ticks again indicate adjacency and dashes indicate nonadjacency.

be a fair number of trivial solutions (i.e. maximal cliques with just a few nodes).

Perhaps oddly, some of the trivial solutions that the maximal clique technique gives rise to are provably *incorrect*. For example, in the above problem, one solution appears to include regions 1 and 2 being forest and sky, respectively, whereas it is clear from the presence of birds that region 1 *cannot* be forest and *must* be sky. Here logic appears to be a stronger arbiter of correctness than an evidence building scheme such as the maximal clique technique. However, on the plus side, the maximal clique technique will take contradictory evidence and produce the best possible response. For example, if noise appears and introduces an invalid "bird" in the forest, it will simply ignore it, whereas a purely logic-based scheme will be unable to find a satisfactory solution: clearly, much depends on the accuracy and universality of the assertions on which these schemes are based.

Logical problems such as the ones outlined above are commonly tackled with the aid of declarative languages such as PROLOG, in which the rules are written explicitly in a standard form not dissimilar to IF statements in English. Space does not permit a full discussion of PROLOG here. However, it should be noted that PROLOG is basically designed to obtain a single "logical" solution where one exists. It can also be instructed to search for all available solutions or for any given number of solutions. When instructed to search for all solutions, it will finally arrive at the same main set of solutions as the maximal clique approach (although, as indicated above, noise can make the situation more complicated). Thus, despite their apparent differences, these approaches are quite similar: it is the implementation of the underlying search problem that is different (PROLOG uses a "depth-first" search strategy—see below). Next, we move on to another scheme—that of relaxation labelling—which has a quite different formulation.

Relaxation labelling is an iterative technique in which evidence is gradually built up for the proper labelling of the solution space—in this case, of the various regions in the scene. Two main possibilities exist. In the first, called *discrete relaxation*, evidence for pairs of labels is examined and a label discarding rule is instigated which eliminates at each iteration those pairs of labels that are currently inconsistent; this technique is applied until no further change in the labelling occurs. The second possibility is that of *probabilistic relaxation*; here the labels are set up numerically to correspond to the probabilities of a given interpretation of each region, i.e. a table is compiled of regions against possible interpretations, each entry being a number representing the probability of that particular interpretation. After providing a suitable set of starting probabilities (possibly all being weighted equally), these are updated iteratively, and in the ideal case they converge on the values 0 or 1 to give a unique interpretation of the scene. Unfortunately, convergence is by no means guaranteed and can depend on the starting probabilities and the updating rule. A prearranged constraint function defines the underlying process, and this could in principle be either better or worse at matching reality than (for example) the logic programming techniques that are used in PROLOG. Relaxation labelling is a complex optimization process and is not discussed further here. Suffice it to say that the technique often runs into problems of excessive computation. The reader is referred to the seminal paper by Rosenfeld *et al.* (1976) and other papers mentioned in Section 14.9 for further details.

14.7 Search

The above sections have shown how the maximal clique approach may be used to locate objects in an image, or alternatively to label scenes according to predefined rules about what arrangements of regions are expected in scenes. In either case, the basic process being performed is that of search for solutions that are compatible with the observed data. This search takes place in assignment space, i.e. a space in which all combinations of assignments of observed features with possible interpretations exist. The problem is that of finding one or more valid sets of observed assignments.

It generally happens that the search space is very large, so that an exhaustive search for all solutions would involve enormous computational effort and would take considerable time. Unfortunately, one of the most obvious and appealing methods of obtaining solutions, the maximal clique approach, is NP-complete and can require impracticably large amounts of time to find all the solutions. It is therefore useful to clarify the nature of the maximal clique approach: to achieve this, we first describe the two main categories of search—breadth-first and depth-first search.

Breadth-first search is a form of search that systematically works down a tree of possibilities, never taking short-cuts to nearby solutions. Depth-first search, in contrast, involves taking as direct a path as possible to individual solutions, stopping the process when a solution is found and backtracking up the tree whenever a wrong decision is found to have been made. It is normal to curtail the depth-first search when sufficient solutions have been found and this means that much of the tree of possibilities will not have been explored. Although breadth-first search can be curtailed similarly when enough solutions have been found, the maximal clique approach as described earlier is in fact a form of breadth-first search that is exhaustive and runs to completion.

In addition to being an exhaustive breadth-first search, the maximal clique approach may be described as being "blind" and "flat"—i.e. it involves neither heuristic nor hierarchical means of guiding the search. In fact, faster search methods involve guiding the search in various ways. First, heuristics are used to specify at various stages in which direction to proceed (which node of the tree to expand), or which paths to ignore (which nodes to prune). Second, the search can be made more "hierarchical", so that it searches first for outline features of a solution, returning later (perhaps in several stages) to fill in the details. Details of these techniques are omitted here. However, an interesting approach was used by Rummel and Beutel (1984): they searched images for industrial components using features such as corners and holes, alternating at various stages between breadth-first and depth-first search by using a heuristic based on a dynamically adjusted parameter: this being computed on the basis of how far the search is still away from its goal, and the quality of the fit so far. Rummel and Beutel noted the existence of a tradeoff between speed and accuracy as a "guide factor", based on the number of features required for recognition, is adjusted—the problem being that trying to increase speed introduces some risk of not finding the optimum solution.

14.8 Concluding remarks

This chapter has discussed the problem of recognizing objects from their features, and has also considered the related task of scene analysis. The maximal clique approach was seen to be capable of finding solutions to both of these tasks, although ultimately these are search problems, so a much greater range of methods is applicable to each. In particular, blind, flat exhaustive breadth-first search (i.e. the maximal clique method) involves considerable computation and is often best replaced by guided depth-first search, with suitable heuristics being devised to guide the search. In addition, it should be noted that languages such as PROLOG can implement depth-first search and that rather different procedural techniques such as relaxation labelling are available and should be considered (although these are subject to their own complexities and computational problems).

It is of interest that the task of recognizing objects from their features tends to involve considerable computation and that the GHT can provide a satisfactory solution to these problems. This is probably because the graph theory representation is not well matched to the relevant real-space template matching task in the way that the GHT is. Here, recall what was noted in Chapter 10—that the GHT is particularly well suited to object detection in real space as it is one type of spatial matched filter. Indeed, the maximal clique approach can be regarded as a rather inefficient substitute for the GHT form of the spatial matched filter. Furthermore, the LFF method takes a short-cut to save computation and this makes it less like a spatial matched filter, thereby adversely affecting its ability to detect objects amongst noise and clutter. Note that for scene analysis, where relational descriptors rather than precise dimensional (binary) attributes are involved, the GHT does not provide any very obvious possibilities—probably because we are here dealing with much more abstract data that are linked to real space only at a very high level (but see the work of Kasif *et al.*, 1983).

Finally, note that, on an absolute scale, the graph matching approach takes very little note of detailed image structure, using at most only pairwise feature attributes. This is adequate for 2-D image interpretation but inadequate for situations such as 3-D image analysis where there are more degrees of freedom to contend with (normally three degrees of freedom for position and three for orientation, for each object in the scene). Hence more specialized and complex approaches need to be taken in such cases: these are examined in the following chapter.

14.9 Bibliographical and historical notes

Graph matching and clique finding algorithms started to appear in the literature around 1970: for an early solution to the graph isomorphism problem, see Corneil and Gotlieb (1970). The subgraph isomorphism problem was tackled soon after by Barrow *et al.* (1972): see also Ullmann (1976). The double subgraph isomorphism (or subgraph–subgraph isomorphism) problem is commonly tackled by seeking maximal cliques in the match graph, and algorithms for achieving this have been described by Bron and Kerbosch (1973), Osteen and Tou (1973) and Ambler *et al.* (1975) (note that in 1989, Kehtarnavaz and Mohan reported preferring the algorithm of Osteen and Tou on the grounds of speed). Improved speed has also been achieved using the minimal match graph concept (Davies, 1991a).

Bolles (1979) applied the maximal clique technique to real-world problems (notably the location of engine covers) and showed how operation could be made more robust by taking additional features into account. By 1982 Bolles and Cain had formulated the local-feature-focus method, which (i) searches for restricted sets of features on an object, (ii) takes symmetry into account to save computation, and (iii) reconsiders the original image data in order to confirm a valid match: the paper gives various criteria for ensuring satisfactory solutions with this type of method.

Not satisfied by the speed of operation of maximal clique methods, other workers have tended to use depth-first search techniques. Rummel and Beutel (1984) developed the idea of alternating between depth-first and breadth-first search as dictated by the data—a powerful approach, although the heuristics that they used for this may well lack generality. Meanwhile, Kasif *et al.* (1983) showed how a modified GHT (the "relational HT") could be used for graph matching, although their paper gives few practical details. A somewhat different application of the GHT to perform 2-D matching (Davies, 1988g) was described in Section 14.4, and has been extended to optimize accuracy (Davies, 1992d) and also to improve speed (Davies, 1993b). Geometric hashing has been developed to perform similar tasks on objects with complex polygonal shapes (Tsai, 1996).

Relaxation labelling in scenes dates from the seminal paper by Rosenfeld *et al.* (1976); for later work on relaxation labelling and its use for matching, see Kitchen and Rosenfeld (1979), Hummel and Zucker (1983) and Henderson (1984); for rule-based methods in image understanding, see for example, Hwang *et al.* (1986); and for preparatory discussion and careful contrasting of these approaches, see Ballard and Brown (1982). Discussion of basic AI search techniques and rule-based systems may be found in Charniak and McDermott (1985). A useful reference on PROLOG is Clocksin and Mellish (1984).

14.10 Problems

1. Find the match graph for a set of features arranged in the form of an isosceles triangle. Find how much simplification occurs by taking account of symmetry and using the symmetry-reduced match graph.

Extend your results to the case of a kite (two isosceles triangles arranged symmetrically base to base).

2. Two lino-cutter blades (trapeziums) are to be located from their corners. Consider images in which two corners of one blade are occluded by the other blade. Sketch the possible configurations, counting the number of corners in each case. If corners are treated like point features with no other attributes, show that the match graph will lead to an ambiguous solution. Show further that the ambiguity can in general be eliminated if proper account is taken of corner orientation. Specify how accurately corner orientation would need to be determined for this to be possible.

3. In problem 14.2, would the situation be any better if the GHT were used?

15

The Three-Dimensional World

15.1 Introduction

In the foregoing chapters, it has on the whole been assumed that objects are essentially flat and are viewed from above in such a way that there are only three degrees of freedom—namely, the two associated with position, and a further one concerned with orientation. While this approach was adequate for carrying out many useful visual tasks, it is totally inadequate for interpreting outdoor or factory scenes or even for helping with quite simple robot assembly and inspection tasks. Indeed, over the past 10 years a considerable amount of quite sophisticated theory has been developed and backed up by experiment, to find how scenes composed of real 3-D objects can be understood in detail.

In general, this means attempting to interpret scenes in which objects may appear in totally arbitrary positions and orientations—corresponding to six degrees of freedom. Interpreting such scenes, and deducing the translation and orientation parameters of arbitrary sets of objects, is still a formidable computing task—partly because of the inherent ambiguity in inferring 3-D information from 2-D images. The parts of this task that are currently solvable typically take many minutes to complete on conventional computers.

A variety of approaches are now available for proceeding with 3-D vision. A single chapter will be unable to describe all of them but the intention here is to provide an overview, outlining the basic principles and classifying the methods according to generality, applicability and so on. While computer vision need not necessarily mimic the capabilities of the human eye–brain system, much research on 3-D vision has been aimed at biological modelling. This type of research shows that the human visual system makes use of a number of different methods simultaneously, taking appropriate cues from the input data and forming hypotheses about the content of a scene, progressively enhancing these hypotheses until a useful working model of what is present is produced. Thus individual methods are not expected to work in isolation but instead merely need to provide the model generator with whatever data become available. Clearly, biological machinery of various types may in principle lie idle for much of the time until triggered by specific input stimuli. Computer vision systems are currently less sophisticated than this and tend to be built on specific processing models, so that they can be applied efficiently to more restricted types of image data. In this chapter, we adopt the pragmatic view that particular methods need to be (or have been) developed for specific types of situation, and that they should be used only when appropriate—although some care is taken to elucidate what the appropriate types of application are.

15.2   Three-dimensional vision—the variety of methods

One of the most obvious characteristics of the human visual system is that it employs two eyes, and it is well known to the layman that binocular (or "stereo") vision permits depth to be discerned within a scene. However, the loss of vision that results when one eye is shut is relatively insignificant and is by no means a disqualification from driving a car or even an aeroplane. On the contrary, depth can readily be deduced in monocular vision from a plethora of cues that are buried in an image. Naturally, to achieve this the eye–brain system is able to call on a huge amount of prestored data about the physical world and about the types of object in it, be they man-made or natural entities. For example, the size of any car being viewed is strongly constrained; likewise, most objects have highly restricted sizes, both absolutely and in their depths relative to their frontal dimensions. Nevertheless, in a single view of a scene, it is normally impossible to deduce absolute sizes—all the objects and their depths can be scaled up or down by arbitrary factors and this cannot be discerned from a monocular view.

While it is clear that the eye–brain system makes use of a huge database relating to the physical world, there is much that can be learnt with negligible prior knowledge, even from a single monocular view. The main key to this is the "shape from shading" concept. For 3-D shape to be deducible from shading information (i.e. from the grey-scale intensities in an image), something has to be known about how the scene is lit—the simplest situation being when the scene is illuminated by a single point light source at a known position: note that indoors a single overhead tungsten light is still the most usual illuminator, while outdoors the sun performs a similar function. In either case an obvious result is that a single source will illuminate one part of an object and not another—which then remains in shadow—and parts that are orientated in various ways relative to the source and to the observer appear with different brightness values, so that orientation can in principle be deduced. In fact, as will be seen below, deduction of orientation and position is not at all trivial and may even be ambiguous. Nevertheless, successful methods have been developed for carrying out this task. One problem that often arises is that the position of the light source is unknown but this information can generally be extracted (at least by the eye) from the scene being examined, so a bootstrapping procedure is then able to unlock the image data gradually and proceed to an interpretation.

While these methods enable the eye to interpret real scenes, it is difficult to say quite to what degree of precision they are carried out. With computer vision, the required precision levels are liable to be higher, although the machine will be aided by knowing exactly where the source of illumination is. However, with computer vision we can go further and arrange artificial lighting schemes that would not appear in nature, so the computer can acquire an advantage over the human visual system. In particular, a set of light sources can be applied in sequence to the scene—an approach known as photometric stereo—which can in certain cases help the computer to interpret the scene more rigorously and efficiently. In other cases structured light may be applied: this means projecting onto the scene a pattern of spots or stripes, or even a grid of lines, and measuring their positions in the resulting image. By this means depth information can be obtained much as for pairs of stereo images.

Finally, a number of methods have been developed for analysing images on the basis of readily identifiable sets of features. These methods are the 3-D analogues of the graph matching and GHT approaches of Chapter 14. However, they are significantly more complex because they generally involve six degrees of freedom in place of the three assumed throughout Chapter 14. It should also be noted that such methods make strong assumptions about the particular objects to be located within the scene. In general situations it is unlikely that such assumptions could be made, and so initial analysis of any images must be made on the basis that the entire scene must be mapped out in 3-D, then 3-D models built up and finally deductions must be made by noticing what relation one part of the scene bears to another part. Note that if a scene is composed from an entirely new set of objects, all that can be done is to *describe* what is present and say perhaps

what the set most closely *resembles*: recognition *per se* cannot be performed. It should be noticed that scene analysis is—at least from a single
monocular image—an inherently ambiguous process: every scene can have
a number of possible interpretations and there is evidence that the eye looks
for the simplest and most probable explanation rather than an absolute
interpretation. Indeed, it is underlined by the many illusions to which the
eye–brain system is subject, that decisions must repeatedly be made
concerning the most likely interpretation of a scene and that there is some
risk that its internal model builder will lock on to an interpretation or part-
interpretation that is suboptimal (see the paintings of Escher!).

This section has indicated that methods of 3-D vision can be categorized
according to whether they start by mapping out the shapes of objects in 3-D
space and then attempt to interpret the resulting shapes, or whether they try
to identify objects directly from their features. In either case a knowledge
base is ultimately called for. It has also been seen that methods of mapping
objects in real space include monocular and binocular methods, although
structured lighting can help to offset the deficiencies of employing a single
"eye". Laser scanning and ranging techniques must also be included in
methods of 3-D mapping, although space precludes detailed discussion of
these techniques in this book.

15.3   Projection schemes for three-dimensional vision

It is common in engineering drawings to provide three views of an object to
be manufactured—the plan, the side view and the elevation. Traditionally
these views are simple orthographic (nondistorting) projections of the
object—i.e. they are made by taking sets of parallel lines from points on the
object to the flat plane on which it is being projected.

However, when objects are viewed by eye or from a camera, rays
converge to the lens and so images formed in this way are subject not only
to change of scale but also to perspective distortions (Fig. 15.1). This type
of projection is called perspective projection, although it includes
orthographic projection as the special case of viewing from a distant point.
Unfortunately, perspective projections have the disadvantage that they
tend to make objects appear more complex than they really are by
destroying simple relationships between their features: thus parallel edges
no longer appear parallel and midpoints no longer appear as such
(although many useful geometric properties still hold—e.g. a tangent line
remains a tangent line and the order of points on a straight line remains
unchanged).

a

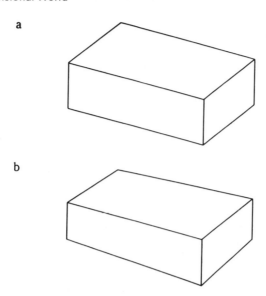

b

Fig. 15.1 (a) Image of a rectangular box taken using orthographic projection; (b) the same box taken using perspective projection. In (b) note that parallel lines no longer appear parallel, although paradoxically the box appears more realistic.

In outdoor scenes it is very common to see lines which are known to be parallel apparently converging towards a vanishing point on the horizon line (Fig. 15.2). In fact, the horizon line is the projection onto the image plane of the line at infinity on the ground plane $\mathcal{G}$: it is the set of all possible vanishing points for parallel lines on $\mathcal{G}$. In general, the vanishing points of a plane $\mathcal{P}$ are defined as the projections onto the image plane corresponding to points at infinity in a given direction on $\mathcal{P}$. Thus any plane $\mathcal{Q}$ within the field of view may have vanishing points in the image plane, and these will lie on a vanishing line which is the analogue of the horizon line for $\mathcal{Q}$.

Figure 15.3(a) shows how an image is projected into the image plane by a convex (eye or camera) lens at the origin. It is inconvenient to have to consider inverted images and it is a commonly used convention in image analysis to set the centre of the lens at the origin $(0, 0, 0)$ and to imagine the image plane to be the plane $Z = f$, f being the focal length of the lens; with this simplified geometry (Fig. 15.3(b)), images in the image plane appear noninverted. Taking a general point in the scene as (X, Y, Z), which appears in the image as (x_1, y_1), perspective projection now gives:

$$(x_1, y_1) = (fX/Z, fY/Z) \tag{15.1}$$

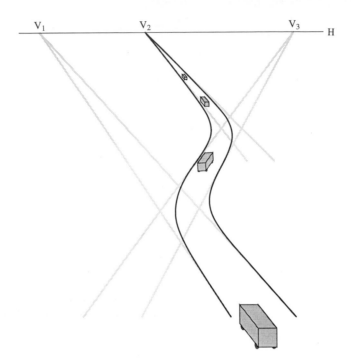

Fig. 15.2 Vanishing points and the horizon line. This figure shows how parallel lines on the ground plane appear, under perspective projection, to meet at vanishing points V_i on the horizon line H. (Note that V_i and H lie in the *image* plane.) If two parallel lines do not lie on the ground plane, their vanishing point will lie on a different vanishing line. Hence it should be possible to determine whether any roads are on an incline by computing all the vanishing points for the scene.

15.3.1 Binocular images

Figure 15.4 shows the situation when two lenses are used to obtain a stereo pair of images. In general, the two optical systems do not have parallel optical axes but exhibit a "vergence" (which may be variable, as it is for human eyes), so that they intersect at some point within the scene. Then a general point (X, Y, Z) in the scene has two different pairs of coordinates, (x_1, y_1) and (x_2, y_2), in its two images, which differ both because of the vergence between the optical axes and because the baseline b between the lenses causes relative displacement or "disparity" of the points in the two images.

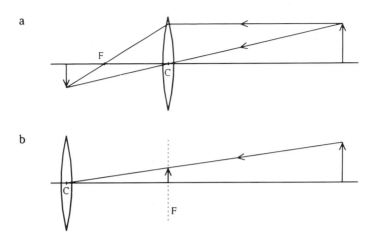

Fig. 15.3 (a) Projection of an image into the image plane by a convex lens; note that a single image plane only brings objects at a single distance into focus but that for far-off objects the image plane may be taken to be the focal plane, a distance f from the lens; (b) a commonly used convention which imagines the projected image to appear noninverted at a focal plane F in front of the lens. The centre of the lens is said to be the centre of projection for image formation.

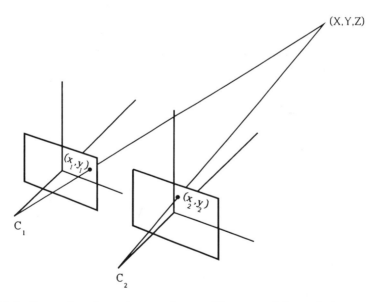

Fig. 15.4 Stereo imaging using two lenses. The axes of the optical systems are parallel, i.e. there is no "vergence" between the optical axes.

For simplicity, we now take the vergence to be zero, i.e. the optic axes are parallel. Then, with suitable choice of Z-axis on the perpendicular bisector of the baseline b, we obtain two equations:

$$x_1 = (X + b/2)f/Z \tag{15.2}$$

$$x_2 = (X - b/2)f/Z \tag{15.3}$$

so that the disparity is

$$D = x_1 - x_2 = bf/Z \tag{15.4}$$

Rewriting this equation in the form:

$$Z = bf/(x_1 - x_2) \tag{15.5}$$

now permits the depth Z to be calculated. In fact, computation of Z only requires the disparity for a stereo pair of image points to be found and parameters of the optical systems to be known. However, confirming that both points in a stereo pair actually correspond to the same point in the original scene is in general not at all trivial, and much of the computation in stereo vision is devoted to this task. In addition, to obtain good accuracy in the determination of depth, a large baseline b is required: unfortunately, as b is increased the correspondence between the images decreases, so it becomes more difficult to find matching points.

15.3.2 The correspondence problem

There are two important approaches to finding pairs of points that match in the two images of a stereo pair. One is that of "light striping" (one form of structured lighting), which encodes the two images so that it is easy to see pairs of corresponding points. If a single vertical stripe is used, for every value of y there is in principle only one light stripe point in each image and so the matching problem is solved. We return to this problem in a later section.

The second important approach is to employ epipolar lines. To understand this approach, imagine that we have located a distinctive point in the first image and that we are marking all possible points in the object field which could have given rise to it. This will mark out a line of points at various depths in the scene and, when viewed in the second image plane, a locus of points can be constructed in that plane. This locus is the *epipolar line* corresponding to the original image point in the alternate image (Fig. 15.5). If we now search along the epipolar line for a similarly distinctive point in the second image, the chance of finding the correct match is significantly enhanced. This method has the advantage not only of cutting down the amount of computation required to find corresponding points, but also of reducing significantly the chance of false alarms. Note that the concept of an epipolar line applies to both images—a point in one image gives an epipolar line in the other image. Note also that in the simple geometry of Fig. 15.4, all epipolar lines are parallel to the x-axis, although this is not so in general (in fact, the general situation is that all epipolar lines in one image plane pass through the point that is the image of the projection point of the alternate image plane).

The correspondence problem is rendered considerably more difficult by the fact that there will be points in the scene which give rise to points in one image but not in the other. Such points are either occluded in the one image, or else are so distorted as not to give a recognizable match in the two images (e.g. the different background might mask a corner point in one image while

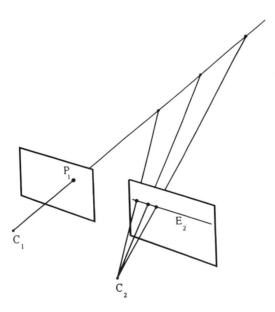

Fig. 15.5 Geometry of epipolar lines. A point P_1 in one image plane may have arisen from any one of a line of points in the scene, and may appear in the alternate image plane at any point on the so-called epipolar line E_2.

permitting it to stand out in the other). Any attempt to match such points can then only lead to false alarms. Thus it is necessary to search for consistent sets of solutions in the form of continuous object surfaces in the scene. For this reason iterative "relaxation" schemes are widely used to implement stereo matching.

Broadly speaking, correspondences are sought by two methods: one is the matching of near-vertical edge points in the two images (near-horizontal edge points do not give the required precision); the other is the matching of local intensity patterns using correlation techniques. Correlation is an expensive operation and in this case is relatively unreliable—principally because intensity patterns frequently appear significantly foreshortened[*] in one or other image and hence are difficult to match reliably. In such cases the most practical solution is to reduce the baseline; as noted earlier, this has the effect of reducing the accuracy of depth measurement. Further details of these techniques are to be found in Shirai (1987).

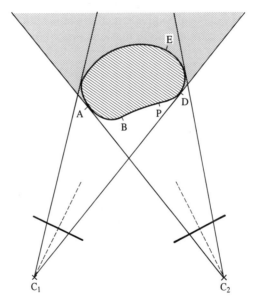

Fig. 15.6   Visibility of feature points in two stereo views. Here an object is viewed from two directions. Only feature points which appear in both views are of value for depth estimation. This eliminates all points in the shaded region, such as E, from consideration.

---

[*] i.e. distorted by the effects of perspective.

Before leaving this topic, it is worth considering in slightly more detail how the problems of visibility mentioned above arise. Figure 15.6 shows a situation in which an object is being observed by two cameras giving stereo images. Clearly, much of the object will not be visible in either image because of self-occlusion, while some feature points will only be visible in one or the other image. Now consider the order in which the points appear in the two images (Fig. 15.7). The points which are visible appear in the same order as in the scene, and the points which are just going out of sight are those for which the order between the scene and the image is just about to change. Points which provide information about the front surface of the object can thus only bear a simple geometrical relation to each other: in particular, for points not to obscure, or be obscured by, a given point P, they must not lie within a double-

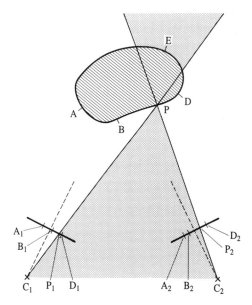

Fig. 15.7 Ordering of feature points on an object. In the two views of the object shown here, the feature points all appear in the same order A, B, P, D as on the surface of the object. Points for which this would not be valid, such as E, are behind the object and are obscured from view. Relative to a given visible feature P, there is a double-ended cone (shaded) in which feature points must not appear if they are not to obscure, or be obscured by, the feature under consideration. An exception to these rules might be if the object had a semitransparent window through which an additional feature T were visible: in that case interpretation would be facilitated by noting that the orderings of the features seen in the two views were different—e.g. A_1, T_1, B_1, P_1, D_1 and A_2, B_2, T_2, P_2, D_2.

ended cone region defined by P and the centres of projection C_1, C_2 of the two cameras: this region is shown shaded in Fig. 15.7. A surface passing through P for which full depth information can be retrieved must lie entirely within the non-shaded region. (Of course, a new double-ended cone must be considered for each point on the surface being viewed.) Note that the possibility of objects containing holes, or having transparent sections, must not be forgotten (such cases can be detected from differences in the ordering of feature points in the two views—see Fig. 15.7); neither must it be ignored that the foregoing figures represent a single horizontal cross-section of an object which can have totally different shapes and depths in different cross-sections.

15.4 Shape from shading

It was mentioned in Section 15.2 that it is possible to analyse the pattern of intensities in a single (monocular) image and to deduce the shapes of objects from the shading information. The principle underlying this technique is that of modelling the reflectance of objects in the scene as a function of the angles of incidence i and emergence e of light from their surfaces. In fact, a third angle is also involved, and it is called the phase g (Fig. 15.8).

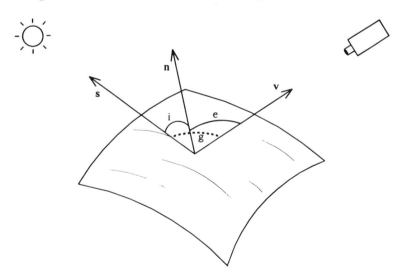

Fig. 15.8 Geometry of reflection. An incident ray from source direction **s** is reflected along the viewer direction **v** by an element of the surface whose local normal direction is **n**; i, e and g are defined respectively as the incident, emergent and phase angles.

A general model of the situation gives the radiance I (light intensity in the image) in terms of the irradiance E (energy per unit area falling on the surface of the object) and the reflectance R:

$$I(x_1, y_1) = E(x, y, z)R(\mathbf{n}, \mathbf{s}, \mathbf{v}) \tag{15.6}$$

It is well known that a number of matt surfaces approximate reasonably well to an ideal Lambertian surface whose reflectance function depends only on the angle of incidence i—i.e. the angles of emergence and phase are immaterial:

$$I = (1/\pi)E \cos i \tag{15.7}$$

For the present purpose E is regarded as a constant and combined with other constants for the camera and the optical system (including, for example, the f-number). In this way a normalized reflectance is obtained, which in this case is:

$$R = R_0 \cos i = R_0 \mathbf{s} \cdot \mathbf{n}$$

$$= \frac{R_0(1 + pp_s + qq_s)}{(1 + p^2 + q^2)^{1/2}(1 + p_s^2 + q_s^2)^{1/2}} \tag{15.8}$$

where the commonly used convention is taken of writing orientations in 3-D in terms of p and q values. These are not direction cosines but correspond to the coordinates of the point $(p, q, 1)$ at which a particular direction vector from the origin meets the plane $z = 1$: hence they need suitable normalization, as in the above equation.

The above equation gives a reflectance map in gradient (p, q) space. We now temporarily put the absolute reflectance value R_0 equal to unity. The reflectance map can be drawn as a set of contours of equal brightness, starting with a point having $R = 1$ at $\mathbf{s} = \mathbf{n}$, and going down to zero for $\mathbf{n}$ perpendicular to $\mathbf{s}$. When $\mathbf{s} = \mathbf{v}$, so that the light source is along the viewing direction (here taken to be the direction $p = q = 0$), zero brightness occurs only for infinite distances on the reflectance map $((p^2 + q^2)^{1/2}$ approaching infinity) (Fig. 15.9(a)). In a more general case, when $\mathbf{s} \neq \mathbf{v}$, zero brightness occurs along a straight line in gradient space (Fig. 15.9(b)). To find the exact shapes of the contours we can set R at a constant value a, which results in:

$$a(1 + p^2 + q^2)^{1/2}(1 + p_s^2 + q_s^2)^{1/2} = 1 + pp_s + qq_s \tag{15.9}$$

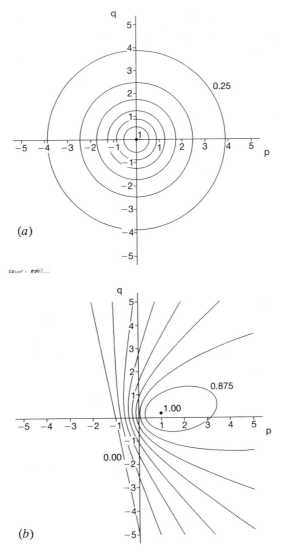

Fig. 15.9 Reflectance maps for Lambertian surfaces: (a) contours of constant intensity plotted in gradient (p, q) space for the case where the source direction **s** (marked by a black dot) is along the viewing direction **v** $(0, 0)$ (the contours are taken in steps of 0.125 between the values shown); (b) the contours that arise where the source direction (p_s, q_s) is at a point (marked by a black dot) in the positive quadrant of (p, q) space: note that there is a well-defined region, bounded by the straight line $1 + pp_s + qq_s = 0$, for which the intensity is zero (the contours are again taken in steps of 0.125).

Squaring this equation clearly gives a quadratic in p and q, which could be simplified by a suitable change of axes. Thus the contours must be curves of conic section, namely circles, ellipses, parabolas, hyperbolas, lines or points (the case of a point arises only when $a = 1$, when we get $p = p_s$, $q = q_s$; and that of a line only if $a = 0$, when we get the equation $1 + pp_s + qq_s = 0$: both of these solutions were implied above).

Unfortunately, object reflectances are not all Lambertian and an obvious exception is for surfaces that approximate to pure specular reflection. In that case $e = i$ and $g = i + e$ (**s**, **n**, **v** are coplanar); the only nonzero reflectance position in gradient space is the point representing the bisector of the angle between the source direction $\mathbf{s}(p, q)$ and the viewing direction $\mathbf{v}(0, 0)$—i.e. **n** is along $\mathbf{s} + \mathbf{v}$—and *very* approximately:

$$p \simeq p_s/2 \tag{15.10}$$

$$q \simeq q_s/2 \tag{15.11}$$

For less perfect specularity, a peak is obtained around this position. A good approximation to the reflectance of many real surfaces is obtained by modelling them as basically Lambertian but with a strong additional reflectance near the specular reflectance position. Using the Phong (1975) model for the latter component gives:

$$R = R_0 \cos i + R_1 \cos^m \theta \tag{15.12}$$

θ being the angle between the actual emergence direction and the ideal specular reflectance direction.

The resulting contours now have two centres around which to peak: the first is the ideal specular reflection direction ($p \simeq p_s/2$, $q \simeq q_s/2$), and the second is that of the source direction ($p = p_s$, $q = q_s$). When objects are at all shiny—such as metal, plastic, liquid, or even wood surfaces—the specular peak is quite sharp and rather intense: casual observation may not even indicate the presence of another peak since Lambertian reflection is so diffuse (Fig. 15.10). In other cases the specular peak can broaden and become more diffuse: hence it may merge with the Lambertian peak and effectively disappear.

Some remarks should be made about the Phong model employed above. First, it is adapted to different materials by adjusting the values of R_0, R_1 and m. Phong remarks that R_1 typically lies between 10% and 80%, while m is in the range 1–10. However, Rogers (1985) indicates that m may be as

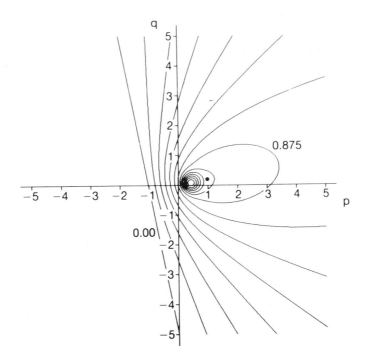

Fig. 15.10 Reflectance map for a non-Lambertian surface: a modified form of Fig. 15.9(b) for the case where the surface has a marked specular component ($R_0 = 1.0$, $R_1 = 0.8$): note that the specular peak can have very high intensity (much greater than the maximum value of unity for the Lambertian component). In this case the specular component is modelled with a $\cos^8 \theta$ variation (the contours are again taken in steps of 0.125).

high as 50. It is important to note that there is no physical significance in these numbers—the model is simply a phenomenological one. This being so, care should be taken to prevent the $\cos^m \theta$ term from contributing to reflectance estimates when $|\theta| > 90°$. The Phong model is reasonably accurate but has been improved upon by Cook and Torrance (1982). This is important in computer graphics applications but probably the improvement is difficult to make use of in computer vision, because of lack of data concerning the reflectances of real objects and because of variability in the current state (cleanliness, degree of polish, etc.) of a given surface. However, the method of photometric stereo gives some possibility of overcoming these problems.

15.5 Photometric stereo

Photometric stereo is a form of structured lighting that increases the information available from surface reflectance variations. Basically, instead of taking a single monocular image of a scene illuminated from a single source, several images are taken, from the same vantage point, with the scene illuminated in turn by separate light sources. Again, these light sources are ideally point sources some distance away in various directions, so that there is in each case a well-defined light source direction from which to measure surface orientation.

The basic idea of photometric stereo is that of cutting down the number of possible positions in gradient space for a given point on the surface of an object. It has already been seen that, for known absolute reflectance R_0, a constant brightness in one image permits the surface orientation to be limited to a curve of conic cross-section in gradient space. This would also be true for a second such image, the curve being a new one if the illuminating source is different. In general two such conic curves meet in two points, so there is now only a single ambiguity in the gradient of the surface at any given point in the image. To resolve this ambiguity a third source of illumination can be employed (this must not be in the plane containing the first two and the surface point being examined), and the third image gives another curve in gradient space which should pass through the appropriate crossing point of the first two curves (Fig. 15.11). If a third source of illumination cannot be used, it is sometimes possible to arrange that the inclination of each of the sources is so high that $(p^2 + q^2)^{1/2}$ on the surface is always lower than $(p_s^2 + q_s^2)^{1/2}$ for each of the sources, so that only one interpretation of the data is possible. This method is prone to difficulty, however, since it means that parts of the surface could be in shadow, thereby preventing the gradient for these parts of the surface from being measured. Another possibility is to assume that the surface is reasonably smooth, so that p and q vary continuously over it. This itself ensures that ambiguities are resolved over most of the surface.

However, there are other advantages to be gained from using more than two sources of illumination. One is that information on the absolute surface reflectance can be obtained. Another is that the assumption of a Lambertian surface can be tested. Thus, three sources of illumination ensure that the remaining ambiguity is resolved *and* permit absolute reflectivity to be measured: this is obvious since if the three contours in gradient space do not pass through the same point, then the absolute reflectivity cannot be unity, so corresponding contours should be sought

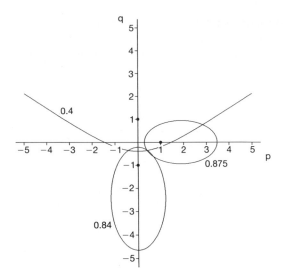

Fig. 15.11 Obtaining a unique surface orientation by photometric stereo. Three contours of constant intensity arise for different light sources of equal strength: all three contours pass through a single point in (p, q) space and result in a unique solution for the local gradient.

which do pass through the same point. In practice the calculation is normally carried out by defining a set of nine matrix components of irradiance, s_{ij} being the jth component of light source vector $\mathbf{s}_i$. Then, in matrix notation:

$$\mathbf{E} = R_0 \mathscr{S} \mathbf{n} \tag{15.13}$$

where

$$\mathbf{E} = (E_1, E_2, E_3)^{\mathrm{T}} \tag{15.14}$$

and

$$\mathscr{S} = \begin{bmatrix} s_{11} & s_{12} & s_{13} \\ s_{21} & s_{22} & s_{23} \\ s_{31} & s_{32} & s_{33} \end{bmatrix} \tag{15.15}$$

Provided that the three vectors s_1, s_2, s_3 are not coplanar, so that $\mathscr{S}$ is not a singular matrix, R_0 and $\mathbf{n}$ can now be determined from the formulae:

$$R_0 = |\mathscr{S}^{-1}\mathbf{E}| \tag{15.16}$$

$$\mathbf{n} = \mathscr{S}^{-1}\mathbf{E}/R_0 \tag{15.17}$$

An interesting special case arises if the three source directions are mutually perpendicular; taking them to be aligned along the respective major axes directions, $\mathscr{S}$ is now the unit matrix, so that:

$$R_0 = [E_1^2 + E_2^2 + E_3^2]^{1/2} \tag{15.18}$$

and

$$\mathbf{n} = (E_1, E_2, E_3)^{\mathrm{T}}/R_0 \tag{15.19}$$

   If four or more images are obtained using further illumination sources, more information can be obtained: for example, the coefficient of specular reflectance, R_1. In practice this coefficient varies somewhat randomly with the cleanness of the surface and it may not be relevant to determine it accurately. More probably it will be sufficient to check whether significant specularity is present, so that the corresponding region of the surface can be ignored for absolute reflectance calculations. Nonetheless, finding the specularity peak can itself give important surface orientation information, as will be clear from the previous section. It should be remarked that although the information from several illumination sources should ideally be collated using least-squares analysis, this method requires significant computation. Hence it seems better to use the images resulting from further illumination sources as confirmatory—or, instead, to select the three that exhibit the least evidence of specularity as giving the most reliable information on local surface orientation.

15.6 The assumption of surface smoothness

It was hinted above that the assumption of a reasonably smooth surface permits ambiguities to be removed in situations where there are two illuminating sources. In fact, this method can be used to help analyse the brightness map even for situations where a single source is employed: indeed, the fact that the eye can perform this feat of interpretation indicates that it is possible to find computer methods for achieving it. Much research has been carried out on this topic and a set of methods is available, although the calculations are complex iterative procedures that cannot be carried out in real time on conventional computers. For this reason they are not studied in depth here: the reader is referred to the volume by Horn (1986) for detailed information on this topic. However, one or two remarks are in order.

First, consider the representation to be employed for this type of analysis. It turns out that normal gradient (p, q) space is not very appropriate for the purpose. In particular, it is necessary to average gradient (i.e. the $\mathbf{n}$-values) locally within the image; however, (p, q)-space is not "linear", in that a simple average of (p, q) values within a window gives biased results. It turns out that a conformal representation of gradient (i.e. one which preserves small shapes) is closer to the ideal, in that the distances between points in such a representation provide better approximations to the relative orientations of surface normals: averaging in such a representation gives reasonably accurate results. The required representation is obtained by a stereographic projection, which maps the unit (Gaussian) sphere onto a plane $(z = 1)$ through its north pole but this time using as a projection point not its centre but its south pole. This projection has the additional advantage that it projects all possible orientations of a surface onto the plane, not merely those from the northern hemisphere. Hence backlit objects can be represented conveniently in the same map as used for frontlit objects.

Second, the relaxation methods used to estimate surface orientation have to be provided with accurate boundary conditions: in principle, the more correct the orientations that are presented initially to such procedures, the more quickly and accurately the iterations proceed. There are normally two sets of boundary conditions that can be applied in such programs. One is the set of positions in the image where the surface normal is perpendicular to the viewing direction. The other is the set of positions in the image where the surface normal is perpendicular to the direction of illumination: this set of positions corresponds to the set of shadow edges (Fig. 15.12). Careful analysis of the image must be undertaken to find each set of positions, but once they have been located they provide valuable cues for unlocking the information content of the monocular image, and mapping out surfaces in detail.

Finally, it should be remarked that all shape from shading techniques provide information which initially takes the form of surface orientation maps. Dimensions are not obtainable directly but these can be computed by integration across the image from known starting points. In practice this tends to mean that absolute dimensions are unknown and that dimensional

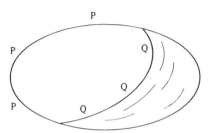

Fig. 15.12 The two types of boundary condition that can be used in shape-from-shading computations of surface orientation: (i) positions P where the surface normal is perpendicular to the viewing direction; (ii) positions Q where the surface normal is perpendicular to the direction of illumination (i.e. shadow boundaries).

maps are obtainable only if the size of an object is given or if its depth within the scene is known.

15.7 Shape from texture

Texture can be very helpful to the human eye in permitting depth to be perceived. Although textured patterns can be very complex, it turns out that even the simplest textural elements can carry depth information. Ohta *et al.* (1981) showed how circular patches on a flat surface viewed more and more obliquely in the distance become first elliptical and then progressively flatter and flatter. At infinite distance, on the horizon line (here defined as the line at infinity in the given plane), they would clearly become very short line segments. To disentangle such textured images sufficiently to deduce depths within the scene, it is first necessary to find the horizon line reliably. This is achieved by taking all pairs of texture elements and deducing from their areas where the horizon line would have to be. To proceed we make use of the rule:

$$d_1^3/d_2^3 = A_1/A_2 \tag{15.20}$$

which applies since circles at various depths would give a square law, although the progressive eccentricity also reduces the area linearly in proportion to the depth. This information is accumulated in a separate image space and a line is then fitted to these data: false alarms are eliminated automatically by this Hough-based procedure. At this stage the original data—the ellipse areas—provide direct information on depth, although some averaging is required to obtain accurate results. Although this type of method has been demonstrated in certain instances, it is in practice highly restricted unless very considerable amounts of computation are performed. Hence it is doubtful whether it can be of general practical use in machine vision applications.

15.8   Use of structured lighting

Structured lighting has already been considered briefly in Section 15.2 as an alternative to stereo for mapping out depth in scenes. Basically, a pattern of light stripes, or other arrangement of light spots or grids, is projected onto the object field. Then these patterns are enhanced in a (generally) single monocular image and analysed to extract the depth information. To obtain the maximum information the light pattern must be close-knit and the received images must be of very high resolution. When shapes are at all complex, the lines can in places appear so close together that they are unresolvable. It then becomes necessary to separate the elements in the projected pattern, trading resolution and accuracy for reliability of interpretation. Even so, if parts of the objects are along the line of sight, the lines can merge together and even cross back and forth, so unambiguous interpretation is never assured. In fact, this is part of a larger problem, in that parts of the object will be obscured from the projected pattern by occluding bodies or perhaps by self-occlusion: the method has this feature in common with the shape from shading technique and with stereo vision, which relies on *both* cameras being able to view various parts of the objects simultaneously. Hence the structured light approach is subject to similar restrictions to those found for other methods of 3-D vision and is not a panacea. Nevertheless, it is a useful technique that is generally simple to set up so as to acquire specific 3-D information which can enable a computer to start the process of cueing into complex images.

Light spots provide perhaps the most obvious form of structured light. However, they are restricted because for each spot an analysis has to be performed to determine which spot is being viewed: connected lines, in contrast, carry a large amount of coding information with them so that ambiguities are less likely to arise. Grids of lines carry even more coding information but do not necessarily give any more depth information. Indeed, if a pattern of light stripes can be projected (for example) from the left of the camera so that they are parallel to the y-axis in the observed image, then it is clear that there is no point in projecting another set of lines parallel to the x-axis, since these merely replicate information that is already available from the rows of pixels in the image—all the depth information is carried by the vertical lines and their horizontal displacements in the image. This analysis assumes that the camera and projected beams are carefully aligned, and that no perspective or other distortions are present. In fact, most practical structured lighting systems in current use appear to employ light stripe patterns rather than spot patterns or full grid patterns.

This section ends with an analysis of the situations that can arise when a single stripe is incident on objects as simple as rectangular blocks. Figure 15.13 shows three types of structure in observed stripes: (i) the effect of a sharp angle being encountered; (ii) the effect of "jump edges" at which light stripes jump horizontally and vertically at the same time; and (iii) the effect of discontinuous edges at which light stripes jump horizontally but not vertically. The reasons for these circumstances will be obvious from Fig. 15.13. Basically, the problem to be tackled with jump and discontinuous edges is to find whether a given stripe end marks an occluding edge or an occluded edge. The importance of this distinction is that occluding edges mark actual edges of the object being observed, whereas occluded edges may be merely edges of shadow regions and are then not *directly* significant[*]. A simple rule is that, if stripes are projected from the left, the left-hand component of a discontinuous edge will be the occluding edge and the right-hand component will be the occluded edge. Angle edges are located by applying a Laplacian type of operator which detects the change in orientation of the light stripe.

The ideas outlined above correspond to possible 1-D operators that interpret light stripe information to locate nonvertical edges of objects. The method provides no direct information concerning vertical edges. To obtain such information it is necessary to analyse the information from sets of light stripes. For this purpose 2-D edge operators are required, which collect sufficient data from at least two or three adjacent light stripes. Further details are beyond the scope of this chapter.

---

[*] More precisely, they involve interactions of light with two objects rather than with one, and are therefore more complex to interpret.

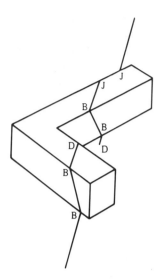

Fig. 15.13   Three of the structures that are observed when a light stripe is incident on even quite simple shapes: bends (B), jumps (J) and discontinuities (D).

Before leaving the topic of light stripe analysis, it should be remarked that stripes provide a very useful means of recognizing planes forming the faces of polyhedra and other types of manufactured object. The characteristic sets of parallel lines can be found and demarcated relatively easily, and the fact that the lines usually give rather strong signals means that line tracking techniques can be applied and that algorithms can operate quite rapidly. However, whole-scene interpretation, including inferring the presence and relative positions of different objects, remains a more complex task, as will be seen below.

15.9   Three-dimensional object recognition schemes

The methods described so far in this chapter employ various means for finding depth at all places in a scene, and are hence able to map out 3-D surfaces in a fair amount of detail. However, on the whole they do not give any clue as to what these surfaces represent. In some situations it may be clear that certain planar surfaces are parts of the background, e.g. the floor and the walls of a room, but in general individual objects will not be inherently identifiable. Indeed, objects tend to merge with each other and with the background, so specific methods are needed to segment the 3-D space map* and finally recognize the objects, giving detailed information on their positions and orientations.

Before proceeding to study this problem, it should be noted that further general processing can be carried out to analyse the 3-D shapes. Agin and Binford (1976) and others have developed techniques for likening 3-D shapes to "generalized cylinders", these being like normal (right circular) cylinders but with additional degrees of freedom so that the axes can bend and the cross-sections can vary, both in size and in detailed shape. Thus even an animal like a sheep can be likened to a distorted cylinder. On the whole this approach is elegant but may not be well adapted to describe many industrial objects, and it is therefore not pursued further here. A simpler approach may be to model the 3-D surfaces as planar, quadratic, cubic and quartic surfaces, and then to try to understand these model surfaces in terms of what is known about existing objects. This approach was adopted by Hall *et al.* (1982) and was found to be viable, at least for certain quite simple objects such as cups. Shirai (1987) has taken the approach even further so that a whole range of objects can be found and identified in quite complex indoor scenes.

At this point it is worth examining the situation anew to find what it is we are trying to achieve. First, can recognition be carried out *directly* on the mapped out 3-D surfaces, just as it could for the 2-D images of earlier chapters? Second, if we can bypass the 3-D modelling process, and still recognize objects, might it not be possible to save even more computation and omit the stage of mapping out 3-D surfaces, instead identifying 3-D objects directly in 2-D images? It might even be possible to locate 3-D objects from a single 2-D image.

Consider the first of these problems. When we studied 2-D recognition, many instances were found where the HT approach was of great help. It turned out to give trouble in more complex cases, particularly when attempts were made to find objects where there were more than two or at most three degrees of freedom. Here, however, we have situations where objects normally have six degrees of freedom—three degrees of freedom for translation and another three for rotation. This doubling of the number of free parameters on going from 2-D to 3-D makes the situation far worse, since the search space is proportional in size not to the number of degrees of

---

* This may be defined as an imagined 3-D map showing, without interpretation, the surfaces of all objects in the scene and incorporating all the information from depth or range images. Note that it will generally include only the front surfaces of objects seen from the vantage point of the camera.

freedom, but to its exponent: for example, if each degree of freedom in translation or rotation can have 256 values, the number of possible locations in parameter space changes from 256^3 in 2-D to 256^6 in 3-D. This will be seen to have a very profound effect on object location schemes and tends to make the HT technique difficult to implement. The problem was examined by Ballard and Sabbah (1983) for the particular case where space maps are available for 3-D scenes.

15.10   The method of Ballard and Sabbah

In their paper Ballard and Sabbah (1983) made great play about the order in which various parameters could be determined, concluding that there is a natural precedence—(i) scale, (ii) orientation, (iii) translation. Thus, if the scale of an object is unknown, it can in principle be determined without first finding the three orientation parameters, and the latter can in turn be found independently of the three translation parameters. It is easy to see that scale can be found irrespective of the other six parameters if a complete space map is available, since the volume of the object[*] can be measured and this immediately determines the scale[†]. Similarly, the orientation and translation parameters can be decoupled merely by ignoring the absolute locations of the features in the space map and hence narrowing down the search problem to that of finding the three object orientation parameters. Once this has been achieved, we can then go back to the original image data and find the three translation parameters, keeping the orientation parameters fixed.

The case Ballard and Sabbah examined was that of a polyhedron with planar faces of known area. To proceed they first considered the case of a polygon all of whose edges have been located and whose lengths l_i have been measured in a 2-D image (Fig. 15.14).

They then considered the difference between the body-centred frame of axes and the viewer-centred frame. It is assumed that once an edge of length l_{vi} is found in the image, its orientation θ_{bi} relative to the body-centred frame can be obtained from a lookup table $\{(l_{bj}, \theta_{bj})\}$. Then its actual orientation θ_{vi} in the image gives an estimate $\theta_{vi} - \theta_{bi}$ for the orientation of the body-centred frame relative to that of the viewer-centred frame. Thus it

---

[*] Or what is temporarily hypothesized to be a object.

[†] However, there is little evidence to support the idea that separating scale from orientation is universally possible as Ballard and Sabbah imply. Indeed, their paper supports the view that scale and orientation generally have to be determined simultaneously, e.g. in the same parameter space.

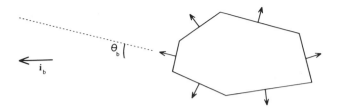

Fig. 15.14 Locating the body axis of a polygon: an estimate of one body axis of a polygon may be obtained from the orientation of a single edge.

is possible to construct a 1-D parameter space for relative orientation $\theta_v - \theta_b$ and to use the individual estimates $\theta_{vi} - \theta_{bi}$ to obtain an accurate measure of the body orientation (similar ideas have already been met in Chapters 12 and 14).

To generalize the concept to 3-D, the lengths l_{vi} are first generalized to the areas A_{vi} of the planar polygonal faces of a polyhedral shape that is to be located in the 3-D space map (Fig. 15.15). The problem now becomes rather more complicated, since each area A_{vi} (and the direction of its normal $\mathbf{n}_{vi}$) leads to an incomplete estimate of the direction of the body-centred axis $\mathbf{i}_b$. In fact, $\mathbf{i}_b$ is assumed to be inclined at an angle θ_{bi} relative to $\mathbf{n}_{vi}$: clearly this defines not a single orientation in space but rather a *cone* of directions whose symmetry axis is $\mathbf{n}_{vi}$ (Fig. 15.15). The parameter space for determining the orientation of body axis $\mathbf{i}_b$ is best pictured on the unit (Gaussian) sphere of direction cosines (l, m, n), where the relation:

$$l^2 + m^2 + n^2 = 1 \qquad (15.21)$$

ensures that this apparently 3-D parameter space is actually only 2-D. Details of how to find peaks in this parameter space via 2-D representations are to be found in Ballard and Sabbah (1983).

Next, note another complication—that so far we have only seen how to determine one of the body axes. Determining a second body axis $\mathbf{j}_b$ normal to the first requires a separate lookup table of orientations φ_{bi} to be stored for areas A_{bi}, and a separate parameter space has to be used to accumulate the cones specifying $\mathbf{j}_b$. Once $\mathbf{i}_b$ and $\mathbf{j}_b$ are known, the third axis of a right-angled set is determined uniquely using the relation

$$\mathbf{k}_b = \mathbf{i}_b \times \mathbf{j}_b \qquad (15.22)$$

Finally, it should be noted that the method may falsely identify as a polyhedron of the relevant type a set of areas that happen to have the correct

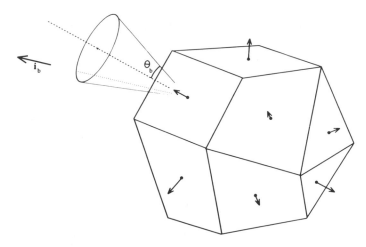

Fig. 15.15 Locating the body axis of a polyhedron: the orientation of one body axis of a polyhedron may be narrowed down to a cone of directions, the precise orientation ultimately being determined by the intersection of a number of such cones.

orientations but not the correct translations. Hence it is necessary to proceed by finding the position of the body using another parameter space, and to check at that stage that consistent results (sufficiently high peaks) are obtained, indicating a valid body of this type.

Clearly, the whole procedure involves more computation for 3-D than for 2-D but it is tractable supposing that the true areas of the polyhedral faces and the directions of their normals can be measured from the space map. It should be noted that the method has the normal robustness against occlusion and noise that is to be expected with the HT approach, although the maximum peak height for a polyhedron of p faces is p and this reduces to p' when only p' faces are unoccluded. Thus the method will fail when p' is less than 3 (for two unoccluded faces, there will be two cones specifying a given body axis and the resulting two circles on the unit sphere intersect in two places, leaving an ambiguity of orientation).

There are many problems of 3-D vision that cannot easily be solved by the above approach. Clearly cases where objects have curved faces are intractable by this method, as are situations where simpler bodies such as cubes (for which $p = 3$ but p' may be lower) need to be examined. In addition, depth maps are somewhat more difficult to obtain than intensity images. The method of Silberberg *et al.* (1984) tackles the problem of locating objects with plane faces in normal 2-D intensity images where depth information is not available; in addition, by working directly from individual surface features it avoids the stage of segmenting surfaces.

15.11   The method of Silberberg *et al.*

This method aims to employ the GHT to locate planar objects in 2-D images by means as close as possible to the normal GHT approach. The problem of too many parameters having to be determined is first minimized (i) by ensuring that the scale of the objects is known in advance, so that six rather than seven parameters have to be determined, and (ii) by arranging that projection is orthographic rather than perspective, so that there is no way in which depth information can affect the image: hence only five parameters are active in the image—i.e. there are two translational and three rotational parameters[*].

Of the remaining parameters, two of the rotations are made implicit by assuming a particular viewpoint over the object; then only its 2-D translation parameters and orientation about the camera axis remain to be found. A normal parameter space is constructed for the latter three parameters; the effect of implicit viewpoint variation is investigated by specifying a fixed large number of viewpoints arranged regularly over the Gaussian sphere, this being achieved by employing a regular icosahedron whose 20 (equilateral) triangular faces are each subdivided into four smaller equilateral triangles. This type of approach has been employed by a number of workers (Koenderink and van Doorn, 1979; Chakravarty and Freeman, 1982).

The overall strategy is to record the height of the main peak in parameter space for each of the 80 viewpoints and then to improve the best set of five viewpoints in several iterations, each time splitting each of the triangles into four smaller equilateral triangles. After two iterations the best viewpoint is obtained and the object is located by the method to a suitable degree of accuracy.

The case investigated by Silberberg was that of a cube on whose faces certain letters appear. The faces of a cube are evidently planar and all relevant features are straight edges. Votes are accumulated in parameter space at every position representing some translation or rotation of a line which leads to a valid position in a model of the object. The criterion for validity also includes matching the lengths of lines in the image and in the projected model,

---

[*] Note that Ballard and Sabbah suggest this idea but then employ space maps which completely specify depth information, thereby bringing the effective parameter count back to six.

although some fragmentation of lines is taken to be acceptable: hence lines must, if anything, appear shorter than the lengths deduced from the model. After the initial run, two iterations were found to permit objects to be positively identified, located within 4 pixels and orientated within 5°. Overall, the method is successful for the restricted problem it tackles—objects with planar faces and straight line features, moderate accuracy requirements, the need to find only five parameters, and the accompanying restriction to orthographic projection (which may often be inconvenient).

With this background of work on orthographic projection, other workers have felt it useful to revert to perspective projection. There are good grounds for doing so since, as stated above, orthographic projection prevents any information on depth within a scene from being deduced using a single monocular view. Unfortunately, there are potential penalties from proceeding with perspective projection: (i) the parameter count goes back up from five to six, so a larger parameter space has to be searched, and (ii) parallel body lines no longer remain parallel in the image, so a valuable set of cues which can help unlock complex image data is lost. We next show how a powerful strategy developed by Horaud (1987) can be employed, under either projection scheme, to facilitate the recognition of polyhedral objects.

15.12   Horaud's junction orientation technique[*]

Horaud's (1987) technique is special in that it uses as its starting point 2-D images of 3-D scenes and "backprojects" them into the scene, with the aim of making interpretations in 3-D rather than 2-D frames of reference. This has the initial effect of increasing mathematical complexity, although in the end useful and supposedly more accurate results emerge.

Initially the boundaries of planar surfaces on objects are backprojected. Each boundary line is thus transformed into an "interpretation plane" defined by the centre of the camera projection system and the boundary line in the image plane: clearly, the interpretation plane must contain the line that originally projected into the boundary line in the image. Similarly, angles between boundary lines in the image are backprojected into two interpretation planes, which must contain the original two object lines. Finally, junctions between three boundary lines are backprojected into three interpretation planes which must contain a corner in the space map (Fig. 15.16). The paper focuses on the backprojection of junctions and shows how

---

[*] This and related techniques are sometimes referred to as "shape from angle".

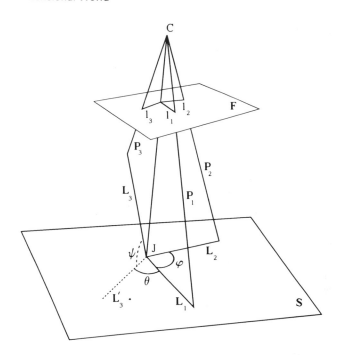

Fig. 15.16 Geometry for backprojection from junctions: a junction of three lines in an image may be backprojected into three planes, from which the orientation in space of the original corner J may be deduced.

measurements of the junction angles in the image relate to those of the original corner; it also shows how the space orientation of the corner can be computed. In fact, it is interesting that the orientation of an object in 3-D can in general be deduced from the appearance of just one of its corners in a single image. This is a powerful result and in principle permits objects to be recognized and located from extremely sparse data.

To understand the method, the mathematics first needs to be set up with some care. Assume that lines L_1, L_2, L_3 meet at a junction in an object, and appear as lines l_1, l_2, l_3 in the image (Fig. 15.16). Take respective interpretation planes containing the three lines and label them by unit vectors P_1, P_2, P_3 along their normals, so that:

$$P_1 \cdot L_1 = 0 \tag{15.23}$$

$$P_2 \cdot L_2 = 0 \tag{15.24}$$

$$P_3 \cdot L_3 = 0 \tag{15.25}$$

In addition, take the space plane containing L_1 and L_2, and label it by a unit vector S along its normal, so that:

$$S \cdot L_1 = 0 \qquad (15.26)$$

$$S \cdot L_2 = 0 \qquad (15.27)$$

Since L_1 is perpendicular to S and to P_1, and L_2 is perpendicular to S and to P_2, it is found that:

$$L_1 = S \times P_1 \qquad (15.28)$$

$$L_2 = S \times P_2 \qquad (15.29)$$

Note that S is not in general perpendicular to P_1 and P_2, so L_1 and L_2 are not in general unit vectors. Defining φ as the angle between L_1 and L_2, we now have:

$$L_1 \cdot L_2 = L_1 L_2 \cos \varphi \qquad (15.30)$$

which can be re-expressed in the form:

$$(S \times P_1) \cdot (S \times P_2) = |S \times P_1||S \times P_2| \cos \varphi \qquad (15.31)$$

Next we need to consider the junction between L_1, L_2, L_3. To proceed it is necessary to specify the relative orientations in space of the three lines. θ is the angle between L_1 and the projection L_3' of L_3 on plane S, while ψ is the angle between L_3' and L_3 (Fig. 15.16). Thus the structure of the junction J is described completely by the three angles φ, θ, ψ. L_3 can now be found in terms of other quantities:

$$L_3 = S \sin \psi + L_1 \cos \theta \cos \psi + (S \times L_1) \sin \theta \cos \psi \qquad (15.32)$$

Applying equation (15.25), it may be deduced that:

$$S \cdot P_3 \sin \psi + L_1 \cdot P_3 \cos\theta \cos \psi + (S \times L_1) \cdot P_3 \sin \theta \cos \psi = 0 \qquad (15.33)$$

Substituting for L_1 from equation (15.28), and simplifying, we finally obtain:

$$(S \cdot P_3)|S \times P_1| \sin \psi + S \cdot (P_1 \times P_3) \cos \theta \cos \psi$$
$$+ (S \cdot P_1)(S \cdot P_3) \sin\theta \cos \psi = (P_1 \cdot P_3) \sin \theta \cos \psi \qquad (15.34)$$

Equations (15.31) and (15.34) now exclude the unknown vectors $\mathbf{L}_1$, $\mathbf{L}_2$, $\mathbf{L}_3$ but they retain $\mathbf{S}$, $\mathbf{P}_1$, $\mathbf{P}_2$, $\mathbf{P}_3$ and the three angles φ, θ, ψ. $\mathbf{P}_1$, $\mathbf{P}_2$, $\mathbf{P}_3$ are known from the image geometry, and the angles φ, θ, ψ are presumed to be known from the object geometry; in addition, only two components (α, β) of the unit vector $\mathbf{S}$ are independent, so the two equations should be sufficient to determine the orientation of the space plane $\mathbf{S}$. Unfortunately, the two equations are highly nonlinear and it is necessary to solve them numerically. Horaud (1987) achieved this by re-expressing the formulae in the forms:

$$\cos \varphi = f(\alpha, \beta) \tag{15.35}$$

$$\sin \theta \cos \psi = g_1(\alpha, \beta) \sin \psi + g_2(\alpha, \beta) \cos \theta \cos \psi$$
$$+ g_3(\alpha, \beta) \sin \theta \cos \psi \tag{15.36}$$

For each image junction, $\mathbf{P}_1$, $\mathbf{P}_2$, $\mathbf{P}_3$ are known and it is possible to evaluate f, g_1, g_2, g_3. Then, assuming a particular interpretation of the junction, values are assigned to φ, θ, ψ and curves giving the relation between α and β are plotted for each equation. Possible orientations for the space plane $\mathbf{S}$ are then given by positions in (α, β) space where the curves cross. Horaud has shown that, in general, 0, 1 or 2 solutions are possible: the case of no solutions corresponds to trying to make an impossible match between a corner and an image junction when totally the wrong angles φ, θ, ψ are assumed; one solution is the normal situation; and two solutions arise in the interesting special case when orthographic or near-orthographic projection permits perceptual reversals—i.e. a convex corner is interpreted as a concave corner or *vice versa*. In fact, under orthographic projection the image data from a single corner are insufficient, taken on their own, to give a unique interpretation: in this situation even the human visual system makes mistakes—as in the case of the well-known Necker cube illusion (see Chapter 16). However, when such cases arise in practical situations, it may be better to take the convex rather than the concave corner interpretation as a working assumption, as it has slightly greater likelihood of being correct.

Horaud has shown that such ambiguities are frequently resolved if the space plane orientation is estimated simultaneously for all the junctions bordering the object face in question, by plotting the α and β values for all such junctions on the same α, β graph. For example, with a cube face on which there are three such junctions, nine curves are coincident at the correct solution, and there are nine points where only two curves cross, indicating false solutions. On the other hand, if the same cube is viewed under conditions approximating very closely to orthographic projection, two solutions with nine coincident curves appear and the situation remains unresolved, as before.

Overall, this technique is important in showing that although lines and angles individually lead to virtually unlimited numbers of possible interpretations of 3-D scenes, junctions lead individually to at most two solutions and any remaining ambiguity can normally be eliminated if junctions on the same face are considered together. As has been seen, the exception to this rule occurs when projection is accurately orthographic, although this is a situation that can often be avoided in practice.

So far we have considered only how a given hypothesis about the scene may be tested: nothing has been said about how assignments of the angles φ, θ, ψ are made to the observed junctions. Horaud's paper discussed this aspect of the work in some depth. In general, the approach is to use a depth-first search technique in which a match is "grown" from the initial most promising junction assignment. In fact, considerable preprocessing of sample data is carried out to find how to rank image features for their utility during depth-first search interpretation. The idea is to order possible alternatives such as linear or circular arcs, convex or concave junctions, short or long lines, etc. In this way the tree search becomes more planned and efficient at run time. Generally, the more frequently occurring types of feature should be weighted down in favour of the rarer types of feature, for greater search efficiency. In addition, it should be remembered that hypothesis generation is relatively expensive in that it demands a stage of backprojection, as described above. Ideally, this stage need be employed only once for each object (in the case that only a single corner is, initially, considered). Subsequent stages of processing then involve hypothesis verification in which other features of the object are predicted and their presence sought in the image: if found they are used to refine the existing match; if the match at any stage becomes worse, then the algorithm backtracks and eliminates one or more features and proceeds with other ones. This process is unavoidable, since more than one image feature may be present near a predicted feature.

One of the factors that has been found to make the method converge quickly is the use of grouped rather than individual features, since this tends to decrease the combinatorial explosion in the size of the search: in the present context this means that attempts should be made to match first all junctions or angles bordering a given object face, and further that a face should be selected that has the greatest number of matchable features around it.

In summary, this approach is claimed to be successful since it back-projects from the image and then uses geometrical constraints and heuristic assumptions for matching in 3-D space. It turns out to be suitable for matching objects that possess planar faces and straight line boundaries, hence giving angle and junction features. However, extending the back-projection technique to situations where object faces are curved and have curved boundaries will almost certainly prove much more difficult.

15.13   The 3DPO system of Bolles and Horaud

The 3-D part orientation (3DPO) system of Bolles and Horaud (1986) has much the same overall philosophy as that of Horaud (1987) with regard to its depth-first heuristic search technique, but takes its data from range images (obtained using a structured light system) and looks for different features of the objects it matches. The data acquisition aspect of the scheme is not considered here—beyond noting that the range maps are initially searched for edges—but consideration is given mainly to how the particular parts being examined lead to particular methods for finding features and matching them.

Figure 15.17 shows in simplified form the type of industrial part being sought in the images. In typical scenes several of these parts may appear jumbled on a worktable, with perhaps three or four being piled on top of each other in some places. In such cases it is vital that the matching scheme be

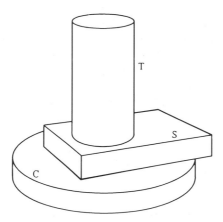

Fig. 15.17 The essential features of the industrial components located by the 3DPO system of Bolles and Horaud (1986). S, C and T indicate respectively straight and circular dihedral edges and straight tangential edges, all of which are searched for by the system.

highly robust if most of the parts are to be found, since even when a part is unoccluded, it appears against a highly cluttered and muddled background. However, the parts themselves have reasonably simple shapes and possess certain salient features. In the particular problem cited, each has a cylindrical base with a concentric cylindrical head, and also a planar shelf attached symmetrically to the base. To locate such objects it is natural to attempt to search for circular and straight dihedral edges. In addition, because of the type of data being used, it is useful to search for straight tangential edges, which appear where the sides of curved cylinders are viewed obliquely.

In general, circular dihedral edges appear elliptical, and parameters for five of the six degrees of freedom of the part can be determined by analysing these edges. The parameter that cannot be determined in this way corresponds to rotation about the axis of symmetry of the cylinder.

Straight dihedral edges also permit five free parameters to be determined, since location of one plane eliminates three degrees of freedom and location of an adjacent plane eliminates a further two degrees of freedom. The parameter that remains undetermined is that of linear motion along the direction of the edge. However, there is also a further ambiguity in that the part may appear either way around on the dihedral edge.

Straight tangential edges determine only four free parameters, since the part is free to rotate about the axis of the cylinder and can also move along the tangential edge. It should be noted that these edges are the most difficult to locate accurately, since range data are subject to greater levels of noise as surfaces curve away from the sensor.

All three of these types of edge are planar. They also provide useful additional information that can help to identify where they are on a part. For example, straight and curved dihedral edges both provide information on the size of the included angle, and the curved edges in addition give radius values. By now it will be plain that curved dihedral edges provide significantly more parametric information about a part than either of the other two types of edge, and therefore they are of most use to form initial hypotheses about the pose (position and orientation) of a part. Having found such an edge, it is necessary to try out various hypotheses about which edge it is, for example by searching for other circular dihedral edges at specific relative positions: this is a vital hypothesis verification step. Next, the problem of how to determine the remaining free parameter is solved by searching for the linear straight dihedral edge features from the planar shelf on the part. In principle this could be done using pure depth-first search but in this case an exception is made, since the available line features are all very similar and they are best located together by the maximal clique (breadth-first exhaustive search) technique.

At this stage hypothesis generation is complete and the part is essentially found, but hypothesis verification is required (i) to confirm that the part is genuine and not an accidental grouping of independent features in the image, (ii) to refine the pose estimate, and (iii) to determine the "configuration" of the part, i.e. to what extent it is buried under other parts (making it difficult for a robot to pick it up). Bolles and Horaud discussed the problem of configuration understanding in some depth. The problem of pose refinement was tackled by Rutkowski and Benton (1984) and by Tabandeh and Fallside (1986), who used an iterative scheme for this purpose. When the most accurate pose has been obtained, the overall degree of fit can be considered and the hypothesis rejected if some relevant criterion is not met.

In common with other researchers (Faugeras and Hebert, 1983; Grimson and Lozano-Perez, 1984), Bolles and Horaud took a depth-first tree search as the basic matching strategy. Their 3DPO scheme uses a minimum number of features to key into the data, first generating hypotheses and then taking care to ensure verification (note that Bolles and Cain (1982) had earlier used this technique in a 2-D part location problem). This contrasts with much work (especially that based on the HT) which makes hypotheses but does not check them (however, it is worth noting that forming the initial hypotheses is the difficult and computationally intensive part of the work: researchers will therefore write about this aspect of their work and perhaps not bother to state the minor amount of computation that went into confirming that objects had indeed been located: note also that in much 2-D work, images can be significantly simpler and the size of the peak in parameter space can be so large as to make it virtually certain that an object has been located—thus rendering verification unnecessary).

15.14 The IVISM system

The last four sections have shown how totally different features have been used by various workers to cue into complex image data. This section presents yet another approach to object location, based again on area information. The IVISM (Intelligent Vision Integrated with a Solid Modeller) system of Tabandeh and Fallside (1986; see also Tabandeh, 1988) is nominally similar to that of Ballard and Sabbah (1983) but uses an intensity image rather than a space map to examine planar object surfaces and to obtain data on their areas. The apparent area of a surface is used to give information on the inclination θ of the surface normal to the viewer-centred frame, using the formula:

$$A_{\text{apparent}} = A_{\text{absolute}} \cos \theta \qquad\qquad (15.37)$$

Thus the orientation of the object may in principle be deduced from the intersections of various circles on the Gaussian sphere. The geometry for this is essentially identical to that given in Section 15.2 on the Ballard and Sabbah work.

This approach is viable since the application is one with strong constraints: the objects are of known size and appear at known range since they lie on a worktable ready to be picked up by a robot. Hence one of the six degrees of freedom for position is missing, and the absolute area of each face of the object is known once it has (at first tentatively) been identified.

One of the important features of the IVISM system is that it incorporates a solid modeller which takes in various hypothesized values for the position and orientation (pose) parameters of the object and deduces its appearance from a given viewpoint: the latter is then compared with the actual appearance, and the pose parameters are refined; then the solid modeller is redeployed to obtain a new prediction. This procedure is iterated until sufficiently accurate agreement is obtained between the predicted and actual views of the object, a process that normally takes only two or three iterations. The advantage of this approach is that at the same time as the object viewpoint is refined, predictions of other object faces that ought to be visible are made and extra matches corresponding to these are incorporated into the interpretation. A maximal clique algorithm is employed to obtain a maximal set of matches for available features: in fact, this is the slowest part of the interpretive process, sometimes taking of the order of one minute over several iterations—although this may be justified in view of the extra reliability and accuracy of the interpretation.

Another feature of IVISM is that skeletons are computed for each of the visible areas and these are used to deduce a set of local coordinate systems. The skeleton computation time is negligible compared with the time required for the rest of the algorithm. A subtle point is that it was found useful to apply the same segmentation and skeletonization procedures to the image obtained from the solid modeller and to the raw input image data before comparisons were made. In this way better matches, for example with small obliquely viewed faces, are obtained. Several other sophisticated techniques are incorporated into the IVISM system, such as a "transformation history" which guides the matching process as it progresses, but these cannot be discussed in detail here.

The wide divergence between the IVISM approach and that due for example to Horaud (1987) confirms that 3-D vision is a vibrant subject. However, the wide divergence is moderated by certain similarities of approach—the use of search techniques incorporating a high degree of intelligence in order to obtain the best interpretation of all available features, coupled with specific mathematical techniques for deducing object pose with high accuracy from a limited number of starting features.

15.15   Lowe's approach

Earlier in this chapter, the question was raised of whether objects could be located within a single 2-D intensity image with no specific preliminary mapping out of depth information. It was later found that certain methods (Horaud 1987; Tabandeh and Fallside, 1986) permit this to be achieved, thereby saving computational effort in the specific cases in which these methods prove possible. Lowe (1987) has made strong statements about the value of this strategy. In particular, he states that object recognition has not been shown to be easier for methods which employ depth maps than for systems which begin only with 2-D images. Nonetheless, he does not deny that depth information may be crucial in some situations, for example for training a vision system, or for making particular precise measurements. Lowe also remarks on the use of extreme forms of perspective projection which make parallel lines in the scene appear strongly convergent, hence leading to strong depth cues: he feels that such approaches are too restrictive and nongeneral (see also Horaud, 1987).

If depth in general and perspective convergence in particular are to be eschewed, other means must be obtained for cueing into complex scenes. Unfortunately, the methods so far outlined for proceeding further are also restricted to objects bounded by plane surfaces, and rely very much on highly specific features of known types of object. In this connection Lowe regards it as "useless to look for features with particular sizes or angles or other properties that are highly dependent upon viewpoint". Instead, he advocates an "initial stage of matching ... based upon the detection of structures in the image that can be formed bottom-up in the absence of domain knowledge". The method he adopts is to find elements of perceptual organization—particularly those based on proximity, parallelism and collinearity (Fig. 15.18). These three properties are preserved over a wide range of viewpoints (although in the case of parallelism there must be some restriction on how close a viewpoint will be acceptable). Finding instances

Fig. 15.18 Elements of perceptual organization: the three types of perceptual element identified by Lowe (1987)—proximity, parallelism and collinearity—which may be used to locate 3-D objects.

of these properties is therefore of high significance for image interpretation. Lowe derives criteria by which to test the levels of significance of groups of lines having these properties, the idea being to disregard instances which are accidental collections of image features.

These considerations lead to an overall scheme for image interpretation in which bottom-up (data-driven) analysis is applied initially to find viewpoint-invariant features, and top-down analysis—which makes use of hypotheses about the specific objects being sought—is applied subsequently to incorporate viewpoint-dependent features. Viewpoint-invariant features are taken as cues because they can reduce markedly the amount of search required to interpret the image. Finally, a viewpoint consistency constraint is taken to be vital for confirming the final interpretation: highly overconstrained solutions make the interpretation robust against missing data and measurement errors of various types. In Lowe's work, correct matches of plastic razors usually had over 20 segments agreeing with the final model, compared with seven or less for incorrect matches.

Lowe's 3-D object recognition scheme is backed up by much rationale outlining what is wrong with certain other strategies for 3-D vision, and describing in some detail what makes his own approach successful. However, it is important to note that his approach is not unique: for example, the feedback scheme required to verify correct interpretation is essentially the same as that of Bolles and Horaud (1986).

15.16 Concluding remarks

The historical approach to 3-D vision follows the rather obvious fact that the human visual system is binocular, and presumes that depth maps are required initially in artificial schemes for scene interpretation. Then it was gradually realized that short-cuts may be taken so that objects can be recognized directly in monocular images. However, it is arguable that when a restricted set of specific objects is being sought in an image, it will not always be worthwhile to search first for the perceptual groupings identified by Lowe (1987). Nonetheless, it is likely that in the wider context of unrestricted outdoor scenes, perceptual groupings will have to be used as bottom-up "trigger" features.

This chapter has studied a number of approaches to 3-D vision. Perhaps the most obvious feature of these is the degree of variation from one to the other. Indeed, within the confines of a single chapter, it seemed useful to illustrate the diversity of existing methods. The fact that several of these approaches (all those concerned with interpretation rather than with derivation of depth maps) arose over the short span of 6 years showed that the subject had by no means settled down and that there was still considerable scope for creative research.

Despite the diversity of methods mentioned above, there are certain common themes: the use of trigger features, the value of combining features into groups that are analysed together, the need for working hypotheses to be generated at an early stage, the use of depth-first heuristic search (combined where appropriate with more rigorous breadth-first evaluation of the possible interpretations), and the detailed verification of hypotheses. All these can be taken as parts of current *methodology*; *details*, however, vary with the dataset. More specifically, if a new type of industrial part is to be considered, some study must be made of its most salient features: then this causes not only the feature detection scheme to vary but also the heuristics of the search employed—and also the mathematics of the hypothesis mechanism. In this last respect, note how different are the mathematics of the Horaud junction analysis and the Tabandeh area-measurement approaches, and how much this difference reflects on the computational load of the overall scheme in each case. The reader is referred to the following chapter for further discussion of object recognition under perspective projection.

The last few sections of this chapter have concentrated on object recognition and have tended to eschew the value of range measurements and depth maps. However, this might give a misleading impression of the situation. There are many situations where recognition is largely irrelevant but where it is mandatory to map out 3-D surfaces in great detail. Turbine blades, automobile body parts or even food products such as fruit may need to be measured accurately in 3-D: in such cases it is known in advance what object is in what position, but some inspection or measurement function has to be carried out and a diagnosis made. In such instances, the methods of structured lighting, stereopsis or photometric stereo come into their own and are highly effective methods. Ultimately also, one might expect that a robot vision system will have to use all the tricks of the human visual system if it is to be as adaptable and useful when operating in an unconstrained environment rather than at a particular worktable.

15.17   Bibliographical and historical notes

As noted earlier in the chapter, the most obvious approach to 3-D perception is to employ a binocular camera system. Burr and Chien (1977) and Arnold (1978) showed how a correspondence could be set up between the two input images by use of edges and edge segments. Forming a correspondence can involve considerable computation: Barnea and Silverman (1972) showed how this problem could be alleviated by passing quickly over unfavourable matches. Likewise, Moravec (1980) devised a coarse-to-fine matching procedure which arrives systematically at an accurate correspondence between images. Marr and Poggio (1979) formulated two constraints—those of uniqueness and continuity—that have to be satisfied in choosing global correspondences: these constraints are important in leading to the simplest available surface interpretation. Ito and Ishii (1986) found that there is something to be gained from three-view stereo in offsetting ambiguity and the effects of occlusions.

The structured lighting approach to 3-D vision was introduced independently by Shirai (1972) and Agin and Binford (1973, 1976), in the form of a single plane of light, while Will and Pennington (1971) developed the grid coding technique. Nitzan *et al.* (1977) employed an alternative LIDAR (light detecting and ranging) scheme for mapping objects in 3-D; here short light pulses were timed as they travelled to the object surface and back.

Meanwhile, other workers were attempting monocular approaches to 3-D vision. Some basic ideas underlying shape-from-shading date from as long ago as 1929, with Fesenkov's investigations of the lunar surface: see also van Digellen (1951). However, the first shape-from-shading problem to be solved both theoretically and in an operating algorithm appears to have been that of Rindfleisch (1966), also relating to lunar landscapes. Thereafter Horn systematically tackled the problem both theoretically and with computer investigations, starting with a notable review (1975) and resulting in prominent papers (e.g. Horn, 1977; Ikeuchi and Horn, 1981; Horn and Brooks, 1986), an important book (Horn, 1986) and an edited work (Horn and Brooks, 1989). Interesting papers by other workers in this area include Blake *et al.* (1985), Bruckstein (1988) and Ferrie and Levine (1989). Woodham (1978, 1980, 1981) must be credited with the photometric stereo idea. Finally, the vital contributions made by workers on computer graphics in this area must not be forgotten—see for example Phong (1975), Cook and Torrance (1982).

The concept of shape-from-texture arose from the work of Gibson (1950) and was developed by Bajcsy and Liebermann (1976), Stevens (1980), and notably by Kender (1981), who carefully explored the underlying theoretical constraints.

The paper by Barrow and Tenenbaum (1981) provides a very readable review of much of this earlier work. It is probably true to say that 1980 marked a turning point, when the emphasis in 3-D vision shifted from mapping out surfaces to interpreting images as sets of 3-D objects. Possibly, this segmentation task could not be tackled easily earlier because basic tools such as the HT were not sufficiently well developed (e.g. the GHT dates from 1981). The work of Koenderink and van Doorn (1979) and Chakravarty and Freeman (1982) was probably also crucial in providing a framework of potential 3-D views of objects enabling interpretation schemes to be developed. The work of Ballard and Sabbah (1983) provided an early breakthrough in segmentation of real objects in 3-D (see Section 15.10) and this was followed by vital further work by Faugeras and Hebert (1983), Silberberg *et al.* (1984), Bolles and Horaud (1986), Horaud (1987), Pollard *et al.* (1987), and many others (see foregoing sections). Lowe's (1987) research and the rationale he provides is of great import, since it sums up the motivation of many groups as well as having its own inherent originality. However, there are signs that it too has been overtaken—see especially the work of Worrall *et al.* (1989), which claims to require less computation through use of the interpretation plane method coupled with least-squares analysis of the resulting nonlinear equations.

The review article of Besl and Jain (1984) is particularly useful but is now somewhat dated, and the volume by Shirai (1987) partly fills the gap. Other interesting work includes that of Horaud *et al.* (1989) on solving the perspective 4-point problem (finding the position and orientation of the camera relative to known points), and complements the work of Section 15.12: for further references on this topic see Section 16.6. The volume of articles edited by Winston (1975) contains several classic papers on 3-D vision—namely, those of Horn, Waltz and Shirai. Waltz's work on understanding line drawings of polyhedra is interesting in rigorously classifying all types of line junction and making full use of information available from shadows: however, it assumes an ideal "blocks world" and for this reason its importance is ultimately limited.

Finally, though finding vanishing points might be thought a well worked-through topic, research on it is still proceeding in the 1990s (Straforini *et al.*, 1993; Lutton *et al.*, 1994). Similarly, stereo correlation matching techniques are still under development, to maintain robustness in real-time applications (Lane *et al.*, 1994); and there is an increasing emphasis on incorporating stereo viewing in real-time visual feedback loops for robot control (Hollinghurst and Cipolla, 1994).

15.18   Problems

1. Prove that all epipolar lines in one image plane pass through the point that is the image of the projection point of the alternate image plane.
2. What is the physical significance of the straight line contour in gradient space (see Fig. 15.9(b))?
3. Find the condition for a contour in gradient space to be an ellipse.
4. Derive a vector formula giving the unit vector **r** of the direction for pure specular reflection in terms of the unit vectors **s** and **n** of the source and surface normal directions (Section 15.4). Hence show how the total reflectance (equation (15.12)) may be computed as a function of p and q.
5. Sketch a curve of the function $\cos^m \theta$. Estimate what value m would have to have for 90% of the R_1 component to be reflected within 10° of the direction for pure specular reflection.

16

Tackling the Perspective *N*-Point Problem

16.1 Introduction

This chapter follows on from the previous introductory chapter, and tackles a problem of central importance in the analysis of images from 3-D scenes. It has been kept separate and fairly short so as to focus carefully on relevant factors in the analysis. First, we look closely at the phenomenon of perspective inversion, which has already been alluded to several times in Chapter 15. Then we refine our ideas on perspective, and proceed to consider the determination of object pose from salient features which are located in the images. It will be of interest to consider how many salient features are required for unambiguous determination of pose.

16.2 The phenomenon of perspective inversion

In this section we study first the phenomenon of *perspective inversion*. This is actually a rather well-known effect which appears in the following "Necker cube" illusion. Consider a wire cube made from 12 pieces of wire welded together at the corners. Looking at it from approximately the direction of one corner, it is difficult to tell which way round the cube is, i.e. which of the opposite corners of the cube is the nearer (Fig. 16.1). Indeed, on looking at the cube for a time, one gradually comes to feel one knows which way round it is, but then it suddenly appears to reverse itself; then that perception remains for some time, until it too reverses

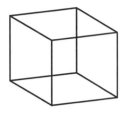

Fig. 16.1   The phenomenon of perspective inversion. This figure shows a wire cube viewed approximately from the direction of one corner. The phenomenon of perspective inversion makes it difficult to see which of the opposite corners of the cube is the nearer: in fact there are two stable interpretations of the cube, either of which may be perceived at any moment.

itself[*]. This illusion reflects the fact that the brain is making various hypotheses about the scene, and even making decisions based on incomplete evidence about the situation (Gregory, 1971, 1972).

The wire cube illusion could perhaps be regarded as somewhat artificial. But consider instead an aeroplane (Fig. 16.2(a)) which is seen in the distance (Fig. 16.2(b)) against a bright sky. The silhouetting of the object means that its surface details are not visible. In that case interpretation requires that a hypothesis be made about the scene, and it is possible to make the wrong one. Clearly (Fig. 16.2(c)), the aeroplane could be at an angle α (as for P), though it could equally well be at an angle $-\alpha$ (as for Q). The two hypotheses about the orientation of the object are related by the fact that the one can be obtained from the other by reflection in a plane R normal to the viewing direction D.

Strictly, there is only an ambiguity in this case if the object is viewed under orthographic or scaled orthographic projection[†]. However, in the distance, perspective projection approximates to scaled orthographic projection, and it is often difficult to detect the difference[‡]. If the aeroplane

---

[*] In psychology, this shifting of attention is known as *perceptual reversal*, which is unfortunately rather similar to the term *perspective inversion*, but is actually a much more general effect which leads to a host of other types of optical illusion—see Gregory (1971) and the many illustrations produced by M.C. Escher.

[†] *Scaled orthographic projection* is orthographic projection with the final image scaled in size by a constant factor.

[‡] In this case, the object is said to be viewed under *weak perspective projection*. For weak perspective, the depth ΔZ within the object has to be much less than its depth Z in the scene. On the other hand, the perspective scaling factor can be different for each object and will depend on its depth in the scene: so the perspective can validly be locally weak and globally normal.

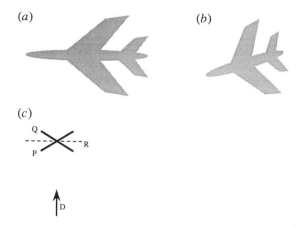

Fig. 16.2 Perspective inversion for an aeroplane. An aeroplane (a) is silhouetted against the sky and appears as in (b); (c) shows the two planes P and Q in which the aeroplane could lie, relative to the direction D of viewing: R is the reflection plane relating the planes P and Q.

in Fig. 16.2 was quite close, it would be obvious that one part of the silhouette was nearer, as the perspective would distort it in a particular way. In general, perspective projection will break down symmetries, so searching for symmetries that are known to be present in the object should reveal which way around it is: however, if the object is in the distance, as in Fig. 16.2(b), it will be virtually impossible to see the breakdown. Unfortunately, short-term study of the motion of the aeroplane will not help with interpretation in the case shown in Fig. 16.2(b). Eventually, however, the aeroplane will appear to become smaller or larger, and this will give the additional information needed to resolve the issue.

16.3 Ambiguity of pose under weak perspective projection

It is instructive to examine to what extent the pose of an object can be deduced under weak perspective projection. We can reduce the above problem to a simplest case in which three points have to be located and identified. Any set of three points is coplanar, and the common plane corresponds to that of the silhouette shown in Fig. 16.2(a) (we assume here that the three points are not collinear, so that they do in fact define a plane). The problem then is to match the corresponding points on the idealized object (Fig. 16.2(a)) with those on the observed object (Fig. 16.2(b)). It is not yet completely obvious that this is possible, or that the solution is unique, even apart from the reflection operation noted earlier. It could be that more than three points will be required—especially if the scale is unknown—or it could be that there are several solutions, even if we ignore the reflection ambiguity. Of particular interest will be the extent to which it is possible to deduce which is which of the three points in the observed image.

To understand the degree of difficulty, let us briefly consider full perspective projection. In this case, any set of three noncollinear points can be mapped into any other three. This means that it may not be possible to deduce much about the original object just from this information: we will certainly not be able to deduce which point maps to which other point. However, we shall see that the situation is rather less ambiguous when viewing the object under weak perspective projection.

Perhaps the simplest approach (due to Huang *et al.* as recently as 1995) is to imagine a circle drawn through the original set of points P_1, P_2, P_3 (Fig. 16.3(a)). We then find the centroid C of the set of points, and draw additional lines through the points, all passing through C and meeting the circle in another three points Q_1, Q_2, Q_3 (Fig. 16.3(a)). Now in common with orthographic projection, scaled orthographic projection maintains ratios of distances on the same straight line, and weak perspective projection approximates to this. Thus the distance ratio $P_iC : CQ_i$ remains unchanged after projection. Thus when we project the whole figure, as in Fig. 16.3(b), we find that the circle has become an ellipse, though all lines remain lines, and all linear distance ratios remain unchanged. The significance of this is as follows. When the points P_1', P_2', P_3' are observed in the image, the centroid C' can be computed, as can the positions of Q_1', Q_2', Q_3'. Thus we have six points by which to deduce the position and parameters of the ellipse (in fact, five are sufficient). Once the ellipse is known, the orientation of its major axis gives the axis of rotation of the object; while the ratio of the lengths of the minor to major axes immediately gives the value of cos α. (Notice how the ambiguity in the sign of α comes up naturally in this calculation.) Finally, the length of the major axis of the ellipse permits the depth of the object in the scene to be deduced.

We have now shown that observing three projected points permits a unique ellipse to be computed passing through them, and when this is backprojected into a circle, the axis of rotation of the object, and the angle of rotation can be deduced, but not the sign of the angle of rotation. There are two important comments to be made about the above calculation. The first is that the three distance ratios must be stored in memory, before interpretation of the observed scene can begin. The second is that the order of the three points apparently has to be known before interpretation can be undertaken: otherwise

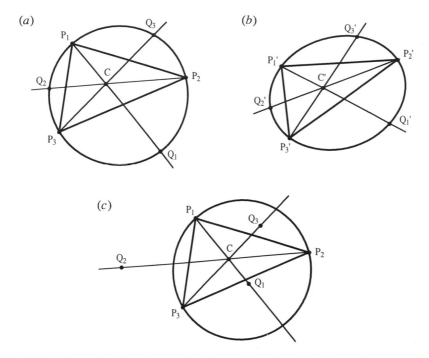

Fig. 16.3 Determination of pose for three points viewed under weak perspective projection. (a) The three feature points P_1, P_2, P_3 lie on a known type of object. The circle passing through P_1, P_2, P_3 is drawn, and lines through the points and their centroid C meet the circle in Q_1, Q_2, Q_3. The ratios $P_iC:CQ_i$ are then deduced. (b) The three points are observed under weak perspective as P'_1, P'_2, P'_3, together with their centroid C', and the three points Q'_1, Q'_2, Q'_3 are located using the original distance ratios. An ellipse drawn through the six points P'_1, P'_2, P'_3, Q'_1, Q'_2, Q'_3 can now be used to determine the orientation of the plane in which P_1, P_2, P_3 must lie, and also (from the major axis of the ellipse) the distance of viewing. (c) An erroneous interpretation of the three points does not permit a circle to be drawn passing through P_1, P_2, P_3, Q_1, Q_2, Q_3 and hence no ellipse can be found which passes through the observed and the derived points P'_1, P'_2, P'_3, Q'_1, Q'_2, Q'_3.

we will have to perform six computations in which all possible assignments of the distance ratios are tried; furthermore, it might appear from the earlier introductory remarks that several solutions are possible. While there are some instances in which feature points might be distinguishable, there are many cases when they are not (especially in 3-D situations where corner features might vary considerably when viewed from different positions). Thus the potential ambiguity is important. However, if we can try out each of the six cases, little difficulty will generally arise. For immediately we deduce the positions Q_1', Q_2', Q_3', we will find that it is not possible in general to fit the six resulting points to an ellipse. The reason is easily seen on returning to the original circle. In that case, if the wrong distance ratios are assigned, the Q_i will clearly not lie on the circle, since the only values of the distance ratios for which the Q_i do lie on the circle are the correctly assigned ones (Fig. 16.3(c)). This means that though computation is wasted testing the incorrect assignments, there appears to be no risk of their leading to ambiguous solutions. Nevertheless, there is one contingency under which things could go wrong. Suppose the original set of points P_1, P_2, P_3 forms an almost perfect equilateral triangle. Then the distance ratios will be very similar, and, taking numerical inaccuracies into account, it may not be clear which ellipse provides the best and most likely fit. This mitigates against taking sets of feature points which form approximately isosceles or equilateral triangles. However, in practice more than three coplanar points will generally be used to optimize the fit, making fortuitous solutions rather unlikely.

Overall, it is fortunate that weak perspective projection requires such weak conditions for the identification of unique (to within a reflection) solutions, especially as full perspective projection demands four points before a unique solution can be found (see below). However, under weak perspective projection additional points lead to greater accuracy but no reduction in the reflection ambiguity: this is because the information content from weak perspective projection is impoverished in the lack of depth cues which could (at least in principle) resolve the ambiguity. To understand this lack of additional information from more than three points under weak perspective projection, note that each additional feature point in the same plane is predetermined once three points have been identified (here we are assuming that the model object with the correct distance ratios can be referred to).

These considerations indicate that we have two potential routes to unique location of objects from limited numbers of feature points. The first is to resort to use of noncoplanar points viewed still under weak perspective projection. The second is to use full perspective projection to view coplanar or noncoplanar sets of feature points. We shall see below that whichever of these options we take, a unique solution demands that a minimum of four feature points be located on any object.

16.4 Obtaining unique solutions to the pose problem

The overall situation is summarized in Table 16.1. Looking first at the case of weak perspective projection, the number of solutions only becomes finite for three or more point features. Once three points have been employed, in the coplanar case there is no further reduction in the number of solutions, since (as noted earlier) the positions of any additional points can be deduced from the existing ones. However, this does not apply when the additional points are noncoplanar since they are able to provide just the right information to eliminate any ambiguity (see Fig. 16.4). (Although this might appear to contradict what was said earlier about perspective inversion, note that we are assuming here that the body is rigid and that all its features are at *known* fixed points on it in three dimensions; hence this particular ambiguity no longer applies, except for objects with special symmetries which we shall ignore here—see Fig. 16.4(d).)

Table 16.1   Ambiguities when estimating pose from point features.

Arrangements of the points	N	WPP	FPP
	$\leqslant 2$	∞	∞
	3	2	4
Coplanar	4	2	1
	5	2	1
	$\geqslant 6$	2	1
	$\leqslant 2$	∞	∞
	3	2	4
Noncoplanar	4	1	2
	5	1	2
	$\geqslant 6$	1	1

This table summarizes the numbers of solutions that will be obtained when estimating the pose of a rigid object from point features located in a single image. It is assumed that N point features are detected and identified correctly and in the correct order. The columns WPP and FPP signify weak perspective projection and full perspective projection, respectively. The upper half of the table applies when all N points are coplanar; the lower half of the table applies when the N points are noncoplanar. Note that when $N \leqslant 3$, the results for the two cases are identical, since for $N \leqslant 3$, all points are necessarily coplanar.

Considering next the case of full perspective projection, the number of solutions again becomes finite only for three or more point features. The lack of information provided by three point features means that four solutions are in principle possible (see the example in Fig. 16.5 and the detailed explanation in Section 16.2.1), but the number of solutions drops to one as soon as four coplanar points are employed (the correct solution can be found by making cross checks between subsets of three points, and eliminating

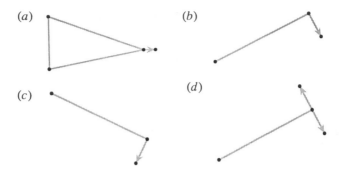

Fig. 16.4 Determination of pose for four points viewed under weak perspective projection. (a) An object containing four noncoplanar points, as seen under weak perspective projection; (b) a side view of the object. If the first three points (connected by non-arrowed grey lines) were viewed alone, perspective inversion would give rise to a second interpretation (c). However, the fourth point gives additional information about the pose which permits only one overall interpretation. This would not be the case for an object containing an additional symmetry as in (d), since its reflection would be identical to the original view (not shown).

inconsistent solutions); when the points are noncoplanar, it is only when six or more points are employed that there is sufficient information to unambiguously determine the pose: there is necessarily no ambiguity with six or more points, as all 11 camera calibration parameters can be deduced from the 12 linear equations that then arise (see Appendix C). Correspondingly, it is deduced that five noncoplanar points will in general be *insufficient* for all 11 parameters to be deduced, so there will still be some ambiguity in this case.

Next, it should be questioned why the coplanar case is at first $(N = 3)$ better[*] under weak perspective projection and then $(N > 3)$ better under full perspective projection; while the noncoplanar case is always better, or as good, under weak perspective projection. The reason must be that intrinsically full perspective projection provides more detailed information, but is frustrated by lack of data when there are relatively few points: however, the exact stage at which the additional information becomes available is different in the coplanar and noncoplanar cases. In this respect it is important to note that when coplanar points are being observed under

---

[*] In this context "better" means less ambiguous, and leading to fewer solutions.

weak perspective projection there is never enough information to eliminate the ambiguity.

It should be emphasized that the above discussion assumes that the correspondences between object and image features are all known, i.e. that *N* point features are detected and identified correctly and in the correct order. If this is not so, the number of possible solutions could increase substantially, considering the number of possible permutations of quite small numbers of points. This makes it attractive to use the minimum number of features for ascertaining the most probable match (Horaud *et al.*, 1989). Other workers have used heuristics to help reduce the number of possibilities. For example, Tan (1995) used a simple compactness measure (see Section 6.9) to determine which geometric solution is the most likely: extreme obliqueness is perhaps unlikely, and the most likely solution is taken to be the one with highest compactness value. This idea follows on from the extremum principle of Brady and Yuille (1984), which states that the most probable solutions are those nearest to extrema of relevant (e.g. rotation) parameters[*]. In this context, it is worth noting that coplanar points viewed under weak or full perspective projection always appear in the same cyclic order: this is not trivial to check given the possible distortions of an object, though if a convex polygon can be drawn through the points, the cyclic order around its boundary will not change on projection[†]. However, for noncoplanar points, the pattern of the perceived points can re-order itself almost randomly: this means that a considerably greater number of permutations of the points have to be considered for noncoplanar points than for coplanar points.

Finally, it should be noted that the above discussion has concentrated on the existence and uniqueness of solutions to the pose problem. The stability of the solutions has not so far been discussed. However, the concept of stability gives a totally different dimension to the data presented in Table 16.1. In particular, noncoplanar points tend to give more stable solutions to the pose problem. For example, if the plane containing a set of coplanar points is viewed almost head-on ($\alpha \simeq 0$), there will be very little information on the exact orientation of the plane, because the changes in lateral displacement of the points will vary as $\cos \alpha$ (see Section 16.1) and there will be no linear term in the Taylor expansion of the orientation dependence.

---

[*] Perhaps the simplest way of understanding this principle is obtained by considering a pendulum, whose extreme positions are also its most probable! However, in this case the extremum occurs when the angle α (see Fig 16.1) is close to zero.

[†] The reason for this is that planar convexity is an invariant of projection.

16.4.1 Solution of the three-point problem

Figure 16.5 showed how four solutions can arise when three point features are viewed under full perspective projection. Here we briefly explore this situation by considering the relevant equations. Figure 16.5 shows that the camera sees the points as three image points representing three directions in space. This means that we can compute the angles α, β, γ between these three directions. If the distances between the three points A, B, C on the object are the known values D_{AB}, D_{BC}, D_{CA}, we can now apply the cosine rule in an attempt to determine the distances R_A, R_B, R_C of the feature points from the centre of projection:

$$D_{BC}^2 = R_B^2 + R_C^2 - 2R_B R_C \cos \alpha \tag{16.1}$$

$$D_{CA}^2 = R_C^2 + R_A^2 - 2R_C R_A \cos \beta \tag{16.2}$$

$$D_{AB}^2 = R_A^2 + R_B^2 - 2R_A R_B \cos \gamma \tag{16.3}$$

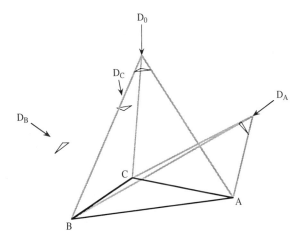

Fig. 16.5 Ambiguity for three points viewed under full perspective projection. Under full perspective projection, the camera sees three points A, B, C as three directions in space, and this can lead to four-fold ambiguity in interpreting a known object. The figure shows the four possible viewing directions and centres of projection of the camera (indicated by the directions and tips of the bold arrows): in each case the image at each camera is indicated by a small triangle. D_A, D_B, D_C correspond approximately to views from the general directions of A, B, C respectively.

Eliminating any two of the variables R_A, R_B, R_C yields an eighth degree equation in the other variable, indicating that eight solutions to the system of equations could be available (Fischler and Bolles, 1981). However, the above cosine rule equations contain only constants and second degree terms: hence, for every positive solution there is another solution which differs only by a sign change in all the variables. These solutions correspond to inversion through the centre of projection and are hence unrealizable. Thus there are at most four realizable solutions to the system of equations. In fact, we can quickly demonstrate that there may sometimes be fewer than four solutions: since in some cases, for one or more of the "flipped" positions shown in Fig. 16.5, one of the features could be on the negative side of the centre of projection, and hence would be unrealizable.

Before leaving this topic, it is worth noting the homogeneity of equations (16.1)–(16.3) implies that observation of the angles α, β, γ permits the orientation of the object to be estimated independently of any knowledge of its scale: in fact, estimation of scale depends directly on estimation of range, and vice versa. Thus knowledge of just one range parameter (e.g. R_A) will permit the scale of the object to be deduced. Alternatively, knowledge of its area will permit the remaining parameters to be deduced. This concept provides a slight generalization of the main results of Sections 16.2–16.3, which start generally with the assumption that all the dimensions of the object are known.

16.4.2 Using symmetric trapezia for estimating pose

One more example will be of interest here. That is the case of four points arranged at the corners of a symmetric trapezium (Tan, 1995). When viewed under weak perspective projection, the midpoints of the parallel sides are easily measured, but under full perspective projection midpoints do not remain midpoints, so the axis of symmetry cannot be obtained in this way. However, producing the skewed sides to meet at S', and forming the intersection I' of the diagonals permits the axis of symmetry to be located as the line I'S' (Fig. 16.6). Thus we now have not four points but six to describe the perspective view of the trapezium. What is more important is that the axis of symmetry has been located and this is known to be perpendicular to the parallel sides of the trapezium. This is a great help in making the mathematics more tractable, and in obtaining solutions quickly so that (for example) object motion can be tracked in real time. Again, this is a case where object orientation can be deduced straightforwardly, even when the situation is one of strong perspective, and even when the size of

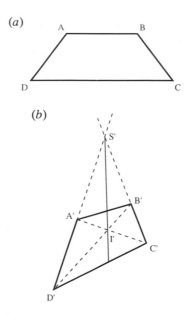

Fig. 16.6 Trapezium viewed under full perspective projection. (a) Symmetrical trapezium; (b) how it appears when viewed under full perspective projection. In spite of the fact that midpoints do not project into midpoints under perspective projection, the two points S' and I' on the symmetry axis can be located unambiguously as the intersection of two nonparallel sides and two diagonals respectively. This gives six points (from which the two midpoints on the symmetry axis can be deduced if necessary), which is sufficient to compute the pose of the object, albeit with a single ambiguity of interpretation (see text).

the object is unknown. This result is a generalization from that of Haralick (1989), who noted that a single view of a rectangle of unknown size is sufficient to determine its pose. In either case, the range of the object can be found if its area is known, or its size can be deduced if a single range value can be found from other data (see also Section 16.3.1).

16.5 Concluding remarks

This chapter has aimed to cover certain aspects of 3-D vision that were not studied in depth in the previous chapter. In particular, it was worth investigating the topic of perspective inversion in some detail, and exploring how it was affected by the method of projection. Orthographic projection, scaled orthographic projection, weak perspective projection and full perspective projection were considered, and the numbers of object points that would lead to correct or ambiguous interpretations were analysed. It was found that scaled orthographic projection and its approximation, weak perspective projection, led to straightforward interpretation when four or more noncoplanar points were considered, though the perspective inversion ambiguity remained when all the points were coplanar. This latter ambiguity was resolved with four or more points viewed under full perspective projection. However, in the noncoplanar case, some ambiguity remained until six points were being viewed. The key to understanding the situation is the fact that full perspective projection makes the situation more complex, though it also provides more information by which, ultimately, to resolve the ambiguity.

Additional problems were found to arise when the points being viewed are indistinguishable, and then a good many solutions may have to be tried before a minimally ambiguous result is obtained. With coplanar points fewer possibilities arise, and this leads to less computational complexity: the key to success here is the natural ordering that can arise for points in a plane—as for example when they form a convex set which can be ordered uniquely around a bounding polygon. In this context the role that can be played by the extremum principle of Brady and Yuille (1984) in reducing the number of solutions is significant (for further insight on the topic, see Horaud and Brady, 1988).

It is of great relevance to devise methods for rapid interpretation in real-time applications. To achieve this it is important to work with a minimal set of points, and to obtain analytic solutions which move directly to solutions without computationally expensive iterative procedures: for example, Horaud *et al.* (1989) found an analytic solution for the perspective four-point problem which works both in the general noncoplanar case and in the planar case. Other low computation methods are still being developed, as with pose determination for symmetrical trapezia (Tan, 1995). It should also be noted that understanding is still advancing, as demonstrated by Huang *et al.*'s (1995) neat geometrical solution to the pose determination problem for three points viewed under weak perspective projection.

16.6 Bibliographical and historical notes

The development of solutions to the so-called *perspective N-point problem* (finding the pose of objects from *N* features under various forms of perspective) has been proceeding for well over a decade, and is by no means complete. The situation was summarized by Fischler and Bolles in 1981, and several new algorithms were described by them. However, they did not discuss pose determination under weak perspective, and perhaps surprisingly, considering its reduced complexity, this is still the subject of much research (e.g. Alter, 1994; Huang *et al.*, 1995). Horaud *et al.* (1989) discussed the problem of finding rapid solutions to the perspective N-point problem by reducing N as far as possible: they also obtained an analytic solution for the case $N = 4$ which should help considerably with real-time implementations. Their solution is related to Horaud's earlier (1987) corner interpretation scheme—described in Section 15.12—while Haralick *et al.* (1984) provided useful basic theory for matching wire frame objects.

Faugeras (1993) provides an excellent general reference for the whole area of 3-D vision; for an interesting viewpoint on the subject, with particular emphasis on pose refinement, see Sullivan (1992). For an early theorem on how a scene can be reconstructed from two projections using a simple algorithm, see Longuet-Higgins (1981).

17

Motion

17.1 Introduction

This chapter is concerned with the analysis of motion in digital images. In the space available it will not be possible to cover the whole subject comprehensively: instead the aim will be to provide the flavour of the subject, airing some of the principles that have proved important over the past 10–15 years. Over much of this time optical flow has been topical and it is appropriate to study it in fair detail, particularly, as we shall see later from some case studies on the analysis of traffic scenes, since it is actually being used seriously in practical situations.

17.2 Optical flow

When scenes contain moving objects, analysis is necessarily more complex than for scenes where everything is stationary, since temporal variations in intensity have to be taken into account. However, intuition suggests that it should be possible—even straightforward—to segment moving objects by virtue of their motion: image differencing over successive pairs of frames should permit this to be achieved. More careful consideration shows that things are not quite so simple, as illustrated in Fig. 17.1. The reason is that regions of constant intensity give no sign of motion, while edges parallel to the direction of motion also give the appearance of not moving: only edges with a component normal to the direction of motion carry information about the motion. In addition, there

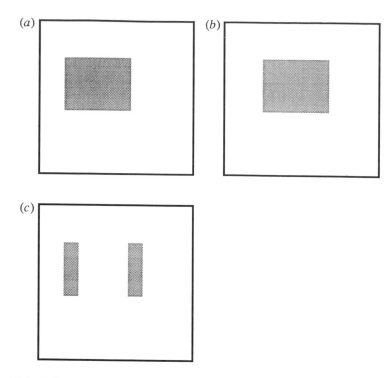

Fig. 17.1 Effect of image differencing. This figure shows an object which has moved between frames (a) and (b); (c) shows the result of performing an image differencing operation. Note that the edges parallel to the direction of motion do not show up in the difference image. Also, regions of constant intensity give no sign of motion.

is some ambiguity in the direction of the velocity vector. This arises partly because there is too little information available within a small aperture to permit the full velocity vector to be computed (Fig. 17.2): this is hence called the *aperture problem*.

These elementary ideas can be taken further, and they lead to the notion of optical flow, wherein a local operator which is applied at all pixels in the image will lead to a motion vector field which varies smoothly over the whole image. The attraction lies in the use of a local operator, with its limited computational burden. Ideally, it would have an overhead comparable to an edge detector in a normal intensity image—though clearly it will have to be applied locally to pairs of images in an image sequence.

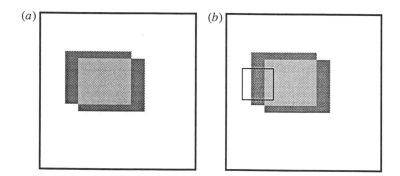

Fig. 17.2 The aperture problem: (a) shows (dark grey) regions of motion of an object whose central uniform region (light grey) gives no sign of motion; (b) shows how little is visible in a small aperture (black border), thereby leading to ambiguity in the deduced direction of motion of the object.

We start by considering the intensity function $I(x, y, t)$ and expanding it in a Taylor series:

$$I(x + dx, y + dy, t + dt) = I(x, y, t) + I_x \, dx + I_y \, dy + I_t \, dt + \cdots \quad (17.1)$$

where second and higher order terms have been ignored. In this formula I_x, I_y, I_t denote respective partial derivatives with respect to x, y and t.

We next set the local condition that the image has shifted by amount (dx, dy) in time dt, so that it is functionally identical at $(x + dx, y + dy, t + dt)$ and (x, y, t):

$$I(x + dx, y + dy, t + dt) = I(x, y, t) \quad (17.2)$$

Hence we can deduce:

$$I_t = -(I_x \dot{x} + I_y \dot{y}) \quad (17.3)$$

Writing the local velocity **v** in the form:

$$\mathbf{v} = (v_x, v_y) = (\dot{x}, \dot{y}) \quad (17.4)$$

we find:

$$I_t = -(I_x v_x + I_y v_y) = -\nabla I \cdot \mathbf{v} \quad (17.5)$$

I_t can be measured by subtracting pairs of images in the input sequence, while ∇I can be estimated by Sobel or other gradient operators. Hence it should be possible to deduce the velocity field $\mathbf{v}(x, y)$ using the above equation. Unfortunately, this equation is a scalar equation, and will not suffice for determining the two local components of the velocity field as we require. There is a further problem with this equation—that the velocity value will depend on the values of both I_t and ∇I, and these quantities are only estimated approximately by the respective differencing operators: in both cases significant noise will arise, and this will be exacerbated by taking the ratio in order to calculate $\mathbf{v}$.

Let us now return to the problem of computing the full velocity field $\mathbf{v}(x, y)$. All we know about $\mathbf{v}$ is that its components lie on the following line in (v_x, v_y)-space (Fig. 17.3):

$$I_x v_x + I_y v_y + I_t = 0 \qquad (17.6)$$

This line is normal to the direction (I_x, I_y), and has a distance from the (velocity) origin which is equal to:

$$|\mathbf{v}| = -I_t / [I_x^2 + I_y^2]^{1/2} \qquad (17.7)$$

Clearly, we need to deduce the component of $\mathbf{v}$ along the line given by equation (17.6). However, there is no purely local means of achieving this

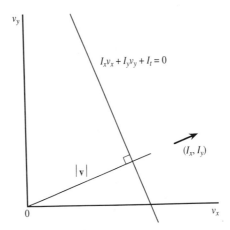

Fig. 17.3   Computation of the velocity field. This graph shows the line in velocity space on which the velocity vector $\mathbf{v}$ must lie. The line is normal to the direction (I_x, I_y) and its distance from the origin is known to be $|\mathbf{v}|$ (see text).

with first derivatives of the intensity function. The accepted solution (Horn and Schunck, 1981) is to use relaxation labelling to arrive iteratively at a self-consistent solution which minimizes the global error. In principle, this approach will also minimize the noise problem indicated earlier.

In fact, there are still problems with the method. Essentially, these arise as there are liable to be vast expanses of the image where the intensity gradient is low. In that case only very inaccurate information is available about the velocity component parallel to ∇I, and the whole problem becomes ill-conditioned. On the other hand, in a highly textured image, this situation should not arise (assuming that the texture has a large enough grain size to give good differential signals).

Finally, we return to the idea mentioned at the beginning of this section— that edges parallel to the direction of motion would not give useful motion information. Such edges will have edge normals that are normal to the direction of motion, so ∇I will be normal to v. Thus, from equation (17.5), I_t will be zero. In addition, regions of constant intensity will have $\nabla I = 0$, so again I_t will be zero. It is interesting and highly useful that such a simple equation (17.5) embodies all the cases that were suggested earlier on the basis of intuition.

In what follows we assume that the optical flow (velocity field) image has been computed satisfactorily, i.e. without the disadvantages of inaccuracy or ill-conditioning. It must now be interpreted in terms of moving objects and perhaps a moving camera. In fact, we shall ignore motion of the camera by remaining within its frame of reference.

17.3 Interpretation of optical flow fields

We start by considering a case where no motion is visible. In that case the velocity field image contains only vectors of zero length (Fig. 17.4(a)). Next we take a case where one object is moving towards the right, with a simple effect on the velocity field image (Fig. 17.4(b)). Next we consider the case where the camera is moving forwards; in this case all the stationary objects in the field of view appear to be diverging from a point which is called the *focus of expansion* (FOE)—see Fig. 17.4(c); this image also shows an object which is moving rapidly past the camera, and which has its own separate FOE. Figure 17.4(d) shows the case of an object moving directly towards the camera: in this case its FOE lies within its outline. Similarly, objects which are receding appear to move towards their respective *foci of contraction*. Next there are objects which are stationary but which are rotating about the line of sight: for these the vector field appears as in Fig.

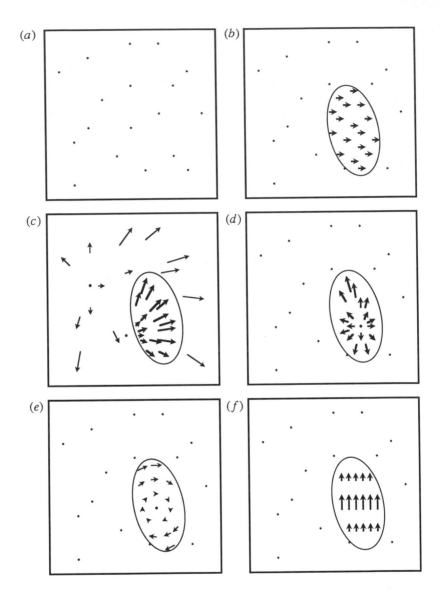

Fig. 17.4 Interpretation of velocity flow fields: (a) the object features all have zero velocity; (b) an object is moving to the right; (c) the camera is moving into the scene, and the stationary object features appear to be diverging from a focus of expansion (FOE), while a single large object is moving past the camera and away from a separate FOE. In (d) an object is moving directly towards the camera which is stationary: the object's FOE lies within its outline. In (e) an object is rotating about the line of sight to the camera, and in (f) the object is rotating about an axis perpendicular to the line of sight. In all cases, the length of the arrow indicates the magnitude of the velocity vector.

17.4(e). There is a final case which is also quite simple—that of an object which is stationary but rotating about an axis normal to the line of sight; if the axis is horizontal, then the features on the object will appear to be moving up or down, while paradoxically the object itself remains stationary (Fig. 17.4(f))—though its outline could oscillate as it rotates.

So far, we have only dealt with cases in which pure translational or pure rotational motion is occurring. If a rotating meteor is rushing past, or a spinning cricket ball is approaching, then both types of motion will occur together. In that case unravelling the motion will be far more complex. We shall not solve this problem here, but refer the reader to more specialized texts (e.g. Maybank, 1992). However, the complexity is due to the way depth (Z) creeps into the calculations. Note that pure rotational motion with rotation about the line of sight does not depend on Z: all we have to measure is the angular velocity, and this can be done quite simply.

17.4 Using focus of expansion to avoid collision

We now take a simple case in which a focus of expansion is located in an image and show how it is possible to deduce the distance of closest approach of the camera to a fixed object of known coordinates. This type of information is valuable for guiding robot arms or robot vehicles and helping to avoid collisions.

In the notation of Chapter 15 and Appendix C, we have the following formulae for the location of an image point (x, y, z) resulting from a world point (X, Y, Z):

$$x = fX/Z \tag{17.8}$$

$$y = fY/Z \tag{17.9}$$

$$z = f \tag{17.10}$$

Assuming that the camera has a motion vector $(-\dot{X}, -\dot{Y}, -\dot{Z}) = (-u, -v, -w)$, fixed world points will have velocity (u, v, w) relative to the camera. Now a point (X_0, Y_0, Z_0) will after a time t appear to move to $(X, Y, Z) = (X_0 + ut, Y_0 + vt, Z_0 + wt)$ with image coordinates:

$$(x, y) = \left(\frac{f(X_0 + ut)}{Z_0 + wt}, \frac{f(X_0 + ut)}{Z_0 + wt} \right) \tag{17.11}$$

and as $t \longrightarrow \infty$ this approaches the focus of expansion $f(fu/w, fv/w)$. This point is in the image, but the true interpretation is that the actual motion of the centre of projection of the imaging system is towards the point:

$$\mathbf{p} = (fu/w, fv/w, f) \tag{17.12}$$

(This is of course consistent with the motion vector (u, v, w) assumed initially.) The distance moved during time t can now be modelled as:

$$\mathbf{X}_c = (X_c, Y_c, Z_c) = \alpha t\, \mathbf{p} = f\alpha t(u/w, v/w, 1) \tag{17.13}$$

where α is a normalization constant. To calculate the distance of closest approach of the camera to the world point $\mathbf{X} = (X, Y, Z)$, we merely specify that the vector $\mathbf{X}_c - \mathbf{X}$ be perpendicular to $\mathbf{p}$ (Fig. 17.5), so that:

$$(\mathbf{X}_c - \mathbf{X}) \cdot \mathbf{p} = 0 \tag{17.14}$$

$$\text{i.e. } (\alpha t\, \mathbf{p} - \mathbf{X}) \cdot \mathbf{p} = 0 \tag{17.15}$$

$$\therefore\ \alpha t\, \mathbf{p} \cdot \mathbf{p} = \mathbf{X} \cdot \mathbf{p} \tag{17.16}$$

$$\therefore\ t = (\mathbf{X} \cdot \mathbf{p})/\alpha(\mathbf{p} \cdot \mathbf{p}) \tag{17.17}$$

Substituting in the equation for $\mathbf{X}_c$ now gives:

$$\mathbf{X}_c = \mathbf{p}(\mathbf{X} \cdot \mathbf{p})/(\mathbf{p} \cdot \mathbf{p}) \tag{17.18}$$

Hence the minimum distance of approach is given by:

$$d_{min}{}^2 = \left[\frac{\mathbf{p}(\mathbf{X} \cdot \mathbf{p})}{(\mathbf{p} \cdot \mathbf{p})} - \mathbf{X}\right]^2 = \frac{(\mathbf{X} \cdot \mathbf{p})^2}{(\mathbf{p} \cdot \mathbf{p})} - \frac{2\,(\mathbf{X} \cdot \mathbf{p})^2}{(\mathbf{p} \cdot \mathbf{p})} + (\mathbf{X} \cdot \mathbf{X})$$

$$= (\mathbf{X} \cdot \mathbf{X}) - \frac{(\mathbf{X} \cdot \mathbf{p})^2}{(\mathbf{p} \cdot \mathbf{p})} \tag{17.19}$$

which is naturally zero when $\mathbf{p}$ is aligned along $\mathbf{X}$. Clearly, avoidance of collisions requires an estimate of the size of the machine (e.g. robot or vehicle) attached to the camera and the size to be associated with the world point feature $\mathbf{X}$. Finally, it should be noted that while $\mathbf{p}$ is obtained from the image data, $\mathbf{X}$ can only be deduced from the image data if the depth Z can be estimated from other information. In fact, this information should be available from time-to-adjacency analysis (see below) if the speed of the camera through space (and specifically w) is known.

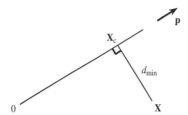

Fig. 17.5 Calculation of distance of closest approach. Here the camera is moving from 0 to X_c in the direction **p**, not in a direct line to the object at **X**. d_{min} is the distance of closest approach.

17.5 Time-to-adjacency analysis

Next we consider the extent to which the depths of objects can be deduced from optical flow. First, note that features on the same object share the same focus of expansion, and this can help us to identify them. But how can we get information on the depths of the various features on the object from optical flow? The basic approach is to start with the coordinates of a general image point (x, y), deduce its flow velocity, and then find an equation linking this with the depth Z.

Taking the general image point (x, y) given in equation (17.11), we find:

$$\dot{x} = f[(Z_0 + wt)u - (X_0 + ut)w]/(Z_0 + wt)^2 = f[Zu - Xw]/Z^2 \qquad (17.20)$$

and

$$\dot{y} = f[Zv - Yw]/Z^2 \qquad (17.21)$$

Hence:

$$\dot{x}/\dot{y} = [Zu - Xw]/[Zv - Yw] = (u/w - X/Z)/(v/w - Y/Z)$$
$$= (x - x_F)/(y - y_F) \qquad (17.22)$$

This result was to be expected, as the motion of the image point has to be directly away from the focus of expansion (x_F, y_F). With no loss of generality, we now take a set of axes such that the image point considered is moving along the x-axis. Then we have:

$$\dot{y} = 0 \qquad (17.23)$$

$$y_F = y = fY/Z \qquad (17.24)$$

Defining the distance from the focus of expansion as Δr (see Fig. 17.6), we find:

$$\Delta r = \Delta x = x - x_F = fX/Z - fu/w = f(Xw - Zu)/Zw \qquad (17.25)$$

$$\therefore \ \Delta r/\dot{r} = \Delta x/\dot{x} = -Z/w \qquad (17.26)$$

What this equation means is that the *time to adjacency*, when the origin of the camera coordinate system will arrive at the object point, is the same (Z/w) when seen in real-world coordinates as when seen in image coordinates $(-\Delta r/\dot{r})$. Hence it is possible to relate the optical flow vectors for object points at different depths in the scene. This is important, as the assumption of identical values of w now allows us to determine the relative depths of object points merely from their apparent motion parameters:

$$\frac{Z_1}{Z_2} = \frac{\Delta r_1/\Delta r_2}{\dot{r}_1/\dot{r}_2} \qquad (17.27)$$

This is thus the first step in the determination of structure from motion. In this context it should be noted how the implicit assumption that the objects under observation are rigid is included—namely that all points on the same object are characterized by identical values of w. The assumption of rigidity underlies much of the work on interpretation of motion in images.

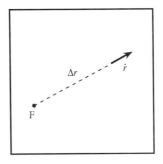

Fig. 17.6 Calculation of time to adjacency. Here an object feature is moving directly away from the focus of expansion F with speed $\dot{r}$. At the time of observation the distance of the feature from F is Δr. These measurements permit the time to adjacency and hence also the relative depth of the feature to be calculated.

17.6 Basic difficulties with the optical flow model

When the optical flow ideas presented above are tried on real images, certain problems arise which are not apparent from the above model. First, not all edge points which should appear in the motion image are actually present. This is due to the contrast between the moving object and the background vanishing locally and limiting visibility. The situation is exactly as for edges which are produced by Sobel or other edge detection operators in non-moving images: the contrast simply drops to a low value in certain localities and the edge peters out. This signals that the edge model, and now the velocity flow model, is limited and such local procedures are *ad hoc* and too impoverished to permit proper segmentation unaided.

Here we take the view that simple models can be useful, but they become inadequate on certain occasions and robust methods are required to overcome the problems that then arise. Some of the problems were noticed by Horn as early as 1986. First, a smooth sphere may be rotating but the motion will not show up in an optical flow (difference) image. We can if we wish regard this as a simple optical illusion, as the rotation of the sphere may well be invisible to the eye too. Second, a motionless sphere may *appear* to rotate as the light rotates around it: the object is simply subject to the laws of Lambertian optics, and again we may regard this effect as an optical illusion. (The illusion is relative to the baseline provided by the *normally correct* optical flow model.)

We next return to the optical flow model and see where it could be wrong or misleading. The answer is at once apparent: we stated in writing equation (17.2) that we were assuming that the *image* is being shifted. Yet it is not images that shift but the objects imaged within them. Thus we ought to be considering the images of objects moving against a fixed background (or a variable background if the camera is moving). This will then permit us to see how sections of the motion edge can go from high to low contrast and back again in a rather fickle way, which we must nevertheless allow for in our algorithms. With this in mind it should be permissible to go on using optical flow and difference imaging, even though these concepts have distinctly limited theoretical validity. (For a more thorough analysis of the underlying theory, see Faugeras, 1993.)

17.7 Stereo from motion

An interesting aspect of camera motion is that over time the camera sees a succession of images that span a baseline in a similar way to binocular (stereo) images. Thus it should be possible to obtain depth information by taking two such images and tracking object features between them. The technique is in principle more straightforward than normal stereo imaging in that feature tracking is possible, so the correspondence problem should be nonexistent. However, there is a difficulty in that the object field is viewed from almost the same direction in the succession of images, so that the full benefit of the available baseline is not obtained (Fig. 17.7). We can analyse the effect as follows.

First, in the case of camera motion, the equations for lateral displacement in the image depend not only on X but also on Y, though we can make a simplification in the theory by working with R, the radial distance of an object point from the optical axis of the camera, where:

$$R = [X^2 + Y^2]^{1/2} \tag{17.28}$$

We now obtain the radial distances in the two images as:

$$r_1 = Rf/Z_1 \tag{17.29}$$

$$r_2 = Rf/Z_2 \tag{17.30}$$

so the disparity is:

$$D = r_2 - r_1 = Rf(1/Z_2 - 1/Z_1) \tag{17.31}$$

Writing the baseline as:

$$b = Z_1 - Z_2 \tag{17.32}$$

and assuming that $b \ll Z_1, Z_2$, and then dropping the suffices, gives:

$$D = Rbf/Z^2 \tag{17.33}$$

While this would appear to mitigate against finding Z without knowing R, we can overcome this problem by observing that:

$$R/Z = r/f \tag{17.34}$$

where r is approximately the mean value $\frac{1}{2}(r_1 + r_2)$. Substituting for R now gives:

$$D = br/Z \tag{17.35}$$

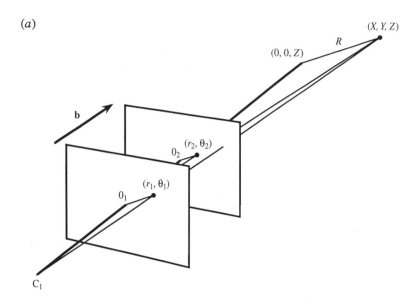

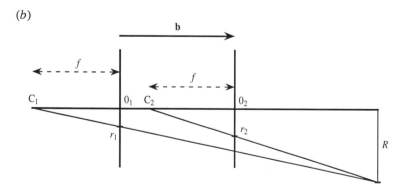

Fig. 17.7 Calculation of stereo from camera motion: (a) shows how stereo imaging can result from camera motion, the vector **b** representing the baseline. (b) The simplified planar geometry required to calculate the disparity. It is assumed that the motion is directly along the optical axis of the camera.

Hence, we can deduce the depth of the object point as:

$$Z = br/D = br/(r_2 - r_1) \qquad (17.36)$$

This equation should be compared with equation (16.5) representing the normal stereo situation. The important point to note is that for motion stereo, the disparity depends on the radial distance r of the image point from the optical axis of the camera, whereas for normal stereo the disparity is independent of r; as a result motion stereo gives no depth information for points on the optical axis, and the accuracy of depth information depends on the magnitude of r.

17.8   Applications to the monitoring of traffic flow

17.8.1   The system of Bascle *et al.*

An important area of 3-D and motion studies is provided by visual analysis of traffic flow. Here it is of interest to see how the principles outlined in the earlier chapters are brought to bear on such practical problems. In one recent study (Bascle *et al.*, 1994) the emphasis was on tracking of complex primitives in image sequences, the primitives in question being vehicles. There are several aspects of the problem which make the analysis easier than the general case: in particular, there is a constraint that the vehicles run on a roadway, and that the motion is generally smooth. Nevertheless, the methods that have to be used to make scene interpretation reliable and robust are nontrivial, and will be outlined in some detail.

First, motion-based segmentation is used to initialize the interpretation of the sequence of scenes. The motion image is used to obtain a rough mask of the object, and then the object outline is refined by classical edge detection and linking. B-splines are used to obtain a smoother version of the outline, which is fed to a snake-based tracking algorithm. The latter updates the fit of the object outline and repeats this for each incoming image.

However, snake-based segmentation concentrates on isolation of the object boundary, and therefore ignores motion information from the main region of the object. It is therefore more reliable to perform motion-based segmentation of the entire region bounded by the snake, and to use this information to refine the description of the motion and to predict the position of the object in the next image. This means that the snake-based tracker is provided with a reliable starting point from which to estimate the position of the object in the next image. The overall process is thus to feed the output of the snake boundary estimator into a motion-based segmenter and position predictor which re-initializes the snake for the next image—so both constituent algorithms perform the operations they are best adapted to. It is especially relevant that the snake has a good starting approximation in each frame, both to help eliminate ambiguities and to save on computation. The motion-based region segmenter operates principally by analysis of optical flow, though in practice the increments between frames are not especially small: this means that while true derivatives are not obtained, the result is not as bedevilled by noise as it might otherwise be.

Various refinements have been incorporated into the basic procedure:

(i)   B-splines are used to smooth the outlines obtained from the snake algorithm.
(ii) The motion predictions are carried out using an affine motion model which works on a point-by-point basis. (The affine model is sufficiently accurate for this purpose if perspective is weak so that motion can be approximated locally by a set of linear equations.)
(iii) A multiresolution procedure is invoked to perform a more reliable analysis of the motion parameters.
(iv) Temporal filtering of the motion is performed over several image frames.
(v)   The overall trajectories of the boundary points are smoothed by a Kalman filter.

The affine motion model used in the algorithm involves six parameters:

$$\begin{bmatrix} x(t+1) \\ y(t+1) \end{bmatrix} = \begin{bmatrix} a_{11}(t) & a_{12}(t) \\ a_{21}(t) & a_{22}(t) \end{bmatrix} \begin{bmatrix} x(t) \\ y(t) \end{bmatrix} + \begin{bmatrix} b_1(t) \\ b_2(t) \end{bmatrix} \tag{17.37}$$

This leads to an affine model of image velocities, also with six parameters:

$$\begin{bmatrix} u(t+1) \\ v(t+1) \end{bmatrix} = \begin{bmatrix} m_{11}(t) & m_{12}(t) \\ m_{21}(t) & m_{22}(t) \end{bmatrix} \begin{bmatrix} u(t) \\ v(t) \end{bmatrix} + \begin{bmatrix} c_1(t) \\ c_2(t) \end{bmatrix} \tag{17.38}$$

Once the motion parameters have been found from the optical flow field, it is straightforward to estimate the following snake position.

An important factor in the application of this type of algorithm is the degree of robustness it permits. In this case, both the snake algorithm and the motion-based region segmentation scheme are claimed to be rather robust to partial occlusions. While the precise mechanism by which this is achieved is not made clear, the abundance of available motion information for each object, the insistence on consistent motion, and the recursive application of smoothing procedures including a Kalman filter, all help to achieve this end. However, no specific nonlinear outlier rejection process is mentioned, and this could ultimately cause problems, e.g. if two vehicles merged together and became separated later on: such a situation should perhaps be treated at a higher level in order to achieve a satisfactory interpretation. Then the problem mentioned by the authors of coping with instances of total occlusion would not arise.

Finally, it is of interest that the initial motion segmentation scheme locates the vehicles with their shadows since these are also moving (see Fig. 17.8); subsequent analysis seems able to eliminate the shadows and arrive at smooth vehicle boundaries. Clearly, shadows can be an encumbrance to practical systems, and systematic rather than *ad hoc* procedures are ultimately needed to eliminate their effects. Likewise, the algorithm appears able to cope with cluttered backgrounds when tracking faces, though in more complex scenes this might not be possible without specific high-level guidance. In particular, snakes are liable to be confused by background structure in an image, so however good the starting approximation, they could diverge from appropriate answers.

Fig. 17.8 Vehicles located with their shadows. In many practical situations, shadows move with the objects that cause them, and simple motion segmentation procedures produce composite objects that include the shadows. Here a snake tracker envelops the car and its shadow.

17.8.2 The system of Koller *et al.*

Another scheme for automatic traffic scene analysis has been described by Koller *et al.* (1994). This contrasts with the system described above in placing heavy reliance on high-level scene interpretation through use of belief networks. The basic system incorporates a low-level vision system employing optical flow, intensity gradient and temporal derivatives. These provide feature extraction, and lead to snake approximations to contours. Since convex polygons would be difficult to track from image to image (because the control points would tend to move randomly), the boundaries are smoothed by closed cubic splines having twelve control points: tracking can then be achieved using Kalman filters. The motion is again approximated by an affine model, though in this case only three parameters are used, one being a scale parameter and the other two being velocity parameters:

$$\Delta \mathbf{x} = s(\mathbf{x} - \mathbf{x}_m) + \Delta \mathbf{x}_m \qquad (17.39)$$

Here the second term gives the basic velocity component of the centre of a vehicle region, and the first term gives the relative velocity for other points in the region, s being the change in scale of the vehicle ($s = 0$ if there is no change in scale). The rationale for this is that vehicles are constrained to move on the roadway, and rotations will be small. In addition, motion with a component towards the camera will result in an increase in size of the object and a corresponding increase in its apparent speed of motion.

Occlusion reasoning is achieved by assuming that the vehicles are moving along the roadway, and are proceeding in a definite order, so that later vehicles (when viewed from behind) may partly or wholly obscure earlier ones. This depth ordering defines the order in which vehicles are able to occlude each other, and appears to be the minimum necessary to rigorously overcome problems of occlusion.

As stated above, belief networks are employed in this system to distinguish between various possible interpretations of the image sequence. Belief networks are directed acyclic graphs in which the nodes represent random variables and arcs between them represent causal connections. In fact, each node has an associated list of the conditional probabilities of its various states corresponding to assumed states of its parents (i.e. the previous nodes on the directed network). Thus observed states for subsets of nodes permit deductions to be made about the probabilities of the states of other nodes. The reason for using such networks is to permit rigorous analysis of probabilities of different outcomes when a limited amount of knowledge is available about the system. Likewise, once various outcomes are known with certainty (e.g. a particular vehicle has passed beneath a bridge), parts of the network will become redundant and can be removed: however, before removal their influence must be "rolled up" by updating the probabilities for the remainder of the network. Clearly, when applied to traffic, the belief network has to be updated in a manner appropriate to the vehicles that are currently being observed; indeed, each vehicle will have its own belief network which will contribute a complete description of the entire traffic scene. However, one vehicle will have some influence on other vehicles, and special note will have to be taken of stalled vehicles or those making lane changes. In addition, one vehicle slowing down will have some influence on the decisions made by drivers in following vehicles. All these factors can be encoded into the belief network and can aid in arriving at globally correct interpretations. General road and weather conditions can also be taken into account.

It is realistic to expect real-time performance of such a system in the near future, with the help of special hardware. Specifically, the time taken by the belief network to cope with each vehicle was around one second (running on a Sun SparcStation 10), and requires a 3- to 10-fold increase in speed if expected levels of traffic are to be dealt with. However, further sophistication is required to enable the vision part of the system to deal with shadows, brake lights and other signals, while it remains to be seen whether the system will be able to cope with a wide enough variety of weather conditions.

Overall, the system is designed in a very similar manner to that of Bascle *et al.* (1994), though its use of belief networks places it in a more sophisticated and undoubtedly more robust category.

In a more recent paper (Malik *et al.*, 1995) the same research group has found that a stereo camera scheme can be set up to use geometrical constraints in a way which improves both speed of processing and robustness. Under a number of conditions, interpretation of the disparity map is markedly simplified. The first step is to locate the ground plane (the roadway) by finding road markers such as white lines. Then the disparity map is modified to give the ground plane zero disparity. After this has been achieved, vehicles on the roadway are located trivially via their positive disparity values, reflecting their positions above the roadway. The effect is essentially one-dimensional (see Fig. 17.9). It only applies if the following conditions are valid:

(i)   The camera parameters are identical (e.g. identical focal lengths and image planes, optical axes parallel, etc.), differing only in a lateral baseline **b**.

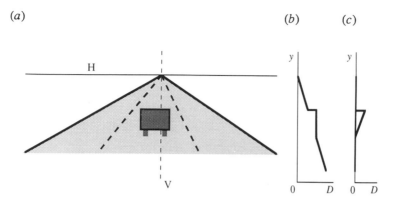

(a)   (b)   (c)

Fig. 17.9 Disparity analysis of a road scene. In scene (a), the disparity D is greatest near the bottom of the image where depth Z is small, and is reduced to zero at the horizon line H. The vertical line V in the image passes up through the back of a vehicle, which is at constant depth, thereby giving constant disparity in (b). In (c) the disparity graph is modified when the ground plane is reduced to zero disparity by subtracting the Helmholtz shear.

(ii)   The baseline vector **b** has a component only along the image x-axes.

(iii) The ground plane normal **n** has no component along the image x-axes.

In that case there is no y-disparity, and the x-disparity on viewing the ground plane reduces to:

$$D_g = (b/h)(y \cos \delta + f \sin \delta) \qquad (17.40)$$

where b is the inter-camera baseline distance, h is the height of the centre of projection of the camera above the ground plane, and δ is the angle of declination of the optical axes below the horizontal. The conditions listed above are reasonable for practical stereo systems, and if they are valid then the disparity will have high values near the bottom of the scene, reducing to $(b/h)f \sin \delta$ for vanishing points on the roadway. If we now stipulate that the disparity of the roadway must everywhere be reduced to zero, this can be achieved by subtracting the quantity D_g from the observed disparity D. Thus we are subtracting what has become known as the Helmholtz shear from the image (it is a shear because the x-disparity and hence the x-values depend on y). Here we refrain from giving a full proof of the above formula, and instead give some insight into the situation by taking the important case $\delta = 0$. In that case we have (cf.

equation 16.4):

$$D_g = bf/Z \qquad (17.41)$$

while from Fig. 17.10:

$$y/f = h/Z \qquad (17.42)$$

Eliminating Z between these equations leads to:

$$D_g = by/h \qquad (17.43)$$

which proves the formula in that case.

Finally, it is interesting to note that whereas the subtraction of D_g from D is certainly valid below the horizon line (where the ground plane disappears), above that level the definition of D_g is open to question: equation (17.40) could provide an invalid extrapolation above the horizon line (would zero perhaps be a more appropriate value?). Assuming that equation (17.40) remains valid, and that D_g is negative above the horizon line, then we will be subtracting a negative quantity from the observed disparity. This means that the deduced disparity will increase with height. By way of example, take two telegraph poles equidistant from the camera, and placed either side of the roadway. Their disparity will now increase with height and as a result they will appear to have a separation which also increases with height. It turns out that this was the form of optical illusion discovered by von Helmholtz (1925) which led him to the hypothesis that a shear (equation (17.40)) is active within the human visual system. In the context of machine

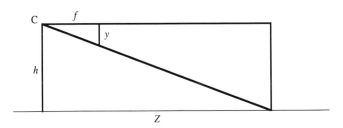

Fig. 17.10 Geometry for subtracting the Helmholtz shear. Here the optical centre of the camera C is a height h above the ground plane, and the optical axis of the camera is horizontal and parallel to the ground plane. The geometry of similar triangles can then be used to relate declination in the image (measured by y) to depth Z in the scene.

vision, it is clearly a construction which has a degree of usefulness, but also carries important warnings about the possibility of making misleading measurements from the resulting sheared images. The situation arises even for the backs of vehicles seen below the horizon line (see Fig. 17.9(c)).

17.9 Concluding remarks

Early in this chapter we described the formation of optical flow fields and showed how a moving object or a moving camera leads to a focus of expansion. In the case of moving objects, the focus of expansion can be used to decide whether a collision will occur. In addition, analysis of the motion taking account of the position of the focus of expansion led to the possibility of determining structure from motion. Specifically, this can be achieved via time-to-adjacency analysis, which yields the relative depth in terms of the motion parameters measurable directly from the image. We then went on to demonstrate some basic difficulties with the optical flow model, which arise since the motion edge can have a wide range of contrast values, making it difficult to measure motion accurately. In practice, this means that larger time intervals may have to be employed to increase the motion signal. Otherwise, feature-based processing related to that of Chapter 14 can be used. Corners are the features which are the most widely used for this purpose, because of their ubiquity and because they are highly localized in 3-D (note that this high degree of localization ensures that corners are not subject to the aperture problem discussed in Section 17.2). Space prevents details of this approach from being described here: details may be found in Barnard and Thompson (1980), Scott (1988), Shah and Jain (1984) and Ullman (1979).

In Section 17.8 a study was made of two vision systems that have recently been developed for monitoring traffic flow. The reason for studying these systems here was to show the extent to which optical flow ideas can be applied in real situations, and to find what other techniques have to be used to build complete vision systems for this type of application. Interestingly, both systems made use of (i) optical flow, (ii) snake approximation to boundary contours, (iii) affine models of the displacements and velocity flows, and (iv) Kalman filters for motion prediction—confirming the value of all these techniques. On the other hand the Koller system also employed belief networks for monitoring the whole process and making decisions about occlusions and other relevant factors. In addition, the later version of the Koller system (Malik *et al.*, 1995) used binocular vision to obtain a disparity map and then applied a Helmholtz shear to reduce the ground plane to zero disparity so that vehicles could be tracked robustly in crowded road scenes. Ground plane location was also a factor in the design of the system.

17.10 Bibliographical and historical notes

Optical flow has been investigated by many workers over a good many years: see for example Horn and Schunck (1981) and Heikkonen (1995). A definitive account of the mathematics relating to FOE appeared in 1980 (Longuet-Higgins and Prazdny, 1980). In fact, foci of expansion can be obtained either from the optical flow field or directly (Jain, 1983). The results of Section 17.5 on time-to-adjacency analysis stem originally from the work of Longuet-Higgins and Prazdny (1980) which provides some deep insights into the whole problem of optical flow and the possibilities of using its shear components. Note that numerical solution of the velocity field problem is not trivial; typically, least-squares analysis is required to overcome the effects of measurement inaccuracies and noise, and to obtain finally the required position measurements and motion parameters (Maybank, 1986). Overall, resolving ambiguities of interpretation is one of the main problems, and challenges, of image sequence analysis (see Longuet-Higgins (1984) for an interesting analysis of ambiguity in the case of a moving plane).

Unfortunately, the substantial and important literature on motion, image sequence analysis and optical flow, which impinges heavily on 3-D vision, could not be discussed in detail here for reasons of space. For seminal work on these topics, see for example, Ullman (1979), Snyder (1981), Huang (1983), Jain (1983), Nagel (1983, 1986) and Hildreth (1984).

Further work on traffic monitoring appears in Fathy and Siyal (1995), and there has been significant effort on automatic visual guidance in convoys (Schneiderman et al., 1995; Stella et al., 1995). Vehicle guidance and ego-motion control cover much further work (Brady and Wang, 1992; Dickmanns and Mysliwetz, 1992), while several papers show that following vehicles can conveniently be tackled by searching for symmetrical moving objects (e.g. Kuehnle, 1991; Zielke et al., 1993).

Work on the application of snakes to tracking has been carried out by Delagnes et al. (1995); on the use of Kalman filters to tracking by Marslin et al. (1991); on the tracking of plant rows in agricultural vehicles by Marchant and Brivot (1995); and on recognition of vehicles on the ground plane by Tan et al. (1994). For details of belief networks see Pearl (1988).

In spite of all these successes, there is probably still only a very limited number of fully automated visual vehicle guidance systems in actual everyday use. The current problems would appear to be (i) lack of the robustness and reliability that would be required to trust the system in an "all hours–all weathers" situation, and (ii) lack of hardware capable of implementing the requisite advanced algorithms reliably in real time. Note that there are also legal implications for a system that is to be used for control rather than merely for vehicle counting functions.

18

Invariants and Their Applications

18.1 Introduction

Pattern recognition is a complex task; as stated in Chapter 1, it involves the twin processes of discrimination and generalization. Indeed, the latter process is in many ways more important than the first—especially in the initial stages of recognition—since there is so much redundant information in a typical image. Thus we need to find ways of helping to eliminate invalid matches. This is where the study of invariants comes into its own.

An invariant is a property of an object or class of objects that does not change with changes of viewpoint or object pose and which can therefore be used to help distinguish it from other objects. The procedure is to search for objects with a specific invariant, so that those which do not possess the invariant can immediately be discarded from consideration. An invariant property can be regarded as a necessary condition for an object to be in the chosen class, though in principle only detailed subsequent analysis will confirm that it is. In addition, if an object is found to possess the correct invariant, it will then be profitable to pursue the analysis further and find its pose, size or other relevant data. Ideally an invariant would uniquely identify an object as being of a particular type or class. Thus an invariant should not merely be a property which leads to further hypotheses being made about the object, but one which fully characterizes it. However, the difference is a subtle one, more a matter of degree and purpose than an absolute criterion. We shall see below the extent of the difference by appealing to a number of specific cases.

Let us first consider an object being viewed from directly overhead at a known distance by a camera whose optical axis is normal to the plane on which the object is lying. We shall assume that the object is flat. Take two point features on the object such as corners or small holes (cf. Chapter 14). If we measure the image distance between these features, then this acts as an invariant, in that:

(i)   it has a value independent of the translation and orientation parameters of the object;

(ii) it will be unchanged for different objects of the same type;

(iii) it will in general be different from the distance parameters of other objects that might be on the object plane.

Thus measurement of distance provides a certain lookup or indexing quality which will ideally identify the object uniquely, though further analysis will be required to fully locate it and ascertain its orientation. Hence distance has all the requirements of an invariant, though it could also be argued that it is only a feature which helps to classify objects. Clearly, we are here ignoring an important factor—the effect of imprecision in measurement, due to spatial quantization (or inadequate spatial resolution), noise, lens distortions, and so on; in addition, the effects of partial occlusion or breakage are also being ignored. Most definitely, there is a limit to what can be achieved with a single invariant measure, though in what follows we attempt to reveal what is possible, and demonstrate the advantages of the way of thinking which involves analysis of invariants.

The above ideas relating to distance as an invariant measure showed it to be useful in suppressing the effects of translations and rotations of objects in 2-D. Hence it is of little direct value when considering translations and rotations in 3-D. Furthermore, it is not even able to cope with scale variations of objects in 2-D. Moving the camera closer to the object plane and refocusing totally changes the situation and all values of the distance invariant residing in the object indexing table must be changed and the old values ignored. However, a moment's thought will show that this last problem could be overcome. All we need to do is to take *ratios* of distances. This requires a minimum of three point features to be identified in the image and the distances between features measured. If we call two of these distances d_1 and d_2, then the ratio d_1/d_2 will act as a scale-independent invariant, i.e. we will be able to identify objects using a single indexing operation whatever their 2-D translation, orientation, or apparent size or scale. An alternative to this idea is to measure the angle between pairs of distance vectors, $\cos^{-1}(\mathbf{d}_1 \cdot \mathbf{d}_2 / |\mathbf{d}_1| \, \|\mathbf{d}_2|)$, which will again be scale invariant.

Of course, this consideration has already been invoked in our earlier work on shape analysis. If objects are subject only to 2-D translations and rotations but not to changes of scale, they can be characterized by their perimeters or areas as well as their normal linear dimensions; furthermore, parameters such as compactness and aspect ratio, which employ dimensionless ratios of image measurements, were acknowledged in Chapter 6 to overcome the size/scale problem.

Nevertheless, the main motivation for using invariants is to obtain mathematical measures of configurations of object features which are carefully designed to be independent of the viewpoint or coordinate system used and indeed not to require specific setup or calibration of the image acquisition system. However, it must be emphasized that camera distortions are assumed to be absent or to have been compensated for by suitable post-camera transformations (see Appendix C).

18.2 Cross ratios: the "ratio of ratios" concept

It would be most useful if we could extend the above ideas to permit indexing for general transformations in 3-D. Indeed, an obvious question is whether finding *ratios* of ratios of distances will provide suitable invariants and lead to such a generalization. The answer is that ratios of ratios do provide useful further invariants, though going further than this leads to considerable complication, and there are restrictions on what can be

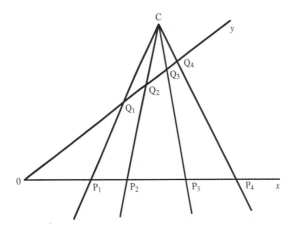

Fig. 18.1 Perspective transformation of four collinear points. This figure shows four collinear points (P_1, P_2, P_3, P_4) and a transformation of them (Q_1, Q_2, Q_3, Q_4) similar to that produced by an imaging system with optical centre C. Such a transformation is called a *perspective transformation*.

achieved with limited computation. In addition, noise ultimately becomes a limiting factor, since so many parameters become involved in the computation of complex invariants that the method ultimately loses steam (it becomes just one of many ways of raising hypotheses and therefore has to complete with other approaches in a manner appropriate to the particular problem application being studied).

We now consider the ratio of ratios approach. Initially, we only examine a set of four collinear points on an object. Figure 18.1 shows such a set of four points (P_1, P_2, P_3, P_4) and a transformation of them (Q_1, Q_2, Q_3, Q_4) such as that produced by an imaging system with optical centre C (c, d). Choice of a suitable pair of oblique axes permits the coordinates of the points in the separate sets to be expressed respectively as:

$$(x_1, 0), \ (x_2, 0), \ (x_3, 0), \ (x_4, 0)$$
$$(0, y_1), \ (0, y_2), \ (0, y_3), \ (0, y_4)$$

Taking points P_i, Q_i, we can write the ratio $CQ_i : P_i Q_i$ both as $c/(-x_i)$ and as $(d - y_i)/y_i$. Hence:

$$\frac{c}{x_i} + \frac{d}{y_i} = 1 \tag{18.1}$$

which must be valid for all i. Subtraction of the ith and jth relations now gives:

$$\frac{c(x_j - x_i)}{x_i x_j} = \frac{-d(y_j - y_i)}{y_i y_j} \tag{18.2}$$

Forming a ratio between two such relations will now eliminate the unknowns c and d. For example, we will have:

$$\frac{x_3(x_2 - x_1)}{x_2(x_3 - x_1)} = \frac{y_3(y_2 - y_1)}{y_2(y_3 - y_1)} \tag{18.3}$$

However, the result still contains factors such as x_3/x_2 which depend on absolute position. Hence it is necessary to form a suitable ratio of such results which cancels out the effects of absolute positions:

$$\left(\frac{x_2 - x_4}{x_3 - x_4}\right) \bigg/ \left(\frac{x_2 - x_1}{x_3 - x_1}\right) = \left(\frac{y_2 - y_4}{y_3 - y_4}\right) \bigg/ \left(\frac{y_2 - y_1}{y_3 - y_1}\right) \tag{18.4}$$

Thus our original intuition that a ratio of ratios type of invariant might exist which would cancel out the effects of a perspective transformation is correct. In particular, four collinear points viewed from any perspective viewpoint yield the same value of the cross ratio, defined as above. The value of the cross ratio of the four points is written:

$$\mathscr{C}(P_1, P_2, P_3, P_4) = \frac{(x_3 - x_1)(x_2 - x_4)}{(x_2 - x_1)(x_3 - x_4)} \tag{18.5}$$

For clarity, we shall write this particular cross ratio as κ in what follows. It should be noted that there are $4! = 24$ possible ways in which four collinear points can be ordered on a straight line, and hence there could be 24 cross ratios. However, they are not all distinct, and in fact there are only six different values. To verify this we start by interchanging pairs of points:

$$\mathscr{C}(P_2, P_1, P_3, P_4) = \frac{(x_3 - x_2)(x_1 - x_4)}{(x_1 - x_2)(x_3 - x_4)} = 1 - \kappa \tag{18.6}$$

$$\mathscr{C}(P_1, P_3, P_2, P_4) = \frac{(x_2 - x_1)(x_3 - x_4)}{(x_3 - x_1)(x_2 - x_4)} = \frac{1}{\kappa} \tag{18.7}$$

$$\mathscr{C}(P_1, P_2, P_4, P_3) = \frac{(x_4 - x_1)(x_2 - x_3)}{(x_2 - x_1)(x_4 - x_3)} = 1 - \kappa \tag{18.8}$$

$$\mathscr{C}(P_4, P_2, P_3, P_1) = \frac{(x_3 - x_4)(x_2 - x_1)}{(x_2 - x_4)(x_3 - x_1)} = \frac{1}{\kappa} \tag{18.9}$$

$$\mathscr{C}(P_3, P_2, P_1, P_4) = \frac{(x_1 - x_3)(x_2 - x_4)}{(x_2 - x_3)(x_1 - x_4)} = \frac{\kappa}{\kappa - 1} \tag{18.10}$$

$$\mathscr{C}(P_1, P_4, P_3, P_2) = \frac{(x_3 - x_1)(x_4 - x_2)}{(x_4 - x_1)(x_3 - x_2)} = \frac{\kappa}{\kappa - 1} \tag{18.11}$$

These cases provide the main possibilities, but of course interchanging more points will yield a limited number of further values—in particular:

$$\mathscr{C}(P_3, P_1, P_2, P_4) = 1 - \mathscr{C}(P_1, P_3, P_2, P_4) = 1 - \frac{1}{\kappa} = \frac{\kappa - 1}{\kappa} \tag{18.12}$$

$$\mathscr{C}(P_2, P_3, P_1, P_4) = \frac{1}{\mathscr{C}(P_2, P_1, P_3, P_4)} = \frac{1}{1 - \kappa} \tag{18.13}$$

This covers all six cases, and a moment's thought (based on trying further interchanges of points) will show there can be no others (we can only repeat κ, $1 - \kappa$, $\kappa/(\kappa - 1)$, and their inverses). Of particular interest is the fact that numbering the points in reverse (which would correspond to viewing the line from the other side) leaves the cross ratio unchanged. Nevertheless, it is inconvenient that the same invariant has six different manifestations, as this implies that six different index values have to be looked up before the class of an object can be ascertained. On the other hand, if points are labelled in order along the line rather than randomly it should generally be possible to circumvent this situation.

So far we have been able to produce only one projective invariant, and this corresponds to the rather simple case of four collinear points. The usefulness of this measure is augmented considerably when it is noted that four collinear points, taken in conjunction with another point, define a pencil* of concurrent coplanar lines passing through the latter point. It will be clear that we can assign a unique cross ratio to this pencil of lines, equal to the cross ratio of the collinear points on any line passing through them. We can clarify the situation by considering the angles between the various lines (Fig. 18.2). Applying the sine rule four times to determine the four distances in the cross ratio $\mathscr{C}(P_1, P_2, P_3, P_4)$ gives:

$$\frac{x_3 - x_1}{\sin \alpha_{13}} = \frac{OP_1}{\sin \beta_3} \tag{8.14}$$

$$\frac{x_2 - x_4}{\sin \alpha_{24}} = \frac{OP_4}{\sin \beta_2} \tag{8.15}$$

$$\frac{x_2 - x_1}{\sin \alpha_{12}} = \frac{OP_1}{\sin \beta_2} \tag{8.16}$$

$$\frac{x_3 - x_4}{\sin \alpha_{34}} = \frac{OP_4}{\sin \beta_3} \tag{8.17}$$

Substituting in the cross ratio formula (equation (18.5)) and cancelling the factors OP_1, OP_4, $\sin \beta_2$ and $\sin \beta_3$ now gives:

$$\mathscr{C}(P_1, P_2, P_3, P_4) = \frac{\sin \alpha_{13} \sin \alpha_{24}}{\sin \alpha_{12} \sin \alpha_{34}} \tag{8.18}$$

---

* It is a common nomenclature of projective geometry to call a set of concurrent lines a *pencil* (e.g. Tuckey and Armistead, 1953).

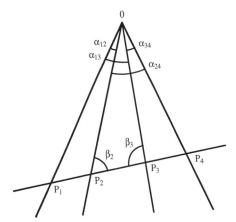

Fig. 18.2 Geometry for calculation of the cross ratio of a pencil of lines in terms of the angles between them.

Thus the cross ratio depends only on the angles of the pencil of lines. It is interesting that appropriate juxtaposition of the sines of the angles gives the final formula invariance under perspective projection: using the angles themselves would not give the desired degree of mathematical invariance. Indeed, we can immediately see one reason for this: inversion of the direction of any line must leave the situation unchanged, so the formula must be tolerant to adding π to each of the two angles linking the line; this could not be achieved if the angles appeared without suitable trigonometric functions.

We can extend this concept to four concurrent planes since the concurrent lines can be projected into four concurrent planes once a separate axis for the concurrency has been defined. As there are infinitely many such axes, there are infinitely many ways in which sets of planes can be chosen. Thus the original simple result on collinear points can be extended to a much more general case.

It is worth noting here that we started by trying to generalize the case of four collinear points, but what we achieved was first to find a dual situation in which points become lines also described by a cross ratio, and then to find an extension in which planes are described by a cross ratio. We now return to the case of four collinear points, and see how we can extend it in other ways.

18.3   Invariants for noncollinear points

First, imagine that not all the points are collinear: specifically let us assume that one point is not in the line of the other three. If this is the case, then

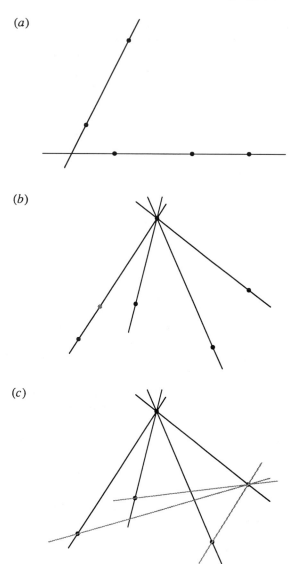

Fig. 18.3 Calculation of invariants for a set of noncollinear points. (a) The addition of a fifth point to a set of four points, one of which is not collinear with the rest, permits the cross ratio to be calculated. (b) The calculation can be extended to any set of noncollinear points; also shown is an additional (grey) point which a single cross ratio fails to distinguish from other points on the same line. (c) Any failure to identify a point uniquely can be overcome by calculating the cross ratio of a second pencil generated from the five original points.

there is not enough information to calculate a cross ratio. However, if a further coplanar point is available, we can draw an imaginary line between the noncollinear points to intersect the common line in a unique point, which will then permit a cross ratio to be computed (Fig. 18.3(a)). Nevertheless, this is some way from a general solution to the characterization of a set of noncollinear points. We might enquire how many point features in general position* on a plane will be required to calculate an invariant. In fact, the answer is five, since the fact that we can form a cross ratio from the angles between four lines immediately means that forming a pencil of four lines from five points defines a cross ratio invariant (Fig. 18.3(b)).

While the value of this cross ratio provides a necessary condition for a match between two sets of five general coplanar points, it could be a fortuitous match, as the condition depends only on the relative directions between the various points and the reference point, i.e. any of the non-reference points is only defined to the extent that it lies on a given line. Clearly, two cross ratios formed by taking two reference points will define the directions of all the remaining points uniquely (Fig. 18.3(c)).

We can now summarize the general result, which stipulates that for five general coplanar points, no three of which are collinear, two different cross ratios are required to characterize the shape. These cross ratios correspond to taking in turn two separate points and producing pencils of lines passing through them and (in each case) the remaining four points (Fig. 18.3(c)). While it might appear that at least five cross ratios result from this sort of procedure, there are only two functionally independent cross ratios—essentially because the position of any point is defined once its direction relative to two other points is known.

Next, we consider the problem of finding the ground plane in practical situations—especially that of egomotion including vehicle guidance (Fig. 18.4). Here a set of four collinear points can be observed from one frame to the next. If they are on a single plane, then the cross ratio will remain constant, but if one is elevated above the ground plane (as for example a bridge or another vehicle) then the cross ratio will vary over time. Taking a larger number of points, it should clearly be possible to deduce by a process of elimination which are on the ground plane and which are not (though the amount of noise and clutter will determine the computational complexity of the task): note that all this is possible without any calibration of the camera, this being perhaps the main value of concentrating attention on projective

---

*Points on a plane which are chosen at random, and which are not collinear or in any special pattern such as a regular polygon, are described as being in *general position*.

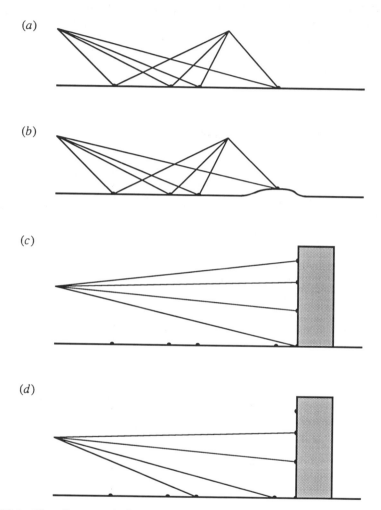

Fig. 18.4 Use of cross ratio for egomotion guidance. (a) The cross ratio for a set of four collinear points can be tracked to confirm that the points are collinear: this suggests that they lie on the ground plane. (b) A case where the cross ratio will not be constant. (c) The cross ratio is constant, although the points actually lie on a plane which is not the ground plane. (d) All four points lie on planes, yet the cross ratio will not be constant.

invariants. Note that there is a potential problem regarding irrelevant planes, such as the vertical faces of buildings. The cross ratio test is so resistant to viewpoint and pose that it merely ascertains whether the points being tested are coplanar. It is only by using a sufficiently large number of independent sets of points that one plane can be discriminated from another (for simplicity we ignore here any subsequent stages of pose analysis that might be carried out).

18.4 Invariants for points on conics

These discussions clearly help to build up an understanding of how geometric invariants can be designed to cope with sets of points, lines and planes in 3-D. Significantly more difficult is the case of curved lines and surfaces, though much headway has now been made in the understanding of conics and certain other surfaces (see Mundy and Zisserman, 1992a). It will not be possible to examine all such cases in depth here. However, it will be useful to consider conic sections and particularly ellipses in more detail.

First, we consider Chasles' theorem, which dates from the nineteenth century. (The history of projective geometry is quite rich and was initially carried out totally independently of the requirements of machine vision.) Suppose we have four fixed coplanar points F_1, F_2, F_3, F_4 on a conic section curve and one variable point P in the same plane (Fig. 18.5). Then the four lines joining P to the fixed points form a pencil whose cross ratio will in general vary with the position of P. Chasles' (1855) theorem states that if P now moves so as to keep the cross ratio constant, then P will trace out a conic section. This clearly provides a means of checking whether a set of points lies on a planar curve such as an ellipse. Note the close analogy with

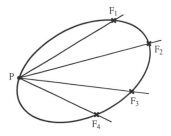

Fig. 18.5 Definition of a conic using a cross ratio. Here P is constrained to move so that the cross ratio of the pencil from P to F_1, F_2, F_3, F_4 remains constant. By Chasles' theorem, P traces out a conic curve.

the problem of ground plane detection already mentioned. Again the amount of computation could become excessive if there were a lot of noise or clutter in the image. When the image contains N boundary features which need to be checked out, the problem complexity is intrinsically $O(N^5)$, since there are $O(N^4)$ ways of selecting the first four points, and for each such selection, $N-4$ points must be examined to determine whether they lie on the same conic. However, choice of suitable heuristics would be expected to limit the computation. Note the problem of ensuring that the first four points are tested in the same order around the ellipse, which is liable to be tedious (i) for point features, and (ii) for disconnected boundary features.

While Chasles' theorem gives an excellent opportunity to use invariants to locate conics in images, it is not at all discriminatory in this. The theorem applies to a general conic: hence it does not immediately permit circles, ellipses, parabolas or hyperbolas to be distinguished, a fact that would sometimes be a distinct disadvantage. This is an example of a more general problem in pattern recognition system design—of deciding exactly how and in what sequence one object should be differentiated from another: space will not permit this point to be considered further here.

Finally, we state without proof that conic section curves can all be transformed under perspective projection to other types of conic section, and thus into ellipses; subsequently they can be transformed into circles. Thus any conic section curve can be transformed projectively into a circle, while the inverse transformation can transform it back again (Mundy and Zisserman, 1992b). This means that simple properties of the circle can frequently be generalized to ellipses or other conic sections. In this context, points to bear in mind are that, after perspective projection, lines intersecting curves do so in

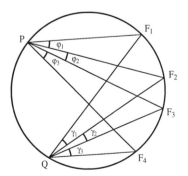

Fig. 18.6 Proof of Chasles' theorem. The four points F_1, F_2, F_3, F_4 subtend the same angles at P as they do at the fixed point Q. Thus the cross ratio is the same for all points on the circle. This means that Chasles' theorem is valid for a circle.

the same number of points, and thus tangents transform into tangents, chords into chords, three-point contact (in the case of nonconic curves) remains three-point contact, and so on. Returning to Chasles' theorem, a simple proof in the case of circles will automatically generalize to more complex conic section curves.

In response to this assertion, we can in fact derive Chasles' theorem almost trivially for a circle. Appealing to Fig. 18.6, we see that the angles φ_1, φ_2, φ_3 are equal to the respective angles γ_1, γ_2, γ_3 (angles in the same segment of a circle). Thus the pencils PF_1, PF_2, PF_3, PF_4 and QF_1, QF_2, QF_3, QF_4 have equal angles, their *relative* directions being superposable. This means that they will have the same cross ratio, defined by equation (18.18). Hence the cross ratio of the pencil will remain constant as P traces out the circle. As stated above, the property will automatically generalize to any other conic.

18.5 Concluding remarks

This chapter has aimed to give some insight into the important subject of invariants and its application in image recognition. The subject takes off when ratios of ratios of distances are considered, and this idea leads in a natural way to the cross ratio invariant. While its immediate manifestation lies in its application to recognition of the spacings of points on a line, it generalizes immediately to angular spacings for pencils of lines, and also to angular separations of concurrent planes. A further extension of the idea is the development of invariants which can describe sets of noncollinear points, and it turns out that just two cross ratios suffice to characterize a set of four noncollinear points. The cross ratio can also be applied to conics: indeed Chasles' theorem describes a conic as the locus of points which maintains a pencil of constant cross ratio with a given set of four points. However, this theorem does not permit one type of conic curve to be distinguished from another.

Many other theorems and types of invariant exist, but space prevents more than a mention being made of them. As an extension to the line and conic examples given in this chapter, invariants have been produced which cover a conic and two coplanar nontangent lines; a conic and two coplanar points; two coplanar conics. Of particular value is the group approach to the design of invariants (Mundy and Zisserman, 1992a). However, certain mathematically viable invariants, such as those which describe local shape parameters on curves, prove to be too unstable to use in practice because of image noise.

Finally, there is the warning of Åström (1995) that perspective transformations can produce such incredible changes in shape that a duck silhouette can be projected arbitrarily closely into something that looks like a rabbit or a circle, hence upsetting invariant-based recognition[*]: while such reports seem absent from the previous literature, Åström's work indicates that care must be taken to regard recognition via invariants as hypothesis formation which is capable of leading to false alarms.

Overall, the value of invariants lies in making computationally efficient checks of whether points or other features might form part of specific objects. In addition, they achieve this without the necessity[†] for camera calibration or knowing the viewpoint of the camera (though there is an implicit assumption that the camera is Euclidean). While invariants have been known within the vision community for well over 20 years, it is only during the last decade that they have been systematically developed and applied for machine vision. Such is their power that they will undoubtedly assume a stronger and more central role in the future.

18.6 Bibliographical and historical notes

The mathematical subject of invariance is very old (cf. the work of Chasles, 1855), but it has only recently been developed systematically for machine vision. Notable in this context is the work of Rothwell, Zisserman and their coworkers, as reported by Forsyth et al. (1991), Mundy and Zisserman (1992a, b), Rothwell et al. (1992a, b), and Zisserman et al. (1990). In particular, the paper by Forsyth et al. (1991) shows the range of available invariant techniques and discusses problems of stability which arise in certain cases. The appendix (Mundy and Zisserman, 1992b) on projective geometry for machine vision, which appears in Mundy and Zisserman (1992a), is especially valuable, and provides the background needed for

---

* It could of course be argued that all recognition methods will be subject to the effects of perspective transformations. However, invariant-based recognition will not flinch from invoking highly extreme transformations which appear to grossly distort the objects in question, whereas more conventional methods are likely to be designed to cope with a *reasonable* range of expected shape distortions.

† Here we assume that the aim is location of specific objects in the image. If the objects are then to be located in the world coordinates, camera calibration, or some use of reference points will of course be needed. However, there are many applications, such as inspection, surveillance and identification (e.g. of faces or signatures) where location of objects in the *image* can be entirely adequate.

understanding the other papers in the volume. On the whole the volume has a theoretical flavour which demonstrates what ought to be possible using invariants, though comparisons between invariants and other approaches to recognition are on the whole lacking. Thus it is only by examining whether workers *choose* to use invariants in real applications that the full story will emerge. In this respect the paper by Kamel *et al.* (1994) on face recognition is of great interest, as it shows how invariants helped to achieve more than had been achieved earlier after many attempts using other approaches— specifically in correcting for perspective distortions during face recognition.

19

Automated Visual Inspection

19.1 Introduction

Over the past few years machine vision has consolidated its early promise and has become a vital component in the design of advanced manufacturing systems. On the one hand it provides a means of maintaining control of quality during manufacture, and on the other it is able to feed the assembly robot with the right sort of information to construct complex products from sets of basic components. In the latter case this is particularly important as it helps to make flexible manufacture a reality, so that costs due to under-use of expensive production lines are virtually eliminated.

These two major applications of vision—automated visual inspection and automated assembly—have much in common and can on the whole be performed by similar hardware vision systems employing closely related algorithms. Perhaps the most obvious use of visual inspection is to check products for quality so that defective ones may be rejected: this is easy to visualize in the case of a line making biscuits or washers at rates of the order of one million per day. Another area of application of inspection is to measure a specific parameter for each product and to feed its value back to an earlier stage in the plant in order to "close the loop" of the manufacturing process: a typical application is to adjust the temperature of jam or chocolate in a biscuit factory, when the coating is found not to be spreading correctly. A third use of inspection is simply to gather statistics on the efficiency of the manufacturing process, finding, for example, how product diameters vary, in order to provide information which will help management with advance planning.

In automated assembly, it is clear that vision can provide feedback to control the robot arm and wrist. For this purpose it needs to provide detailed information on the positions and orientations of objects within the field of view: clearly, it also needs to be able to distinguish individual components within the field of view. In addition, a well-designed vision system will be able to check components before assembly, for example so as to prevent the robot trying to fit a screw into a nonexistent hole.

It will now be clear that inspection and assembly require virtually identical vision systems, the most notable difference often being that a linescan camera is used for inspecting components on a conveyor, whereas a vidicon or other area (whole picture) camera is required for assembly operations on a worktable. In what follows discussion is centred on inspection although, because of the similarity of the two types of application, many of the concepts that are developed are also useful for automated assembly tasks.

19.2 The process of inspection

Inspection is the process of comparing individual manufactured items against some pre-established standard with a view to maintenance of quality. Before proceeding to study inspection tasks in detail, it is useful to note that the process of inspection commonly takes place in three definable stages:

(i)   image acquisition;
(ii) object location;
(iii) object scrutiny and measurement.

We defer detailed discussion of image acquisition until Chapter 23 and comment here on the relevance of separating the processes of location and scrutiny. This is important because (either on a worktable or on a conveyor) large numbers of pixels usually have to be examined before a particular product is found, whereas once it has been located, its image frequently contains relatively few pixels and so rather little computational effort need be expended in scrutinizing and measuring it. For example, on a biscuit line, products may be separated by several times the product diameter in two dimensions, so that some 100 000 pixels may need to be examined to locate products occupying say 5000 pixels. This means that product location is likely to be a much more computationally intensive problem than product scrutiny. While this is generally true, sampling techniques may permit object location to be performed with much increased efficiency (Chapter 9): under these circumstances it is possible that location may be faster than scrutiny, since the latter process, though straightforward, tends to permit no short-cuts and requires all pixels to be examined.

19.3 Review of the types of object to be inspected

Before studying methods of visual inspection (including those required for assembly applications), it is worth considering the types of object with which such systems may have to cope. The situation is illustrated by taking two rather opposing categories: (i) goods such as food products which are subject to wide variation during manufacture but for which physical appearance is an important factor, and (ii) those such as precision metal parts which are needed in the electronics and automotive industries. The problems that are specific to each category are discussed first; then size measurement and the problem of 3-D inspection are considered briefly.

19.3.1 Food products

Food products are a particularly wide category, ranging from chocolate cream biscuits to pizzas, and from frozen food packs to complete set meals (as provided by airlines). In the food industry the trend is towards products of high added value. Logically, such products should be inspected at every stage of manufacture, so that further value is not added to products that are already deficient. However, inspection systems are still quite expensive and there is a tendency to inspect only at the end of the product line: this at least ensures (i) that the final appearance is acceptable and (ii) that the size of the product is within the range required by the packaging machine[*]. In fact, this strategy is reasonable for many products—as for some types of biscuit where a layer of jam is clearly discernable underneath a layer of chocolate. Pizzas exemplify another category wherein many additives appear on top of the final product, all of which are in principle detectable by a vision system at the end of the product line.

When checking the shapes of chocolate products (chocolates, chocolate bars, chocolate biscuits, etc.), a particular complication that arises is the "footing" around the base of the product. This footing is often quite jagged and makes it difficult to recognize the product or to determine its orientation.

---

[*] On many food lines, jamming of packaging machines due to oversize products is one of the major problems.

However, the eye generally has little difficulty with this task, and hence if a robot is to place a chocolate in its proper place in a box it will have to emulate the eye and employ a full grey-scale image: short-cuts with silhouettes in binary images are unlikely to work well. In this context it should be remarked that chocolate is one of the more expensive ingredients of biscuits and cakes. A frequently recurring inspection problem is to check that chocolate cover is sufficient to please the consumer while low enough to maintain adequate profit margins.

Returning to packaged meals, these present both an inspection and an assembly problem. A robot or other mechanism has to place individual items on the plastic tray and it is clearly preferable that every item should be checked to ensure, for example, that each salad contains an olive or that each cake has the requisite blob of cream.

19.3.2 Precision components

Many other parts of industry have also progressed to the automatic manufacture and assembly of complex products. It is clearly necessary for items such as washers and O-rings to be tested for size and roundness, for screws to be checked for the presence of a thread, and for mains plugs to be examined for the appropriate pins, fuses and screws. Engines and brake assemblies also have to be checked for numerous possible faults. Perhaps the worst problems arise when items such as flanges or slots are missing so that further components cannot be fitted properly. It cannot be emphasized enough that what is missing is at least as important as what is present: missing holes and threads can effectively prevent proper assembly. It is sometimes stated that checking the pitch of a screw thread is unnecessary —

Table 19.1   Features to be checked on precision components.

Dimensions within specified tolerances
Correct positioning, orientation and alignment
Correct shape, especially roundness, of objects and holes
Whether corners are misshapen, blunted or chipped
Presence of holes, slots, screws, rivets, etc.
Presence of a thread in screws
Presence of burr and swarf
Pits, scratches, cracks, wear and other surface marks
Quality of surface finish and texture
Continuity of seams, folds, laps and other joins

if a thread is present it is bound to be correct. However, the author has come upon at least one industrial application where this is not true.

Table 19.1 summarizes some of the common features that need to be checked when dealing with individual precision components. Note that measurement of the extent of any defect, together with knowledge of its inherent seriousness, should permit components to be graded according to quality, thereby saving money for the manufacturer (rejecting all defective items is a very crude option).

19.3.3 Differing requirements for size measurement

Size measurement is important both in the food industry and in the automotive and small-parts industry. However, the problems in the two cases are often rather different. For example, the diameter of a biscuit can vary within quite wide limits (~5%) without giving rise to undue problems but when it gets outside this range there is a serious risk of jamming the packing machine, and the situation must be monitored carefully. In contrast, for mechanical parts, the required precision can vary from 1% for objects such as O-rings to 0.01% for piston heads. This variation clearly makes it difficult to design a truly general-purpose inspection system. However, the manufacturing process often permits little variation in size from one item to the next. Hence it may be adequate to have a system that is capable of measuring to an accuracy of rather better than 1%, so long as it is capable of checking all the characteristics mentioned in Table 19.1.

For cases where high precision is vital, it is important to note that accuracy of measurement is proportional to the resolution of the input image. Currently images of up to 512×512 pixels are common, so accuracy of measurement is basically of the order of 0.2%. Fortunately, grey-scale images provide a means of obtaining significantly greater accuracy than indicated by the above arguments, since the exact transition from dark to light at the boundary of an object can be estimated more closely. In addition, averaging techniques (e.g. along the side of a rectangular block of metal) permit accuracies to be increased even further—by a factor $\sqrt{N}$ if N pixel measurements are made. These factors permit measurements to be made to subpixel resolution, sometimes even down to 0.1 pixels.

19.3.4 Three-dimensional objects

Next it is noted that all real objects are 3-D, although the cost of setting up an inspection station frequently demands that they are examined from one viewpoint in a single 2-D image. This is clearly highly restrictive, and in many cases over-restrictive. Nevertheless, it is generally possible to do an enormous amount of useful checking and measurement from one such image. The clue that this is possible lies in the prodigious capability of the human eye—for example, to detect at a glance from the play of light on a surface whether or not it is flat. Furthermore, in many cases products are essentially flat and the information that we are trying to find out about them is simply expressible via their shape or via the presence of some other feature which is detectable in a 2-D image. In cases where 3-D information is required, methods exist for obtaining it from one or more images, for example via binocular vision or structured lighting, as has already been seen in Chapter 15. More is said about this below.

19.3.5 Other products and materials for inspection

This subsection briefly mentions a few types of product and material that are not fully covered in the foregoing discussion. First, electronic components are increasingly having to be inspected during manufacture, and of these, printed circuit boards (PCBs) and integrated circuits are subject to their own special problems which are currently receiving considerable attention. Second, steel strip and wood inspection are also of great importance. Third, bottle and glass inspection has its own particular intricacies, because of the nature of the material, glints being a relevant factor—as also in the case of inspection of cellophane-covered foodpacks. In this chapter, space permits only a short discussion of some of these topics (see Sections 19.7–19.8).

19.4 Summary—the main categories of inspection

The above sections have given a general review of the problems of inspection but have not shown how they might be solved. This section takes the analysis a stage further. First, note that the items in Table 19.1 may be classified as *geometrical* (measurement of size and shape—in 2-D or 3-D as necessary), *structural* (whether there are any missing items or foreign objects) and *superficial* (whether the surface has adequate quality). It should be evident from Table 19.1 that these three categories are not completely distinct but they are useful for the following discussion.

Start by noting that the methods of object location are also inherently capable of providing geometrical measurements. Distances between relevant edges, holes and corners can be measured; shapes of boundaries can be checked both absolutely and via their salient features; and alignments can usually be checked quite simply, for example by finding how closely various straight line segments fit to the sides of a suitably placed rectangle. In addition, shapes of 3-D surfaces can be mapped out by binocular vision, photometric stereo, structured lighting or other means (see Chapter 15), and subsequently checked for accuracy.

Structural tests can also be made once objects have been located, assuming that a database of the features they are supposed to possess is available. In the latter case, it is necessary merely to check whether the features are present in predicted positions. As for foreign objects, these can be looked for via unconstrained search as objects in their own right. Alternatively, they may be found as differences between objects and their idealized forms, as predicted from templates or other data in the database. In either case, the problem is very data-dependent and an optimal solution needs to be found for each situation. For example, scratches may be searched for directly as straight line segments.

Tests of surface quality are perhaps more complex. In Chapter 23 methods of lighting are described which illuminate flat surfaces uniformly, so that variations in brightness may be attributed to surface blemishes. For curved surfaces it might be hoped that the illumination on the surface would be predictable, and then differences would again indicate surface blemishes. However, in complex cases, there is probably no alternative but to resort to the use of switched lights coupled with rigorous photometric stereo techniques (see Chapter 15). Finally, it should be noted that the problem of checking quality of surface finish is akin to that of ensuring an attractive physical appearance, and this can be highly subjective; this means that inspection algorithms need to be "trained" in some way to make the right judgements (note that judgements are decisions or classifications and so the methods of Chapter 20 are appropriate).

Overall, accurate object location is a prerequisite to reliable object scrutiny and measurement, for all three main categories of inspection. If a CAD system is available, then providing location information permits an image to be generated which should closely match the observed image, and template matching (or correlation) techniques should in many cases permit the remaining inspection functions to be fulfilled. However, this will not always work without trouble—as in the case where object surfaces have a random or textured component. This means that preliminary analysis of the texture may have to be carried out before relevant templates can be applied—or at least checks made of the maximum and minimum pixel intensities within the product area. More generally, in order to solve this and other problems, some latitude in the degree of fit should be permitted.

It is interesting that the same general technique—that of template matching—arises in both the measurement and scrutiny phase and the object location phase. However (as remarked earlier) this need not consume as much computational effort as in object location. The underlying reason for this is that template matching is difficult when there are many degrees of freedom inherent in the situation, since comparisons with an enormous number of templates may be required. However, when the template is in a standard position relative to the product and when it has been orientated correctly template matching is much more likely to constitute a practical solution to the inspection task, although the problem is very data-dependent.

Despite these considerations, there is a need to find computationally efficient means of performing the necessary checks of parts. The first possibility is to use suitable algorithms to model the image intensity and then to employ the model to check relevant surfaces for flaws and blemishes. Another useful approach is to convert 2-D to 1-D intensity profiles. This approach leads to the lateral histogram (Chapter 12) and radial histogram techniques; the latter can conveniently be applied for inspecting the very many objects possessing circular symmetry, as will be seen below. However, we first consider a simple but useful means of checking shapes.

19.5   Shape deviations relative to a standard template

For food and certain other products, an important factor in 2-D shape measurement is the deviation relative to a standard template. Maximum deviations are important because of the need (already referred to) of fitting the product into a standard pack. Another useful measure is the area of overflow or underfill relative to the template (Fig. 19.1). For simple shapes that are bounded by circular arcs or straight lines (a category that includes many types of biscuit or bracket) it is straightforward to test whether a particular pixel on or near the boundary is inside the template or outside it. For straight line segments this is achieved in the following way. Taking the pixel as (x_1, y_1) and the line as having equation:

$$lx + my + n = 0 \tag{19.1}$$

the coordinates of the pixel are substituted in the expression:

$$f(x, y) = lx + my + n \tag{19.2}$$

The sign will be positive if the pixel lies on one side of the line, negative on

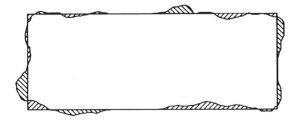

Fig. 19.1 Measurement of product area relative to a template: in this example two measurements are taken, indicating respectively the areas of overflow and underfill relative to a prespecified rectangular template.

the other, and zero if it is on the idealized boundary. Furthermore, the distance on either side of the line is given by the formula:

$$d = (lx_1 + my_1 + n)/(l^2 + m^2)^{1/2} \qquad (19.3)$$

The same observation about the signs applies to any conic curve if appropriate equations are used: for example, for an ellipse:

$$f(x, y) = (x - x_c)^2/a^2 + (y - y_c)^2/b^2 - 1 \qquad (19.4)$$

where $f(x, y)$ changes sign on the ellipse boundary. For a circle the situation is particularly simple, since the distance to the circle centre need only be calculated and compared with the idealized radius value. For more complex shapes, deviations need to be measured using centroidal profiles or the other methods described in Chapter 7. However, the method outlined above is useful as it is simple, quick and reasonably robust, and does not need to employ sequential boundary tracking algorithms: a raster scan over the region of the product is sufficient for the purpose.

19.6 Inspection of circular products

Since circular products are so common, it has proved beneficial to develop special techniques for scrutinizing them efficiently. The "radial histogram" technique was developed for this purpose (Davies, 1984c, 1985b). It involves plotting a histogram of intensity as a function of radial distance from the centre. Varying numbers of pixels at different radial distances complicate the problem but smooth histograms result when suitable normalization procedures are applied (Figs 19.2 and 19.3). These give

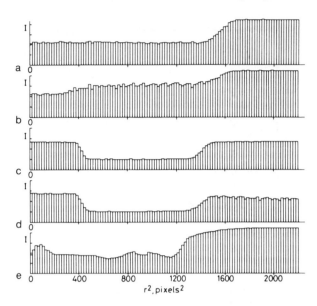

Fig. 19.2   Practical applications of the radial histogram approach. In all cases an r^2 base variable is used and histogram columns are individually normalized: this corresponds to method (iv) in the text (Section 19.6.1). These histograms were generated from the original images of Fig. 19.3.

accurate information on all three major categories of inspection. In particular, they provide values of all relevant radii: for example, both radii of a washer can be estimated accurately using this technique. Surprisingly, accuracy can be as high as 0.3% even when objects have radii as low as 40 pixels (Davies, 1985c), with correspondingly higher accuracies at higher resolutions.

Radial histograms are particularly well suited to the scrutiny of symmetrical products that do not exhibit a texture, or for which texture is not prominent and may validly be averaged out. In addition, the radial histogram approach ignores correlations between pixels in the dimension being averaged (i.e. angle); where such correlations are significant it is not possible to use the approach. An obvious example is in the inspection of components such as buttons, where angular displacement of one of four button holes will not be detectable unless the hole encroaches on the space of a neighbouring hole. Similarly, the method is not able to check the detailed shape of each of the small holes. Clearly, the averaging involved in finding the radial histogram mitigates against such detailed inspection, which is then best carried out by separate direct scrutiny of each of the holes.

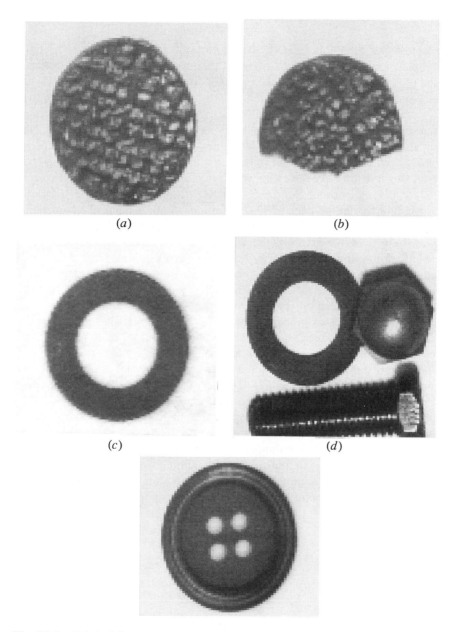

(a)

(b)

(c)

(d)

Fig. 19.3 Original images used in generating the radial histograms of Figs 19.2 and 19.5.

19.6.1   Computation of the radial histogram: statistical problems

In computing the radial intensity histogram of a region in an image, one of the main variables to be considered is the radial step size s. If s is small, then few pixels contribute to the corresponding column of the histogram and little averaging of noise and related effects takes place: for textured products this could be a disadvantage unless it is specifically required to analyse texture by an extension of the method. Ignoring the latter possibility, and increasing s in order to obtain a histogram having a statistically significant set of ordinates, there is bound to be a loss in radial resolution. To find more about this tradeoff between radial resolution and accuracy, it is necessary to analyse the statistics of interpixel (radial) distances. Assuming a Poisson distribution for the number of pixels falling within a band of given radius and variable s, it is found that histogram column height or "signal" varies, at given radius, as the step size s, whereas "noise" varies as $\sqrt{s}$ giving an overall signal-to-noise ratio varying (for given radius) as $\sqrt{s}$. Radial resolution is of course inversely proportional to s. For small radii, of less than about 20 pixels, this simple model is inaccurate and it is better to follow the actual detailed statistics of a discrete square lattice. Under these circumstances, it seems prudent to examine the pixel statistics plotted in Fig. 19.4 for various values of s and from these to select a value of s which is the best compromise for the application being considered.

For many applications it is inconvenient to have radial intensity distributions such as that depicted in Fig. 19.5, in which the column height varies inherently as the radius because the number of pixels in a band of radius r and given s is (approximately) proportional to r. It is more convenient to normalize the distribution so that regions of uniform intensity give rise to a uniform radial intensity distribution. There are two means by which this may be achieved. The first is to multiply each column of the radial histogram by a factor which brings it to an inherently uniform value. The other is to plot the histogram as a function of some function $u(r)$ other than r which gives the same effect.

Taking the latter approach first, changing variables from r to $u(r)$ results in

$$\int_{r}^{r+\delta r} \int_{0}^{2\pi} I(r)r \, dr \, d\theta = I(r) \times 2\pi r \, \delta r$$
$$= I(u)\delta u = \text{constant} \tag{19.5}$$

Hence

$$\delta u = 2\pi r \delta r \tag{19.6}$$

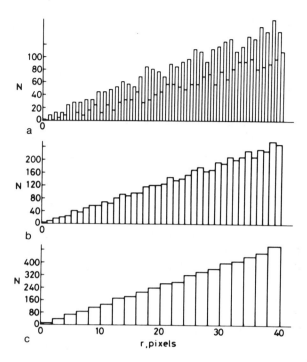

Fig. 19.4 Pixel statistics for various radial step sizes: the number of pixels, as a function of radius, that fall into bands of given radial step size. The chosen step sizes are (a) 0.50, (b) 1.00 and (c) 2.00 pixels.

We now require that $I(u) =$ constant for fixed step size δu. Setting δu equal to a constant γ leads to

$$\delta r = \gamma/2\pi r \qquad (19.7)$$

so that

$$\delta r \propto 1/r \qquad (19.8)$$

as might indeed have been expected. In addition, note that

$$u = \int_0^r 2\pi r \, dr = \pi r^2 \qquad (19.9)$$

making the histogram base parameter equal to r^2 (within a constant of proportionality).

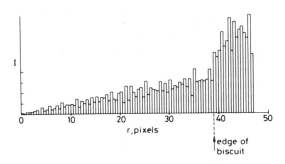

Fig. 19.5 Basic form of radial intensity distribution, which varies approximately as the radius. The product being inspected here is the biscuit of Fig. 19.3(a).

Unfortunately, this approach is still affected by the pixel statistics problem, as shown in Fig. 19.6. Thus we are effectively being forced to apply the approach of normalizing the histogram by multiplying by suitable numerical factors, as suggested earlier. There are now four alternatives:

(i)   use the basic radial histogram;
(ii)   use the modified radial histogram, with $u(r) = r^2$ base parameter;
(iii) use the basic histogram, with individual column normalization;
(iv)   use the $u(r) = r^2$ histogram, with individual column normalization.

As a rule, it appears to be necessary to normalize each column of the histogram individually to obtain the most accurate results. For ease of interpretation, it would also be preferable to use the modified base parameter $u(r) = r^2$ rather than r.

Note that the square-law approach is trivial to implement, since r^2 rather than r is the value that is actually available during computation (r has to be

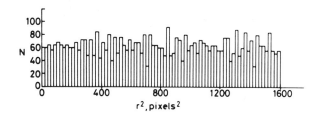

Fig. 19.6 Pixel statistics for an r^2 histogram base parameter: the pixel statistics are not exactly uniform even when the radial histogram is plotted with an r^2 base parameter.

derived from r^2 if it is needed), and further the r^2 base parameter gives a nonlinear stretch to the histogram which matches the increased radial accuracy that is actually available at high values of r. This means that the $u(r) = r^2$ histogram permits greater accuracy in the estimation of radius at large r than it does at small r, thus making $u(r) = r^2$ the natural base variable to employ (Davies, 1985b).

These considerations make method (iv) the most generally suitable one, although method (ii) is useful when speed of processing is a particular problem. Finally, it is relevant to note that, to a first approximation, all types of radial intensity histogram mentioned above are able to eliminate the effects of a uniform gradient of background illumination over an object.

19.6.2 Application of radial histograms

Figure 19.2 shows practical applications of the above theory to various situations, depicted in Fig. 19.3: see also Fig. 19.7, which relates to the biscuits shown in Fig. 9.1. In particular, notice that the radial histogram approach is able to give vital information on various types of defect: the presence or absence of holes in a product such as a washer or a button; whether circular objects are in contact or overlapping; broken objects; "show-through" of biscuit where there are gaps in a chocolate or other coating; and so on. In addition, it is straightforward to derive dimensional measurements from radial histograms. In particular, radii of discs or washers

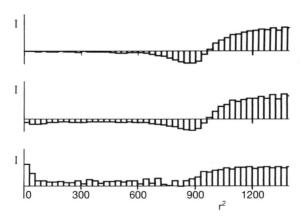

Fig. 19.7 Radial intensity histograms for the three biscuits of Fig. 9.1: the order is from top to bottom in both figures. Intensity is here measured relative to that at the centre of an ideal product.

can be obtained to significantly better than 1 pixel accuracy because of the averaging effect of the histogram approach. However, the method is limited here by the accuracy with which the centre of the circular region is first located: this underlines the value of the high-accuracy centre finding technique of Chapter 9. To a certain extent, accuracy is limited by the degree of roundness of the product feature being examined: radius can be measured only to the extent that it is meaningful!

It might be imagined that the radial histogram technique is applicable only for symmetrical objects. However, it is also possible to use radial histograms as signatures of intensity patterns in the region of specific salient features. Small holes are suitable salient features but corners are less suitable unless the background is saturated out at a constant value; otherwise, too much variation arises from the background and the technique does not prove viable.

It is worth underlining that the combination of techniques described above is not only accurate but also sparing in computation. Thus processing times for a 128×128 image are typically a few seconds in software and considerably faster in special-purpose hardware (see Chapter 24).

Finally, note that the radial histogram approach has the useful characteristic that it is trainable, since the relevant 1-D templates may be accumulated, averaged over a number of ideal products and stored in memory, ready for comparisons with less ideal products. This characteristic is valuable, not only for convenience in setting up, but also because it permits inspection to be adaptable to cover a range of products.

19.7   Inspection of printed circuits

Over the past 10 years machine vision has been used increasingly in the electronics industry, notably for inspecting PCBs. First, PCBs may be inspected before components are inserted; second, they may be inspected to check that the correct components have been inserted the correct way round; and third, all the soldered joints may be scrutinized. The faults that have to be checked for on the boards include touching tracks, whisker bridges, broken tracks (including hairline cracks) and mismatch between pad positions and holes drilled for component insertion. Controlled illumination is required to eliminate glints from the bare metal and to ensure adequate contrast between the metal and its substrate. With adequate control over the lighting, most of the checks (e.g. apart from reading of any print on the substrate) may be carried out on a binarized image and the problem devolves into that of checking shape. This may be tackled by gross template matching—using a logical exclusive-or operation—but this approach requires large data storage and precise registration has to be achieved.

Difficulties with registration errors can largely be avoided if shape analysis is performed by connectedness measurement (using thinning techniques) and morphological processing: for example, if a track disappears or becomes broken after too few erosion operations, then it is too narrow; a similar procedure will check whether tracks are too close together. Likewise, hairline cracks may be detected by dilations followed by tests to check for changes in connectedness.

Alignment of solder pads with component holes is customarily checked by employing a combination of back and front lighting. Powerful back lighting (i.e. from behind the PCB) gives bright spots at the hole positions, while front lighting gives sufficient contrast to show the pad positions; it is then necessary to confirm that each hole is, within a suitable tolerance, at the centre of its pad. Counting of the bright spots from the holes, plus suitable measurements around hole positions (e.g. via radial histogram signatures), permits this process to be performed satisfactorily.

Overall, the main problem with PCB inspection is the resolution required: typically, images have to be digitized to at least 4000×4000 pixels—as when a 20×20 cm board is being checked to an accuracy of $50\,\mu\text{m}$. In addition, a suitable inspection system normally has to be able to check each board fully in less than one minute; it also has to be trainable, to allow for upgrades in the design of the circuit or improvements in the layout; and it should cost no more than around £40000. However, considerable success has been attained with these aims.

To date, the bulk of the work in PCB inspection has concerned the checking of tracks. Nevertheless, useful work has also been carried out on the checking of soldered joints. Here, each joint has to be modelled in 3-D by structured light or other techniques. In one such case light stripes were used (Nakagawa, 1982), and in another surface reflectance was measured with a fixed lighting scheme (Besl et al., 1985). Note that surface brightness says something about the quality of the soldered joint. This type of problem is probably completely solvable (at least up to the subjective level of a human inspector) but detailed scrutiny of each joint at a resolution of say 64×64 pixels may well be required to guarantee that the process is successful, and this implies an enormous amount of computation to cope with the several hundreds (or in some cases thousands) of joints on most PCBs. Hence special architectures are needed to handle the information in the time available. Indeed, knowledge of how to devise suitable algorithms must currently be well ahead of the development of economic hardware for high-speed implementation.

Similar work is under way on the inspection of integrated circuit masks and die bonds but space does not permit discussion of this rapidly developing area. For a useful review, see Chin (1988).

19.8   Steel strip and wood inspection

The problem of inspecting steel strip is one that is very exacting for human operators. First, it is virtually impossible for the human eye to focus on surface faults when the strip is moving past the observer at rates in excess of 20 m/s; second, several years of experience are required for this sort of work; and third, the conditions in a steel mill are far from congenial, with considerable heat and noise constantly being present. Hence much work has been done to automate the inspection process (Browne and Norton-Wayne, 1986). At its simplest this requires straightforward optics and intensity thresholding, although special laser scanning devices have also been developed to facilitate the process (Barker and Brook, 1978).

The problem of wood inspection is more complex, since this natural material is very variable in its characteristics. For example, the grain varies markedly from sample to sample. As a result of this variation the task of wood inspection is still in its infancy and many problems remain. However, the purpose of wood inspection is reasonably clear: first, to look for cracks, knots, holes, bark inclusions, embedded pine needles, miscoloration and so on; and ultimately to make full use of this natural resource by identifying regions of the wood where strength or appearance is substandard. In addition, the timber may have to be classified as appropriate for different categories of use—furniture, building, outdoor, etc. Overall, wood inspection is something of an art—i.e. it is a highly subjective process— although valiant attempts have been made to solve the problems (e.g. Sobey, 1989).

19.9   Inspection of products with high levels of variability

In the above sections we have concentrated on certain aspects of inspection —particularly dimensional checking of components ranging from precision parts to food products, and the checking of complex assemblies to confirm the presence of holes, nuts, springs, and so on. These could be regarded as the geometrical aspects of inspection. For the more imprecisely made products such as foodstuffs and textiles, there are greater difficulties as the template against which any product has to be compared is not fixed.

Broadly, there are two ways of tackling this problem: one is the use of a range of templates, each of which is acceptable to some degree; the other is the specification of a variety of descriptive parameters. In either case there will be a number of numerical measurements whose values have to be within prescribed tolerances. Overall, it seems inescapable that variable products demand greater amounts of checking and computation, and that inspection is significantly more demanding. Nowhere is this clearer than for food and textiles, for which the relevant parameters are largely textural. However, "fuzzy" inspection situations can also occur for certain products which might initially be considered as precision components: e.g. for electric lamps the contour of the element and the solder pads on the base have significant variability. Thus this whole area of inspection involves checking that a range of parameters do not fall outside certain prespecified limits on some relevant distribution which *may* be reasonably approximated by a Gaussian.

We have seen above that the inspection task is made significantly more complicated by natural valid variability in the product; in the end it seems best to regard inspection as a process of making measurements which have to be checked statistically. Defects can be detected relative to the templates, either as gross mismatches or else as numerical deviations. And missing parts can likewise be detected since they do not appear at the appropriate positions relative to the templates. Foreign objects also appear to fit into this pattern, being essentially defects under another name. However, this view is rather too simple for several reasons:

(i) Foreign objects are frequently unknown in size, shape, material or nature.
(ii) Foreign objects may appear in the product in a variety of unpredictable positions and orientations.
(iii) Foreign objects may have to be detected in a background of texture which is so variable in intensity that they will not stand out.

Overall, it is the unpredictability of foreign objects that can make them difficult to see, especially in textured backgrounds (Fig. 19.8). If one knew their nature in advance, then a special detector could perhaps be designed to locate them. But in many practical situations, the only means of detecting them is to look for the unusual. In fact, the human eye is well tuned to search for the unusual. On the other hand there are few obvious techniques which can be used to seek it out automatically in digital images. Clearly, simple thresholding would work in a variety of practical cases, especially where plain surfaces have to be inspected for scratches, holes, swarf or dirt. However, looking for extraneous vegetable matter (such as leaves, twigs or

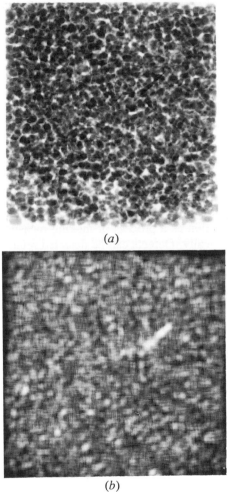

(a)

(b)

Fig. 19.8 Foreign object detection in a packet of frozen vegetables (in this case sweetcorn). (a) The original X-ray image; (b) an image in which texture analysis procedures have been applied to enhance any foreign objects; (c) and (d), the respective thresholded images. Notice the false alarms that are starting to arise in (c), and the increased confidence of detection of the foreign object in (d): for further details, see Patel *et al.* (1995).

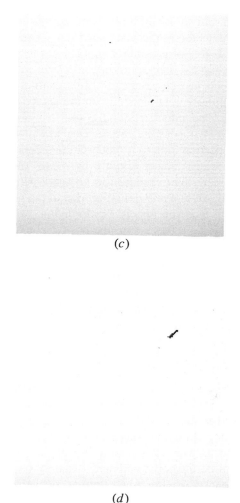

(c)

(d)

Fig. 19.8 (continued).

pods) amongst a sea of peas on a conveyor may be less easy, as the contrast levels may be similar, and the textural cues may not be able to distinguish the shapes sufficiently accurately. Of course, in the latter example, it could be imagined that every pea could be identified by its intrinsic circularity. However, the incidence of occlusion, and the very computation intensive nature of this approach to inspection, inhibits such an approach. In any case, a method which would detect pods amongst peas might not detect round stones or small pieces of wood—especially in a grey-scale image.

Ultimately, the problem is difficult because the paradigm means of designing sensitive detectors—the matched filter—cannot be used, simply because of the high degree of variability in what has to be detected. With so many degrees of freedom—shape dimensions, size, intensity, texture, and so on—foreign objects can be extremely difficult to detect successfully in complex images. Naturally, a lot depends on the nature of the substrate, and while a plain background might render the task trivial, a textured product substrate may render it impossible, or at least practically impossible in a real-time factory inspection milieu. In general the solution devolves into not trying to detect the foreign object directly by means of carefully designed matched filters, but in trying to model the intensity pattern of the substrate sufficiently accurately, so that any deviation due to the presence of a foreign object is detected and rendered visible. As hinted above, the approach is to search for the unusual. To achieve this, the basic technique is to identify the 3σ or other appropriate points on all available measures, and initiate rejection when they are exceeded.

There is a fundamental objection to this procedure: if some limit (e.g. 3σ) is assumed, this cannot easily be optimized, since the proper method for achieving this is to find both the distributions—for the background substrate and for the foreign objects—and to obtain a minimum error decision boundary between them (ideally, this would be implemented in a multidimensional feature space appropriate to the task). However, in this case we do not have the distribution corresponding to the foreign objects, so we have to fall back on "reasonable" acceptance limits based on the substrate distribution.

It might appear that this argument is flawed in that the proportion of foreign objects coming along the conveyor is well known. While this might occasionally be so, the levels of detectability of the foreign objects in the received images will be unknown and will certainly be less than the actual occurrence levels: hence arriving at an optimal decision level will be very difficult.

However, a far worse problem often exists in practice. The occurrence rates for foreign objects might be almost totally unknown, because of (i) their intrinsic variability and (ii) their rarity. We might well ask "How often will an elastic band fall onto a conveyor of peas?", but this is a question which it is virtually impossible to answer. Maybe it is possible to answer somewhat more accurately the more general question of how often a foreign object of some sort will fall onto the conveyer, but even then the response may well be that somewhere between 1 in 100 000 and 1 in 10 million of bags of peas contain a foreign object. With low levels of risk, the probabilities are extremely difficult to estimate, and indeed there is little available data or other basis on which to calculate them. Consumer complaints can indicate the possible levels of risk, but these arise as individual items, and in any case many customers will not make any fuss and by no means all instances come to light.

With food products the penalties for not detecting individual foreign objects are not usually especially great. Glass in baby food may be more apocryphal than real, and may be unlikely to cause more than alarm. Similarly, small stones amongst the vegetables are more of a nuisance than a harm, though cracked teeth could perhaps result in litigation in the £1000 bracket. Far more serious are problems with electric lamps, where a wire emerging from the solder pads is potentially lethal and is a substantive worry for the manufacturer. Litigation for deaths arising from this source could run to a million pounds or so (corresponding to the individual's potential lifetime earnings).

This discussion reveals very clearly that it is the cost rate[*] rather than the error rate which is the important parameter when there is even a remote risk to life and limb. Indeed, it concerned Rodd and his coworkers so much in relation to the inspection of electric lamps than they decided to develop special techniques for ensuring that their algorithms were tested sufficiently (Thomas and Rodd, 1994; Thomas *et al.*, 1995): computer graphics techniques were used to produce large numbers of images with automatically generated variations on the basic defects, and it was checked that the inspection algorithms would always locate them.

19.10 X-ray inspection

In inspection applications, there is a tension between inspecting products early on, before significant value has been added to a potentially defective product, and at the end of the line, so that the quality of the final products is guaranteed. This consideration applies particularly with food products, where additives such as chocolate can be very expensive and constitute substantial waste if the basic product is broken or misshapen. In addition, inspection at the end of the line is especially valuable as oversized products (which may arise if two normal products become stuck together) can jam packing machines. Ideally, it would be beneficial if two inspection stations were placed on the line in appropriate positions, but if only one can be

---

[*] In fact, the *perceived* cost rate may be even more important, and this can change markedly with reports appearing in the daily press.

afforded (the usual situation) it will generally have to be placed at the end of the line.

With many products it is useful to be absolutely sure about the final quality as the customer will receive it. Thus there is especial value in inspecting the packaged products. Since the packets will usually be opaque, it will be necessary to inspect them under X-radiation. This results in significant expense, since the complete system will include not only the X-ray source and sensing system but also various safety features including heavy shielding. As a result commercial X-ray food inspection systems rarely cost less than £40 000, and £100 000 is a more typical figure. Such figures do not take account of maintenance costs, and it is also important to note that the X-ray source and the sensors deteriorate with time, so that sensitivity falls, and special calibration procedures have to be invoked.

Fortunately, X-ray sensors can nowadays take the form of linear photodiode packages constructed using integrated circuit technology[*]. These are placed end to end to span the width of a food conveyor which may be 30–40 cm wide. They act as a linescan camera which grabs images as the product moves along the conveyor. The main adjustments to be made in such a system are the voltage across the X-ray tube and the current passing through it. In food inspection applications, the voltage will be in the range 30–100 kV, and the current will be in the range 3–10 mA. A basic commercial system will include thresholding and pixel counting hardware, permitting the detection of small pieces of metal or other hard contaminants, but not soft contaminants. In general, the latter can only be detected if more sophisticated algorithms are used which examine the contrast levels over various regions of the image and arrive at a consensus that foreign objects are present—typically with the help of texture analysis procedures.

The sensitivity of an X-ray detection system depends on a number of factors. Basically, it is highly dependent on the number of photons arriving at the sensors, and this number is proportional to the electron current passing through the X-ray tube. There are stringent rules on the intensities of X-radiation to which food products may be exposed, but in general these limits are not approached because of good sensitivity at moderate current levels. However, we shall not delve into these matters further here.

Sensitivity also depends critically on the voltages that are applied to X-ray tubes. It is well known that the higher the voltage, the higher the electron energies, the higher the energies of the resulting photons, and the greater the

---

[*]The X-ray photons are first converted to visible light by passage through a layer of scintillating material.

penetrating power of the X-ray beam. However, greater penetrating power is not necessarily an advantage, as the beam will tend to pass through the food without attenuation, and therefore without detecting any foreign objects. While a poorly set up system may well be able to detect quite small pieces of metal without much trouble, detection of small stones and other hard contaminants will be less easy, and detection of soft contaminants will be virtually impossible. Thus it is necessary to optimize the contrast in the input images.

Unfortunately, X-ray sources provide a wide range of wavelengths, all of which are scattered or absorbed to varying degrees by the intervening substances. In a thick sample the X-rays may be scattered so that the X-radiation that arrives at the detector has passed through material that is not in a direct line between the X-ray tube and the sensors. This makes a complete analysis of sensitivity rather complicated: in what follows we will ignore this effect and assume that the bulk of the radiation reaching the sensors follows the direct path from the X-ray source. We will also assume that the radiation is gradually absorbed by the intervening substances, in proportion to its current strength. Thus we obtain the standard exponential formula for the decay of radiation through the material which we shall temporarily take to be homogeneous and of thickness z:

$$I = I_0 \exp\left[-\int \mu \, dz\right] = I_0 \exp\left[-\mu \int dz\right] = I_0 e^{-\mu z} \qquad (19.10)$$

Here μ depends on the type of material and the penetrating power of the X-radiation. For monochromatic radiation of energy E, we have:

$$\mu = (\rho N/A)[k_P Z^a / E^b + k_C Z/E] \qquad (19.11)$$

where ρ is the density of the material, A is its atomic weight, N is Avogadro's number, a and b are numbers depending on the type of material, and k_P and k_C are decay constants resulting from photoelectric and Compton scattering respectively (see for example Eisberg, 1961). It will not be appropriate to examine all the implications of this formula. Instead we proceed with a rather simplified model which nevertheless shows how to optimize sensitivity:

$$\mu = \alpha/E \qquad (19.12)$$

Substituting in equation (19.10), we find:

$$I = I_0 \exp(-\alpha z/E) \qquad (19.13)$$

If a minute variation in thickness, or a small foreign object, is to be detected sensitively, we need to consider the change in intensity resulting from a change in z or in αz (ultimately it is the integral of $\mu \, dz$ which is important—see equation (19.10)). It will be convenient to relabel the latter quantity as a generalized distance X, and the inverse energy factor as f:

$$I = I_0 \exp (-Xf) \tag{19.14}$$

$$\therefore \ dI/dX = -I_0 f \exp (-Xf) \tag{19.15}$$

so that:

$$\Delta I = -\Delta X \, I_0 f \exp (-Xf) \tag{19.16}$$

The contrast due to the variation in generalized distance can now be expressed as:

$$\Delta I/I = -\Delta X \, I_0 f \exp (-Xf)/I_0 \exp (-Xf) = -\Delta X \, f = -\Delta X/E \tag{19.17}$$

This calculation shows that contrast should improve as the energy of the X-ray photons decreases. However, this result appears wrong, as reducing the photon energy will reduce the penetrating power, and in the end no radiation will pass through the sample. First, the sensors will not be sufficiently sensitive to detect the radiation: specifically, noise (including quantization noise) will become the dominating factor. Second, we have ignored the fact that the X-radiation is not monochromatic. We shall content ourselves here with modelling the situation to take account of the latter factor. Assume that the beam has two energies, one fairly low (as above), and one rather high and penetrating. This high-energy component will add a substantially constant value to the overall beam intensity, and will result in a modified expression for the contrast:

$$\Delta I/I = -\Delta X \, I_0 f \exp (-Xf)/[I_1 + I_0 \exp (-Xf)]$$
$$= -\Delta X \, I_0 f/[I_1 \exp (Xf) + I_0] \tag{19.18}$$

To optimize sensitivity, we differentiate with respect to f:

$$d(\Delta I/I)/df = -\Delta X \, I_0 \{ [I_1 \exp (Xf) + I_0]$$
$$- Xf I_1 \exp (Xf) \}/[I_1 \exp (Xf) + I_0]^2 \tag{19.19}$$

This is zero when:

$$XfI_1 = I_1 + I_0 \exp(-Xf) \qquad (19.20)$$

i.e. $\qquad\qquad E = X/[1 + (I_0/I_1) \exp(-X/E)] \qquad (19.21)$

When $I_1 \ll I_0$ we have the previous result, that optimum sensitivity occurs for low E. However, when $I_1 \gg I_0$ we have the result that optimum sensitivity occurs when $E = X$. In general, this formula gives an optimum X-ray energy which is above zero, in accordance with intuition. In passing we note that graphical or iterative solutions of equation (19.21) are easily obtained.

Finally, we consider the exponential form of the signal given by equations (19.10), (19.13) and (19.14). These are nonlinear in X (i.e. αz), and meaningful image analysis algorithms would tend to require signals which are linear in the relevant physical quantity, namely X. Thus it is appropriate to take the logarithm of the signal from the input sensor, before proceeding with texture analysis or other procedures:

$$I' = \log I = \log [I_0 \exp(-Xf)] = A - Xf \qquad (19.22)$$

where

$$A = \log I_0 \qquad (19.23)$$

In this way, doubling the width of the sample doubles the change in intensity, and subsequent (e.g. texture analysis) algorithms can once more be designed on an intuitive basis. (In fact, there is a more fundamental reason for performing this transformation—that it performs an element of noise whitening, which should ultimately help to optimize sensitivity.)

19.11 Bringing inspection to the factory

The relationship between the producer of vision systems and the industrial user is more complex than might appear at first sight. The user has a need for an inspection system and states his need in a particular way. However, subsequent tests in the factory may show that the initial statement was inaccurate or imprecise; for example, the line manager's requirements[*] may

---

[*] In many factories line managers have the brief of maintaining production at a high level on an hour-by-hour basis, while at the same time keeping track of quality: the tension between these two aims, and particularly the underlying economic constraints, means that on occasion quality is bound to suffer.

not exactly match those envisaged by the factory management board. Part of the problem lies in the relative importance given to the three disparate functions of inspection mentioned earlier. Another lies in the change of perspective once it is seen exactly what defects the vision system is able to detect: it may be found immediately that one or more of the major defects that a product is subject to may be eliminated by modifications to the manufacturing process; in that case the need for vision is greatly reduced, and indeed the very process of trying out a vision system may end in its value being undermined and its not being taken up after a trial period. Clearly, this does not detract from the inherent capability of vision systems to perform 100% untiring inspection and to help maintain strict control of quality. However, it must not be forgotten that vision systems are not cheap and that they can in some cases be justified only if they replace a number of human operators: frequently a payback period of 2–3 years is specified for installing a vision system.

Textural measurements on products are an attractive proposition for applications in the food and textile industries. Often textural analysis is written into the prior justification for, and initial specification of, an inspection system. However, what a vision researcher understands by texture and what a line inspector in either of these industries means by it tend to be different. Indeed, what is required of textural measurements varies markedly with the application. The vision researcher may have in mind higher-order[*] statistical measures of texture, such as would be useful with a rough irregular surface of no definite periodicity[†]—as in the case of sand or pebbles on a beach, or grass or leaves on a bush. However, the textile manufacturer would be very sensitive to the periodicity of his fabric, and to the presence of faults or overly large gaps in the weave. Similarly, the food manufacturer might be interested in the number and spatial distribution of pieces of pepper on a pizza, while for fish coatings (e.g. batter or bread-crumbs) uniformity will be important and "texture measurement" may end by being interpreted as determining the number of holes per unit area of the coating. Thus it is clear that texture may be characterized not by

---

[*] The zero-order statistic is the mean intensity level; first-order statistics such as variance and skewness are derived from the histogram of intensities; second- and higher-order statistics take the form of grey-level co-occurrence matrices, showing the number of times particular grey values appear at two or more pixels in various relative positions. For more discussion on textures and texture analysis, see Chapter 22.

[†] More rigorously, the fabric is intended to have a long-range periodic order that does not occur with sand or grass: in fact, there is a close analogy here with the long- and short-range periodic order for atoms in a crystal and in a liquid, respectively.

higher-order statistics but instead by rather obvious counting or uniformity checks. In such cases a major problem is likely to be that of reducing the amount of computation to the lowest possible level, so that considerable expanses of fabric, or large numbers of products, can be checked with minimum expense on special hardware. In addition, it is frequently important to keep the inspection system flexible by training on samples, so that maximum utility of the production line can be maintained.

With this backcloth to factory requirements, it is clearly vital for the vision researcher to be sensitive to actual rather than idealized needs or the problem as initially specified. There is no substitute for detailed consultation with the line manager *and* close observation in the factory before setting up a trial system. Then the results from trials need to be considered very carefully to confirm that the system is producing the information that is really required.

19.12 Concluding remarks

This chapter has been concerned with the application of computer vision to industrial tasks, and notably to automated visual inspection. The number of relevant applications is exceptionally high and for that reason it has been necessary to concentrate on principles and methods rather than on individual cases. The repeated mention of hardware implementation has been a necessary one, since the economics of real-time implementation is often the factor that ultimately decides whether a given system will be installed on a production line. However, speeds of processing are also heavily dependent on the specific algorithms employed, and these in turn depend on the nature of the image acquisition system—including both the lighting and the camera setup (indeed, the decision of whether to inspect products on a moving conveyor or to bring them to a standstill for more careful scrutiny is perhaps the most fundamental one for implementation). Hence image acquisition and real-time electronic hardware systems are the main topics of later chapters (Chapters 23 and 24).

More fundamentally, the reader will have noticed that a major purpose of inspection systems is to make instant decisions on the adequacy of products. Related to this purpose are the often fluid criteria for making such decisions and the need to train the system in some way, so that the decisions that are made are at least as good as those that would have arisen with human inspectors. In addition, training schemes are valuable in making inspection systems more general and adaptive, particularly with regard to change of product. Hence the pattern recognition techniques of Chapter 20 are highly relevant to the process of inspection.

Ideally, the present chapter would have appeared after Chapters 20 and 23–24. However, it was felt better to include it as early as possible, in order to maintain the motivation of the reader; also Chapters 23–24 are perhaps more relevant to implementation than to the rationale of image analysis.

On a different tack, it is worth noting that automated visual inspection falls under the general heading of computer-aided manufacture (CAM), of which computer-aided design (CAD) is another part. Nowadays many manufactured parts can in principle be designed on a computer, visualized on a computer screen, made by computer-controlled (e.g. milling) machines, and inspected by computer—all without human intervention in handling the parts themselves. There is much sense in this computer integrated manufacture (CIM) concept, since the original design dataset is stored in the computer, and therefore it might as well be used (i) to aid the image analysis process that is needed for inspection, and (ii) as the actual template by which to judge the quality of the goods. After all, why key in a separate set of templates to act as inspection criteria when the specification already exists on a computer? However, some augmentation of the original design information to include valid tolerances is necessary before the dataset is sufficient for implementing a complete CIM system. It should also be remarked that the purely dimensional input to a numerically controlled milling machine is not generally sufficient—as the frequent references to surface quality in the present chapter indicate.

19.13 Bibliographical and historical notes

It is very difficult to provide a bibliography of the enormous number of papers on applications of vision in industry or even in the more restricted area of automated visual inspection. In any case, it can be argued that a book on research ought to concentrate on principles and to a lesser extent on detailed applications and "mere" history. In fact, there are a number of useful and (overall) fairly comprehensive review articles on these topics, and the reader is referred first to these: Kruger and Thompson (1981), Chin and Harlow (1982), Chin (1988), Wallace (1988), Newman and Jain (1995). For a more recent review of PCB inspection, see Moganti *et al.* (1996).

The overall history of industrial applications of vision has been one of relatively slow beginnings as the potential for visual control became clear, followed only in recent years by explosive growth as methods and techniques evolved and as cost-effective implementations became possible as a result of cheaper computational equipment. A key development was the instigation of important conferences and symposia, notably that on "Computer Vision and Sensor-Based Robots" held at General Motors Research Laboratories during 1978 (see Dodd and Rossol, 1979), and the ROVISEC (Robot Vision and Sensory Controls) series of conferences organized annually by IFS (Conferences) Ltd, UK, from 1981. In addition, useful compendia of papers have been published (e.g. Pugh, 1983; Billingsley, 1985), and books outlining relevant principles and practical details (e.g. Batchelor *et al.*, 1985; Browne and Norton-Wayne, 1986). Also relevant is the volume of company profiles and (brief) case-studies produced by Braggins and Hollingham (1986). The reader is also referred to a special issue of IEEE *Transactions on Pattern Analysis and Machine Intelligence* (now somewhat dated) which is devoted to industrial machine vision and computer vision technology (Sanz, 1988).

Noble (1995) presents an interesting and highly relevant view of the use of machine vision in manufacturing. Davies (1995) develops the same topic by presenting several case-studies together with a discussion of some major problems that remain to be tackled in this area.

The past few years have seen considerable interest in X-ray inspection techniques, particularly in the food industry (Boerner and Strecker, 1988; Wagner, 1988; Chan *et al.*, 1990; Penman *et al.*, 1992; Graves *et al.*, 1994; Noble *et al.*, 1994). In the case of X-ray inspection of food, the interest is almost solely in the detection of foreign objects, which could in some cases be injurious to the consumer. Indeed, this has been a prime motivation for much work in the author's laboratory (Patel *et al.*, 1994, 1995; Patel and Davies, 1995).

Another topic of growing interest is the automatic visual control of materials such as lace during manufacture; in particular, high-speed laser scalloping of lace is set to become fully automated (King and Tao, 1995).

It is often the case that details become less relevant with time, and the fundamental theory of vision limits what is practically possible in real applications. Hence it does not seem fruitful to dwell further on applications here, beyond noting that the possibilities are continually expanding and should eventually cover most aspects of inspection, assembly automation and vehicle guidance that are presently handled by humans (note, however, that for safety, legal and social reasons, it may not prove acceptable for machines to take over control in all cases which they could theoretically handle). It should be pointed out that applications are limited particularly by available techniques for image acquisition and (as hinted above) by possibilities for cost-effective hardware implementation. For this reason, it is important that Chapters 23 and 24 should be read in conjunction with the present chapter. In this respect it is odd how many books on image analysis almost or even totally ignore one or other of these aspects (especially the former) as if they had no real importance.

20

Statistical Pattern Recognition

20.1 Introduction

The earlier chapters of this book have tackled the task of interpreting images on the basis that when suitable cues have been found, the positions of various objects and their identities will emerge in a natural and straightforward way. When the objects that appear in an image have simple shapes, just one stage of processing may be required—as in the case of circular washers on a conveyor. For more complex objects, location requires at least two stages—as when graph matching methods are used. For situations where the full complexity of three dimensions occurs, more subtle procedures are usually required, as has already been seen in Chapter 15. Indeed, the very ambiguity involved in interpreting the 2-D images from a set of 3-D objects generally requires cues to be sought and hypotheses to be made before any serious attempt can be made at the task. Thus cues are vital to keying into the complex data structures of many images. However, for simpler situations, concentration on small features is valuable in permitting image interpretation to be carried out efficiently and rapidly; neither must it be forgotten that in many applications of machine vision the task is made simpler by the fact that the main interest lies in specific types of object, e.g. the widgets to be inspected on a widget line.

However, the fact that it is often expedient to start by searching for cues means that the vision system could lock on to erroneous interpretations. For example, it is worth noting that Hough-based methods lead to interpretations based on the most prominent peaks in parameter space and that the maximal clique approach supports interpretations based on the largest number of feature matches between parts of an image and the object model. Clearly, noise, background clutter and other artefacts make it possible that an object will be seen (or its presence hypothesized) when none is there, while occlusions, shadows, etc. may cause an object to be missed altogether.

Hence a rigorous interpretation strategy would ideally try to find all possible feature and object matches, and should have some means of distinguishing between them. An attractive idea is that of not being satisfied with an interpretation until the *whole* of any image has been understood. However, enough has been seen in previous chapters (e.g. Chapter 10) to demonstrate that the computational load of such a strategy generally makes it impracticable, at least for real (e.g. outdoor) scenes. Yet there are some more constrained types of situation in which the strategy is worth considering and in which the various possible solutions can be evaluated carefully before a final interpretation is reached. These situations arise practically where small relevant parts of images can be segmented and interpreted in isolation. One such case is that of optical character recognition (OCR): a commonly used approach for tackling it is that of statistical pattern recognition (SPR).

The following sections study some of the important principles of SPR. A complete description of all the work that has been carried out in this area would take several volumes to cover and would be out of place here. Fortunately, SPR has been researched for more than three decades and has settled down sufficiently so that an overview chapter can serve a useful purpose. The reader is referred to several existing excellent texts for further details (Duda and Hart, 1973; Tou and Gonzalez, 1974; Devijver and Kittler, 1982). We start by describing the nearest neighbour approach to SPR and then go on to consider Bayes' decision theory which provides a more general model of the underlying process.

20.2   The nearest neighbour algorithm

The principle of the nearest neighbour (NN) algorithm is that of comparing input image[*] patterns against a number of paradigms and then classifying the input pattern according to the class of the paradigm that gives the closest match (Fig. 20.1). An instructive but rather trivial example is that shown in

---

[*] Note that a number of the methods discussed in this chapter are very general and can be applied to the recognition of widely different datasets, including for example speech and electrocardiograph waveforms.

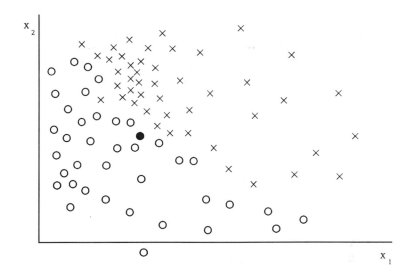

Fig. 20.1 Principle of the nearest neighbour algorithm for a two-class problem: O, class 1 training set patterns; ×, class 2 training set patterns; ●, test pattern.

Fig. 1.1. Here a number of binary patterns are presented to the computer in the training phase of the algorithm: then the test patterns are presented one at a time and compared bit by bit against each of the training patterns. It is clear that this gives a generally reasonable result, the main problems arising when (i) training patterns of different classes are close together in Hamming distance (i.e. they differ in too few bits to be readily distinguishable), and (ii) minor translations, rotations or noise cause variations that inhibit accurate recognition. More generally, problem (ii) means that the training patterns are insufficiently representative of what will appear during the test phase. The latter statement encapsulates an exceptionally important principle and it implies that there must be sufficient patterns in the training set for the algorithm to be able to generalize over all possible patterns of each class. However, problem (i) implies that patterns of two different classes may in some cases be so close as to be indistinguishable by any algorithm, and then it is inevitable that erroneous classifications will be made. It is seen below that this is because the underlying distributions in feature space overlap.

The example of Fig. 1.1 is rather trivial but nevertheless carries important lessons. It is to be noted that general images have many more pixels than those in Fig. 1.1 and also are not merely binary. However, since it is pertinent to simplify the data as far as possible to save computation, it is usual to concentrate on various features of a typical image and classify on the basis of these. One example is provided by a more realistic version of the OCR problem where characters have linear dimensions of at least 32 pixels (although we continue to assume that the characters have been located *reasonably* accurately so that it remains only to classify the subimages containing them). We can start by thinning the characters to their skeletons and making measurements on the skeleton nodes and limbs (see also Chapters 1 and 6): this gives (i) the numbers of nodes and limbs of various types, (ii) the lengths and relative orientations of limbs, and perhaps (iii) information on curvatures of limbs. Thus we arrive at a set of numerical features that describe the character in the subimage.

The general technique is now to plot the characters in the training set in a multidimensional feature space and to tag the plots with the classification index. Then test patterns are placed in turn in the feature space and classified according to the class of the nearest training set pattern. Clearly this generalizes the method adopted in Fig. 20.1. Note that in the general case the distance in feature space is no longer Hamming distance but some more general measure such as Mahalanobis distance (Duda and Hart, 1973). In fact, a problem arises since there is no reason why the different dimensions in feature space should contribute equally to distance: rather, they should each have different weights in order to match the physical problem more closely. The problem of weighting cannot be discussed in detail here and the reader is referred to other texts such as that by Duda and Hart (1973). Suffice it to say that with an appropriate definition of distance, the generalization of the method outlined above is adequate to cope with a variety of problems.

It will be clear that in order to achieve a suitably low error rate, large numbers of training set patterns are normally required. This then leads to significant storage and computation problems. Means have been found for reducing these problems by several important strategies. Notable amongst these is that of pruning the training set by eliminating patterns which are not near the boundaries of class regions in feature space, since such patterns do not materially help in reducing the misclassification rate.

An alternative strategy for obtaining equivalent performance at lower computational cost is to employ a piecewise linear or other functional classifier instead of the original training set. It will be clear that the NN method itself can be replaced, with no change in performance, by a set of planar decision surfaces which are the perpendicular bisectors (or their analogues in multidimensional space) of the lines joining pairs of training patterns of different classes that are on the boundaries of class regions. If this system of planar surfaces is simplified by any convenient means then the computational load may be reduced further (Fig. 20.2). This may be

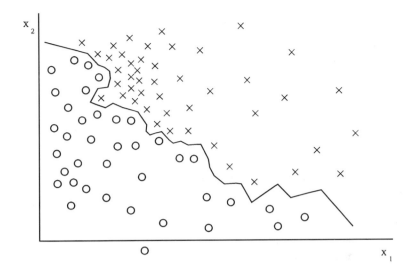

Fig. 20.2 Use of planar decision surfaces for pattern classification: in this example the "planar decision surface" reduces to a piecewise linear decision boundary in two dimensions. Once the decision boundary is known, the training set patterns themselves need no longer be stored.

achieved either indirectly by some smoothing process, as implied above, or directly by finding training procedures that act to update the positions of decision surfaces immediately on receipt of each new training set pattern. The latter approach is in many ways more attractive, since it drastically cuts down storage requirements—although it must be confirmed that a training procedure is selected which converges sufficiently rapidly. Again, discussion of this well-researched topic is left to other texts (Nilsson, 1965; Duda and Hart, 1973; Devijver and Kittler, 1982).

We now turn to a more generalized approach—that of Bayes' decision theory—since this underpins all the possibilities thrown up by the NN method and its derivatives.

20.3 Bayes' decision theory

The basis of Bayes' decision theory will now be examined. If we are trying to get a computer to classify objects, a sound approach is to get it to measure some prominent feature of each object such as its length and to use this feature as an aid to classification. Sometimes such a feature may give very little indication of the pattern class—perhaps because of the effects of manufacturing variation. For example, a handwritten character may be so ill-formed that its features are of little help in interpreting it; it then becomes much more reliable to make use of the known relative frequencies of letters, or to invoke context: in fact, either of these strategies can give a greatly increased probability of correct interpretation. In other words, when feature measurements are found to be giving an error rate above a certain threshold, it is more reliable to employ the *a priori* probability of a given pattern appearing.

The next step in improving recognition performance is to combine the information from feature measurements and from *a priori* probabilities: this is achieved by applying Bayes' rule. For a single feature x this takes the form:

$$P(C_i | x) = p(x | C_i)P(C_i)/p(x) \tag{20.1}$$

where

$$p(x) = \sum_j p(x | C_j)P(C_j) \tag{20.2}$$

Mathematically, the variables here are (i) the *a priori* probability of class C_i, $P(C_i)$; (ii) the probability density for feature x, $p(x)$; (iii) the class-conditional probability density for feature x in class C_i, $p(x | C_i)$—i.e. the probability that feature x arises for objects known to be in class C_i; and (iv) the *a posteriori* probability of class C_i when x is observed, $P(C_i | x)$.

The notation $P(C_i | x)$ is by now a standard one, being defined as the probability that the class is C_i when the feature is known to have the value x. Bayes' rule now says that to find the class of an object we need to know two sets of information about the objects that might be viewed: the first is the basic probability $P(C_i)$ that a particular class might arise; the second is the distribution of values of the feature x for each class. Fortunately, each of these sets of information can be found straightforwardly by observing a sequence of objects, for example as they move along a conveyor. Such a sequence of objects is again called the training set.

Many common image analysis techniques give features that may be used to help identify or classify objects. These include the area of an object, its perimeter, the numbers of holes it possesses, and so on. It is important to note that classification performance may be improved not only by making use of the *a priori* probability but also by employing a number of features simultaneously. Generally, increasing the number of features helps to resolve object classes and reduce classification errors (Fig. 20.3): however, the error rate is rarely reduced to zero merely by adding more and more features, and indeed the situation eventually deteriorates for reasons explained in Section 20.5.

Bayes' rule can be generalized to cover the case of a generalized feature x, in multidimensional feature-space, by using the modified formula:

$$P(C_i|\mathbf{x}) = p(\mathbf{x}|C_i)P(C_i)/p(\mathbf{x}) \tag{20.3}$$

where $P(C_i)$ is the *a priori* probability of class C_i, and $p(\mathbf{x})$ is the overall probability density for feature vector $\mathbf{x}$:

$$p(\mathbf{x}) = \sum_j p(\mathbf{x}|C_j)P(C_j) \tag{20.4}$$

The classification procedure is then to compare the values of all the $P(C_j|\mathbf{x})$ and to classify an object as class C_i if:

$$P(C_i|\mathbf{x}) > P(C_j|\mathbf{x}) \quad \text{for all} \quad j \neq i \tag{20.5}$$

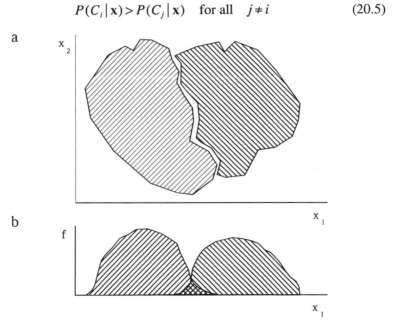

Fig. 20.3 Use of several features to reduce classification errors: (a) the two regions to be separated in 2-D (x_1, x_2) feature space; (b) frequencies of occurrence of the two classes when the pattern vectors are projected onto the x_1-axis. Clearly, error rates will be high when either feature is used on its own but will be reduced to a low level when both features are employed together.

20.4   Relation of the nearest neighbour and Bayes' approaches

When Bayes' theory is applied to simple pattern recognition tasks, it is immediately clear that *a priori* probabilities are important in determining the final classification of any pattern, since these probabilities arise explicitly in the calculation. However, this is not so for the NN type of classifier. Indeed, the whole idea of the NN classifier appears to be to get away from such considerations, instead classifying patterns on the basis of training set patterns that lie nearby in feature space. However, there must be a definite answer to the question of whether *a priori* probabilities are or are not taken into account *implicitly* in the NN formulation, and therefore of whether an adjustment needs to be made to the NN classifier to minimize the error rate. Since it is clearly important to have a categorical statement of the situation, the next subsection is devoted to providing such a statement, together with necessary analysis.

20.4.1   Mathematical statement of the problem

This subsection considers in detail the relation between the NN algorithm and Bayes' theory. For simplicity (and with no ultimate loss of generality), we here take all dimensions in feature space to have equal weight, so that the measure of distance in feature space is not a complicating factor.

For greatest accuracy of classification, many training set patterns will be used and it will be possible to define a density of training set patterns in feature space, $D_i(\mathbf{x})$, for position $\mathbf{x}$ in feature space and class C_i. Clearly, if $D_k(\mathbf{x})$ is high at position $\mathbf{x}$ in class C_k, then training set patterns lie close together and a test pattern at $\mathbf{x}$ will be likely to fall in class C_k. More particularly, if

$$D_k(\mathbf{x}) = \max_i D_i(\mathbf{x}) \tag{20.6}$$

then our basic statement of the NN rule implies that the class of a test pattern $\mathbf{x}$ will be C_k.

However, according to the outline given above, this analysis is flawed in not showing explicitly how the classification depends on the *a priori* probability of class C_k. To proceed, note that $D_i(\mathbf{x})$ is closely related to the conditional probability density $p(\mathbf{x} \mid C_i)$ that a training set pattern will appear at position $\mathbf{x}$ in feature space if it is in class C_i. Indeed, the $D_i(\mathbf{x})$ are merely non-normalized values of the $p(\mathbf{x} \mid C_i)$:

$$p(\mathbf{x} \mid C_i) = D_i(\mathbf{x}) \Big/ \int D_i(\mathbf{x}) \, d\mathbf{x} \tag{20.7}$$

The standard Bayes' formulae (equations (20.3) and (20.4)) can now be used to calculate the *a posteriori* probability of class C_i.

So far it has been seen that the *a priori* probability should be combined with the training set density data before valid classifications can be made using the NN rule: as a result it seems invalid merely to take the nearest training set pattern in feature space as an indicator of pattern class. However, note that when clusters of training set patterns and the underlying within-class distributions scarcely overlap, there is anyway a rather low probability of error in the overlap region, and the result of using $p(\mathbf{x} \mid C_i)$ rather than $P(C_i \mid \mathbf{x})$ to indicate class often introduces only a very small bias in the decision surface. Hence, although totally invalid *mathematically*, the error introduced need not always be disastrous.

We now consider the situation in more detail, finding how the need to multiply by the *a priori* probability affects the NN approach. In fact, multiplying by the *a priori* probability can be achieved either *directly*, by multiplying the densities of each class by the appropriate $P(C_i)$, or *indirectly*, by providing a suitable amount of additional training for classes with high *a priori* probability. It may now be seen that the amount of additional training required is *precisely* the amount that would be obtained if the training set patterns were allowed to appear with their natural frequencies (see equations below). For example, if objects of different classes are moving along a conveyor, we should not first separate them and then train with equal numbers of patterns from each class: we should instead allow them to proceed normally and train on them all at their normal frequencies of occurrence in the training stream. Clearly, if training set patterns do not appear for a time with their proper natural frequencies, this will introduce a bias into the properties of the classifier. Thus we must make every effort to permit the training set to be representative not only of the *types* of pattern of each class but also of the *frequencies* with which they are presented to the classifier during training.

The above ideas for *indirect* inclusion of *a priori* probabilities may be expressed as follows:

$$P(C_i) = \frac{\int D_i(\mathbf{x}) \, d\mathbf{x}}{\sum_j \int D_j(\mathbf{x}) \, d\mathbf{x}} \tag{20.8}$$

Hence

$$P(C_i|\mathbf{x}) = \frac{D_i(\mathbf{x})}{\left(\sum_j \int D_j(\mathbf{x})\,d\mathbf{x}\right)p(\mathbf{x})} \tag{20.9}$$

where

$$p(\mathbf{x}) = \frac{\sum_k D_k(\mathbf{x})}{\sum_j \int D_j(\mathbf{x})\,d\mathbf{x}} \tag{20.10}$$

Substituting for $p(\mathbf{x})$ now gives:

$$P(C_i|\mathbf{x}) = D_i(\mathbf{x}) \Big/ \sum_k D_k(\mathbf{x}) \tag{20.11}$$

so the decision rule to be applied is to classify an object as class C_i if

$$D_i(\mathbf{x}) > D_j(\mathbf{x}) \quad \text{for all} \quad j \neq i \tag{20.12}$$

The following conclusions have now been arrived at:

(i)   The NN classifier may well not include *a priori* probabilities and hence could give a classification bias.

(ii) It is in general wrong to train a NN classifier in such a way that an equal number of training set patterns of each class are applied.

(iii) The correct way to train a NN classifier is to apply training set patterns at the natural rates at which they arise in raw training set data.

The third conclusion is perhaps the most surprising and the most gratifying. Essentially it adds further fire to the principle that training set patterns should be representative of the class distributions from which they are taken, although we now see that it should be generalized to the following: *training sets should be fully representative of the populations from which they are drawn*, where "fully representative" includes ensuring that the frequencies of occurrence of the various classes are representative of those in the whole population of patterns. Phrased in this way, the principle becomes a general one which is relevant to many types of trainable classifier (including, in particular, the MLP type of Ann–see Chapter 21).

20.4.2 The importance of the nearest neighbour classifier

The NN classifier is important in being perhaps the simplest of all classifiers to implement on a computer: in addition, it has the advantage of being guaranteed to give an error rate within a factor of two of the ideal error rate (obtainable with a Bayes' classifier). By modifying the method to base classification of any test pattern on the most commonly occurring class amongst the k nearest training set patterns (giving the "k-NN" method), the error rate can be reduced further until it is arbitrarily close to that of a Bayes' classifier (note that equation (20.11) can be interpreted as covering this case too). However, both the NN and (*a fortiori*) the k-NN methods have the disadvantage that they often require enormous storage to record enough training set pattern vectors, and correspondingly large amounts of computation to search through them to find an optimal match for each test pattern—hence necessitating the pruning and other methods mentioned earlier for cutting down the load.

20.5 The optimum number of features

It was stated in Section 20.3 that error rates can be reduced by increasing the number of features used by a classifier, but that there is a limit to this, after which performance actually deteriorates. We here consider why this should happen. Basically, the reason is similar to the situation where many parameters are used to fit a curve to a set of D data points. As the number of parameters P is increased, the fit of the curve becomes better and better, and in general becomes perfect when $P = D$. However, by that stage the significance of the fit is poor, since the parameters are no longer overdetermined and no averaging of their values is taking place. Essentially, all the noise in the raw input data is being transferred to the parameters. The same thing happens with training set patterns in feature space. Eventually, training set patterns are so sparsely packed in feature space that the test patterns have reduced probability of being nearest to a pattern of the same class, so error rates become very high. This situation can also be regarded as due to a proportion of the features having negligible statistical significance, i.e. they add little additional information and serve merely to add uncertainty to the system.

However, an important factor is that the optimum number of features depends on the amount of training a classifier receives. If the number of training set patterns is increased, more evidence is available to support the determination of a greater number of features and hence to provide more accurate classification of test patterns. Indeed, in the limit of very large numbers of training set patterns, performance continues to increase as the number of features is increased.

This situation was first clarified by Hughes (1968) and verified in the case of n-tuple pattern recognition (a variant of the NN classifier due to Bledsoe and Browning (1959)) by Ullmann (1969). Both workers produced clear curves showing the initial improvement in classifier performance as the number of features increased, this improvement being followed by a fall in performance for large numbers of features.

Before leaving this topic, note that the above arguments relate to the number of features that should be used but not to their selection. Clearly some features are more significant than others, the situation being very data-dependent. It is left as a topic for experimental tests to determine in a particular case which subset of features will minimize classification errors (see also Chittineni, 1980).

20.6   Cost functions and error–reject tradeoff

In the foregoing sections it has been implied that the main criterion for correct classification is that of maximum *a posteriori* probability. However, although probability is always a relevant factor, in a practical engineering environment it is often more important to minimize costs. Hence it is necessary to compare the costs involved in making correct or wrong decisions. Such considerations can be expressed mathematically by invoking a loss function $L(C_i | C_j)$ which represents the cost involved in making a decision C_i when the true class for feature $\mathbf{x}$ is C_j.

To find a modified decision rule based on minimizing costs we first define a function known as the conditional risk:

$$R(C_i | \mathbf{x}) = \sum_j L(C_i | C_j) P(C_j | \mathbf{x}) \tag{20.13}$$

This function expresses the expected cost of deciding on class C_i when $\mathbf{x}$ is observed. As it is wished to minimize this function, we now decide on class C_i only if:

$$R(C_i | \mathbf{x}) < R(C_j | \mathbf{x}) \quad \text{for all} \quad j \neq i \tag{20.14}$$

If we were to choose a particularly simple cost function, of the form:

$$L(C_i|C_j) = \begin{cases} 0 & \text{for} \quad i = j \\ 1 & \text{for} \quad i \neq j \end{cases} \qquad (20.15)$$

then the result would turn out to be identical to the previous probability-based decision rule, relation (20.5). Clearly, it is only when certain errors lead to relatively large (or small) costs that it pays to deviate from the normal decision rule. Such cases arise when we are in a hostile environment and must, for example, give precedence to the sound of an enemy tank over that of other vehicles—it is better to be oversensitive and risk a false alarm than to retain a small chance of not noticing the hostile agent. Similarly, on a production line it may in some circumstances be better to reject a small number of good products than to risk selling a defective product. Cost functions therefore permit classifications to be biased in favour of a safe decision in a rigorous, predetermined and controlled manner, and the desired balance of properties obtained from the classifier.

Another way of minimizing costs is to arrange for the classifier to recognize when it is "doubtful" about a particular classification, because two or more classes are almost equally likely. Then one solution is to make a safe decision, the decision plane in feature space being biased away from its position for maximum probability classification. An alternative is to reject the pattern, i.e. place it into an "unknown" category: in that case some other means can be employed for making an appropriate classification. Such a classification could be made by going back to the original data and measuring further features but in many cases it is more appropriate for a human operator to be available to make the final decision. Clearly, the latter approach is more expensive and so introducing a "reject" classification can incur a relatively large cost factor. A further problem is that the error rate[*] is reduced only by a fraction of the amount that the rejection rate[*] is increased. Indeed, in a simple two-class system, the initial decrease in error rate is only one-half the initial increase in reject rate (i.e. a 1% decrease in error rate is obtained only at the expense of a 2% increase in reject rate), and the situation gets rapidly worse as progressively lower error rates are attempted (Fig. 20.4). Thus very careful cost analysis of the error–reject tradeoff curve must be made before an optimal scheme can be developed.

---

[*] All error and reject rates are assumed to be calculated as proportions of the total number of test patterns to be classified.

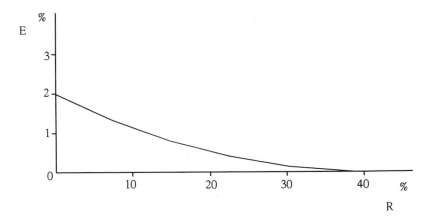

Fig. 20.4 An error–reject tradeoff curve (E, error rate; R, reject rate). In this example, the error rate E drops substantially to zero for a reject rate R of 40%. More usually E cannot be reduced to zero until R is 100%.

Finally, it should be noted that the overall error rate of the classification system depends on the error rate of the classifier that examines the rejects (e.g. the human operator), and this needs to be taken into account in determining the exact tradeoff to be used.

20.7 Cluster analysis

20.7.1 Supervised and unsupervised learning

In the earlier parts of this chapter, we made the implicit assumption that the classes of all the training set patterns are known, and in addition that they should be used in training the classifier. Indeed, this assumption might be thought of as inescapable. However, classifiers may actually use two approaches to learning—*supervised learning* (in which the classes are known and used in training) and *unsupervised learning* (in which they are either unknown or else known and not used in training). Unsupervised learning can frequently be advantageous in practical situations. For example, a human operator is not required to label all the products coming along a conveyor, as the computer can find out for itself both how many classes of product there are and which categories they fall into: in this way considerable operator effort is eliminated; in addition, it is not unlikely that a number of errors would thereby be circumvented. Unfortunately, unsupervised learning involves a number of difficulties, as will be seen in the following subsections.

Before proceeding, it is worth stating two other reasons why unsupervised learning is useful. First, when the characteristics of objects vary with time—for example beans changing in size and colour as the season develops—it will be necessary to track these characteristics within the classifier, and unsupervised learning provides an excellent means of approaching this task. Second, when setting up a recognition system, the characteristics of objects, and in particular their most important parameters (e.g. from the point of view of quality control) may well be unknown, and it will be useful to gain some insight into the nature of the data. Thus types of fault will need to be logged, and permissible variants on objects will need to be noted. As an example, many OCR fonts (such as Times Roman) have a letter *a* with a stroke bent over the top from right to left, though other fonts (such as Monaco) do not have this feature. An unsupervised classifier will be able to flag this up by locating a cluster of training set patterns in a totally separate part of feature space (see Fig. 20.5). In general, unsupervised learning is about the location of clusters in feature space.

20.7.2   Clustering procedures

As indicated above, an important reason for performing cluster analysis is characterization of the input data. However, the underlying motivation is normally to classify test data patterns reliably. To achieve these aims, it will

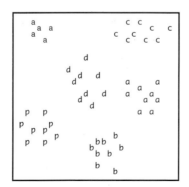

Fig. 20.5   Location of clusters in feature space. The letters correspond to samples of characters taken from various fonts. The small cluster of *a*'s with strokes bent over the top from right to left appear at a separate location in feature space: this type of deviation should be detectable by cluster analysis.

be necessary both to partition feature space into regions corresponding to significant clusters, and to label each region (and cluster) according to the type of data involved. In practice, this can happen in two ways:

(i) By performing cluster analysis, and then labelling the clusters by specific queries to human operators on the classes of a small number individual training set patterns.

(ii) By performing supervised learning on a small number of training set patterns, and then performing unsupervised learning to expand the training set to realistic numbers of examples.

In either case, we see that ultimately we do not escape from the need for supervised classification. However, by placing the main emphasis on unsupervised learning we save on tedium and on the possibility of pre-conceived ideas on possible classes from affecting the final recognition performance.

Before proceeding further, it will be useful to notice that there are cases where we may have absolutely no idea in advance about the number of clusters in feature space: this occurs in classifying the various regions in satellite images. Such cases are in direct contrast with applications such as OCR or recognizing chocolates being placed in a chocolate box.

Unfortunately, cluster analysis involves a number of very significant problems. Not least is the visualization problem. First, in one, two, or even three dimensions, we can easily visualize and decide on the number and location of any clusters, but this capability is misleading: we cannot extend this capability to feature spaces of many dimensions. Second, computers do not visualize as we do, and special algorithms will have to be provided to enable them to do so. Computers could be made to emulate our capability in low-dimensional feature spaces, but a combinatorial explosion would occur if we attempted this for high-dimensional spaces. This means that we will have to develop algorithms which operate on *lists* of feature vectors, if we are to produce automatic procedures for cluster location.

Available algorithms for cluster analysis fall into two main groups— *agglomerative* and *divisive*. Agglomerative algorithms start by taking the individual feature points (training set patterns, excluding class) and progressively grouping them together according to some similarity function until a suitable target criterion is reached. Divisive algorithms start by taking the whole set of feature points as a single large cluster, and progressively dividing it until some suitable target criterion is reached. Let us assume that there are P feature points. Then, in the worst case, the number of comparisons between pairs of individual feature point positions which will be required to decide whether to combine a pair of clusters in an agglomerative algorithm will be:

$$\binom{P}{2} = \frac{1}{2} P(P-1) \tag{20.16}$$

while the number of iterations required to complete the process will be of order $P - k$ (here we are assuming that the final number of clusters to be found is k, where $k \leqslant P$. On the other hand, for a divisive algorithm, the number of comparisons between pairs of individual feature point positions will be reduced to:

$$\binom{k}{2} = \frac{1}{2} k(k-1) \tag{20.16}$$

while the number of iterations required to complete the process will be of order k.

Although it would appear that divisive algorithms require far less computation than agglomerative algorithms, this is not so. This is because any cluster containing p feature points will have to be examined for a huge number of potential splits into subclusters, the actual number being of order:

$$\sum_{q=1}^{p} \binom{p}{q} = \sum_{q=1}^{p} \frac{p!(p-q)!}{q!} \tag{20.18}$$

This means that in general the agglomerative approach will have to be adopted. In fact, the type of agglomerative approach outlined above is exhaustive and rigorous, and a less exacting, iterative approach can be used. First, a suitable number k of cluster centres are set (these can be decided from *a priori* considerations, or by making arbitrary choices). Second, each feature vector is assigned to the closest cluster centre. Third, the cluster centres are recalculated. This process is repeated if any feature points have moved from one cluster to another during the iteration, though termination can also be instituted if the quality of clustering ceases to improve. The overall algorithm, which was originally due to Forgy (1965), is given in Table 20.1.

It will be apparent that the effectiveness of this algorithm will be highly data-dependent—in particular, with regard to the order in which the data points are presented. In addition, the result could be oscillatory or nonoptimal (in the sense of not arriving at the best solution). This could happen if at

Table 20.1 Basis of Forgy's algorithm for cluster analysis.

```
choose target number k of clusters;
set initial cluster centres;
calculate quality of clustering;
repeat
  assign each data point to the closest cluster centre;
  recalculate cluster centres;
  recalculate quality of clustering;
until no further change in the clusters or the quality of the clusters;
```

any stage a single cluster centre arose near the centre of a pair of small clusters. In addition, the method gives no indication of the most appropriate number of clusters. Accordingly, a number of variant and alternative algorithms have been devised. One such algorithm is the ISODATA algorithm (Ball and Hall, 1966): this is similar to Forgy's method, but is able to merge clusters which are close together, and to split elongated clusters.

Another disadvantage of iterative algorithms is that it may not be obvious when to get them to terminate: as a result, they are liable to be too computation intensive. Thus there has been some support for noniterative algorithms. MacQueen's *k-means* algorithm (MacQueen, 1967) is one of the best known noniterative clustering algorithms: it involves two runs over the data points, one being required to find the cluster centres and the other to finally classify the patterns (see Table 20.2). Again, the choice of which data points are to act as the initial cluster centres can be either arbitrary or on some more informed basis.

Noniterative algorithms are, as indicated earlier, very dependent on the order of presentation of the data points. With image data this is especially

Table 20.2 Basis of MacQueen's *k-means* algorithm.

```
choose target number k of clusters;
set the k initial cluster centres at k data points;
for all other data points do (* first pass *)
  begin
    assign data point to closest cluster centre;
    recalculate relevant cluster centre;
  end;
for all data points do (* second pass *)
  reassign data point to closest cluster centre;
```

problematic, as the first few data points are quite likely to be similar (e.g. all derived from sky or other background pixels). A useful way of overcoming this problem is to randomize the choice of data points, so that they can arise from anywhere in the image. In general, noniterative clustering algorithms are less effective than iterative algorithms, because they are overinfluenced by the order of presentation of the data.

Overall, the main problem with the algorithms described above is the lack of indication they give of the most appropriate value of k. However, if a range of possible values for k is known, all of them can be tried (using one of the above algorithms), and the one giving the best performance in respect of some suitable target criterion can be taken as providing an optimal result. In that case, we will have found the set of clusters which, in some specified sense, gives the best overall description of the data. Alternatively, some method of analysing the data to determine k can be used before final cluster analysis, e.g. using the k-means algorithm: the Zhang and Modestino (1990) approach falls into this category.

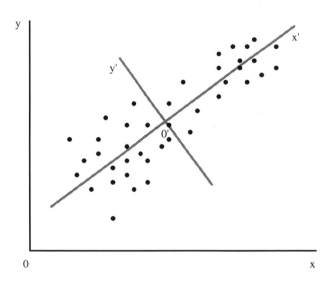

Fig. 20.6 Illustration of principal components analysis. The dots represent patterns in feature space and are initially measured relative to the x- and y-axes. Then the sample mean is located at $0'$, and the direction $0'x'$ of the first principal component is found as the direction along which the variance is maximized. The direction $0'y'$ of the second principal component is normal to $0'x'$; in a higher dimensional space it would be found as the direction normal to $0'x'$ along which the variance is maximized.

20.8 Principal components analysis

Closely related to cluster analysis is the concept of data representation. One powerful way of approaching this task is that of principal components analysis. This involves finding the mean of a cluster of points in feature space, and then finding the principal axes of the cluster in the following way. First an axis is found which passes through the mean position and which gives the maximum variance when the data are projected onto it. Then a second such axis is found which maximizes variance in a direction normal to the first. This process is carried out until a total of N principal axes have been found for an N-dimensional feature space. The process is illustrated in Fig. 20.6. In fact, the process is entirely mathematical and need not be undertaken in the strict sequence indicated above: it merely involves finding a set of orthogonal axes which diagonalizes the covariance matrix.

The covariance matrix for the input population is defined as:

$$C = E\{(\mathbf{x}_{(p)} - \mathbf{m})(\mathbf{x}_{(p)} - \mathbf{m})^{\mathrm{T}}\} \tag{20.19}$$

where $\mathbf{x}(p)$ is the location of the pth data point, and $\mathbf{m}$ is the mean of the P data points; $E\{\cdots\}$ indicates expectation value for the underlying population. We can estimate C from the equations:

$$C = \frac{1}{P} \sum_{p=1}^{P} \mathbf{x}_{(p)} \mathbf{x}_{(p)}^{\mathrm{T}} - \mathbf{m}\mathbf{m}^{\mathrm{T}} \tag{20.20}$$

$$\mathbf{m} = \frac{1}{P} \sum_{p=1}^{P} \mathbf{x}_{(p)} \tag{20.21}$$

Since C is real and symmetric, it is possible to diagonalize it using a suitable orthogonal transformation matrix $\mathbf{A}$, obtaining a set of N orthonormal eigenvectors $\mathbf{u}_i$ with eigenvalues λ_i given by:

$$C\mathbf{u}_i = \lambda_i \mathbf{u}_i \qquad (i = 1, 2, \ldots, N) \tag{20.22}$$

The vectors $\mathbf{u}_i$ are derived from the original vectors $\mathbf{x}_i$ by:

$$\mathbf{u}_i = \mathbf{A}(\mathbf{x}_i - \mathbf{m}) \tag{20.23}$$

and the inverse transformation needed to recover the original data vectors is:

$$\mathbf{x}_i = \mathbf{m} + \mathbf{A}^{\mathrm{T}}\mathbf{u}_i \tag{20.24}$$

Here we have recalled that, for an orthogonal matrix:

$$\mathbf{A}^{-1} = \mathbf{A}^{\mathrm{T}}$$

(20.25)

In fact it may be shown that $\mathbf{A}$ is the matrix whose rows are formed from the eigenvectors of $\mathbf{C}$, and that the diagonalized covariance matrix $\mathbf{C}'$ is given by:

$$\mathbf{C}' = \mathbf{A}\mathbf{C}\mathbf{A}^{\mathrm{T}}$$

(20.26)

so that:

$$\mathbf{C}' = \begin{bmatrix} \lambda_1 & & & & 0 \\ & \lambda_2 & & & \\ & & \cdot & & \\ & & & \cdot & \\ 0 & & \cdot & \cdot & \lambda_N \end{bmatrix}$$

(20.27)

Note that in an orthogonal transformation, the trace of a matrix remains unchanged. Thus the trace of the input data is given by:

$$\text{trace } \mathbf{C} = \text{trace } \mathbf{C}' = \sum_{i=1}^{N} \lambda_i = \sum_{i=1}^{N} \sigma_i^2$$

(20.28)

where we have interpreted the λ_i as the variances of the data in the directions of the principal component axes (note that for a real symmetric matrix, the eigenvalues are all real and positive).

In what follows we shall assume that the eigenvalues have been placed in an ordered sequence, starting with the largest. In that case, λ_1 represents the most significant characteristic of the set of data points, with the later eigenvalues representing successively less significant characteristics. We could even go so far as to say that, in some sense, λ_1 represents the most interesting characteristic of the data, while λ_N would be largely devoid of "interest". More practically, if we ignored λ_N, we might not lose much useful information, and indeed the last few eigenvalues would frequently represent characteristics which are not statistically significant and are essentially noise. For these reasons, principal components analysis is commonly used for reduction in the dimensionality of the feature space from N to some lower value N'. In some applications, this would be taken as leading to a useful amount of data compression. In other applications, it would be taken as providing a reduction in the enormous redundancy present in the input data.

We can quantify these results by writing the variance of the data in the reduced dimensionality space as:

$$\text{trace } (\mathbf{C}')_{\text{reduced}} = \sum_{i=1}^{N'} \lambda_i = \sum_{i=1}^{N'} \sigma_i^2 \qquad (20.29)$$

Not only is it now clear why this leads to reduced variance in the data, but also we can see that the mean square error obtained by making the inverse transformation (equation (20.24)) will be:

$$\overline{\varepsilon^2} = \sum_{i=1}^{N} \sigma_i^2 - \sum_{i=1}^{N'} \sigma_i^2 = \sum_{i=N'+1}^{N} \sigma_i^2 \qquad (20.30)$$

One application in which principal components analysis has become especially important is the analysis of multispectral images, e.g. from earth-orbiting satellites. Typically, there will be six separate input channels (e.g. three colour and three infrared), each providing an image of the same ground region. If these images are 512×512 pixels in size, there will be about a quarter of a million data points and these will have to be inserted into a six-dimensional feature space. After finding the mean and covariance matrix for these data points, the latter is diagonalized and a total of six principal component images can be formed. Commonly, only two or three of these will contain immediately useful information, and the rest can be ignored. (For example, the first three of the six principal component images may well possess 95% of the variance of the input images.) Ideally, the first few principal component images in such a case will highlight such areas as fields, roads and rivers—precisely the information that is required for map-making or other purposes. In general, the vital pattern recognition tasks can be aided *and* considerable savings in storage can be achieved on the incoming image data by attending to just the first few principal components.

Finally, it is as well to note that principal components analysis really provides a particular form of data representation. In itself it does not deal with pattern classification, and methods which are required to be useful for the latter type of task must possess useful discrimination. Thus selection of features simply because they possess the highest variability does not mean that they will necessarily perform well in pattern classifiers. Another important factor which is relevant to the whole study of data analysis in feature space is the scales of the various features. Often, these will be an extremely variegated set, including length, weight, colour, numbers of holes, and so on. Clearly, such features will have no special comparability and are unlikely even to be measurable in the same units. This means that placing them in the same feature space and assuming that the scales on the various axes should have the same weighting factors must be invalid. One way of tackling this problem is to normalize the individual features to some standard scale given by measuring their variances. Such a procedure will naturally totally change the results of principal components calculations, and further mitigates against principal components methodology being used thoughtlessly. On the other hand, there are some occasions on which different features can be compatible, and where principal components analysis can be performed without such worries: one such situation is where all the features are pixel intensities in the same window (this case is discussed in Section 22.5).

20.9   The relevance of probability in image analysis

Having seen the success of Bayes' theory in pointing to apparently absolute answers in the interpretation of certain types of image, it is attractive to consider complex scenes in terms of the probabilities of various interpretations and the likelihood of a particular interpretation being the correct one. Given a sufficiently large number of such scenes and their interpretations, it seems that it ought to be possible to use them to train a suitable classifier. However, practical interpretation in real time is quite another matter. The eye–brain system, that well-known real-time scene understanding machine, does not appear to operate in a manner corresponding to the algorithms we have studied. Instead, it appears to pay attention to various parts of an image in a nonpredetermined sequence of "fixations" of the eye, interrogating various parts of the scene in turn and using the newly acquired information to work out where the next piece of relevant information is to come from. Clearly it is employing a process of *sequential pattern recognition*, which saves effort overall by progressively building up a store of knowledge about relevant parts of the scene—and at the same time forming and testing hypotheses about its structure.

   The above process can be modelled as one of modifying and updating the *a priori* probabilities as analysis progresses. This is an inherently powerful process, since the eye is thereby not tied to "average" *a priori* probabilities for *all* scenes but is able to use information in a particular scene to improve on the average *a priori* probabilities. Another factor making this a powerful process is that it is not possible to know the *a priori* probabilities for sequences of real, complex scenes at all accurately. Hence the methods of this chapter are unlikely to be applicable in more complex situations. In such cases, the concept of probability is a nice idea but should probably be treated lightly.

However, the concept of probability is useful when it can validly be applied: at present, we can sum up the possible applications as those where a restricted range of images can arise, such that the consequent image description contains very few bits of information—namely those forming the various pattern class names. Thus the image data are particularly impoverished, containing relatively little structure: structure can here be defined as relationships between relevant parts of an image which have to be represented in the output data. If such structure is to be present, then it implies recognition of several parts of an image, coupled with recognition of their relation: when only rudimentary structure is present, the output data will contain only a few times the amount of output data of a set of simple classes, whereas more complicated cases will be subject to a combinatorial explosion in the amount of output data. SPR is a topic that is characterized by probabilistic interpretation when there is a many-to-few relation between image and interpretation rather than a many-to-many relation. Thus the input data need to be highly controlled before probabilistic interpretation is possible and SPR can be applied. It is left to other works to consider situations where images are best interpreted as verbal descriptions of their contents and structures. Note, however, that the nature of such descriptions is influenced strongly by the *purpose* for which the information is to be used.

20.10 Concluding remarks

The methods of this chapter make it rather surprising that so much of image processing and analysis is possible without any reference to *a priori* probabilities. It seems likely that this situation is due to several factors: (i) expediency, and in particular the need for speed of interpretation; (ii) the fact that algorithms are designed by humans who have knowledge of the types of input data and thereby incorporate *a priori* probabilities implicitly, for example via the application of suitable threshold values; and (iii) tacit recognition of the situation outlined in the last section, that probabilistic methods have limited validity and applicability. In practice (following on from the ideas of the last section) it is only at the stage of simple image structure and contextual analysis that probabilistic interpretations come into their own. This explains why SPR is not covered in more depth in this book.

Nonetheless, SPR is extremely valuable within its own range of utility. This includes identifying objects on conveyors and making value judgements of their quality, reading labels and codes, verifying signatures, checking fingerprints, and so on. Indeed, the number of distinct applications of SPR is impressive and it forms an essential counterpart to the other methods described in this book.

This chapter has concentrated mainly on the supervised learning approach to SPR. However, unsupervised learning is also vitally important, particularly when training on huge numbers of samples (e.g. in a factory environment) is involved. The section on this topic should therefore not be ignored as a minor and insignificant perturbation: much the same comments apply to the subject of principal components analysis which has had an increasing following in many areas of machine vision—see Section 20.8. Neither should it go unnoticed that these topics link in strongly with those of the following chapter on biologically motivated pattern recognition methods, with artificial neural networks playing a very powerful role.

20.11   Bibliographical and historical notes

Although the subject of statistical pattern recognition tends not to be at the centre of attention in image analysis work[*], it provides an important background—especially in the area of automated visual inspection where decisions continually have to be made on the adequacy of products. Most of the relevant work on this topic was already in place by the early 1970s, including the work of Hughes (1968) and Ullmann (1969) relating to the optimum number of features to be used in a classifier. At that stage a number of important volumes appeared—see for example, Duda and Hart (1973), Ullmann (1973), Batchelor (1974) and Tou and Gonzalez (1974), of which the first is still used very widely. Much later and more up to date is the volume by Devijver and Kittler (1982), although the intervening years do not seem to have changed the "shape" of the subject markedly.

In fact, the use of SPR for image interpretation dates from the 1950s. For example, in 1959 Bledsoe and Browning developed the n-tuple method of pattern recognition, which turned out (Ullmann, 1973) to be a form of NN classifier; however, it has been useful in leading to a range of simple hardware machines based on RAM (n-tuple) lookups (see for example,

---

[*]Note, however, that it is vital to the analysis of multispectral data from satellite imagery: see for example, Landgrebe (1981).

Aleksander *et al.*, 1984), thereby demonstrating the importance of marrying algorithms and readily implementable architectures.

Many of the most important developments in this area have probably been those comparing the detailed performance of one classifier with another, particularly with respect to cutting down the amount of storage and computational effort. Papers in these categories include those by Hart (1968) and Devijver and Kittler (1980). Oddly, there appeared to be no overt mention in the literature of how *a priori* probabilities should be used with the NN algorithm, until the author's paper on this topic (Davies, 1988f): see Section 20.4.

On the unsupervised approach to SPR, Forgy's (1965) method for clustering data was soon followed by the now famous ISODATA approach of Ball and Hall (1966), and then by MacQueen's (1967) k-means algorithm. Much related work ensued, and this has been summarized by Jain and Dubes (1988), which has by now become a classic text. However, cluster analysis is an exacting process and various workers have felt the need to push the subject further forward: e.g. Postaire and Touzani (1989) required more accurate cluster boundaries; Jolion and Rosenfeld (1989) wanted better detection of clusters in noise; Chauduri (1994) needed to cope with time-varying data; and Juan and Vidal (1994) required faster k-means clustering. It should be noted that all this work can be described as conventional, and did not involve the use of robust statistics *per se*. However, elimination of outliers is central to the problem of reliable cluster analysis: for a discussion of this aspect of the problem, see Appendix D and the references listed therein.

For a more complete view of the modern state of the subject, it is necessary to take into account recent work on artificial neural networks (see Chapter 21).

20.12 Problems

1. Show that if the cost function of equation (20.15) is chosen, then the decision rule (20.14) can be expressed in the form of relation (20.5).
2. Show that in a simple two-class system, introducing a reject classification to reduce the number of errors by R in fact requires $2R$ test patterns to be rejected, assuming that R is small. What is likely to happen as R increases?

21

Biologically Inspired Recognition Schemes

21.1 Introduction

The artificial intelligence (AI) community has long been intrigued by the capability of the human mind for apparently effortless perception of complex scenes. Of even greater significance is the fact that perception does not occur as a result of overt algorithms, but rather as a result of hardwired processes, coupled with extensive training of the neural pathways in the infant brain. The hardwiring pattern is largely the result of evolution, both of the neurons themselves and of their interconnections, though it seems that training and use can have some effect on the latter. By and large, however, it is probably correct to say that the overall architecture of the brain is the result of evolution, while its specific perceptual and intellectual capabilities are the result of training. It also seems possible that complex thought processes can themselves move towards solutions on an evolutionary basis, on timescales of milliseconds rather than millennia.

These considerations leave us with two interesting possibilities for constructing artificially intelligent systems: the first involves building or simulating networks of neurons, and training them to perform appropriate nontrivial tasks; the second involves designing algorithms which can search for viable solutions to complex problems by "evolutionary" or "genetic" algorithms (GAs). In this chapter we shall only give an introductory study of GAs, since their capabilities for handling vision problems are only at a preliminary stage. On the other hand, we shall study artificial neural networks (ANNs) more carefully, since they are closely linked to quite old ideas on statistical pattern recognition, and in addition, they are starting to be used quite widely in vision applications.

It is well known that neurons operate in ways which are quite different from those of modern electronic circuits. Electronic circuits may be of two main types—analogue and digital—though the latter type is predominant in the present generation of computers. Specifically, waveforms at the outputs of logic gates and flip-flops are strongly synchronized with clock pulses, whereas those at the outputs of biological neurons exhibit 'firing' patterns which are asynchronous, often being considered as random pulse streams for which only the frequency of firing carries useful information. Thus the representation of the information is totally different from that of electronic circuits. Some have argued that the representation used by biological neurons is a vital feature which leads to efficient information processing, and have even built stochastic computing units on this basis. We shall not pursue this line here as there seems to be no reason why the vagaries of biological evolution should have led to the most efficient representation for artificial information processing: i.e. the types of neuron that are suitable for biological systems are not necessarily appropriate for machine vision systems. Nevertheless, in designing machine vision systems, it seems reasonable to take *some* hints from biological systems on what methodologies might be useful.

This chapter will describe ANNs and give some examples of their application in machine vision, and will also give a brief outline of work on GAs. In considering ANNs, it should be borne in mind that there are a number of possible architectures with various characteristics, some being appropriate for supervised learning, and others being better adapted for unsupervised learning. We start by studying the perceptron type of classifier designed by Rosenblatt in the 1960s.

21.2 Artificial neural networks

The concept of an ANN which could be useful for pattern recognition started in the 1950s and continued right through the 1960s. For example, Bledsoe and Browning (1959) developed the "n-tuple" type of classifier which involved bit-wise recording and lookup of binary feature data, leading to the "weightless" or "logical" type of ANN. Although this type of classifier has had some following right through to the present day, it is probably no exaggeration to say that it is Rosenblatt's "perceptron" (1958, 1962) which has had the greatest influence on the subject.

The simple perceptron is a linear classifier that classifies patterns into two classes. It takes a feature vector $\mathbf{x} = (x_1, x_2, \ldots, x_N)$ as its input, and produces a single scalar output

$$\sum_{i=1}^{N} w_i x_i$$

the classification process being completed by applying a threshold (Heaviside step) function at θ (see Fig. 21.1). The mathematics is simplified by writing $-\theta$ as w_0, and taking it to correspond to an input x_0 which is maintained at a constant value of unity. The output of the linear part of the classifier is then written in the form:

$$d = \sum_{i=1}^{N} w_i x_i - \theta = \sum_{i=1}^{N} w_i x_i + w_0 = \sum_{i=0}^{N} w_i x_i \tag{21.1}$$

and the final output of the classifier is given by:

$$y = f(d) = f\left(\sum_{i=0}^{N} w_i x_i\right) \tag{21.2}$$

This type of neuron can be trained using a variety of procedures, such as the *fixed increment rule* given in Table 21.1. (The original fixed increment rule used a learning rate coefficient η equal to unity.) The basic concept of this algorithm was to try to improve the overall error rate by moving the linear discriminant plane a fixed distance towards a position where no misclassification would occur—but only doing this when a classification error had occurred:

$$w_i(k+1) = w_i(k) \qquad\qquad y(k) = \omega(k) \tag{21.3}$$
$$w_i(k+1) = w_i(k) + \eta[\omega(k) - y(k)]x_i(k) \qquad y(k) \neq \omega(k) \tag{21.4}$$

In these equations, the parameter k represents the kth iteration of the classifier and $\omega(k)$ is the class of the kth training pattern. It is clearly important to know whether this training scheme is effective in practice. In fact, it is possible to show that, if the algorithm is modified so that its main loop is applied sufficiently many times, *and* if the feature vectors are linearly separable, then the algorithm will converge on a correct error-free solution.

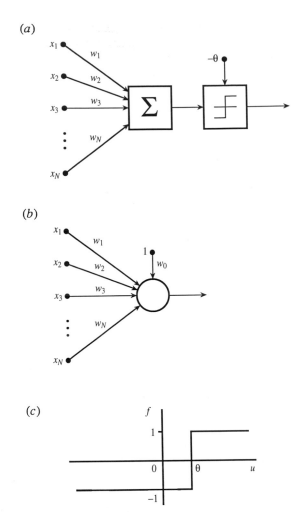

Fig. 21.1 Simple perceptron. (a) The basic form of a simple perceptron: input feature values are weighted and summed, and the result fed via a threshold unit to the output connection. (b) A convenient shorthand notation for the perceptron. (c) The activation function of the threshold unit.

Unfortunately, most sets of feature vectors are not linearly separable: thus it is necessary to find an alternative procedure for adjusting the weights. This is achieved by the Widrow–Hoff delta rule which involves making changes in the weights in proportion to the error $\delta = \omega - d$ made by the classifier. (Note that the error is calculated *before* thresholding to determine the actual

Table 21.1   Perceptron *fixed increment* algorithm.

---

initialize weights with small random numbers;
select suitable value of learning rate coefficient η in the range 0 to 1;
`repeat`
 `for` all patterns in the training set `do`
  `begin`
   obtain feature vector **x** and class ω;
   compute perceptron output y;
   `if` $y \neq \omega$ `then` adjust weights according to $w_i := w_i + \eta(\omega - y)x_i$;
  `end;`
`until` no further change;

---

class: i.e. δ is calculated using d rather than $f(d)$.) Thus we obtain the Widrow–Hoff delta rule in the form:

$$w_i(k+1) = w_i(k) + \eta\delta x_i(k) = w_i(k) + \eta[\omega(k) - d(k)]x_i(k) \quad (21.5)$$

There are two important ways in which the Widrow–Hoff rule differs from the fixed increment rule:

(i)   An adjustment is made to the weights whether or not the classifier makes an actual classification error.

(ii) The output function d used for training is different from the function $y = f(d)$ used for testing.

These differences underline the revised aim of being able to cope with non-linearly separable feature data. Figure 21.2 clarifies the situation by appealing to a 2-D case. Figure 21.2(a) shows separable data which are straightforwardly fitted by the fixed increment rule. However, the fixed increment rule is not designed to cope with nonseparable data of the type shown in Fig. 21.2(b), and results in instability during training, and inability to arrive at an optimal solution. On the other hand the Widrow–Hoff rule copes satisfactorily with this type of data. An interesting addendum to the case of Fig. 21.2(a) is that although the fixed increment rule apparently reaches an optimal solution, the rule becomes "complacent" once a zero error situation has occurred, whereas an ideal classifier would arrive at a solution which minimizes the probability of error. Clearly the Widrow–Hoff rule goes some way to solving this problem.

So far we have considered what can be achieved by a simple perceptron. Clearly, though it is only capable of dichotomizing feature data, a suitably

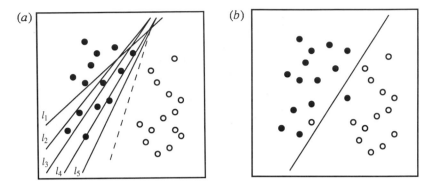

Fig. 21.2 Separable and nonseparable data. (a) Two sets of pattern data: lines l_1–l_5 indicate possible successive positions of a linear decision surface produced by the fixed increment rule. Note that the latter is satisfied by the final position l_5. The dashed line shows the final position which would have been produced by the Widrow–Hoff delta rule. (b) The stable position which would be produced by the Widrow–Hoff rule in the case of nonseparable data: in this case the fixed increment rule would oscillate over a range of positions during training.

trained array of simple perceptrons—the "single-layer perceptron" of Fig. 21.3—should be able to divide feature space into a large number of subregions bounded (in multidimensional space) by hyperplanes. However, in a multiclass application this approach would require a very large number of simple perceptrons—up to

$$\binom{c}{2} = c(c-1)/2$$

for a c-class system. Hence there is a need to generalize the approach by other means. In particular, multilayer perceptron (MLP) networks (see Fig. 21.4)—which would emulate the neural networks in the brain—seem poised to provide a solution since they should be able to recode the outputs of the first layer of simple perceptrons.

Rosenblatt himself proposed such networks, but was unable to propose general means for training them systematically. In 1969 Minsky and Papert published their famous monograph, and in discussing the MLP raised the spectre of "the monster of vacuous generality"; they drew attention to certain problems which apparently would never be solved using MLPs. For example, diameter-limited perceptrons (those that view only small regions of an image within a restricted diameter) would be unable to measure large-

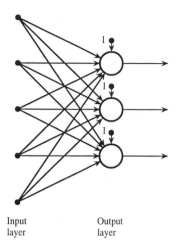

Fig. 21.3 Single-layer perceptron. The single-layer perceptron employs a number of simple perceptrons in a single layer. Each output indicates a different class (or region of feature space). In more complex diagrams, the bias units (labelled "1") are generally omitted for clarity.

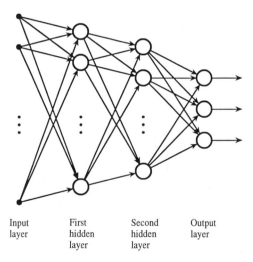

Fig. 21.4 Multilayer perceptron. The multilayer perceptron employs several layers of perceptrons. In principle, this topology permits the network to define more complex regions of feature space, and thus perform much more precise pattern recognition tasks. Finding systematic means of training the separate layers becomes the vital issue. For clarity, the bias units have been omitted from this and later diagrams.

scale connectedness within images. These considerations discouraged effort in this area, and for many years attention was diverted to other areas such as artificial intelligence. It was not until 1986 that Rumelhart *et al.* were successful in proposing a systematic approach to the training of MLPs. Their solution is known as the back-propagation algorithm.

21.3   The back-propagation algorithm

The problem of training a MLP can be simply stated: a general layer of a MLP obtains its feature data from the lower layers, and receives its class data from higher layers. Hence, if all the weights in the MLP are potentially changeable, the information reaching a particular layer cannot be relied upon: there is no reason why training a layer in isolation should lead to overall convergence of the MLP towards an ideal classifier (however defined). In addition, it is not evident what the optimal MLP architecture should be. While it might be thought that this is a rather minor difficulty, in fact this is not so: indeed, this is but one example of the so-called "credit assignment problem"*.

One of the main difficulties in predicting the properties of MLPs and hence of training them reliably is the fact that neuron outputs swing suddenly from one state to another as their inputs change by infinitesimal amounts. Hence we might consider removing the thresholding functions from the lower layers of MLP networks to make them easier to train. Unfortunately, this would result in these layers acting together as larger linear classifiers, with far less discriminatory power than the original classifier (in the limit we would have a set of linear classifiers each with a single thresholded output connection, so the overall MLP would act as a single-layer perceptron!).

The key to solving these problems was to modify the perceptrons composing the MLP by giving them a less "hard" activation function than the Heaviside function. As we have seen, a linear activation function would be of little use, but one of "sigmoid" shape, such as the $\tanh(u)$ function (Fig. 21.5) is effective, and indeed is almost certainly the most widely used

---

*This is not a good first example by which to define the credit assignment problem (in this case it would appear to be more of a deficit assignment problem). The credit assignment problem is the problem of correctly determining the local origins of global properties and making the right assignments of rewards, punishments, corrections and so on, thereby permitting the whole system to be optimized systematically.

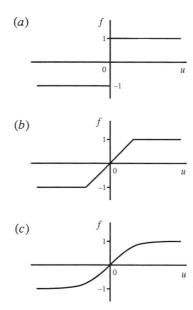

Fig. 21.5 Symmetric activation functions: (a) the Heaviside activation function used in the simple perceptron; (b) a linear activation function, which is, however, limited by saturation mechanisms; (c) a sigmoidal activation function which approximates to the hyperbolic tangent function.

of the available functions[*]. Once these softer activation functions were used, it became possible for each layer of the MLP to "feel" the data more precisely and thus training procedures could be set up on a systematic basis. In particular, the rate of change of the data at each individual neuron could be communicated to other layers which could then be trained appropriately —though only on an incremental basis. We shall not go through the detailed

---

[*] We do not here make a marked distinction between symmetrical activation functions and alternatives which are related to them by shifts of axes, though the symmetrical formulation seems preferable as it emphasizes bidirectional functionality. In fact, the tanh(u) function, which ranges from -1 to 1, can be expressed in the form:

$$\tanh(u) = (e^u - e^{-u})/(e^u + e^{-u}) = 1 - 2/(1 + e^{2u})$$

and is thereby closely related to the commonly used function $(1 + e^{-v})^{-1}$. It can now be deduced that the latter function is symmetrical, though it ranges from 0 to 1 as v goes from $-\infty$ to ∞.

mathematical procedure, or proof of convergence. Instead, we give an outline of the back-propagation algorithm (see Table 21.2). Nevertheless, some notes on the algorithm are in order:

(i)   The outputs of one node are the inputs of the next, and an arbitrary choice is made to label all variables as output (y) parameters rather than as input (x) variables; all output parameters are in the range 0 to 1.

(ii) The class parameter ω has been generalized as the target value t of the output variable y.

(iii) For all except the final outputs, the quantity δ_j has to be calculated using the formula

$$\delta_j = y_j(1 - y_j)\left(\sum_m \delta_m w_{jm}\right)$$

the summation having to be taken over all the nodes in the layer *above* node j.

(iv) The sequence for computing the node weights involves starting with the output nodes and then proceeding downwards one layer at a time.

Table 21.2   The back-propagation algorithm.

---

initialize weights with small random numbers;
select suitable value of learning rate coefficient η in the range 0 to 1;
`repeat`
    `for` all patterns in the training set do
       `for` all nodes j in the MLP do
         `begin`
            obtain feature vector $\mathbf{x}$ and target output value t;
            compute MLP output y;
            `if` node is in output layer
               `then` $\delta_j = y_j(1 - y_j)(t_j - y_j)$

            `else` $\delta_j = y_j(1 - y_j)\left(\sum_m \delta_m w_{jm}\right)$

            adjust weights i of node j according to $w_{ij} := w_{ij} + \eta \delta_j y_i$;
         `end`;
`until` changes are reduced to some predetermined level;

(v) If there are no hidden nodes, the formula reverts to the Widrow–Hoff delta rule, except that the input parameters are now labelled y_i, as indicated above.

(vi) It is important to initialize the weights with random numbers to minimize the chance of the system becoming stuck in some possibly symmetrical state from which it might be difficult to recover.

(vii) Choice of value for the learning rate coefficient η will be a balance between achieving a high rate of learning and avoidance of over-shoot: normally a value of around 0.8 is selected.

When there are many hidden nodes, convergence of the weights can be very slow; indeed, this is one disadvantage of MLP networks. Many attempts have been made to speed convergence, and a method that is almost universally used is to add a "momentum" term to the weight update formula, it being assumed that weights will change in a similar manner during iteration k to the change during iteration $k - 1$:

$$w_{ij}(k + 1) = w_{ij}(k) + \eta\delta_j y_i + \alpha[w_{ij}(k) - w_{ij}(k - 1)] \qquad (21.6)$$

where α is the momentum factor. Primarily, this technique is intended to prevent networks dallying temporarily around, or even becoming stuck permanently at, local minima of the energy surface.

21.4 MLP architectures

The preceding sections gave the motivation for designing a MLP and for finding a suitable training procedure, and then outlined a general MLP architecture and the widely used back-propagation training algorithm. However, having a general solution is only one part of the answer. The next question is how best to adapt the general architecture to specific types of problem. We shall not give a full answer to this question here. However, Lippmann attempted to answer this problem in 1987. He showed that a two-layer (single hidden layer) MLP can implement arbitrary convex decision boundaries, and indicated that a three-layer (two hidden layer) network is required to implement more complex decision boundaries. It was subsequently found that it should never be necessary to exceed two hidden layers, as a three-layer network can tackle quite general situations if sufficient neurons are used (Cybenko, 1988). Subsequently, Cybenko (1989) and Hornik et al. (1989) showed that a two-layer MLP can approximate any continuous function, though

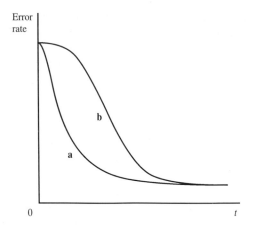

Fig. 21.6 Learning curves: (a) for a single-layer perceptron; (b) for a multilayer perceptron. Note that the multilayer perceptron takes considerable time to get going, since initially each layer receives relatively little useful training information from the other layers. Note also that the lower part of the diagram has been idealized to the case of identical asymptotic error rates—though this situation would seldom occur in practice.

nevertheless there may sometimes be advantages in using more than two layers.

It should be remarked that although the back-propagation algorithm can train MLPs of any number of layers, in practice, training one layer "through" several others introduces an element of uncertainty which is commonly reflected in increased training times (see Fig. 21.6). Thus, there is some advantage to be gained from using a minimal number of layers of neurons. In this context, the above findings on the necessary numbers of hidden layers are especially welcome.

21.5   Overfitting to the training data

When training MLPs and many other types of ANN there is a problem of overfitting the network to the training data. One of the fundamental aims of statistical pattern recognition is for the learning machine to be able to generalize from the particular set of data it is trained on to other types of data it might meet during testing. In particular, the machine should be able to cope with noise, distortions and fuzziness in the data, though clearly not to the extent of being able to respond correctly to types of data different from that on which it has been trained. The main points to be made here are (i) that the machine should learn to respond to the underlying population from which the training data has been drawn, and (ii) that it must not be so well adapted to the specific training data that it responds less well to other data from the same population. Figure 21.7 shows in a 2-D case both a fairly ideal degree of fit and a situation where every nuance of the set of data has been fitted, thereby achieving a degree of overfit.

Typically, overfitting can arise if the learning machine has more adjustable parameters than are strictly necessary for modelling the training data: with too few parameters such a situation should not arise. However, if the learning machine has enough parameters to ensure that relevant details of the underlying population are fitted, there may be overmodelling of part of the training set and the overall fit may become excessively accurate; thus the overall recognition performance will deteriorate. Ultimately, the reason for this is that recognition is a delicate balance between capability to discriminate and capability to generalize, and it is most unlikely that any complex learning machine will get the balance right for all the features it has to take account of.

Be this as it may, we clearly need to have some means of preventing overadaptation to the training data. One way of achieving this is to curtail the training process before overadaptation can occur[*]. However, this is not difficult, since we merely need to test the system periodically during training to ensure that the point of overadaptation has not been reached. Figure 21.8 shows what happens when testing is carried out simultaneously on a separate dataset: at first performance on the test data closely matches that on the training data, being slightly superior for the latter because a small degree of overadaptation is already occurring. But after a time, performance starts deteriorating on the test data while that on the training data appears to go on improving. This is the point where serious overfitting is occurring, and the training process should be curtailed. The aim, then, is to make the whole training process far more rigorous by splitting the original training set into two parts—the first being retained as a normal training set, and the second being called the *validation set*. Note that the latter is actually part of the training set in the sense that it is not part of the eventual test set.

---

[*] It is often stated that this procedure aims to prevent overtraining. However, the term "overtraining" is ambiguous. On the one hand it can mean recycling through the *same* set of training data until eventually the learning machine is overadapted to it. On the other hand it can mean using more and more *totally new* data—a procedure which cannot produce overadaptation to the data, and on the contrary is almost certain to improve performance. In view of this ambiguity it seems better not to use the term.

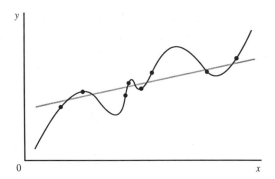

Fig. 21.7 Overfitting of data. The data points are rather too well fitted by the solid curve, which matches every nuance exactly. Unless there are strong theoretical reasons why the solid curve should be used, the grey line will give a higher confidence level.

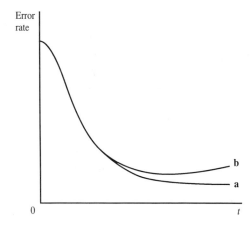

Fig. 21.8 Cross-validation tests. This diagram shows the learning curve for a multilayer perceptron (a) when tested on the training data and (b) when tested on a special validation set. Curve (a) tends to go on improving even when overfitting is occurring. However, this situation is detected when (b) starts deteriorating. To offset the effects of noise (not shown on curves (a) and (b)), it is usual to allow 5–10% deterioration relative to the minimum in (b).

The process of checking the degree of training by use of a validation set is called *cross-validation*, and is vitally important to proper use of an ANN. The training algorithm should include cross-validation as a fully integrated part of the whole training schedule: it should not be regarded as an optional extra.

It is useful to speculate how overadaptation could occur when the training procedure is completely determined by the back-propagation (or other) provably correct algorithm. In fact, there are mechanisms by which overadaptation can occur. For example, when the training data do not control particular weights sufficiently closely, some could drift to large positive or negative values, while retaining a sufficient degree of cancellation so that no problems appear to arise with the training data; yet, when test or validation data are employed, the problems become all too clear. The fact that the form of the sigmoid function will permit some nodes to become "saturated out" does not apparently help the situation, as it inactivates parameters and hides certain aspects of the incoming data. Yet it is intrinsic to the MLP architecture and the way it is trained that some nodes are *intended* to be saturated out in order to ignore irrelevant features of the training set. The problem is whether inactivation is inadvertent or designed. The answer probably lies in the quality of the training set and how well it covers the available or potential feature space.

Finally, let us suppose a MLP is being set up, and it is initially unknown how many hidden layers will be required or how many nodes there will have to be on each layer: it will also be unknown how many training set patterns will be required or how many training iterations will be required—or what values of the momentum or learning parameters will be appropriate. A quite substantial number of tests will be required to decide all the relevant parameters. There is therefore a definite risk that the final system will be overadapted not only to the training set, but also to the validation set. In such circumstances what we need is a second validation set which can be used after the whole network has been finalized and final training is being undertaken.

21.6 Optimizing the network architecture

In the last section we mentioned that the ANN architecture will have to be optimized: in particular, for a MLP the optimum number of hidden layers and numbers of nodes per layer will have to be decided; the optimal degree of connectivity of the network will also have to be determined (it cannot be assumed that full connectivity is necessarily advantageous—it may be better to have some connections which bypass certain hidden layers).

Broadly, there are two approaches to the optimization of network architectures. In the first, the network is built up slowly node by node. In the second, an overly complex network is built and this is gradually pruned until performance is optimized. This latter approach clearly has the disadvantage of requiring excessive computation until the final architecture is reached. It is therefore tempting to consider only the first of the two options—an approach which would also tend to improve generalization. A number of alternative schemes have been described for attempting this; currently, there appears to be no universally best method and research on this is still proceeding (e.g. Patel and Davies, 1995).

Instead of dwelling on network generation or pruning aspects of architecture optimization, it is helpful to consider an alternative based on use of GAs (see below). In this approach, various trials are made in a space of possible architectures. To apply GAs to this task, two requirements must be fulfilled: some means of selecting new or modified candidate architectures is needed; and a suitable fitness function for deciding which architecture is optimal must be chosen. The first requirement fits neatly into the GA formalism if each architecture is defined by a binary codeword which can be modified suitably with a meaningful one-to-one correspondence between codeword and architecture. (This is bound to be possible as each actual connection of a potentially completely connected network can be determined by a single bit in the codeword.) In principle, the fitness function could be as simple as the mean-square error obtained using the validation set. However, some additional weighting factors designed to minimize the number of nodes and the numbers of interconnections can also be useful. While this approach has been tested by a number of workers, the computational requirements are rather high: hence GAs are unlikely to be useful for optimizing very large networks.

21.7 Hebbian learning

In Chapter 20 we found how principal components analysis can help with data representation and dimensionality reduction. In fact, principal components analysis is an especially useful procedure, and it is not surprising that there have been a number of attempts to perform it using different types of ANN. In particular, Oja (1982) was able to develop a method for determining the principal component corresponding to the largest eigenvalue λ_{max} using a single neuron with linear weights.

The basic idea is that of Hebbian learning. To understand this process we must imagine a large network of biological neurons which are firing according to various input stimuli and producing various responses elsewhere in the network. Hebb (1949) considered how a given neuron could learn from the data it receives[*]. His conclusion was that good pathways should be rewarded so that they become stronger over time; or more precisely, synaptic weights should be strengthened in proportion to the correlation between the firing of pre- and postsynaptic neurons. Here we can visualize the process as "rewarding" the neuron inputs and outputs in proportion to the numbers of input patterns that arrive at them, and modelling this by the equation[†]:

$$\Delta w_i = \eta y x_i \qquad (21.7)$$

The problem with this approach is that the weights will grow in an unconstrained manner, and therefore a constraining or normalizing influence is needed. Oja achieved this by adding a weight decay proportional to y^2:

$$\Delta w_i = \eta y x_i - \eta y^2 w_i = \eta y (x_i - y w_i) \qquad (21.8)$$

Linsker (1986, 1988) produced an alternative rule based on clipping at certain maximum and minimum values; and Yuille *et al.* (1989) used a rule which was designed to normalize weight growth according to the magnitude of the overall weight vector $\mathbf{w}$:

$$\Delta w_i = \eta (y x_i - |\mathbf{w}|^2 w_i) \qquad (21.9)$$

It is not possible to give a complete justification for these rules here: the only really satisfactory justification requires rather nontrivial mathematical proofs of convergence. However, these ideas led the way to full determination of principal components by ANNs. Both Oja and Sanger developed such methods, which are quite similar and merit close comparison (Oja, 1982; Sanger, 1989).

First, we consider Oja's training rule, which applies to a single-layer feedforward linear network with N input nodes, M processing nodes, and transfer functions:

$$y_i = \sum_{j=1}^{N} w_{ij} x_j \qquad (i = 1 \text{ to } M, \quad M < N) \qquad (21.10)$$

---

[*] This corresponds to a totally different type of solution to the credit assignment problem than that provided by the back-propagation algorithm.

[†] In what follows we suppress the iteration parameter k, and write down merely the increment Δw_i in w_i over a given iteration.

The rule takes the form:

$$\Delta w_{ij} = \eta y_i \left(x_j - \sum_{k=1}^{N} w_{kj} y_k \right) \qquad (i = 1 \text{ to } M, \quad j = 1 \text{ to } N) \qquad (21.11)$$

This is clearly an extension of the earlier rule for finding the principal component corresponding to $\lambda_{\max}$ (equation (21.8)). In particular, it demands that the M nodes, which are all in the same layer, be connected laterally as shown in Fig. 21.9(a), and it is very plausible that if a single node can determine one principal component, then M nodes can find M principal components. However, the symmetry of the situation indicates that it is not clear which node should correspond to the largest eigenvalue, and which to any of the others. In fact, the situation is less definite than this, and all we can say is that the output vectors span the space of the largest M eigenvalues: the vectors are mutually orthogonal but are not guaranteed to be orientated along the principal axes directions. It should also be noted that the results obtained by the network will vary

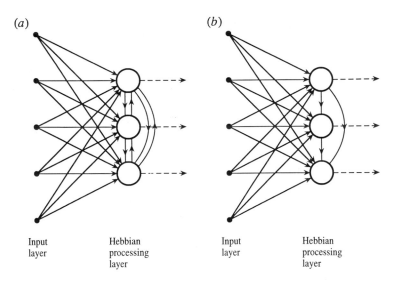

Fig. 21.9 Networks employing Hebbian learning. These two networks employ Hebbian learning and evaluate principal components. During training, both networks learn to project the N-dimensional input feature vectors into the space of the M largest principal components. Unlike the Oja network (a), the Sanger network (b) is able to order the output eigenvalues. For clarity, the output connections are shown as dashed lines.

significantly with the particular data seen during training, and its order of presentation. However, the rule is sufficiently good for certain applications such as data compression, since the principal components corresponding to the smallest $N - M$ eigenvalues are systematically eliminated.

Nevertheless, in many applications it is necessary to force the outputs to correspond to principal components and also to know the order of the eigenvalues. The Sanger network achieves this by making the first node independent of the others; the second dependent only on the first; the third dependent only on the first two; and so on (see Fig. 21.9(b)). Thus orthogonality of the principal components is guaranteed, while independence is maintained at the right level for each node to be sure of producing a specific principal component. It could be argued that the Sanger network is an M-layer network, though this is arguable as the architecture is obtained from the Oja network by *eliminating* some of the connections. The Sanger training rule is obtained from the Oja rule by making the upper limit in the summation i rather than N:

$$\Delta w_{ij} = \eta y_i \left(x_j - \sum_{k=1}^{i} w_{kj} y_k \right) \qquad (i = 1 \text{ to } M, \quad j = 1 \text{ to } N) \qquad (21.12)$$

At the end of the computation, the outputs of the M processing nodes contain the eigenvalues, while the weights of the M nodes provide the M eigenvectors. An interesting point arises since the successive outputs of the Sanger network have successively lower importance as the eigenvalues (data variances) decrease. Hence it is valid and sometimes convenient for images or other output data to be computed and presented with progressively fewer bits for the lower eigenvalues (this applies especially to real-time hardware implementations of the algorithms).

It might be questioned whether there is any value in techniques such as principal components analysis being implemented using ANNs when a number of perfectly good conventional methods exist for computing eigenvalues and eigenvectors. In fact, when the input vectors are rather large, matrix diagonalization involves significant computation, and ANN approaches can arrive at workable approximations very quickly by iterative application of simple linear nodes which are straightforwardly implemented in hardware. Such considerations are especially important in real-time applications, such as those which frequently occur in automated visual inspection or interpretation of satellite images. There is also the possibility that the approximations made by ANNs will permit them to include other relevant information, hence leading to increased reliability of data classification, e.g. by taking into account higher-order correlations (Taylor and Coombes, 1993): however, such possibilities will have to be confirmed by future research.

We end this section with a simple alternative approach to the determination of principal components. Like the Oja approach, this method projects the input vectors onto a subspace which is spanned by the first M principal components, discarding only minimal information. In this case the network architecture is a two-layer MLP with N inputs, M hidden nodes and N output nodes, as shown in Fig. 21.10. The special feature is that the network is trained using the back-propagation algorithm to produce the same outputs as the inputs—a scheme commonly known as "self-supervised back-propagation". Note that, unusually, the outputs are taken from the hidden layer. Interestingly, it has been found both experimentally and theoretically that nonlinearity of the neuron activation function is of no help in finding principal components (Cottrell *et al.*, 1987; Bourland and Kamp, 1988; Baldi and Hornik, 1989).

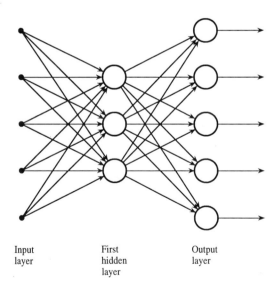

Input   First   Output
layer   hidden   layer
               layer

Fig. 21.10   Self-supervised back-propagation. This network is a normal two-layer perceptron which is trained by back-propagation so that the output vector tracks the input vector. Once this is achieved, the network can provide information on the M largest principal components. Oddly, this information appears at the outputs of the hidden layer.

21.8 Case-study: noise suppression using ANNs

It is pertinent to ask in what ways ANNs would be useful in applications such as those typical of machine vision. There are three immediate answers to this question. The first is that ANNs are learning machines, and their use should result in greater adaptability than that normally available with conventional algorithms. The second is that their concept is that of simple replicable hardware-based computational elements; hence a parallel machine built using them should in principle be able to operate extremely rapidly, thereby emulating some of the intellectual and perceptual capabilities of the human brain. The third is that their parallel processing capability should lead to greater redundancy and robustness in operation which should again mimic that of the human brain. While there is some indication that the last two of these possibilities will be realized in practice, the position is still not clear, and there is no sign of conventional algorithmic approaches being ousted for these reasons. However, the first point—that of improved adaptability—seems to be at the stage of serious test and practical implementation. It is therefore appropriate to consider it more carefully, in the context of a suitable case-study. Here we choose image noise suppression as the vehicle for this study. The reason for using image noise suppression is that it is a well-researched topic about which there is a body of relevant theory, and any advances made using ANNs should immediately be clear.

As we have seen in Chapter 3, there are a number of useful conventional approaches to image noise suppression, the most widely used being the median filter. However, the latter is not without its problems, one being that it causes image distortion through the shifting of curved edges. In addition, it has no adjustable parameters other than neighbourhood size, and so cannot be adapted to the different types of noise it might be called upon to eliminate. Indeed, its performance for noise removal is quite limited, as it is largely an *ad hoc* technique which happens to be best suited to the suppression of impulse noise.

These considerations have prompted a number of researchers to try using ANNs for eliminating image noise (Nightingale and Hutchinson, 1990; Lu and Szeto, 1991; Pham and Bayro-Corrochano, 1992; Yin *et al.*, 1993). Since ANNs are basically recognition tools, they could be used for recognizing noise structures in images, and once a noise structure has been located, it should be possible to replace it by a more appropriate pattern. Greenhill and Davies (1994a, b) have reported a more direct approach, in which they trained the ANN how to respond to a variety of input patterns using various example images.

In this work it did not seem appropriate to use an exotic ANN architecture or training method, but rather to test one of the simpler and more widely used forms of ANN. A normal MLP was therefore selected: this was trained using the standard back-propagation algorithm. Preliminary tests were made to find the most appropriate topology for processing the 25 pixel intensities of each 5×5 square neighbourhood (the idea being to apply this in turn at all positions in the input images, as with a more conventional filter). The result was a three-layer network with 25 inputs, 25 nodes in the first hidden layer, 5 nodes in the second hidden layer, and one output node, there being 100% connectivity between adjacent layers of the network. All the nodes employed the same sigmoidal transfer function, and a grey-scale output was thus available at the output of the network.

Training was achieved using a normal off-camera image as the low-noise "target" image and a noisy input image obtained by adding artificial noise to the target image. Three experiments were carried out using three different types of noise: (i) impulse noise with intensities 0 and 255; (ii) impulse noise with random pixel intensities (taken from a uniform distribution over the range 0 to 255); and (iii) Gaussian noise (truncated at 0 and 255). Noise

(a)

Fig. 21.11 Noise suppression by an ANN filter. (a) An image with added type (i) impulse noise; (b) the effect of suppressing the noise using an appropriately trained ANN filter; and (c) the effect of suppressing the noise using a 5×5 median filter. Notice the "softening" effect and the partial filling in of corners in (c), neither of which is apparent in (b).

(b)

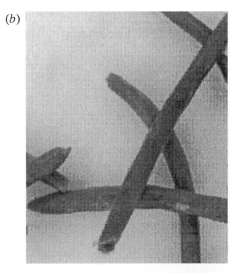

(c)

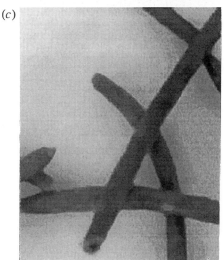

Fig. 21.11 (*continued*).

types (i) and (ii) were applied to random pixels, while noise type (iii) was applied to all pixels in the input images. In the experiments, the performances of mean, median, hybrid median (2LH+) and ANN filters were compared, taking particular note of the average absolute deviations in intensity between the output and target images. Both 3×3 and 5×5 versions of all the filters were used, except that only a 5 × 5 version of the ANN filter was tried, since it should automatically adapt itself to 3 × 3 if required by the data seen during training.

As expected, the mean filters produced gross blurring of the images, and "blotchy" behaviour as a result of partially averaging impulse noise into the surroundings. However, although the median filters eliminated noise

Table 21.3 Properties of the various filters investigated.

Mean filter	Median filter	Hybrid median filter	ANN-based filter
Produces blotches and blur	Does not produce blotches and blur	Does not produce blotches and blur	Does not produce blotches and blur
Does not produce spottiness	Does not produce spottiness	*Produces spottiness*	Does not produce spottiness
Bumps not visible because of blur	*Produces bumps*	Does not produce bumps	Does not produce bumps
Produces corner-filling	*Produces corner-filling*	Produces slight corner-filling	**Does not produce corner-filling**
Cannot adapt to special types of noise	*Cannot adapt to special types of noise*	*Cannot adapt to special types of noise*	**Adapts to the noise seen during training**
Cannot adapt to special types of image data	Cannot adapt to special types of image data	Cannot adapt to special types of image data	**Adapts to the image data seen during training**
Cannot adapt to drifts in the image data	Cannot adapt to drifts in the image data	Cannot adapt to drifts in the image data	Cannot adapt to drifts from the data seen during training
Unlikely to be optimal	Unlikely to be optimal	Unlikely to be optimal	**Potentially optimal**

For clarity of presentation, distinctly good behaviour is marked in bold and distinctly poor behaviour is marked in italics. In the case of the median and hybrid median filters, these are known to be reasonably well adapted to removal of impulse noise, and in the latter case to coping with straight edges. However, as indicated here, they cannot be adapted to arbitrary types of noise or data.

effectively, they were found to soften these particular images to an unusual degree (Fig. 21.11), making the beans appear rather unnatural and less like beans than those in the original image. Of the conventional filters, the hybrid median filters gave the best overall performance with these images. However, they allowed a small number of "spotty" noise structures to remain, since these essentially emulated the straight line and corner structures which this type of filter is designed to retain intact (see Section 3.5.3). These particular images seemed prone to another artefact: the fairly straight sides of the beans became "bumpy" after processing by the median filters, this being (predictably) less evident on the images processed by the hybrid median filters. Finally, the median filters gave significant filling in of the corners between the beans, the effect being present but barely discernible after processing by the hybrid median filters. The situation is summarized in Table 21.3.

In general, the ANN filters[*] performed better than the conventional filters. First, each was found to perform best with the type of noise it had been trained on, and to be more effective than the conventional filters at coping with that particular type of noise. Thus it exhibited the adaptive capability expected of it. In addition, the ANN filters produced images without the remanent spottiness, bumpiness and corner-filling behaviour exhibited by the conventional filters.

Nevertheless, the performance of the ANNs was not always exemplary. In one case an ANN performed less well than a conventional filter when it was tested on an image dissimilar to the ones it had been trained on. This must be due to disobeying a basic rule of pattern recognition—namely that training sets should be fully representative of the classes from which they are drawn (see Chapter 20). However, in this case it could be deemed to have stemmed from the filter being more adapted to the particular type of noise than had been thought possible. Even more surprising was the finding that the training was sensitive not only to the type of noise, but also to the type of data in the underlying image. In particular, closer examination of the original images showed that there were faint vertical camera-induced striations which the ANNs did not remove, but which the median and other filters had eliminated. However, this was because the ANN filters had effectively been "instructed" to keep these striations, since the target images were the original off-camera images, and the input images were produced by adding extra noise to them. Thus the ANN interpreted the striations as valid image

---

[*] At this stage it is easier to imagine there to be three ANNs: in fact, there was one ANN architecture and three different training processes.

data! Clearly, a more appropriate training procedure is called for in this problem.

Unfortunately, it is difficult to see how a more appropriate training procedure—or equivalently, a more appropriate set of training images— could be produced. A *priori* it might have been thought that a median filtered image could be used as a target image. However, if this were done, the ANN would if anything exhibit worse performance than the median filter: it would hardly be able to *improve* on the median filter and adapt effectively to the particular type of noise it is supposed to remove. Such an approach would therefore negate the main aim of using an ANN for noise suppression.

Overall, the results of these experiments are quite significant. They underline that ANN-based filters will be limited by the quality of training. It is also useful to notice that any drift in the characteristics of the incoming image data will tend to cause a mismatch between the ANN filter and its training set, so its performance will be liable to deteriorate if this happens. While this is now apparent for an ANN-based filter, it must also be true to some extent for other more conventional filters, which are by their nature nonadaptable, For example, as noted earlier, a median filter cannot be adapted conveniently, and certainly cannot be matched to its data—except by crudely changing the size of its support region. In fact, we must now reflect the question back on conventional algorithms: are they general-purpose, or are they adapted to some specific conditions or data (which are in general not known in detail)? For by and large, conventional algorithms stem from the often fortuitous or even opportunistic inspirations of individuals, and their specifications are decided less in advance than in retrospect. Thus it is often only later that it is found by chance that they have certain undesirable properties that make them unsuitable for certain applications. Table 21.3 summarizes the properties of the various filters examined in these experiments, and attempts to highlight their successes and failures.

21.9 Genetic algorithms

The introduction to this chapter also included GAs as one of the newer biologically inspired methodologies. In fact, GAs are now starting to have significant impact on image recognition, and there is a growing number of papers which reflect their use in this area. We start this section by outlining how GAs operate.

Basically, a GA is a procedure which mimics biological systems in allowing solutions (including complete algorithms) to evolve until they become optimally adapted to their situation. The first requirement is a means of generating random solutions. The second requirement is a means of judging the effectiveness of these solutions. Such a means will generally take the form of a "fitness" function which should be minimized for optimality, though initially very little information will be available on the optimum value. Next, an initial set of solutions is generated, and the fitness function is evaluated for each of them. The initial solutions are likely to be very poor, but the best of them are selected for "mating" and "procreation". Typically, a pair of solutions is taken, and they are partitioned in some way, normally in a manner that allows corresponding parts to be swapped over (the process is called *crossover*), thereby creating two combination or "infant" solutions. The infants are then subjected to the same procedures— evaluation of their effectiveness using the fitness function, and placement in an order of merit amongst the set of available solutions.

It will be clear from this explanation that, if both the infant and the original parent solutions are retained, there will be an increasing population of solutions, those at the top of the list being of increasing effectiveness. Clearly, the storage of solutions will eventually get out of hand, and the worst solutions will have to be killed off (i.e. erased from memory). However, the population of solutions will continue to improve, but possibly at a decreasing rate. If evolution stagnates, further random solutions may have to be created, though another widely used mechanism is to *mutate* a few existing solutions. This mimics such occurrences as copying errors or radiation damage to genes in biological systems. However, it corresponds to asexual rather than sexual selection, and may well be less effective at improving the system at the required rate. Normally crossover is taken to be the basic mechanism for evolution in GAs. While stagnation remains a possibility, the binary cutting and recombination intrinsic to the basic mechanism is extremely powerful, and can radically improve the population over a limited number of generations. Thus any stagnation is often only temporary.

Overall, the approach can arrive at local minima in a suitably chosen solution space, though it may take considerable computation to improve the situation so that a global minimum is reached. Unfortunately, there is no universal formula for guaranteeing finding global minima over a limited number of iterations. Each application is bound to be different, and there are a good many parameters to be found and operational decisions to be made. These include:

(i) The number of solutions to be generated initially.
(ii) The maximum population of solutions to be retained.

(iii) The number of solutions that are to undergo crossover in each generation.

(iv) How often mutation should be invoked.

(v) Any *a priori* knowledge or rules that can be introduced to guide the system towards a rapid outcome.

(vi) A termination procedure which can make judgements on (a) the likelihood of further significant improvement and (b) the tradeoff between computation and improvement.

This approach can be made to operate impressively on such tasks as timetable generation. In such cases the major decisions to be made are (i) the process of deciding a code so that any solution is expressible as a vector in solution space, and (ii) a suitable method for assessing the effectiveness of any solution (i.e. the fitness function). Once these two decisions have been made, the situation becomes that of managing the GA, which involves making the operational decisions outlined above.

It turns out that GAs are well suited to setting up the weights of ANNs. The GAs essentially perform a coarse search of weight space, finding solutions which are relatively near to minima. Then back-propagation or other procedures can be used to refine the weights and produce optimal solutions. There are good reasons for not having the GAs go all the way to these optimal solutions. Basically, the crossover mechanism is rather brute-force, since binary chopping and recombination is a radical procedure which will tend to make very significant changes. Thus, in the final stages of convergence, if only one bit is crossed over, there is no certainty that this will produce only a minor change in a solution: indeed, it is the whole value of a GA that such changes can have rather large effects commensurate with searching the whole of solution space efficiently and quickly.

As stated earlier, GAs are now starting to have significant impact on image recognition, and more and more papers which reflect their use in this area are appearing (see Section 21.11). While unfortunately algorithms run slowly and only infrequently lend themselves to real-time implementation, they have the potential for helping with the design and adaptation of algorithms for which long evolution times are less important: what this means is that they can help to build masks or other processing structures which will subsequently be used for real-time processing, even if they are not themselves capable of real-time processing (Siedlecki and Sklansky, 1989; Harvey and Marshall, 1994; Katz and Thrift, 1994). Indeed, adaptation is currently the weak point of image recognition—not least in the area of automated visual inspection, where automatic setup and adaptation procedures are desperately needed to save programming effort and provide the capability for development in new application areas.

21.10 Concluding remarks

In the recent past there has been much euphoria and glorification of ANNs: while this was probably inevitable considering the wilderness years before the back-propagation algorithm became widely known, the situation now appears to be settling down and ANNs are starting to achieve a balanced role in vision and other application areas. In this chapter, space has demanded a concentration on three main topics—MLPs trained using back-propagation, Hebbian networks which can emulate principal components analysis, and GAs. MLPs are almost certainly the most widely used of all the biologically inspired approaches to learning, and inclusion here seems amply justified on the basis of wide utility in vision—with the potential for far wider application in the future. Much the same applies to GAs, though these are at a relatively earlier stage of development and application; in addition, they tend to involve very considerable amounts of computation, and their eventual role in machine vision is not too clear: however, it seems most likely (as indicated in Section 21.9) that they will be more useful for setting up vision algorithms than for carrying out the actual processing in real-time systems.

Returning to ANNs which emulate principal components analysis: it is not clear that they can yet perform a role which would provide a serious threat to the conventional approach. What will be important here in determining actual use will be the additional adaptivity provided by the ANN solution, and the possibility that this will, via higher-order correlations (e.g. Taylor and Coombes, 1993) or nonlinear principal component analysis (e.g. Oja and Karhunen, 1993), lead to far more powerful procedures which *are* able to supersede conventional methodologies.

Overall, ANNs have the potential for including automatic setup and adaptivity into vision systems to a far wider degree than at present, though the resulting systems then become dependent on the precise training set that is employed[*]. Thus the function of the system designer changes from code writer to code trainer, raising quite different problems. While this could be thought of as introducing difficulties, it should be noticed that these

---

[*] In this context, note especially that the comments of Section 20.4 on the frequencies of occurrence of the various classes in the training set apply equally to ANNs such as the MLP.

difficulties are to a large extent already present. For who can write effective code without taking in (i.e. being trained on) a large quantity of visual and other data, all of which will be subject to diurnal, seasonal or other changes which are normally totally inexplicit? Similarly, for GAs, who can design a fully effective fitness function appropriate to a given application without taking in similar quantities of data?

21.11   Bibliographical and historical notes

ANNs have had a chequered history. After a promising start in the 1950s and 1960s, they fell into disrepute (or at least, disregard) during the 1970s following the pronouncements of Minsky and Papert in 1969; they picked up again in the early 1980s; were subjected to an explosion in interest after the announcement of the back-propagation algorithm by Rumelhart *et al.* in 1986; and it is only in the mid-1990s that they are settling into the role of normal tools for vision and other applications. In addition, it should not be forgotten that the back-propagation algorithm was invented several times (Werbos, 1974; Parker, 1985) before its relevance was finally recognized. In parallel with these MLP developments, Oja (1982) led the field with his Hebbian principal components network.

Useful recent general references on ANNs include the volumes by Hertz *et al.* (1991), Haykin (1994) and Ripley (1996); and the highly useful review article by Hush and Horne (1993). The reader is also referred to two journal special issues (Backer, 1992; Lowe, 1994) for work specifically related to Machine Vision. In particular, ANNs have been applied to image segmentation by Toulson and Boyce (1992) and Wang *et al.* (1992); and to object location by Ghosh and Pal (1992), Spirkovska and Reid (1992) and Vaillant *et al.* (1994). For work on contextual image labelling, see Mackeown *et al.* (1994). For references related to visual inspection, see Chapter 19. For work on histogram-based thresholding in which the ANN is trained to emulate the maximal likelihood scheme of Chow and Kaneko (1972), see Greenhill and Davies (1995).

Genetic algorithms were invented in the early 1970s, and have only gained wide popularity in the 1990s. A useful early reference is Holland (1975) and a more recent reference is Michalewicz (1992). A short but highly useful review of the subject appears in Srinivas and Patnaik (1994). Although these references are general and do not refer to vision applications of GAs, since the late 1980s GAs have been applied quite regularly to machine vision. For example, they have been used by Siedlecki and

Sklansky (1989) for feature selection, and by Harvey and Marshall (1994) and Katz and Thrift (1994) for filter design; by Lutton and Martinez (1994) and Roth and Levine (1994) for detection of geometric primitives in images; by Pal *et al.* (1994) for optimal image enhancement; by Ankenbrandt *et al.* (1990) for scene recognition; and by Hill and Taylor (1992) and Bhattacharya and Roysam (1994) for image interpretation tasks. The IEE Colloquium on *Genetic Algorithms in Image Processing and Vision* (IEE, 1994) presented a number of relevant approaches; see also Gelsema (1994) for a special issue of *Pattern Recognition Letters* on the subject.

22

Texture

22.1 Introduction

In the foregoing chapters many aspects of image analysis and recognition have been studied. At the core of these matters has been the concept of segmentation, which involves the splitting of images into regions which have some degree of uniformity, whether in intensity, colour, texture, depth, motion or other relevant attributes. Care was taken in Chapter 4 to emphasize that such a process will be largely *ad hoc*, since the boundaries produced will not necessarily correspond to those of real objects. Nevertheless, it is important to make the attempt, either as a preliminary to more accurate or iterative demarcation of objects and their facets, or else as an end in itself—for example to judge the quality of surfaces.

In this chapter we move on to the study of texture and its measurement. Texture is a difficult property to define: indeed, in 1979 Haralick reported that no satisfactory definition of it had up till then been produced. Perhaps we should not be surprised by this, as the concept has rather separate meanings in the contexts of vision, touch and taste, the particular nuances being understood by different people also being highly individual and subjective. Nevertheless we require a working definition of texture, and in vision the particular aspect we focus on is the variation in intensity[*] of a particular surface or region of an image. Even with this statement we are being indecisive about whether it is the physical object being observed which is being described or the image derived

---

[*] We could at this point generalize the definition to cover variation in colour, but this would complicate matters unnecessarily, and would not add substantially to the coverage of the subject.

from it. This reflects the fact that it is the roughness of the surface or the structure or composition of the material which originally gives rise to its visual properties. However, in this chapter we are mainly interested in the interpretation of images, and so we define texture as the characteristic variation in intensity of a region of an image which should allow us to recognize and describe it and to outline its boundaries (Fig. 22.1).

This definition of texture implies that texture is nonexistent in a surface of uniform intensity, and does not say anything about how the intensity might be expected to vary or how we might recognize and describe it. In fact, there are very many ways in which intensity might vary, but if the variation does not have sufficient uniformity, the texture may not be characterized sufficiently closely to permit recognition or segmentation.

We next consider ways in which intensity might vary. Clearly, it can vary rapidly or slowly, markedly or with low contrast, with a high or low degree of directionality, and with greater or lesser degrees of regularity. This last characteristic is often taken as key: either the textural pattern is regular, as for a piece of cloth, or it is random, as for a sandy beach or a pile of grass cuttings. However, this ignores the fact that a regular textural pattern is often not wholly regular (again, as for a piece of cloth), or not wholly random (as for a mound of potatoes of similar size). Thus the degrees of randomness and of regularity will have to be measured and compared when characterizing a texture.

There are more profound things to say about the textures mentioned above. Often the textures are derived from tiny objects or components which are themselves similar, but which are placed together in ways ranging from purely random to purely regular—be they bricks in a wall, grains of sand, blades of grass, strands of material, stripes on a shirt, wickerwork on a basket, or a host of other items. In texture analysis it is useful to have a name for the similar textural elements that are replicated over a region of the image: such textural elements are called *texels*. These considerations lead us to characterize textures in the following ways:

   (i)   The texels will have various sizes and degrees of uniformity.
  (ii)   The texels will be orientated in various directions.
 (iii)   The texels will be spaced at varying distances in different directions.
 (iv)   The contrast will have various magnitudes and variations.
  (v)   Various amounts of background may be visible between texels.
 (vi)   The variations composing the texture may each have varying degrees of regularity *vis-à-vis* randomness.

It is quite clear from this discussion that a texture is a complicated entity to measure. The reason is primarily that many parameters are likely to be required to characterize it: in addition, when so many parameters are involved,

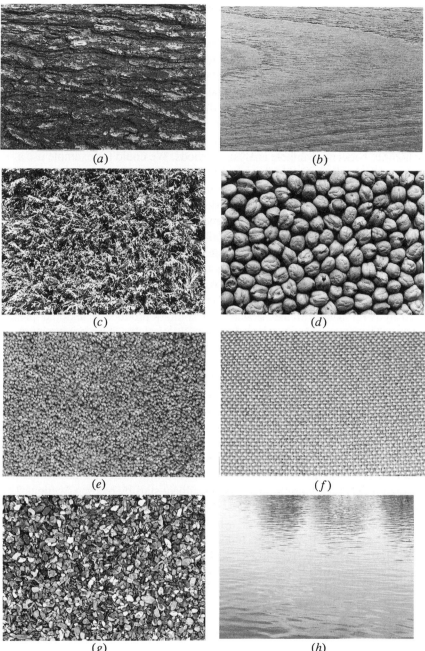

Fig. 22.1 A variety of textures. These textures demonstrate the wide variety of familiar textures which are easily recognized from their characteristic intensity patterns: (a) bark, (b) wood grain, (c) cypress hedge, (d) chickpeas, (e) carpet, (f) canvas, (g) gravel path, (h) water.

it is difficult to disentangle the available data and measure the individual values or decide the ones that are most relevant for recognition. And of course, the statistical nature of many of the parameters is by no means helpful. However, we have so far only attempted to show how complex the situation can be. In what follows we attempt to show that quite simple measures can be used to recognize and segment textures in practical situations.

Before proceeding, it is useful to recall that in the analysis of shape there is a dichotomy between available analysis methods. We could for example use a set of measures such as circularity, aspect ratio and so on which would permit a description of the shape, but which would not allow it to be reconstructed; or else we could use descriptors such as skeletons with distance function values, or moments, which would permit full and accurate reconstruction—though the set of descriptors might have been curtailed so that only limited but predictable accuracy was available. In principle, such a reconstruction criterion should be possible with texture. However, in practice there are two levels of reconstruction. In the first, we could reproduce a pattern which, to human eyes, would be indistinguishable from the off-camera texture until one compared the two on a pixel-by-pixel basis. In the second, we could reproduce a textured pattern exactly. The point is that textures are normally partially statistical in nature, so it will be difficult to obtain a pixel-by-pixel match in intensities: neither, in general, will it be worth aiming to do so. Thus texture analysis generally only aims at obtaining accurate statistical descriptions of textures, from which *apparently* identical textures can be reproduced if desired.

As the final message of this introduction, it ought to be said that very many workers have contributed and used a wide range of approaches for texture analysis over a period of more than 20 years. This fact alone makes detailed study tedious, and the statistical nature of the material raises considerable further tedium and difficulty. It is therefore recommended that those interested in obtaining a quick working view of the subject start by reading Sections 22.2 and 22.4, looking over Sections 22.5 and 22.6, and then proceeding to the end of the chapter: in this way they will bypass much of the literature review material which is part and parcel of a full study of the subject. (Section 22.4 is particularly relevant to practitioners, as it describes the Laws' texture energy approach which is intuitive, straightforward to apply in both software and hardware, and highly effective in many application areas.)

22.2 Some basic approaches to texture analysis

In Section 22.1 we defined texture as the characteristic variation in intensity of a region of an image which should allow us to recognize and describe it and to outline its boundaries. In view of the likely statistical nature of textures, this prompts us to characterize texture by the variance in intensity values taken over the whole region of the texture[*]. However, such an approach will not give a rich enough description of the texture for most purposes, and will certainly not provide any possibility of reconstruction: it will also be especially unsuitable in cases where the texels are well defined, or where there is a high degree of periodicity in the texture. On the other hand, for highly periodic textures such as arise with many textiles, it is natural to consider the use of Fourier analysis. Indeed, in the early days of image analysis, this approach was tested thoroughly, though the results were not always encouraging.

Bajcsy (1973) used a variety of ring and orientated strip filters in the Fourier domain to isolate texture features—an approach that was found to work successfully on natural textures such as grass, sand and trees. However, there is a general difficulty in using the Fourier power spectrum in that the information is more scattered than might at first be expected. In addition, strong edges and image boundary effects can prevent accurate texture analysis by this method, though Shaming (1974) and Dyer and Rosenfeld (1976) tackled the relevant image aperture problems. Perhaps more important is the fact that the Fourier approach is a global one which is difficult to apply successfully to an image that is to be segmented by texture analysis (Weszka *et al.*, 1976).

Autocorrelation is another obvious approach to texture analysis, since it should show up both local intensity variations and also the repeatability of the texture (see Fig. 22.2). An early study was carried out by Kaizer (1955). He examined how many pixels an image has to be shifted before the autocorrelation function drops to $1/e$ of its initial value, and produced a subjective measure of coarseness on this basis. However, Rosenfeld and Troy (1970a, b) later showed that autocorrelation is not a satisfactory measure of coarseness. In addition, autocorrelation is not a very good discriminator of isotropy in natural textures. Hence workers were quick to take up the co-occurrence matrix approach introduced by Haralick *et al.* in 1973: in fact, this approach not only replaced the use of autocorrelation but during the 1970s became to a large degree the "standard" approach to texture analysis.

---

* We defer for now the problem of finding the region of a texture so that we can compute its characteristics in order to perform a segmentation function. However, some preliminary training of a classifier may clearly be used to overcome this problem for supervised texture segmentation tasks.

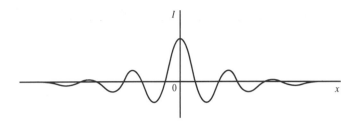

Fig. 22.2 Use of autocorrelation function for texture analysis. This diagram shows the possible 1-D profile of the autocorrelation function for a piece of material in which the weave is subject to significant spatial variation: notice that the periodicity of the autocorrelation function is damped down over quite a short distance.

22.3 Grey-level co-occurrence matrices

The grey-level co-occurrence matrix approach[*] is based on studies of the statistics of pixel intensity distributions. As hinted above with regard to the variance in pixel intensity values, single pixel statistics do not provide rich enough descriptions of textures for practical applications. Thus it is natural to consider second-order statistics obtained by considering *pairs* of pixels in certain spatial relations to each other. Hence, co-occurrence matrices are used, which express the relative frequencies (or probabilities) $P(i, j \mid d, \theta)$ with which two pixels having relative polar coordinates (d, θ) appear with intensities i, j. The co-occurrence matrices provide raw numerical data on the texture, though these data must be condensed to relatively few numbers before they can be used to classify the texture. The early paper by Haralick *et al.* (1973) gave 14 such measures, and these were used successfully for classification of many types of material (including, for example, wood, corn, grass and water). However, Conners and Harlow (1980a) found that only five of these measures were normally used, namely "energy", "entropy", "correlation", "local homogeneity" and "inertia" (note that these names do not provide much indication of the modes of operation of the respective operators).

To obtain a more detailed idea of the operation of the technique, consider the co-occurrence matrix shown in Fig. 22.3. This corresponds to a nearly uniform image containing a single region in which the pixel intensities are subject to an approximately Gaussian noise distribution, the attention being

---

[*] This is also frequently called the spatial grey-level dependence matrix (SGLDM) approach.

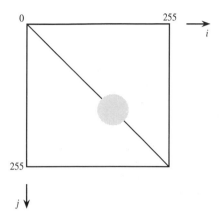

Fig. 22.3 Co-occurrence matrix for a nearly uniform grey-scale image with superimposed Gaussian noise. Here the intensity variation is taken to be almost continuous: normal convention is followed by making the j index increase downwards, as for a table of discrete values (cf. Fig. 22.5).

on pairs of pixels at a constant vector distance $\mathbf{d} = (d, \theta)$ from each other. Next consider the co-occurrence matrix shown in Fig. 22.4, which corresponds to an almost noiseless image with several nearly uniform image regions. In this case the two pixels in each pair may correspond either to the same image regions or to different ones, though if d is small they will only correspond to the same or adjacent image regions. Thus we have a set of N

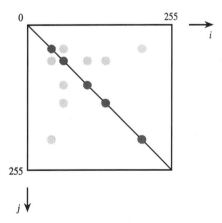

Fig. 22.4 Co-occurrence matrix for an image with several distinct regions of nearly constant intensity. Again, the leading diagonal of the diagram is from top left to bottom right (cf. Figs 22.3 and 22.5).

on-diagonal patches in the co-occurrence matrix, but only a limited number L of the possible number M of off-diagonal patches linking them, where

$$M = \binom{N}{2}$$

and $L \leqslant M$ (typically L will be of order N rather than N^2). With textured images, if the texture is not too strong, it may by modelled as noise, and the $N + L$ patches will be larger but still not overlapping. However, in more complex cases the possibility of segmentation using the co-occurrence matrices will depend on the extent to which $\mathbf{d}$ can be chosen to prevent the patches from overlapping. Since many textures are directional, careful choice of θ will clearly help with this task, though the optimum value of d will depend on several other characteristics of the texture.

As a further illustration, we consider the small image shown in Fig. 22.5(a). To produce the co-occurrence matrices for a given value of $\mathbf{d}$, we merely need to calculate the numbers of cases for which pixels a distance $\mathbf{d}$ apart have intensity values i and j. Here, we content ourselves with the two cases $\mathbf{d} = (1, 0)$ and $\mathbf{d} = (1, \pi/2)$. We thus obtain the matrices shown in Fig. 22.5(b) and (c).

This simple example demonstrates that the amount of data in the matrices is liable to be many times more than in the original image—a situation which is exacerbated in more complex cases by the number of values of d and θ that are required to accurately represent the texture. In addition, the number of grey levels will normally be closer to 256 than to 6, and the amount of matrix data varies as the square of this number. Finally, we should notice that the co-occurrence matrices merely provide a new representation: they do not themselves solve the recognition problem.

These factors mean that the grey-scale has to be compressed into a much smaller set of values, and careful choice of specific sample d, θ values must be made: in most cases it is not at all obvious how such a choice should be made, and it is even more difficult to arrange for it to be made automatically. In addition, various functions of the matrix data must be tested before the texture can be properly characterized and classified.

These problems with the co-occurrence matrix approach have been tackled in many ways: just two are mentioned here. The first is to ignore the distinction between opposite directions in the image, thereby reducing storage by 50%. The second is to work with *differences* between grey levels; this amounts to performing a summation in the co-occurrence matrices along axes parallel to the main diagonal of the matrix. The result is a set of *first*-order *difference*

(a)

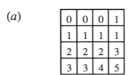

(b)

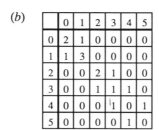

(c)

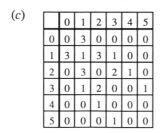

Fig. 22.5 Co-occurrence matrices for a small image. (a) The original image; (b) the resulting co-occurrence matrix for $\mathbf{d} = (1, 0)$; (c) the matrix for $\mathbf{d} = (1, \pi/2)$. Note that even in this simple case the matrices contain more data than the original image.

statistics. While these modifications have given some additional impetus to the approach, the 1980s saw a highly significant diversification of methods for the analysis of textures. Of these, Laws' approach (1979, 1980a, b) is important in that it has led to other developments which provide a systematic, adaptive means of tackling texture analysis. This approach is covered in the following section.

22.4 Laws' texture energy approach

In 1979 and 1980 Laws presented his novel texture energy approach to texture analysis (1979, 1980a, b). This involved the application of simple filters to digital images. The basic filters he used were common Gaussian, edge detector and Laplacian-type filters, and were designed to highlight points of high "texture energy" in the image. By identifying these high energy points, smoothing the various filtered images, and pooling the information from them, he was able to characterize textures highly efficiently and in a manner compatible with pipelined hardware implementations. As remarked earlier, Laws' approach has strongly influenced much subsequent work and it is therefore worth considering it here in some detail.

The Laws' masks are constructed by convolving together just three basic 1×3 masks:

$$\textbf{L3} = [1 \ 2 \ 1] \tag{22.1}$$

$$\textbf{E3} = [-1 \ 0 \ 1] \tag{22.2}$$

$$\textbf{S3} = [-1 \ 2 \ -1] \tag{22.3}$$

The initial letters of these masks indicate *L*ocal averaging, *E*dge detection and *S*pot detection. In fact, these basic masks span the entire 1×3 subspace and form a complete set. Similarly, the 1×5 masks obtained by convolving pairs of these 1×3 masks together form a complete set[*]:

$$\textbf{L5} = [1 \ 4 \ 6 \ 4 \ 1] \tag{22.4}$$

$$\textbf{E5} = [-1 \ -2 \ 0 \ 2 \ 1] \tag{22.5}$$

$$\textbf{S5} = [-1 \ 0 \ 2 \ 0 \ -1] \tag{22.6}$$

$$\textbf{R5} = [1 \ -4 \ 6 \ -4 \ 1] \tag{22.7}$$

$$\textbf{W5} = [-1 \ 2 \ 0 \ -2 \ 1] \tag{22.8}$$

(Here the initial letters are as before, with the addition of *R*ipple detection and *W*ave detection.) We can also use matrix multiplication (see also Section 3.6) to combine the 1×3 and a similar set of 3×1 masks to obtain nine 3×3 masks—for example:

$$\begin{bmatrix} 1 \\ 2 \\ 1 \end{bmatrix} [-1 \ 2 \ -1] = \begin{bmatrix} -1 & 2 & -1 \\ -2 & 4 & -2 \\ -1 & 2 & -1 \end{bmatrix} \tag{22.9}$$

The resulting set of masks also forms a complete set (Table 22.1): note that

---

[*] In principle nine masks can be formed in this way, but only five of them are distinct.

two of these masks are identical to the Sobel operator masks. The corresponding 5×5 masks are entirely similar but are not considered in detail here as all relevant principles are illustrated by the 3×3 masks.

Table 22.1 The nine 3×3 Laws' masks.

$L3^TL3$			$L3^TE3$			$L3^TS3$		
1	2	1	-1	0	1	-1	2	-1
2	4	2	-2	0	2	-2	4	-2
1	2	1	-1	0	1	-1	2	-1

$E3^TL3$			$E3^TE3$			$E3^TS3$		
-1	-2	-1	1	0	-1	1	-2	1
0	0	0	0	0	0	0	0	0
1	2	1	-1	0	1	-1	2	-1

$S3^TL3$			$S3^TE3$			$S3^TS3$		
-1	-2	-1	1	0	-1	1	-2	1
2	4	2	-2	0	2	-2	4	-2
-1	-2	-1	1	0	-1	1	-2	1

All such sets of masks include one whose components do not average to zero. Thus it is less useful for texture analysis since it will give results dependent more on image intensity than on texture. The remainder are sensitive to edge points, spots, lines and combinations of these.

Having produced images that indicate local edginess, etc., the next stage is to deduce the local magnitudes of these quantities. These magnitudes are then smoothed over a fair-sized region rather greater than the basic filter mask size (e.g. Laws used a 15×15 smoothing window after applying his 3×3 masks): the effect of this is to smooth over the gaps between the texture edges and other microfeatures. At this point the image has been transformed into a vector image, each component of which represents energy of a different type. While Laws (1980b) used both squared magnitudes and absolute magnitudes to estimate texture energy, the former corresponding to true energy and giving a better response, the latter are useful in requiring less computation:

$$E(l, m) = \sum_{i=l-p}^{l+p} \sum_{j=m-p}^{m+p} |F(i, j)| \tag{22.10}$$

$F(i, j)$ being the local magnitude of a typical microfeature which is smoothed at a general scan position (l, m) in a $(2p + 1) \times (2p + 1)$ window.

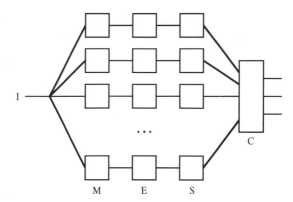

Fig. 22.6 Basic form for a Laws' texture classifier. Here I is the incoming image, M represents the microfeature calculation, E the energy calculation, S the smoothing, and C the final classification.

A further stage is required to combine the various energies in a number of different ways, providing several outputs which can be fed into a classifier to decide upon the particular type of texture at each pixel location (Fig. 22.6): if necessary, principal components analysis is used at this point to help select a suitable set of intermediate outputs.

Laws' method resulted in excellent classification accuracy quoted at (for example) 87% compared with 72% for the co-occurrence matrix method, when applied to a composite texture image of grass, raffia, sand, wool, pigskin, leather, water and wood (Laws, 1980b). He also found that the histogram equalization normally applied to images to eliminate first-order differences in texture field grey-scale distributions gave little improvement in this case.

Research was undertaken by Pietikäinen *et al.* (1983) to determine whether the precise coefficients used in the Laws' masks are responsible for the performance of his method. They found that so long as the general forms of the masks were retained, performance did not deteriorate, and could in some instances be improved. They were able to confirm that Laws' texture energy measures are more powerful than measures based on pairs of pixels (i.e. co-occurrence matrices), though Unser (1986) later questioned the generality of this result.

22.5 Ade's eigenfilter approach

In 1983 Ade investigated the theory underlying the Laws' approach, and developed a revised rationale in terms of eigenfilters. He took all possible pairs of pixels within a 3×3 window, and characterized the image intensity data by a 9×9 covariance matrix. He then determined the eigenvectors required to diagonalize this matrix. These correspond to filter masks similar to the Laws' masks, i.e. use of these "eigenfilter" masks produces images which are principal component images for the given texture. Furthermore, each eigenvalue gives that part of the variance of the original image that can be extracted by the corresponding filter. Essentially, the variances give an exhaustive description of a given texture in terms of the texture of the images from which the covariance matrix was originally derived. Clearly, the filters that give rise to low variances can be taken to be relatively unimportant for texture recognition.

It will be useful to illustrate the technique for a 3×3 window. Here we follow Ade (1983) in numbering the pixels within a 3×3 window in scan order:

1	2	3
4	5	6
7	8	9

This leads to a 9×9 covariance matrix for describing relationships between pixel intensities within a 3×3 window, as stated above. At this point we recall that we are describing a texture, and assuming that its properties are not synchronous with the pixel tessellation, we would expect various coefficients of the covariance matrix $\mathbf{C}$ to be equal: for example, C_{24} should equal C_{57}; in addition, C_{57} must equal C_{75}. It is worth pursuing this matter, as a reduced number of parameters will lead to increased accuracy in determining the remaining ones. In fact, there are

$$\binom{9}{2} = 36$$

Table 22.2 Spatial relationships between pixels in a 3×3 window.

a	b	c	d	e	f	g	h	i	j	k	l	m
9	6	6	4	4	3	3	1	1	2	2	2	2

This table shows the number of occurrences of the spatial relationships between pixels in a 3×3 window. Note that a is the diagonal element of the covariance matrix $\mathbf{C}$, and that all others appear twice as many times in $\mathbf{C}$ as indicated in the table.

ways of selecting pairs of pixels, but there are only 12 distinct spatial relationships between pixels if we disregard translations of whole pairs—or 13 if we include the null vector in the set (see Table 22.2). Thus the covariance matrix (see Section 20.8) takes the form:

$$
\mathbf{C} = \begin{bmatrix}
a & b & f & c & d & k & g & m & h \\
b & a & b & e & c & d & l & g & m \\
f & b & a & j & e & c & i & l & g \\
c & e & j & a & b & f & c & d & k \\
d & c & e & b & a & b & e & c & d \\
k & d & c & f & b & a & j & e & c \\
g & l & i & c & e & j & a & b & f \\
m & g & l & d & c & e & b & a & b \\
h & m & g & k & d & c & f & b & a
\end{bmatrix}
\tag{22.11}
$$

$\mathbf{C}$ is symmetric, and the eigenvalues of a real symmetric covariance matrix are real and positive, and the eigenvectors are mutually orthogonal (see Section 20.8). In addition, the eigenfilters thus produced reflect the proper structure of the texture being studied, and are ideally suited to characterizing it. For example, for a texture with a prominent highly directional pattern, there will be one or more high-energy eigenvalues with eigenfilters having strong directionality in the corresponding direction.

22.6 Appraisal of the Laws and Ade approaches

At this point, it will be worthwhile to compare the Laws and Ade approaches more carefully. In the Laws approach standard filters are used, texture energy images are produced, and *then* principal component analysis may be applied to lead to recognition; whereas in the Ade approach, special filters (the eigenfilters) are applied, incorporating the results of principal component analysis, following which texture energy measures are calculated and a suitable number of these are applied for recognition.

The Ade approach is superior to the extent that it permits low-value energy components to be eliminated early on, thereby saving computation. For example, in Ade's application, the first five of the nine components contain 99.1% of the total texture energy, so the remainder can definitely be ignored; in addition, it would appear that another two of the components containing respectively 1.9% and 0.7% of the energy could also be ignored, with little loss of recognition accuracy. However, in some applications textures could vary continually, and it may well *not* be advantageous to fine-tune a method to the particular data pertaining at any one time[*]. In addition, to do so may prevent an implementation from having wide generality or (in the case of hardware implementations) being so cost-effective. There is therefore still a case for employing the simplest possible complete set of masks, and using the Laws approach. However, Unser and Ade (1984) have adopted the Ade approach and report encouraging results. More recently, the work of Dewaele *et al.* (1988) is built solidly upon the Ade eigenfilter approach.

In 1986, Unser developed a more general version of the Ade technique which also covered the methods of Faugeras (1978), Granlund (1980), Wermser and Liedtke (1982). In this approach not only is performance optimized for texture classification but also it is optimized for discrimination between two textures by simultaneous diagonalization of two covariance matrices. The method was developed further by Unser and Eden (1989, 1990): this work makes a careful analysis of the use of nonlinear detectors. As a result, two levels of nonlinearity are employed, one immediately after the linear filters and designed (by employing a specific Gaussian texture model) to feed the smoothing stage with genuine variance or other suitable measures, and the other after the spatial smoothing stage to counteract the effect of the earlier filter, and aiming to provide a feature value that is in the same units as the input signal. In practical terms this means having the capability for providing an RMS texture signal from each of the linear filter channels.

Overall, the originally intuitive Laws approach emerged during the 1980s as a serious alternative to the co-occurrence matrix approach. It is as well to note that alternative methods that are potentially superior have also been devised—see for example the local rank correlation method of Harwood *et al.* (1985), and the forced-choice method of Vistnes (1989) for finding edges between different textures which apparently has considerably better accuracy than the Laws approach. Vistnes's (1989) investigation concludes that the Laws approach is limited by (i) the small scale of the masks, which can miss larger-scale textural structures, and (ii) the fact that the texture energy smoothing operation blurs the texture feature values

---

[*] For example, these remarks apply (i) to textiles, for which the degree of stretch will vary continuously during manufacture, (ii) to raw food products such as beans, whose sizes will vary with the source of supply, and (iii) to processed food products such as cakes, for which the crumbliness will vary with cooking temperature and water content.

across the edge. The latter finding (or the even worse situation where a third class of texture appears to be located in the region of the border between two textures) has also been noted by Hsiao and Sawchuk (1989, 1990) who applied an improved technique for feature smoothing; they also used probabilistic relaxation for enforcing spatial organization on the resulting data.

22.7 Fractal-based measures of texture

An important new approach to texture analysis that arose in the 1980s is that of fractals. This incorporates the observation due to Mandelbrot (1982) that measurements of the length of a coastline (for example) will vary with the size of the measuring tool used for the purpose, since details smaller than the size of the tool will be missed. If the size of the measuring tool is taken as λ, the measured quantity will be $M = n\lambda^D$, where D is known as the *fractal dimension* and must in general be larger than the immediate geometric dimension if correct measurements are to result (for a coastline we will thus have $D > 2$). Thus, when measurements are being made of 2-D textures, it is found that D can take values from 2.0 to at least 2.8 (Pentland, 1984). Interestingly, these values of D have been found to correspond roughly to subjective measures of the roughness of the surface being inspected (Pentland, 1984). It is important to note that the concept of fractal dimension is a mathematical one, which can only apply over a range of scales for physical surfaces. Nevertheless, if the concept is *not* applied over such a range of scales (as is the case for most existing texture analysis methods), then the measures of texture that result cannot be scale invariant: this means that the texture measure will have to be recalibrated for every possible scale at which it is applied. Clearly, this would be highly inconvenient, so fractal concepts strike at the heart of the texture measurement problem (seemingly, ignoring them would only be valid if for example, textiles or cornfields were being inspected from a fixed height). At the same time, it is fortunate that fractal dimension appears to represent a quantity as useful as surface roughness; furthermore, the fact that it reduces the number of textural attributes from the initially large numbers associated with certain other methods of textural measurement (such as co-occurrence matrices) is also something of an advantage.

Since some of these arguments were put forward by Pentland (1984), other workers have had problems with the approach. For example, reducing all textural measurements to the single measure D clearly cannot permit all textures to be distinguished (Keller *et al.*, 1989). Hence there have been moves to define further fractal-based measures. Mandelbrot himself brought in the concept of *lacunarity* and in 1982 provided one definition, while Voss (1986) and Keller *et al.* (1989) provided further definitions. These definitions differ in mathematical detail, but all three appear to measure a sort of texture mark–space ratio that is distinct from average surface roughness. Lacunarity seems to give good separation for most materials having similar fractal dimensions, though Keller *et al.* proposed further research to combine these measures with other statistical and structural measurements. Ultimately, an important problem of texture analysis will be to determine how many features are required to distinguish different materials and surfaces reliably, while not making the number so large that computation becomes excessive, or (perhaps more important) so that the raw data is unable to support so many features (i.e. so that the features that are chosen are not all statistically significant—see Section 20.5). Meanwhile, it should be noted that the fractal measures currently being used require considerable computation which may not be justified for applications such as automated inspection.

Finally, note that Gårding (1988) found that fractal dimension is not always equivalent to subjective judgements of roughness: in particular he found that a region of Gaussian noise of low amplitude superimposed on a constant grey level will have a fractal dimension that approaches 3.0—a rather high value, which is contrary to our judgement of such surfaces as being quite smooth. (An interpretation of this result is that highly noisy textures appear exactly like 3-D landscapes in relief!)

22.8 Shape from texture

This is another topic in texture analysis that has developed strongly during the 1980s. After early work by Bajcsy and Liebermann (1976) for the case of planar surfaces, Witkin (1981) and Kender (1981) significantly extended this work and at the same time laid the foundations for general development of the whole subject. Many papers have followed (e.g. Aloimonos and Swain, 1985; Aloimonos, 1988; Blostein and Ahuja, 1989; Katatani and Chou, 1989; Stone, 1990) and there is no space to cover all this work here. In general, workers have studied how an assumed standard texel shape is distorted and its size changed by 3-D projections; they then relate this to the local orientation of the surface. Since the texel distortion varies as the cosine of the angle between the line of sight and the local normal to the surface plane, essentially similar "reflectance map" analysis is required as in the case of shape-from-shading estimation. An alternative approach adopted by

Chang *et al.* (1987) involves texture discrimination by projective invariants (see Chapter 18). More recently, Singh and Ramakrishna (1990) exploited shadows and integrated the information available from texture and from shadows: the problem is that in many situations insufficient data arises from either source taken on its own. Though the results of Singh and Ramakrishna are only preliminary, they indicate the direction that texture work must develop if it is to be used in practical image understanding systems.

22.9   Markov random field models of texture

Markov models have long been used for texture synthesis, to help with the generation of realistic images. However, they have also proved increasingly useful for texture analysis. In essence a Markov model is a one-dimensional construct in which the intensity at any pixel depends only upon the intensity of the previous pixel in a chain and upon a transition probability matrix. For images this is too weak a characterization, and various more complex constructs have been devised. Interest in such models dates from as early as 1965 (Abend *et al.*, 1965) (see also Woods, 1972), though such work is accorded no great significance in Haralick's extensive review of 1979. However, by 1982 the situation was starting to change (Hansen and Elliott, 1982), and a considerable amount of further work was soon being published (e.g. Cross and Jain, 1983; Geman and Geman, 1984; Derin and Elliott, 1987) as Gibbs distributions came to be used for characterizing Markov random fields: in fact Ising's much earlier work in statistical mechanics (1925) was the starting point for these developments.

Available space does not permit details of these algorithms to be given here. However, by 1987 impressive results for texture segmentation of real scenes were being achieved using this approach (Derin and Elliott, 1987; Cohen and Cooper, 1987). In particular, it appears to be able to cope with highly irregular boundaries, local boundary estimation errors normally lying between 0 and 3 pixels. Unfortunately, these algorithms depend on iterative processing of the image, and hence tend to require very considerable amounts of computation. It is stated by Cohen and Cooper (1987) that little processing is required, though there is an inbuilt assumption that a suitable highly parallel processor will be available. On the other hand it was also shown by Cohen and Cooper (1987) that a hierarchical segmentation algorithm can achieve considerable computational savings, these being greatest when the texture fields have spatially constant parameters.

Finally, it is of interest that stochastic models are not only useful for modelling texture and segmenting one textured region from another: they can be useful for more general segmentation purposes, including cases where boundary edges are quite noisy (Geman, 1987). Indeed, the fact that a method such as this now effectively takes texture as a mere special case demonstrates the power of the whole approach. Note, however, the comments (Geman, 1987) that parallel hardware improvement rates are consistently underestimated, and (Murray *et al.*, 1986) that the relevant stochastic relaxation processes which take some hours on a VAX take under 1 minute on a suitable parallel array processor: *both* of these figures give some idea of the scale of the computational problem!

22.10 Structural approaches to texture analysis

It has already been remarked that textures approximate to a basic textural element or primitive that is replicated in a more or less regular manner. Structural approaches to texture analysis aim to discern the textural primitive and to determine the underlying gross structure of the texture. Early work (e.g. Pickett, 1970) suggested the structural approach, though little research on these lines was carried out until the late 1970s. Work of this type has been described by Davis (1979), Conners and Harlow (1980b), Matsuyama *et al.* (1982), Vilnrotter *et al.* (1986), and Kim and Park (1990). An unusual and interesting paper by Kass and Witkin (1987) shows how orientated patterns from wood grain, straw, fabric and fingerprints, and also spectrograms and seismic patterns, can be analysed: the method adopted involves building up a flow coordinate system for the image, though the method rests more on edge pattern orientation analysis than on more usual texture analysis procedures. A similar statement may be made about the topologically invariant texture descriptor method of Eichmann and Kasparis (1988), which relies on Hough transforms for finding line structures in highly structured textiles. More recently, pyramidal approaches have been applied to structural texture segmentation (Lam and Ip, 1994).

22.11 Concluding remarks

In this chapter we have seen the difficulties of analysing textures: these arise from the potential, and in many cases the frighteningly real complexities of textures—not least from the fact that their properties are often largely statistical in nature. The erstwhile widely used grey-scale co-occurrence matrix approach has been seen to have distinct computational shortcomings. First, many co-occurrence matrices are in principle required (with different values of d and θ) in order to adequately describe a given texture; second, the co-occurrence matrices can be very large and, paradoxically, may hold more data than the image data they are characterizing—especially if the range of grey-scale values is large. In addition, many sets of co-occurrence matrices may be needed to allow for variation of the texture over the image, and if necessary to initiate segmentation. Hence co-occurrence matrices need to be significantly compressed though in most cases it is not at all obvious *a priori* how this should be achieved, and it is even more difficult to arrange for it to be carried out automatically. This probably explains why attention shifted during the 1980s to other approaches, including particularly Laws' technique and its variations (especially that of Ade). Other natural developments were fractal-based measures and the attention given to Markov approaches; a further important development, which could not be discussed here for space reasons, is the Gabor filter technique (see for example Jain and Farrokhnia, 1991).

In the past few years, considerable attention has also been paid to neural network methodologies, since these are able to extract the rules underlying the construction of textures, apparently without the need for analytic or statistical effort. Although neural networks have been able to achieve certain practical goals in actual applications, there is little firm evidence whether they are actually superior to Markov or other methods because conclusive comparative investigations have still not been performed: however, they do seem to offer solutions requiring minimal computational load.

22.12 Bibliographical and historical notes

Early work on texture analysis was carried out by Haralick *et al.* (1973), and in 1976 Weska and Rosenfeld applied textural analysis to materials inspection. The area was reviewed by Zucker (1976a) and by Haralick (1979), and excellent accounts appear in the books by Ballard and Brown (1982) and Levine (1985).

At the end of the 1970s the Laws technique (1979, 1980a, b) emerged upon the scene (which had up till then been dominated by the co-occurrence matrix approach), and led to the principal components approach of Ade (1983) which was further developed by Dewaele *et al.* (1988), Unser and Eden (1989, 1990) and others; for related work on the optimization of convolution filters see Benke and Skinner (1987). The direction taken by Laws was particularly valuable as it showed how texture analysis could be implemented straightforwardly and in a manner consistent with real-time applications such as inspection.

The 1980s also saw other new developments, such as the fractal approach led by Pentland (1984), and a great amount of work on Markov random field models of texture. Here the work of Hansen and Elliott (1982) was very formative, though the names Cross, Derin, D. Geman, S. Geman and Jain come up repeatedly in this context (see Section 22.9). Bajcsy and Lieber-mann (1976), Witkin (1981) and Kender (1981) pioneered the *shape from texture* concept, which has received much attention ever since. Finally, the last decade has seen much work on the application of neural networks to texture analysis, for example Greenhill and Davies (1993) and Patel *et al.* (1994).

A number of reviews and useful comparative studies have been made, including Van Gool *et al.* (1985), Du Buf *et al.* (1990), Ohanian and Dubes (1992) and Reed and Du Buf (1993). For further work on texture analysis related to inspection for faults and foreign objects, see Chapter 19.

23

Image Acquisition

23.1 Introduction

When implementing a vision system, nothing is more important than image acquisition. Any deficiencies of the initial images can cause great problems with image analysis and interpretation. An obvious example is that of lack of detail due to insufficient contrast or poor focusing of the camera: this can have the effect at best that the dimensions of objects will not be accurately measurable from the images, and at worst that the objects will not even be recognizable, so the purpose of vision cannot be fulfilled. This chapter examines the problems of image acquisition.

Before proceeding it is as well to note that vision algorithms are of use in a variety of areas where visual pictures are not directly input. For example, vision techniques (image processing, image analysis, recognition and so on) can be applied to seismographic maps, to pressure maps (whether these arise from handwriting on pressure pads or from weather data), infrared, ultraviolet, X-ray and radar images, and a variety of other cases. There is no space here to consider methods for acquisition in any of these instances and attention is concentrated on purely optical methods. In addition, space does not permit a detailed study of methods for obtaining range images using laser scanning and ranging techniques, while other methods that are specialized for 3-D work will also have to be passed by. Instead, we concentrate on (i) lighting systems for obtaining intensity images, (ii) technology for receiving and digitizing intensity images, and (iii) basic theory such as the Nyquist sampling theorem which underlies this type of work.

First we consider how to set up a basic system that might be suitable for the thresholding and edge detection work of Chapters 2–5.

23.2   Illumination schemes

The simplest and most obvious arrangement for acquiring images is that shown in Fig. 23.1. A single source provides light over a cluster of objects on a worktable or conveyor, and this scene is viewed by a camera directly overhead. The source is typically a tungsten light which approximates to a point source. Assuming for now that the light and camera are some distance away from the objects, and are in different directions relative to them, it may be noted that:

(i)   different parts of the objects are lit differently, because of variations in the angle of incidence, and hence have different brightnesses as seen from the camera;

(ii)   the brightness values also vary because of the differing absolute reflectivities* of the object surfaces;

(iii)   the brightness values vary with the specularities* of the surfaces in places where the incident, emergent and phase angles are compatible with specular reflection (Chapter 15);

(iv)   parts of the background and of various objects are in shadow and this again affects the brightness values in different regions of the image;

(v)   other more complex effects occur because light reflected from some objects will cast light over other objects—factors that can lead to very complicated variations in brightness over the image.

Clearly, even in this apparently simple case—one point light source and one camera—the situation can become quite complex. However, (v) is normally a reasonably marginal effect and is ignored in what follows. In addition, effect (iii) can often be ignored except in one or two small regions of the image where sharply curved pieces of metal give rise to glints. This still leaves considerable scope for complication due to factors (i), (ii) and (iv).

There are two important reasons for viewing the surfaces of objects: the first is when we wish to locate objects and their facets, and the second is

---

* Referring to equation (15.12), R_0 is the absolute surface reflectivity and R_1 is the specularity.

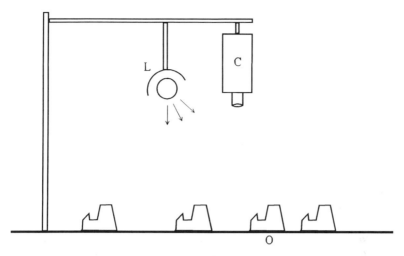

Fig. 23.1 Simple arrangement for image acquisition: C, camera; L, light with simple reflector; O, objects on worktable or conveyor.

when we wish to scrutinize the surfaces themselves. In the first instance it is important to try to highlight the facets by arranging that they are lit differently, so that their edges stand out clearly. In the second instance it might be preferable to do the opposite—i.e. to arrange that the surfaces are lit very similarly, so that any variations in reflectivity caused by defects or blemishes stand out plainly. The existence of effects (i) and (ii) implies that it is difficult to achieve both of these things at the same time: one set of lighting conditions is required for optimum segmentation and location, and another set for optimum surface scrutiny. In most of this book object location has been regarded as the more difficult task and therefore the one that needs the most attention. Hence we have imagined that the lighting scheme is set up for this purpose. In principle, a point source of light is well adapted to this situation. However, it is easy to see that if a very diffuse lighting source is employed, then angles of incidence will tend to average out and effect (ii) will dominate over (i) so that, *to a first approximation*, the observed brightness values will represent variations in surface reflectance. In fact, "soft" or diffuse lighting also subdues specular reflections (effect (iii)), so that for the most part they can be ignored.

Returning to the case of a single point source, recall (effect (iv)) that shadows can become important. There is one special case when this is not so, and that is when the light is projected from exactly the same direction as the camera: we return to this case below. Shadows are a persistent cause of complications in image analysis. One problem is that it is not a trivial task to identify them, so they merely contribute to the overall complexity of any image and in particular add to the number of edges that have to be examined in order to find objects. They also make it much more difficult to use simple thresholding. (However, note that shadows can sometimes provide information that is of vital help in interpreting complex 3-D images—see, for example, Sections 15.6 and 15.18.)

23.2.1 Eliminating shadows

The above considerations suggest that it would be highly convenient if shadows could be eliminated. A strategy for achieving this is to lower their contrast by using several light sources. Then the region of shadow from one source will be a region of illumination from another, and shadow contrast will be lowered dramatically. Indeed, if there are n lights, many positions of shadow will be illuminated by $n-1$ lights and their contrast will be so low that they can be eliminated by straightforward thresholding operations. However, if objects have sharp corners or concavities, there may still be small regions of shadow which are illuminated by only one light or perhaps no light at all; these regions will be immediately around the objects, and if the objects appear dark on a light background, shadows could make the objects appear enlarged, or cause shadow lines immediately around them. For light objects on a dark background this is normally less of a problem.

Clearly, it seems best to aim for large numbers of lights so as to make the shadows more diffuse and less contrasting, and in the limit it appears that we are heading for the situation of soft lighting discussed earlier. However, this is not quite so. What is often required is a form of diffuse lighting that is still directional—as in the case of a diffuse source of restricted extent directly overhead: this can be provided very conveniently by a continuous ring light around the camera. This technique is found to eliminate shadows highly effectively, while retaining sufficient directionality to permit a good measure of segmentation of object facets to be achieved, i.e. it is an excellent compromise although it is certainly not ideal. For these reasons it is worth describing its effects in some detail. In fact, it is clear that it will lead to good segmentation of facets whose boundaries lie in horizontal planes but to poor segmentation of those whose boundaries lie in vertical planes.

The situation just described is very useful for analysing the shape profiles of objects with cylindrical symmetry. It is worth noting the case shown in Fig. 23.2, which involves a special type of chocolate biscuit with jam underneath the chocolate. If this is illuminated by a continuous ring light

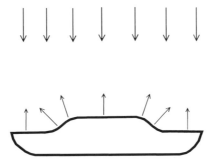

Fig. 23.2 Illumination of a chocolate-and-jam biscuit. This figure shows the cross-section of a particular type of round chocolate biscuit with jam underneath the chocolate. The arrows show how light arriving from vertically overhead is scattered by the various parts of the biscuit.

fairly high overhead (the proper working position), the region of chocolate above the edge of the jam reflects the light obliquely and appears darker than the remainder of the chocolate. However[*], if the ring light is lowered to near the worktable, the region above the edge of the jam appears *brighter* than the rest of the chocolate because it scatters light upwards rather than sideways. Clearly there is also[*] a particular height at which the ring light can make the jam boundary disappear (Fig. 23.3), this height being dictated by the various angles of incidence and reflection and by the relative direction of the ring light. In comparison, if the lighting were made completely diffuse, these effects would tend to disappear and the jam boundary would always have very low contrast.

Suitable fluorescent ring lights are readily available and straightforward to use, and provide a solution that is more practicable than the alternative means of eliminating shadows mentioned above—that of illuminating objects directly from the camera direction, for example via a half-silvered mirror.

Earlier, the one case we did not completely solve arose when we were attempting to segment facets whose joining edges were in vertical planes. There appears to be no simple way of achieving a solution to this problem without recourse to switched lights (see Chapter 15): this is not discussed further here.

We have now identified various practical forms of lighting that can be used to highlight various object features and which eliminate complications

---

[*] The latter two situations are described for interest only.

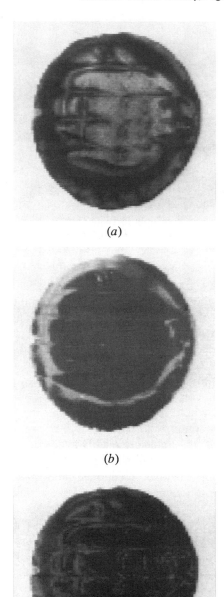

(a)

(b)

(c)

as far as possible. These types of lighting are restricted in what they can achieve (as would clearly be expected from the shape-from-shading ideas of Chapter 15). However, they are exceedingly useful in a variety of applications. A final problem is that two lighting schemes may have to be used in turn, the first for locating objects and the second for inspecting their surfaces. However, this problem can largely be overcome by *not* treating the latter case as a special one requiring its own lighting scheme, but rather noting the direction of lighting and *allowing* for the resulting variation in brightness values by taking account of the known shape of the object. The opposite approach is generally of little use unless other means are used for locating the object. However, the latter situation frequently arises in practice: imagine that a slab of concrete or a plate of steel is to be inspected for defects. In that case the position of the object is known and it is clearly best to set up the most uniform lighting arrangement possible, so as to be most sensitive to small variations in brightness at blemishes. This is, then, an important practical problem, to which we now turn.

23.2.2   Arranging a region of uniform illumination

To arrange a region of uniform illumination on a surface, an analogy can be made with Helmholtz coils in elementary electricity: use two coils symmetrically arranged about the test position, so that where the intensity of one is falling, the increase in intensity from the other cancels it to first order (in fact, this arrangement eliminates *all* odd orders); then arrange that the distance apart is such that the second-order variation at the test position also cancels. In the case of uniform lighting of an area, we are trying to optimize uniformity in two dimensions rather than in one, and there is the added problem that we do not know how specular the reflection from the surface will be. To simplify matters we take the specularity R_1 to be zero. In addition, we assume that the angle of incidence must be taken into account, via Lambert's cosine law. However, the system is still not fully specified, since we may have a ring of point sources, a set of straight fluorescent tubes arranged in a square, a ring light or a number of other arrangements. In any

---

Fig. 23.3   Appearance of the chocolate-and-jam biscuit of Fig. 23.2: (a) how the biscuit appears to a camera directly overhead when illuminated as in Fig. 23.2; (b) appearance when the lights are lowered to just above table level; (c) appearance when the lights are raised to an intermediate level making the presence of the jam scarcely detectable.

of these cases, a symmetrical arrangement will cancel all odd orders in the intensity variation directly under the camera; then raising the set of lights will give a single peak, while lowering it will give peaks under the individual lights; getting the height just right will ensure a maximally flat intensity pattern in which the second-order variation is exactly cancelled out. Simple calculations show that the height should in general be of the order of the linear dimensions of the lighting scheme (e.g. it should be about three-quarters of the diameter for a ring light). In addition, it should be noticed that the size of the region of uniform illumination is also directly proportional to the linear dimensions of the lighting scheme.

The variety of lighting arrangements that are compatible with this general principle makes it not worthwhile to give any detailed dimensional specifications—especially as the lights might be highly directional rather than approximating to point or line sources. However, it is worth underlining that adjustment of just one parameter (the height) permits uniform illumination to be achieved over a reasonable region. Note that (as for a Chebyshev filter) it may be better to arrange a slightly less uniform brightness over a larger region than absolutely uniform brightness over a small region (as is achieved by exact cancellation of the second spatial derivatives of intensity). It is left to empirical tests to finalize the details of the design.

Finally, it should be reiterated that such a lighting scheme is likely to be virtually useless for segmenting object facets from each other—or even for discerning relatively low curvatures on the surface of objects: its particular value lies in the scrutiny of surfaces via their absolute reflectivities, without the encumbrance of switched lights (see Chapter 15). It should also be emphasized that the aim of the discussion in the past few sections has been to achieve as much as possible with a simple static lighting scheme set up systematically. Naturally such solutions are compromises and again no substitute for the full rigour of switched lighting schemes.

23.2.3   Use of linescan cameras

Throughout the above discussion it has been assumed implicitly that a conventional "area" camera is employed to view the objects on a worktable. However, when products are being manufactured in a factory they are very frequently moved from one stage to another on a conveyor. Stopping the conveyor to acquire an image for inspection would impose unwanted design problems: for this reason use is made of the fact that the speed of the conveyor is reasonably uniform, and an area image is built up by taking successive linear snapshots. This is achieved with a linescan camera which consists of a row of photocells on a single integrated circuit sensor; the orientation of the line of photocells must of course be normal to the direction of motion. More will be said below about the internal design of linescan and other cameras. However, we concentrate here on the lighting arrangement to be used with such a camera.

When using a linescan camera it is natural to select a lighting scheme that embodies the same symmetry as the camera: indeed, the most obvious such scheme is a pair of long fluorescent tubes parallel to the line of the camera (and perpendicular to the motion of the conveyor). We here caution against this "obvious" scheme, since a small round object (for example) will not be lit symmetrically. Of course there are difficulties in considering this problem in that different parts of the object are viewed by the linescan camera at different moments, but it should be noted that for small objects a linear lighting scheme will not be isotropic: this could lead to small distortions being introduced in measurements of object dimensions. This means that in practice the ring and other symmetrical lighting schemes described above are likely to be more closely optimal even when a linescan camera is used. For larger objects much the same situation applies, although the geometry is more complex to work out in detail.

Finally, the comment above that conveyor speeds are "reasonably uniform" should be qualified. The author has come across cases where this is true only as a first approximation. As with many mechanical systems, conveyor motion can be unreliable: for example, it can be jerky, and in extreme cases not even purely longitudinal! Such circumstances frequently arise through a variety of problems which cause slippage relative to the driving rollers—the effects of wear or of an irregular join in the conveyor material, misalignment of the driving rollers, and so on. Furthermore, the motors controlling the rollers may not operate at constant speed, either in the short term (e.g. because of varying load) or in the longer term (e.g. because of varying mains frequency and voltage). While, therefore, it cannot be assumed that a conveyor will operate in an ideal way, careful mechanical design can minimize these problems. However, when high accuracy is required, it will be necessary to monitor the conveyor speed, perhaps by using the optically coded disc devices that are widely available, and feeding appropriate distance marker pulses to the controlling computer. Even with this method, it will be difficult to match in the longitudinal direction the extremely high accuracy[*] available from the

---

[*] A number of linescan cameras are now available with 4096 or greater numbers of photocells in a single linear array. In addition, these arrays are fabricated using very high-precision technology (see Section 23.3), so considerable reliance can be placed on the data they provide.

linescan camera in the lateral direction. However, images of 512×512 pixels that are within 1 pixel accuracy in each direction should normally be available.

23.3 Cameras and digitization

For a good many years the camera that was normally used for image acquisition was the TV camera with a vidicon or related type of vacuum tube (Fig. 23.4): even today, many such cameras are still in use and hence it is worth describing their main features and mode of operation. Basically, light is focused by a lens onto a layer of semiconducting material on the back surface of a glass faceplate. Light falling on the semiconductor generates electron–hole pairs, the electrons being attracted towards the target electrode—a thin, high-conductivity transparent film between the glass faceplate and the semiconductor—which is connected to a small positive potential. Thus the free surface of the semiconductor acquires a positive charge that varies in a manner roughly proportional to the local light intensity. An electron beam is scanned over the surface of the target in a regular raster scan—this generally being achieved by means of currents in a set of scanning coils. Where the electron beam hits the surface it discharges the positive charge built up by the action of light, and the remaining electrons in the beam pass through a mesh close to (and on the gun side of) the target, which is connected to a positive potential, and are collected by the wall anode. The current in the mesh circuit is monitored, giving a waveform containing temporally coded information on the light intensity falling on various parts of the target: hence suitable processing can be used to reconstruct the image, or to display it visually on a TV monitor which employs the same spatiotemporal scanning arrangement.

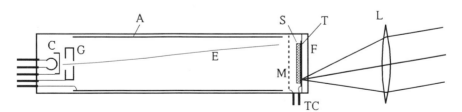

Fig. 23.4 Cross-section of vidicon camera tube: A, wall anode; C, cathode; E, electron beam; F, transparent glass faceplate; G, arrangement of grids (part of electron gun); L, lens; M, mesh (final grid for decelerating electron beam); S, layer of semiconducting material; T, thin transparent target; TC, target connection. Magnetic focusing, scanning and alignment coils are not shown.

The scanning arrangements of such cameras have become standardized, first to 405 lines, then to the current 625 lines (or 525 lines in the USA). In addition, it is usual to interlace the image—i.e. to scan odd lines in one frame and even lines in the next frame, then repeat the process, each full scan taking 1/25 second (1/30 second in the USA). There are also standardized means for synchronizing cameras and monitors, using line and frame "sync" pulses.

It is important to notice that the output of these cameras is inherently analogue, consisting of a continuous voltage variation, although this applies only along the line directions: the scanning action is discrete in that lines are used, making the output of the camera part analogue and part digital. Hence, before the image is available as a set of discrete pixels, the analogue waveform has to be sampled. Since some of the line scanning time is taken up with frame synchronization pulses, only about 550 lines are available for actual picture content. In addition, the aspect ratio of a standard TV image is 4:3 and it is common to digitize TV pictures as 512×768 or as 512×512 pixels. Note also that after the analogue waveform has been sampled and pixel intensity values have been established, it is still necessary to digitize the intensity values. The noise output of vidicons is relatively high and hence only 5–6 bits of useful grey-scale information are commonly available at each pixel location. Finally, it should be noted that the electron beam is scanned electromagnetically—a process that is decidedly nonlinear—so only a certain restricted region of the picture near the centre will be undistorted.

A variety of such cameras have been developed: these include the standard vidicon, silicon vidicon, newvicon, chalnicon, saticon, image isocon and many more. These have various characteristics such as high sensitivity for use in dark studios, high speed for use with fast-moving pictures, and so on. However, modern solid-state cameras are much more compact and robust, and generate less noise; a very important additional advantage is that they are not susceptible to distortion[*], because the pixel pattern is fabricated very accurately by the usual integrated circuit photolithography techniques. These have been gradually replacing vacuum tube cameras except in special situations.

Most solid-state cameras currently available are of the self-scanned CCD type and attention is concentrated on these in what follows. In a solid-state CCD camera, the target is a piece of silicon semiconductor which possesses

---

However, this does not prevent distortions from being introduced by other mechanisms—poor optics, poor lighting arrangements, perspective effects, and so on.

an array of photocells at the pixel positions. Hence this type of camera digitizes the image from the outset, although in one respect—that signal amplitude represents light intensity—the image is still analogue. The analogue voltages (or more accurately the charges) representing the intensities are systematically passed along analogue shift registers to the output of the instrument, where they may be digitized to give 6–8 bits of grey-scale information (the main limitation here being lack of uniformity amongst the photosensors rather than noise *per se*).

The scanning arrangement of a basic CCD camera is not fixed, being fired at any desired rate by externally applied pulses. CCD cameras have been introduced which simulate conventional TV (e.g. vidicon) cameras, with their line and frame sync pulses. In such cases the camera initially digitizes the image into pixels and these are shifted in the standard way to an output connector, being passed en route through a low-pass filter in order to remove the pixellation. When the resulting image is presented on a TV monitor, it appears exactly like a normal TV picture. Some problems are occasionally noticeable, however, when the depixellated image is resampled by a new digitizer for use in a computer frame store; such problems are due to interference or beating between the two horizontal sampling rates to which the information has been subjected, and take the form of faint vertical lines. Clearly, more effective low-pass filters should permit these problems to be eliminated.

Another important problem relating to cameras should be discussed briefly here—that of colour response. The tube of the conventional TV camera has a spectral response curve that peaks at much the same position as the spectral pattern of ("daylight") fluorescent lights—which itself matches the response of the human eye (Table 23.1). However, CCD cameras have significantly lower response to the spectral pattern of fluorescent tubes (Table 23.1). In general this may not matter too greatly but when objects are moving the integration time of the camera is limited and sensitivity can suffer. In such cases the spectral response is an important factor and may dictate against use of fluorescent lights (this is particularly relevant where CCD linescan cameras are used with fast-moving conveyors).

An important factor in the choice of cameras is the delay lag that occurs before a signal disappears. This clearly causes problems with moving images. Fortunately, the effect is entirely eliminated with CCD cameras, since the action of reading an image wipes the old image. However, moving images require frequent reading and this implies loss of integration time and therefore loss of sensitivity—a factor that normally has to be made up by increasing the power of illuminating sources. Camera "burn-in" is another effect that is absent with CCD cameras but which causes severe problems

Table 23.1   Spectral responses.

Device	Band (nm)	Peak (nm)
Vidicon	200–800	~550
CCD	400–1000	~800
Fluorescent tube	400–700	~600
Human eye	400–700	~550

In this table, the response of the human eye is included for reference. Note that the CCD response peaks at much higher wavelength than the vidicon or fluorescent tube, and therefore is often at a disadvantage when used in conjunction with the latter.

with certain types of conventional camera: it is the long-term retention of picture highlights in the light-sensitive material, which makes it necessary to protect the camera against bright lights and to take care to make use of lens covers whenever possible. Finally, "blooming" is the continued generation of electron–hole pairs even when the light-sensitive material is locally saturated with carriers, with the result that the charge spreads and causes highlights to envelop adjacent regions of the image. Both CCD and conventional camera tubes are subject to this problem, although it is inherently worse for CCDs, and this has led to the production of antiblooming structures in these devices: space precludes detailed discussion of the situation here.

23.3.1   Digitization

The remaining important item to be studied in this context is that of digitization, i.e. conversion of the original analogue signals into digital form. There are many types of analogue-to-digital converter (ADC) but the ones that are used for digitizing images have so much data to process—usually in a very short time if real-time analysis is called for—that special types have to be employed. The only one considered here is the "flash" ADC, so called because it digitizes all bits simultaneously, in a flash. In fact, it possesses $n-1$ analogue comparators to separate n grey levels, followed by a priority encoder to convert the $n-1$ results into normal binary code. Such devices produce a result in a very few nanoseconds and their specifications are generally quoted in megasamples per second (typically in the range

50–200 megasamples/second). For some years these were available only in 4-, 5- or 6-bit versions (apart from some very expensive parts) but nowadays it is possible to obtain 8-bit versions for a modest cost (£20 or so): such 8-bit devices are probably sufficient for most needs considering that a certain amount of sensor noise, or variability, is usually present below these levels and that it is *very* difficult to engineer lighting to this accuracy.

23.4 The sampling theorem

The Nyquist sampling theorem underlies all situations where continuous signals are sampled and is especially important where patterns are to be digitized and analysed by computers. This makes it highly relevant both with visual patterns and with acoustic waveforms: hence it is described briefly in this section.

Consider the sampling theorem first in respect of a 1-D time-varying waveform. The theorem states that a sequence of samples (Fig. 23.5) of such a waveform contains all the original information and can be used to regenerate the original waveform exactly, but only if (i) the bandwidth W of the original waveform is restricted and (ii) the rate of sampling f is at least twice the bandwidth of the original waveform—i.e. $f \geq 2W$. Assuming that samples are taken every T seconds, this means that $1/T \geq 2W$.

At first it may be somewhat surprising that the original waveform can be reconstructed exactly from a set of discrete samples. However, the two conditions for achieving this are very stringent. What they are demanding in effect is that the signal must not be permitted to change unpredictably (i.e. at too fast a rate) or else accurate interpolation between the samples will not

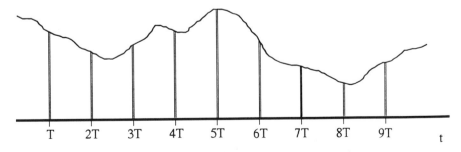

Fig. 23.5 The process of sampling a time-varying signal: a continuous time-varying 1-D signal is sampled by narrow sampling pulses at a regular rate $f_r = 1/T$ which must be at least twice the bandwidth of the signal.

prove possible (the errors that arise from this source are called "aliasing" errors).

Unfortunately, the first condition is virtually unrealizable, since it is close to impossible to devise a low-pass filter with a perfect cut-off. Recall from Chapter 3 that a low-pass filter with a perfect cut-off will have infinite extent in the time domain, so any attempt at achieving the same effect by time domain operations must be doomed to failure. However, acceptable approximations can be achieved by allowing a "guard-band" between the desired and actual cut-off frequencies. This means that the sampling rate must be higher than the Nyquist rate (in telecommunications, satisfactory operation can generally be achieved at sampling rates around 20% above the Nyquist rate—see Brown and Glazier, 1974).

One way of recovering the original waveform is by applying a low-pass filter. This approach is intuitively correct, since it acts in such a way as to broaden the narrow discrete samples until they coalesce and sum to give a continuous waveform. Indeed, this method acts in such a way as to eliminate the "repeated" spectra in the transform of the original sampled waveform (Fig. 23.6): this in itself shows why the original waveform has to be narrow-banded before sampling—so that the repeated and basic spectra of the waveform do not cross over each other and become impossible to separate with a low-pass filter. The idea may be taken further because the Fourier transform of a square cut-off filter is the sinc ($\sin u / u$) function (Fig. 23.7). Hence the original waveform may be recovered by convolving the samples with the sinc function (which in this case means replacing them by sinc functions of corresponding amplitudes). This has the effect of broadening out the samples as required, until the original waveform is recovered.

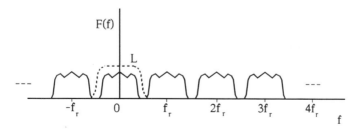

Fig. 23.6 Effect of low-pass filtering to eliminate repeated spectra in the frequency domain (f_r, sampling rate; L, low-pass filter characteristic). This diagram shows the repeated spectra of the frequency transform $F(f)$ of the original sampled waveform. It also demonstrates how a low-pass filter can be expected to eliminate the repeated spectra to recover the original waveform.

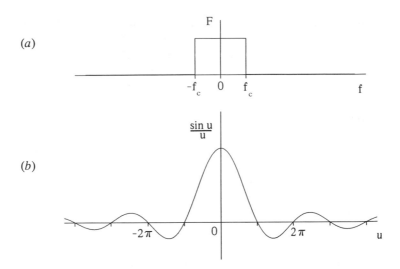

Fig. 23.7   The sinc ($\sin u/u$) function shown in (b) is the Fourier transform of a square pulse (a) corresponding to an ideal low-pass filter. In this case, $u = 2\pi f_c t$, f_c being the cut-off frequency.

So far we have considered the situation only for 1-D time-varying signals. However, recalling that there is an exact mathematical correspondence between time and frequency domain signals on the one hand and spatial and spatial frequency signals on the other, the above ideas may all be applied immediately to each dimension of an image (although the condition for accurate sampling now becomes $1/X \geqslant 2W_X$, where X is the spatial sampling period and W_X is the spatial bandwidth). Here we accept this correspondence without further discussion and proceed to apply the sampling theorem to image acquisition.

Consider next how the signal from a TV camera may be sampled rigorously according to the sampling theorem. First, it is plain that the analogue voltage comprising the time-varying line signals must be narrow-banded, e.g. by a conventional electronic low-pass filter. However, how are the images to be narrow-banded in the vertical direction? The same question clearly applies for both directions with a solid-state area camera. Initially, the most obvious solution to this problem is to perform the process optically, perhaps by defocusing the lens; however, the optical transform function for this case is frequently (i.e. for extreme cases of defocusing) very odd, going negative for some spatial frequencies and causing contrast reversals; hence this solution is far from ideal (Pratt, 1978). Alternatively, we could use a diffraction-limited optical system or perhaps pass the focused beam through some sort of patterned or frosted glass to reduce the spatial bandwidth artificially. None of these techniques will be particularly easy to apply nor (apart possibly from the second) will it give accurate solutions. However, this problem is not as serious as might be imagined. If the sensing region of the camera (per pixel) is reasonably large, and close to the size of a pixel, then the averaging inherent in obtaining the pixel intensities will in fact perform the necessary narrow-banding (Fig. 23.8). To analyse the situation in more detail, note that a pixel is essentially square with a sharp cut-off at its borders: thus its spatial frequency pattern is a 2-D sinc function, which (taking the central positive peak) approximates to a low-pass spatial frequency filter: this approximation improves somewhat as the border between pixels becomes more fuzzy.

The point here is that the worst case from the point of view of the sampling theorem is that of extremely narrow discrete samples, but clearly this worst case is most unlikely to occur with most cameras. However, this does not mean that sampling is automatically ideal—and indeed it is not, since the spatial frequency pattern for a sharply defined pixel shape has (in principle) infinite extent in the spatial frequency domain. The review by Pratt (1978) clarifies the situation and shows that there is a tradeoff between aliasing and resolution error. Overall, it is underlined here that quality of sampling will be one of the limiting factors if the greatest precision in image

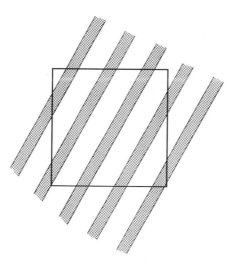

Fig. 23.8 Low-pass filtering carried out by averaging over the pixel region: an image with local high-frequency banding is to be averaged over the whole pixel region by the action of the sensing device.

measurement is aimed for: if the bandwidth of the presampling filter is too low, resolution will be lost; if it is too high, aliasing distortions will creep in; and if its spatial frequency response curve is not suitably smooth, a guard band will have to be included and performance will again suffer.

23.5 Concluding remarks

This chapter has aimed to give some background to the problems of acquiring images, particularly for inspection applications. Methods of illumination were deemed to be worthy of considerable attention since they furnish means by which the practitioner can help to ensure that an inspection system operates successfully—and indeed that its vision algorithms are not unnecessarily complex, thereby necessitating excessive hardware expense for real-time implementation. Means of arranging reasonably uniform illumination and freedom from shadows have been taken to be of significant relevance and allotted fair attention (it is of interest that these topics are scarcely mentioned in most books on this subject—a surprising fact which perhaps indicates the importance that most authors ascribe to this vital aspect of the work).

By contrast, camera systems and digitization techniques have been taken to be purely technical matters to which little space could be devoted (to be really useful—and considering that most workers in this area buy commercial cameras and associated frame-grabbing boards ready to plug into a variety of computers—whole chapters would have been required for each of these topics). Because of its theoretical importance, it appeared to be relevant to give some background to the sampling theorem and its implications although, considering the applications covered in this book, further space devoted to this topic did not appear to be justified (see Rosie, 1966 and Pratt, 1978 for more details).

23.6 Bibliographical and historical notes

It is a regrettable fact that very few papers and books give details of lighting schemes that are used for image acquisition, and even fewer give any rationale or background theory for such schemes. Hence some of the present chapter appears to have broken new ground in this area. However, Batchelor *et al.* (1985) and Browne and Norton-Wayne (1986) give much useful information on light sources, filters, lenses, light guides and so on, thereby complementing the work of this chapter (indeed, the former of these books gives a wealth of detail on how unusual inspection tasks, such as those involving internal threads, may be carried out).

Details of various types of scanning system, camera tube and solid-state (e.g. CCD) device are widely available—see for example, Biberman and Nudelman (1971), Beynon and Lamb (1980), Batchelor *et al.* (1985), Browne and Norton-Wayne (1986), and also various manufacturers' catalogues. Note that much of the existing CCD imaging device technology dates from as recently as the mid-1970s (Barbe, 1975; Weimer, 1975) and is still undergoing development.

Flash (bit-parallel) ADCs are by now used almost universally for digitizing pixel intensities: for a time TRW were pre-eminent in this area and the reader is referred to the catalogues of this and other electronics manufacturers for details of these devices.

The sampling theorem is well covered in very many books on signal processing (see for example, Rosie, 1966). However, details of how band-limiting should be carried out prior to sampling are not so readily available. A brief treatment only is given in Section 23.4: for further details the reader is referred to Pratt (1978) and references contained therein.

24

The Need for Speed: Real-Time Electronic Hardware Systems

24.1 Introduction

In Chapter 1 we started by pointing out that of the five senses, vision has the advantage of providing enormous amounts of information at very fast rates: this was observed to be useful to humans and should also be of great value with robots and other machines. However, the input of large quantities of data necessarily implies large amounts of data processing, and this can create problems—especially in real-time applications. Hence there is no wonder that speed of processing has been alluded to on numerous occasions in earlier chapters of this book.

It is now necessary to examine how serious this situation is and to suggest what can be done to alleviate the problem. Consider a simple situation where it is necessary to examine products of size 64×64 pixels moving at rates of 10–20 per second along a conveyor: this amounts to a requirement to process up to $100\,000$ pixels per second—or typically four times this rate if space between objects is taken into account. In fact, the situation can be significantly worse than indicated by these figures. First, even a basic process such as edge detection generally requires a neighbourhood of at least 9 pixels to be examined before an output pixel value can be computed: thus the number of pixel memory accesses is already ten times that given by the basic pixel processing rate. Second, functions such as skeletonization or size filtering require a number of basic processes to be applied in turn: for example, eliminating objects up to 20 pixels wide requires ten erosion operations, while thinning similar objects using simple "north–south–east–west" algorithms requires at least 40 whole-image operations. Third, typical inspection applications require a number of tasks—edge detection, object location, surface scrutiny, etc.—to be carried out. All these factors mean that the overall processing task may involve anything between 1 and 100 million pixel or other memory accesses per second. Finally, this analysis ignores the complex computations required for some types of 3-D modelling, or for certain more abstract processing operations (see Chapters 14 and 15).

These formidable processing requirements imply a need for very carefully thought out algorithm strategies. However, remembering that most conventional computers operate at around 10 million instructions per second (MIPS), special hardware will ultimately be vital for almost all real-time applications (the exception might be in those tasks where performance rates are governed by the speed of a robot, vehicle or other slow mechanical device). Broadly speaking, there are two main strategies for improving processing rates. The first involves employing a single very fast processor which is fabricated using advanced technology, for example with gallium arsenide semiconductor devices, Josephson junction devices or even optical processing elements. Such techniques can be expected to yield speed increases by a factor of 10 or so, which might be sufficiently rapid for certain applications. However, it is more likely that a second strategy will have to be invoked—that of parallel processing: this involves employing N processors working in parallel, thereby giving the possibility of enhancing speed by a factor N. This second strategy is particularly attractive: to achieve a given processing speed, it should be necessary only to increase the number of processors appropriately—although it has to be accepted that cost will be increased by a factor of around N, as for the speed. Clearly, it is partly a matter of economics which of the two strategies will be the better choice but at any time the first strategy will be technology-limited whereas parallel processing seems more flexible and capable of giving the required speed in any circumstances. Hence, for the most part, parallel processing is considered in what follows.

24.2 Parallel processing

There are two main approaches to parallel processing: in the first, the computational *task* is split into a number of functions which are then implemented by different processors; in the second, the *data* are split into several parts and different processors handle the different parts. These two approaches are sometimes called *algorithmic parallelism* and *data parallelism*, respectively. Notice that if the data are split, different parts of the data are likely to be nominally similar so there is no reason to make the processing elements (PEs) different from each other. However, if the task is split functionally, the functions are liable to be very different and it is most unlikely that the PEs should be identical.

The inspection example cited in Section 24.1 involved a fixed sequence of processes being applied in turn to the input images. On the whole, this type of task is well adapted to algorithmic parallelism and indeed to being implemented as a pipelined processing system, each stage in the pipeline performing one task such as edge detection or thinning (Fig. 24.1) (note that each stage of a thinning task will probably have to be implemented as a single stage in the pipeline). Clearly, such an approach lacks generality but it is cost-effective in a large number of applications, since it is capable of providing the required speedup factor of around two orders of magnitude without undue complexity. Unfortunately, this approach is liable to be inefficient in practice. This is because the speed at which the pipeline operates is dictated by the speed of the slowest device on the pipeline—faster speeds of the other stages constitute wasted computational capability. Variations in the data passing along the pipeline add to this problem: for example, a wide object would require many passes of a thinning operation, so either thinning would not proceed to completion (and the effect of this would have to be anticipated and allowed for), or else the pipeline would have to be run at a slower rate. Obviously, it is necessary for such a system to be designed in accordance with worst-case rather than "average" conditions—although additional buffering between stages can help to reduce this latter problem.

Clearly, the design and control of a reliable pipelined processor is not trivial but, as mentioned above, it gives a generally cost-effective solution in many types of application. However, both with pipelined processors and with other machines that use algorithmic parallelism there are significant difficulties in dividing tasks into functional partitions that match well the PEs on which they are to run. For this and other reasons there have been

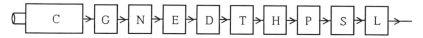

Fig. 24.1 Typical pipelined processing system: C, input from camera; G, grab image; N, remove noise; E, enhance edge; D, detect edge; T, thin edge; H, generate Hough transform; P, detect peaks in parameter space; S, scrutinize products; L, log data and identify products to be rejected.

many attempts at the alternative approach of data parallelism. Indeed, image data are, on the whole, reasonably homogeneous so it is evidently worth searching for solutions incorporating data parallelism. Further consideration then leads to the SIMD type of machine in which each pixel is processed by its own PE. This method is described below.

24.3   SIMD systems

In the SIMD (single instruction stream, multiple data stream) architecture, a 2-D array of PEs is constructed which maps directly onto the image being processed; thus each PE stores its own pixel value, processes it, and stores the processed pixel value. Furthermore, all PEs run the same program and indeed are subject to the same clock; this means that they execute the same instruction simultaneously (hence the existence of a single instruction stream). An additional feature of SIMD machines that are used for image processing is that each PE is connected to those immediately around it, so that neighbourhood operations can conveniently be carried out—the required input data are always available. Typically, each PE is connected to eight others in a square array, as for the CLIP series of machines (Fountain, 1987), although in the DAP computers each PE is connected to only four others (Hunt, 1981). Such machines therefore have the advantage not only of *image parallelism* but also of *neighbourhood parallelism*—data from neighbouring pixels are available immediately and several sequential memory accesses per pixel process are no longer required (for a useful review of these and other types of parallelism, see Danielsson and Levialdi, 1981).

The SIMD architecture is extremely attractive in principle since its processing structure seems closely matched to the requirements of many tasks, such as noise removal, edge detection, thinning, size analysis and so on (although we return to this point below). However, in practice it suffers from a number of disadvantages. Some of these are due to the compromises needed to keep costs at reasonable levels. For example, the PEs are often not powerful floating-point processors and may not contain much memory (this is because available cost is expended on including more PEs rather than making them more powerful); in addition, the processor array may be too small to handle the whole image at once, and problems of continuity and overlap arise when trying to process subimages separately; this can also lead to difficulties when global operations (such as finding an accurate convex hull) have to be performed on the whole image. Finally, getting the data in and out of the array can be a relatively slow process.

Although SIMD machines may appear to operate efficiently on image data, this is not always the case in practice, since many processors may be "ticking over" and not doing anything useful. For example, if a thinning algorithm is being implemented, much of the image may be bare of detail for most of the time, since most of the objects will have shrunk to a fraction of their original area. Thus the PEs are not being kept *usefully* busy. Here the topology of the processing scheme is such that these inactive PEs are unable to get data they can act on, and efficiency drops off markedly. Hence it is not obvious that a SIMD machine can always carry out the *overall* task any faster than a more modest MIMD machine (see below), or a specially fast but significantly cheaper single processor (SISD) machine.

A more important characteristic is that while the SIMD machine is reasonably well adapted for image processing, it is quite restricted in its capabilities for image analysis. For example, it is virtually impossible to use *efficiently* for implementing Hough transforms, especially when these demand mapping image features into an abstract parameter space. In addition, most serial (SISD) computers are much more efficient at operations such as simple edge tracking, since their single processors are generally much faster than costs will permit for the many processors in a SIMD machine. Overall, these problems should be expected since the SIMD concept is designed for image-to-image transformations via local operators and does not map well to (i) image-to-image transformations that demand *nonlocal* operations, (ii) image-to-abstract data transforms (intermediate-level processing), or (iii) abstract-to-abstract data (high-level) processing (note that some would classify (i) as being a form of intermediate-level processing where *deductions* are made about what is happening in distant parts of an image—i.e. higher-level interpretive data are being marked in the transformed image). This means that unaided SIMD machines are unlikely to be well suited for practical inspection work or for related applications.

Before leaving the topic of SIMD machines, recall that they incorporate two types of parallelism—image parallelism and neighbourhood parallelism. Both of these contribute to high processing rates. Although it might at first appear that image parallelism contributes mainly through the high processing bandwidth* it offers, it also contributes through the high data accessing bandwidth: in contrast, neighbourhood parallelism contributes only through the latter mechanism. However, what is important is that this

---

*In this context it is conventional to use the term *bandwidth* to mean the maximum rate realizable via the stated mechanism.

type of parallel machine, in common with any successful parallel machine, incorporates both features. It is of little use to attend to the problem of achieving high processing bandwidth only to run into data bottlenecks through insufficient attention to data structures and data access rates: i.e. it is necessary to match the data access and processing bandwidths if full use is to be made of available processor parallelism.

24.4 The gain in speed attainable with N processors

It is interesting to speculate whether the gain in processing rate could ever be greater than N, say $2N$ or even N^2. It could in principle be imagined that two robots used to make a bed would operate more efficiently than one, or four more efficiently than two, for a square bed. Similarly, N robots welding N sections of a car body would operate more efficiently than a single one. The same idea should apply to N processors operating in parallel on an N-pixel image. At first sight it does appear that a gain greater than N could result. However, closer study shows that any task is split between data organization and actual processing. Thus the maximum gain that could result from the use of N processors is (exactly) N: any other factor is due to the difficulty, either for low or for high N, of getting the right data to the right processor at the right time. Thus in the case of the bed-making robots, there is an overhead for $N = 1$ of having to run around the bed at various stages because the data (the sheets) are not presented correctly. More usually, it is at large N that the data are not available at the right place at the right moment. An immediate practical example of these ideas is that of accessing all eight neighbours in a 3×3 neighbourhood where only four are directly connected, and the corner pixels have to be accessed via these four: then a *three-fold* speedup in data access may be obtained by *doubling* the number of local links from four to eight.

There have been many attempts to model the utilization factor of both SIMD and pipelined machines when operating on branching and other algorithms. Minsky's conjecture (Minsky and Papert, 1971), that the gain in speed from a parallel processor is proportional to $\log_2 N$ rather than N, can be justified on this basis, and leads to an efficiency $\eta = \log_2 N / N$. Hwang and Briggs (1984) produced a more optimistic estimate of efficiency in parallel systems: $\eta = 1/\log_2 N$.

Following Chen (1971), the efficiency of a pipelined processor is usually estimated as $\eta = P/(N + P - 1)$, where there are on average P consecutive data points passing through a pipeline of N stages—the reasoning being based on the proportion of stages that are usefully busy at any one time. For imaging applications, such arguments are often somewhat irrelevant since the total delay through the pipelined processor is unimportant compared with the cycle time between successive input or output data values. This is because a machine that does not keep up with the input data stream will be completely useless, whereas one that incorporates a fixed time delay may be acceptable in some cases (such as a conveyor belt inspection problem) though unacceptable in others (such as a missile guidance system).

Broadly speaking, the situation that is being described here involves a speedup factor N coupled with an efficiency η, giving an overall speedup factor of $N' = \eta N$. The loss in efficiency is often due ultimately to frustrated algorithm branching processes but presents itself as under-utilization of resources, which cannot be reduced because the incoming data are of variable complexity.

24.5 Flynn's classification

Early in the development of parallel processing architectures, Flynn (1972) developed a now well-known classification which has already been referred to above: architectures are either SISD, SIMD or MIMD. Here SI (single instruction stream) means that a single program is employed for all the PEs in the system, whereas MI (multiple instruction stream) means that different programs can be used; SD (single data stream) means that a single stream of data is sent to all the PEs in the system, whereas MD (multiple data stream) means that the PEs are fed with data independently of each other.

The SISD machine is a single processor and is normally taken to refer to a conventional von Neumann computer. However, the definitions given above imply that SISD falls more naturally under the heading of a Harvard architecture, whose instructions and data are fed to it through separate channels: this gives it a degree of parallelism and makes it generally faster than a von Neumann architecture (in fact, there is almost invariably *bit parallelism* also, the data taking the form of words of data holding several bits of information, and the instructions being able to act on all bits simultaneously: however, this possibility is so universal that it is ignored in what follows).

The SIMD architecture has already been described reasonably thoroughly, although it is worth reiterating that the multiple data stream arises in imaging work through the separate pixels being processed by their PEs independently as separate, though similar, data streams. Note that the PEs of SIMD machines invariably embody the Harvard architecture.

The MISD architecture is notably absent from the above classification, although it is possible to envisage that pipelined processors fall into this category since a single stream of data is fed through all processors in turn, albeit being modified as it proceeds so the same *data* (as distinct from the same data stream) do not pass through each PE. However, many parties take the MISD category to be null (e.g. Hockney and Jesshope, 1981).

The MIMD category is a very wide one, containing all possible arrangements of separate PEs that get their data and their instructions independently: it even includes the case where none of the PEs are connected together in any way. However, such a wide interpretation does not solve practical problems. We can therefore envisage linking the PEs together by a common memory bus, or every PE being connected to every other one, or linkage by some other means. A common memory bus would tend to cause severe contention problems in a fast-operating parallel system, whereas maintaining separate links between all pairs of processors is clearly at the opposite extreme but would run into a combinatorial explosion as systems become larger. Hence a variety of other arrangements are used in practice. Crossbar, star, ring, tree, pyramid and hypercube structures have all been used (Fig. 24.2). In the crossbar arrangement, half of the processors can communicate directly with the other half via N links (and $N^2/4$ switches), although all processors can communicate with each other *indirectly*. In the star there is one central PE so the maximum communication path has

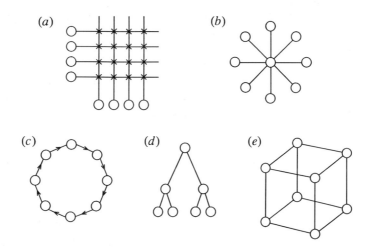

Fig. 24.2   Possible arrangements for linking processing elements: (a) crossbar; (b) star; (c) ring; (d) tree; (e) hypercube. All links are bidirectional except where arrows indicate otherwise.

length 2. In the ring all N PEs are placed symmetrically and the maximum communication path is of length $N-1$ (note that this figure assumes unidirectional rings, which are easier to implement, and a number of notable examples of this type exist). In the tree or pyramid, the maximum path length is of order $2(\log_2 N - 1)$, assuming that there are two branches for each node of the tree. Finally, for the hypercube of n dimensions (built in an n-dimensional space with two positions per dimension), the shortest path length between any two PEs is at most n; in this case there are 2^n processors, so the shortest path length is at most $\log_2 N$. Overall, the basic principle must be to minimize path length while at the same time cutting down the possibility of data bottlenecks: this shows that the star configuration is very limited and explains the considerable recent attention to the hypercube topology. In view of what was said earlier about the importance of matching the data bandwidth to the processing bandwidth, it will be clear that a very careful choice needs to be made concerning a suitable architecture—and there is no lack of possible candidates (the above set of examples is by no means exhaustive).

Finally, it is as well to note that factors other than speed enter into the choice of an architecture. For example, Uhr (1978, 1980) has been a strong advocate of the pyramid architecture, on the basis that it matches well the hierarchical data structures most appropriate for scene interpretation.

Much of the above discussion about architectures has been in the realm of the possible and the ideal (many of the existing systems being expensive experimental ones), and it is now necessary to consider more practical issues. How, for example, are we to match the architecture to the data? More particularly, how do we lay down general guidelines for partitioning the tasks and implementing them on practical architectures? In the absence of general guidelines of this type, it is useful to look in detail at a given practical problem to see how to design an optimal hardware system to implement it: this is done in the next section.

24.6 Optimal implementation of an image analysis algorithm

The particular algorithm considered in this section is one examined earlier by Davies and Johnstone (1986, 1989). It involves the inspection of round food products (see also Chapter 19). The purpose of the analysis is to show how to make a systematic selection between available hardware modules (including computers), so that it can be guaranteed that the final hardware configuration is optimal in specific ways and in particular with regard to relevant cost–speed tradeoffs.

In the particular food product application considered, biscuits are moving at rates of up to 20 per second along a conveyor (Davies and Johnstone, 1986). Since the conveyor is moving continuously, it is natural to use a linescan camera to obtain images of the products. Before they can be scrutinized for defects and their sizes measured, they have to be located accurately within the images. Since the products are approximately circular, it is straightforward to employ the Hough transform technique for the purpose (see Chapter 9): it is also appropriate to use the radial intensity histogram approach to help with the task of product scrutiny (Chapter 19). In addition, simple thresholding can be used to measure the amount of chocolate cover and certain other product features. The main procedures in the algorithm are summarized in Table 24.1. Note that the Hough transform approach requires the rapid and accurate location of edge pixels, which is achieved using the Sobel operator and thresholding the resulting edge enhanced image. Edge detection is in fact the only 3×3 neighbourhood operation in the algorithm and hence it is relatively time-consuming; rather less processing is required by the 1×1 neighbourhood operations (Table 24.1); then come various 1-D processes such as analysis of radial histograms. The fastest operations are those such as logging variables which are neither 1-D nor 2-D processes.

Table 24.1 Breakdown of the inspection algorithm.

Function	Description	Time (s)	Cost (£)	c/t (£/ms)
1. Acquire image	1×1	—	1000	—
2. Clear parameter space	1×1	0.017	200	11.8
3. Find edge points	3×3	4.265	3000	0.7
4. Accumulate points in parameter space	1×1	0.086	2000	23.3
5. Find averaged centre	—	0.020	2000	100.0
6. Find area of product	1×1	0.011	100	9.1
7. Find light area (no chocolate cover)	1×1	0.019	200	10.5
8. Find dark area (slant on product)	1×1	0.021	200	9.5
9. Compute radial intensity histogram	1×1	0.007	400	57.1
10. Compute radial histogram correlation	1-D	0.013	400	30.8
11. Overheads for functions 6–10	—	0.415	1200	2.9
12. Calculate product radius	1-D	0.047	4000	85.1
13. Track parameters and log	—	0.037	4000	108.1
14. Decide if rejection is warranted	—	0.002	4000	2000.0
Time for whole algorithm		4.960		

All timings are relevant to an LSI-11/23 host processor.

24.6.1   Hardware specification and design

On finalization of the algorithm strategy, the overall execution time was found to be about 5 seconds per product on a 16-bit minicomputer (Davies and Johnstone, 1986). With product flow rates of the order of 20 per second, and software optimization subject to severely diminishing returns, a further gain in speed by a factor of around 100 could clearly be obtained only by using special electronic hardware. In this application a compromise was sought with a single CPU linked to a set of suitable hardware accelerators. In this case the latter would have had to be designed specially for the purpose, and indeed some were produced in this way. However, in the discussion below it is immaterial whether the hardware accelerators are made specially or purchased—the object of the discussion is to present rigorous means for deciding which software functions should be replaced by hardware modules.

As a prerequisite to the selection procedure, Table 24.1 lists execution times and hardware implementation costs of all the algorithm functions: the figures are somewhat notional since it is difficult to divide the algorithm rigorously into completely independent sections. However, they are sufficiently accurate to form the basis for useful decisions on the cost-effectiveness of hardware. Since the aim is to examine the principles underlying cost–speed tradeoffs, the figures presented in Table 24.1 are taken as providing a concrete example of the sort of situation that can arise in practice.

24.6.2   Basic ideas on optimal hardware implementation

The basic strategy that was adopted for deciding on a hardware implementation of the algorithm is as follows:

(i)   prioritize the algorithm functions so that it is clear in which order they should be implemented;

(ii) find some criterion for deciding when it is not worth proceeding to implement further functions in hardware (Davies and Johnstone, 1986).

From an intuitive point of view, function prioritization seems a simple process: basically functions should be placed in order of cost-effectiveness, i.e. those saving the most execution time per unit cost (when implemented in hardware) should be placed first, and those giving lesser savings should be placed later. Thus we arrive at the c/t (cost/time) criterion function. Then with limited expenditure we achieve the maximum saving in execution time, i.e. the maximum speed of operation.

To decide at what stage it is not worth implementing further functions in hardware is arguably more difficult. Excluding here the practical possibility of strict cost or time limits, the ideal solution results in the optimal balance between total cost and total time. Since these parameters are not expressible in the same units, it is necessary to select a criterion function such as $C * T$ (total cost $*$ total time) which, when minimized, allows a suitable balance to be arrived at automatically.

The procedure outlined above is simple and does not take account of hardware which is common to several modules. This "overhead" hardware must be implemented for the first such module and is then available at zero cost for subsequent modules. In many cases a speed advantage results from the use of overhead hardware. In the example system, it is found that significant economies are possible when implementing functions 6–10, since common pixel scanning circuitry may be used. In addition, note that any of these functions that are *not* implemented in hardware engender a time overhead in software. This, and the fact that the time overhead is much greater than the sum of the software times for functions 6–10, means that once the initial cost overhead has been paid it is best (in this particular example) to implement all these functions in hardware (Davies and Johnstone, 1986).

Table 24.2   Speed–cost tradeoff figures.

Function (see Table 24.1)	c/t (£/ms)	t (s)	c (£)	T (s)	C (£)	$C * T$ (£–s)	$C' * T'$ (£–s)
—		—	6000	4.990	6000	29940	15080
3	0.7	4.265	3000	0.725	9000	6530	3190
6–11	5.1	0.486	2500	0.239	11500	2750	1400
2	11.8	0.017	200	0.222	11700	2600	1350
4	23.3	0.086	2000	0.136	13700	1860	1040
12	85.1	0.047	4000	0.089	17700	1580	1010
5	100.0	0.020	2000	0.069	19700	1360	860
13	108.1	0.037	4000	0.032	23700	760	770
14	2000.0	0.002	4000	0.030	27700	830	860

The first entry corresponds to the base system cost, including computer, camera, frame store, backplane and power supply. The other entries are derived from Table 24.1 by ordering the c/t values (see text). (However, to allow for a minimum realistic value of T of 0.03 s, all values of T have been increased by this amount.) The final column shows a comparable set of figures for an LSI-11/73 host processor, which is ~2.5 times faster than an LSI-11/23.

Trying out the design strategy outlined above gives the c/t ratio sequence shown in Table 24.2. A set of overall times and costs resulting from implementing in hardware all functions down to and including the one indicated is now deduced. Examination of the column of $C \times T$ products then shows where the tradeoff between hardware and software is optimized: this occurs here when the first 13 functions are implemented in hardware.

The simple analysis presented in this section clearly gives only a general indication of the required hardware–software tradeoff. Indeed, minimizing $C * T$ indicates an overall "bargain package", whereas in practice the system might well have to meet certain cost or speed limits (Davies and Johnstone, 1986). In this food product application it was necessary to aim at an overall cost of around £10 000. By implementing functions 1, 3, 6–11 in hardware, it was found possible to get to within a factor 3.6 of the optimal $C * T$ product. Interestingly, using an upgraded host processor it is possible to get within a much smaller factor (1.8) of the optimal tradeoff, with the same number of functions implemented in hardware (Table 24.2): indeed, a further advantage of using a criterion function is that the choice of which processor to use becomes automatic.

The paper by Davies and Johnstone (1989) goes into some depth concerning the choice of criterion function, showing that more general functions are available and that there is a useful overriding geometrical interpretation—that global concavities are being sought on the chosen curve in $f(C)$, $g(T)$ space. The paper also places the problem of overheads and relevant functional partitions on a more rigorous basis. Finally, a later paper (Davies *et al.*, 1995) emphasizes the value of software solutions, achieved, for example, with the aid of arrays of DSP chips.

24.7   Board-level processing systems

Over the past few years a great number of image processing systems have come onto the market and the number of board-level products that plug into standard computers (such as IBM personal computers) has recently grown very rapidly. Early board-level products incorporated a rather haphazard set of functions: most could be classed as pure image processing systems (i.e. they performed very few intermediate-level vision functions) and most concentrated on binary operations (such as erosion, dilation, thresholding and so on); however, they also carried out Sobel and other edge detection operations, and grey-scale noise removal and histogramming operations. Systems are now available that tackle Hough transforms for line detection and the situation with regard to intermediate-level operations is changing rapidly. In addition, a number of boards have been built which perform quite impressive "brute-force" correlation functions using relatively large templates; this should not be too surprising as correlation is an extremely widely used operation, and such boards facilitate real-time search for many types of object and feature. Further comment is curtailed as the situation is developing so rapidly. However, it is probably true that commercial boards still tend to have restricted capability and are unable to perform in real time more than a fraction of the inspection and interpretation tasks outlined in this book.

24.8 VLSI

Very large scale integration (VLSI) will probably provide the key to much more widespread use of vision systems, since in the long run it will increase the reliability of these systems, lower the cost, and permit the incorporation of a greater degree of sophistication. However, VLSI is itself an expensive undertaking, because mask design, layout and testing is manpower-intensive and can take many weeks of effort: indeed, custom VLSI design is characterized by relatively high capital cost (say £40 000 per mask) but rather low recurrent cost (say £20 per chip) (note, however, that it is often possible to employ "semi-custom" techniques, resulting in costs of a few thousand pounds per design). Hence fabricating a circuit in VLSI is not to be undertaken lightly. So what are the barriers to progress in this area?

First, a great deal of confidence in the utility of a function must be attained before it can be worth committing the design to VLSI. In a fast-moving area such as image analysis, such confidence is not easily come by, since at no moment is the situation static. *Ideally*, we would now be at the stage when low-level functions were fully understood, so these could be implemented as VLSI, while intermediate-level functions should be nearing the same stage, and high-level functions should all be known within a 5–10-year time scale. However, reality is not like that (as earlier chapters of this book should have indicated). As higher-level and 3-D functions become specified more clearly, there is a tendency to find that the earlier low- and intermediate-level functions are inadequate in certain ways—particularly with regard to accuracy. Hence yet another cycle of improvement of low-level functions has to be instigated.

Nonetheless, certain low- and intermediate-level functions have been known about for a long time and by now have probably achieved fairly universal acceptance—at least within a certain range of applicability. Such functions are the Sobel operator, the median filter, correlation, thresholding and the HT for line detection. Hence it is not surprising that several of these functions have been incorporated into commercial VLSI chips (somewhat surprisingly, in view of the discussion of Chapter 5, a commercially available edge detector chip (Plessey Semiconductors, 1986) employs a template matching scheme rather than a true Sobel function). Considering what can nowadays be encapsulated in a VLSI chip (take the Pentium processor or the 16 Mbit dynamic RAM as examples), these imaging functions are rather modest—no doubt reflecting both the current lack of confidence referred to above in the value of these functions, and the much lower likely demand for such circuits when compared with dynamic RAM chips.

However, this represents only the beginning of what is undoubtedly a growth area, and as the usefulness of each new function is proved and VLSI and WSI (wafer-scale integration) technology advance further, the hardware scene for vision will change very radically indeed. These developments will certainly have a considerable impact on those working on machine vision.

There are two particularly important outlets for VLSI circuits. The first is to incorporate them into pipelined processing structures which then feed them with their image data, covering many necessary low- and intermediate-level processing functions. Combining such pipelined structures with powerful multiprocessor systems (able to perform the higher-level "abstract" processing) will provide the desired cost-effective solutions in applications such as inspection. The second outlet is in systolic array processors: these form a specialized type of architecture in which PEs are arranged in 2-D arrays such that input data can pulse in in several (typically two) directions and output data can emerge from another direction. Thus the data are gradually processed as they progress across the array, all cells continuously being kept busy and maintaining a high processing efficiency. The systolic array is a generalization of the pipeline architecture and one that is especially well adapted to integrated circuit implementation. A significant difficulty is that of delivering the data continuously to the array at such a rate as to keep it usefully busy. How to achieve this, and how to design systolic array PEs, is the concern of much current research in this area.

24.9 Concluding remarks

This chapter has studied the means available for implementing image analysis and inspection algorithms in real time. Fast sequential and various parallel processing architectures have been considered, as well as other potentially useful approaches such as systolic arrays with cells implemented in VLSI. In addition, means of selecting between the various realizable schemes have been studied, these being based on criterion function optimizations.

The field is characterized on the one hand by elegant parallel architectures that would cost an excessive amount in most inspection applications, and on the other by hardware solutions that are optimized to the application yet which are bound to lack generality. In fact, it is easy to gain too rosy a view of the elegant parallel architectures that are available: their power and generality are virtually guaranteed to make them an overkill in dedicated applications, while reductions in generality seem likely to ensure that their PEs spend much of their time waiting for data or performing null operations. Overall, the subject of image analysis is quite variegated and it is difficult to find a set of "typical" algorithms for which an "average" architecture can be designed whose PEs will always be kept usefully busy. Thus, the subject of mapping algorithms to architectures—or, better, of matching algorithms *and* architectures, i.e. designing them together as a system—is a truly complex one for which obvious solutions turn out to be less efficient in reality than when initially envisioned. Perhaps the main problem lies in envisaging the nature of image analysis, which in the author's experience turns out to be a far more abstract process than image processing (i.e. image-to-image transformation) ideas would indicate (see especially Chapter 6).

Finally, a topic which has been omitted for reasons of space is that of carrying out sufficiently rigorous timing analysis of image analysis and related tasks so that the overall process is guaranteed to run in real time. In this context, a particular problem is that unusual image data conditions could arise which might engender additional processing, thereby setting the system behind schedule: this situation could arise because of excessive noise, because more than the usual number of products appear on a conveyor, because the heuristics in a search tree prove inadequate in some situation, or for a variety of other reasons. One way of eliminating this type of problem is to include a watchdog timer, so that the process is curtailed at a particular stage (with the product under inspection being rejected, or other suitable action being taken). However, this type of solution is crude, and sophisticated timing analysis methods now exist, based on Quirk analysis: Môtus and Rodd are protagonists of this type of rigorous approach. The reader is referred to Thomas *et al.* (1995) for further details and to Môtus and Rodd (1994) for an in-depth study.

24.10 Bibliographical and historical notes

The problem of implementing vision in real time presents exceptional difficulties since in many cases megabytes of data have to be handled in times of the order of hundredths of a second, and often repeated scrutiny of the data is necessary. Probably it is no exaggeration to say that hundreds of architectures have been considered for this purpose, all having some degree of parallelism. Of these, many have highly structured forms of parallelism, such as SIMD machines. The idea of a SIMD machine for image parallel operations goes back to Unger (1958), and strongly influenced later work both in the UK (where it gave rise to the CLIP series of machines at University College, London) and in the USA (where it led to the Westinghouse SOLOMON computer and to the University of Illinois ILLIAC series of machines)—see for example, Duff *et al.* (1973), Gregory and McReynolds (1963) and Barnes *et al.* (1968). For an overview of these machines, see Fountain (1987).

The SIMD concept was generalized by Flynn (1972) in his well-known classification of computing machines (Section 24.5) and there have been several attempts to devise a more descriptive and useful classification (see Hockney and Jesshope, 1981), although none has so far really caught on (however, see Skillicorn (1988) for an interesting attempt). Arguably, this is because it is all too easy for structures to become too lacking in generality to be useful for a wide range of algorithmic tasks: in fact, there has been a great proliferation of interconnection networks (see Feng, 1981). Reeves (1984) reviewed a number of such schemes that might be useful for image processing. An important problem is that of performing an optimal mapping between algorithms and architectures but for this to be possible a classification of parallel algorithms is needed to complement that of architectures: Kung (1980) made a start with this task. This work has been extended by Cantoni and Levialdi (1983) and Chiang and Fu (1983).

Perhaps the most challenging work in the future will be to make the most of the capability for integration, in VLSI or WSI. Mead and Conway (1980) has become a standard (but now dated) background text in this area, with Offen (1985) showing more clearly the possibilities for the future, with particular regard to image processing and image analysis. Also of interest in this area are the volumes by Kittler and Duff (1985) and Fountain (1987). Within the purview of VLSI processing is the concept of systolic arrays, introduced by Kung and Leiserson (1979); this technique is still being developed (see for example, Kung, 1984) and shows much promise for certain image processing tasks.

At a more down-to-earth level, it must be remembered that many of the existing image processing and analysis systems do not employ sophisticated parallel architectures but "merely" very fast serial processing techniques with hardwired functions—often capable of processing at video rates (i.e. giving results within one TV frame time). Ullmann (1981) reviewed this area.

During the late 1980s, the Transputer provided system implementers with a powerful processor with good communication facilities (four bidirectional links each with the capability for independent communication at rates of 10 Mbits per second *and* operation independent of the CPU), enabling them to build SIMD and MIMD machines on a do-it-yourself basis. The T414 device had a 32-bit processor operating at 10 MIPS with a 5 MHz clock and had 2 kB of internal memory, while the T800 device had 4 kB of internal memory and a floating-point processor (Inmos, 1987) yet cost only around £200. These devices have been incorporated in a number of commercially available board-level products plus a few complete rather powerful systems, and vision researchers have developed parallel algorithms for them, using mainly the Occam parallel processing language (Inmos, 1988; Pountain and May, 1987).

Other solutions include the use of bit-slice types of device (Edmonds and Davies, 1991), and DSP chips (Davies *et al.*, 1995), while special purpose multiprocessor designs are required to implement multiresolution and other Hough transforms (Atiquzzaman, 1994).

Part 4
Perspectives on Vision

Part 4 of the book aims to draw together and broaden the topics of the previous chapters. In particular, attention is paid to systematizing the subject of machine vision and to summarizing the tradeoffs among the parameters that are relevant in vision algorithms. In addition, the significance and ultimate value of the Hough transform technique is discussed. Appendix D is highly relevant to the work of Part 4, as it brings into relief the essential dichotomy between interpretation and accurate measurement (or between robustness "breakdown point" and measurement "efficiency").

25

Machine Vision, Art or Science?

25.1 Introduction

The preceding chapters have covered many topics relating to vision: how images may be processed to remove noise, how features may be detected, how objects may be located from their features, how to set up lighting schemes, how to design hardware systems for automated visual inspection, and so on. The subject is one that has developed over a period of some 30 years and it has clearly come a long way. However, it has developed piecemeal rather than systematically. Often development is motivated by the particular interests of small groups of workers and is relatively *ad hoc*. Coupled with this is the fact that algorithms, processes and techniques are all limited by the creativity of the various researchers: the process of design tends to be intuitive rather than systematic, and so again the *ad hoc* tends to creep in from time to time. As a result, sometimes no means has yet been devised for achieving particular aims, but more usually a number of imperfect methods is available and there is limited scientific basis for choosing between them.

All this poses the problem of how the subject may be placed on a firmer foundation. Time may help but time can also have the effect of making things more difficult as more methods and results arise which have to be considered; in any case, there is no short-cut to intellectual analysis of the state of the art. This book has aimed to carry out a degree of analysis at every stage but in this last chapter it is worth trying to tie it all together, to make some general statements on methodology and to indicate the direction that might be taken in the future.

Machine vision is an engineering discipline and, like all such disciplines, it has to be based on science and an understanding of fundamental processes. However as an engineering discipline it should involve design based on specifications. Once the specifications for a vision system have been laid down, it can be seen how they match up against the constraints provided by nature and technology. In what follows we consider first the parameters of relevance for the specification of vision systems; then we consider constraints and their origins. This leads to some clues as to how the subject could be developed further.

25.2   Parameters of importance in machine vision

The first thing that can be demanded of any engineering design is that it should work! This applies as much to vision systems as to other parts of engineering. Clearly, there is no use in devising edge detectors that do not find edges, corner detectors that do not find corners, thinning algorithms that do not thin, 3-D object detection schemes that do not find objects, and so on. But in what way could such schemes fail? First, ignore the possibility of noise or other artefacts preventing algorithms from operating properly. Then what remains is the possibility that at any stage important fundamental factors have not been taken into account.

For example, a boundary tracking algorithm can go wrong because it encounters a part of the boundary that is one pixel wide and crosses over instead of continuing. A thinning algorithm can go wrong because every possible local pattern has not been taken into account in the design and hence it disconnects a skeleton. A 3-D object detection scheme can go wrong because proper checks have not been made to confirm that a set of observed features is not coplanar. Of course, these types of problem may arise very rarely (i.e. only with highly specific types of input data), which is why the design error is not noticed for a time. Often, mathematics or enumeration of possibilities can help to eliminate such errors, so problems can be removed systematically. However, being absolutely sure no error has been made is difficult—and it must not be forgotten that transcription errors in computer programs can contribute to the problems. These factors mean that algorithms should be put to extensive tests with large datasets in order to ensure that they are correct. In the author's experience, there is no substitute for subjecting algorithms to variegated tests of this type to check out ideas that are "evidently" correct. Although all this is obvious, it is still worth stating, since silly errors continually arise in practice.

At this stage imagine that we have a range of algorithms that all achieve the same results on ideal data, and that they really work. The next problem is to compare them critically and, in particular, to find how they react to *real* data and the nasty realities such as noise that accompany it. These nasty realities may be summed up as follows:

   (i)   noise;
  (ii)   background clutter;
 (iii)   occlusions;
 (iv)   object defects and breakages;
  (v)   optical and perspective distortions;
 (vi)   nonuniform lighting and its consequences;
(vii)   effects of stray light, shadows and glints.

In general, algorithms need to be sufficiently robust to overcome these problems. However, things are not so simple in practice. For example, HT and many other algorithms are capable of operating properly and detecting objects or features despite considerable degrees of occlusion. But how much occlusion is permissible? Or how much distortion, noise, or how much of any other of the nasty realities can be tolerated? In each specific case we could state some figures that would cover the possibilities. For example, we may be able to state that a line detection algorithm must be able to tolerate 50% occlusion, and so a particular HT implementation is (or is not) able to achieve this. However, at this stage we arrive at a lot of numbers that may mean very little on their own: in particular, they seem different and incompatible. In fact, this latter problem can largely be eliminated: each of the defects can be imagined to obliterate a definite proportion of the object (in the case of impulse noise this is obvious; with Gaussian noise the equivalence is not so clear but we suppose here that an equivalence can at least in principle be computed). Hence we end up by establishing that artefacts in a particular dataset eliminate a certain proportion of the area and perimeter of all objects, or a certain proportion of all small objects*. This is a sufficiently clear statement to proceed with the next stage of analysis.

To go further it is necessary to set up a complete specification for the design of a particular vision algorithm. The specification can be listed as follows (but generality is maintained by not stating any particular algorithmic function):

1. the algorithm must work on ideal data;
2. the algorithm must work on data that is $x\%$ corrupted by artefacts;

---

* Clearly, certain of the nasty realities (such as optical distortions) tend to act in such a way as to cut down accuracy but we concentrate here on robustness of object detection.

3. the algorithm must work to p pixels accuracy;
4. the algorithm must operate within s seconds;
5. the algorithm must be trainable;
6. the algorithm must be implemented with failure rate less than 1 per d days:
7. the hardware needed to implement the algorithm must cost less than £L.

No doubt other specifications will arise in practice but these are ignored here.

Before proceeding, note the following points: (i) specification 2 is a statement about the robustness of algorithms, as explained earlier; (ii) the accuracy of p pixels in specification 3 may well be fractional; (iii) in specification 5, trainability is a characteristic of certain types of algorithm and in any instance we can only accept the fact that it is or is not achievable: hence it is not discussed further here; (iv) the failure rate referred to in specification 6 will be taken to be a hardware characteristic and is ignored in what follows.

The set of specifications listed above may at any stage of technological (especially hardware) development be nonachievable; this is because they are phrased in a particular way, so they are not compromisable. However, if a given specification is getting near to its limit of achievability, a switch to an alternative algorithm might be possible[*]; alternatively, an internal parameter might be adjusted which keeps that specification within range, while pushing another specification closer to the limits of its range. In general, there will be some hard (non-negotiable) specifications and others for which a degree of compromise is acceptable. As has been seen in various chapters of the book, this leads to the possibility of tradeoffs—a topic which is reviewed in the next section.

25.3 Tradeoffs

Tradeoffs form one of the most important features of algorithms, since they permit a degree of flexibility subject only to what is possible in the nature of things. Ideally, the tradeoffs that are enunciated by theory provide absolute statements about what is possible, so that if an algorithm approaches these limits it is then probably as "good" as it can possibly be.

---

[*] But note that several, or all, relevant algorithms may be subject to almost identical limitations, because of underlying technological or natural constraints.

Next, there is the problem about where on a tradeoff curve an algorithm should be made to operate. The type of situation was examined carefully in Chapter 24 in a particular context—that of cost–speed tradeoffs of inspection hardware. Generally, the tradeoff curve (or surface) is bounded by hard limits. However, once it has been established that the optimum working point is somewhere within these limits, in a continuum, then it is appropriate to select a criterion function whereby an optimum can be located uniquely. Details will vary from case to case but the crucial point is that an optimum must exist on a tradeoff curve, and that it can be found systematically once the curve is known. Clearly, all this implies that the science of the situation has been studied sufficiently so that relevant tradeoffs have been determined.

25.3.1 Some important tradeoffs

Earlier chapters of this book have revealed some quite important tradeoffs that are more than just arbitrary relations between relevant parameters. Here, a few examples will have to suffice by way of summary.

First, in Chapter 5 the DG edge operators were found to have only one underlying design parameter—that of operator radius r. Ignoring here the important matter of the effect of a discrete lattice in giving preferred values of r, it was found that:

  (i)   signal-to-noise ratio varies linearly with r, because of underlying signal and noise averaging effects;
  (ii) resolution varies inversely with r, since relevant linear features in the image are averaged over the active area of the neighbourhood: the scale at which edge positions are measured is given by the resolution;
  (iii) the accuracy with which edge position (at the current scale) may be measured depends on the square root of the number of pixels in the neighbourhood, and hence varies as r;
  (iv) computational load, and associated hardware cost, is proportional to the number of pixels in the neighbourhood, and hence varies as r^2.

Thus operator radius carries with it four properties which are intimately related—signal-to-noise ratio, resolution (or scale), accuracy and hardware/computational cost.

Another important problem was that of fast location of circle centres (Chapter 9); in this case robustness was seen to be measurable as the amount of noise or signal distortion that can be tolerated. For HT-based schemes, noise, occlusions, distortions, etc. all reduce the peak height in parameter space, thereby reducing the signal-to-noise ratio and impairing accuracy. Furthermore, if a fraction β of the original signal is removed, leaving a fraction $\gamma = 1 - \beta$, either by such distortions or occlusions or else by deliberate sampling procedures, then the number of independent measurements of the centre location drops to a fraction γ of the optimum. This means that the accuracy of estimation of the centre location drops to a fraction around $\sqrt{\gamma}$ of the optimum (note, however, that Gaussian noise can act in another way to reduce accuracy—by broadening the peak instead of merely reducing its height—but this effect is ignored here).

What is important is that the effect of sampling is substantially the same as that of signal distortion, so that the more distortion that must be tolerated, the higher α, the fraction of the total signal sampled, has to be. This means that as the level of distortion increases, the capability for withstanding sampling decreases, and therefore the gains in speed achievable from sampling are reduced—i.e. for fixed signal-to-noise ratio and accuracy, a definite robustness–speed tradeoff exists. Alternatively, the situation can be viewed as a three-way relation between accuracy, robustness and speed of processing. This provides an interesting insight into how the edge operator tradeoff considered earlier might be generalized.

To underline the value of studying such tradeoffs, note that any given algorithm will have a particular set of perhaps rather arbitrary adjustable parameters which are found to control—and hence lead to tradeoffs between—the important quantities such as speed of processing, signal-to-noise ratio and attainable accuracy already mentioned. Ultimately, such *practically* realizable tradeoffs (i.e. arising from the given algorithm) should be considered against those that may be deduced on purely theoretical grounds. Such considerations would then indicate whether a better algorithm might exist than the one currently being examined.

25.3.2 Tradeoffs for two-stage template matching

Two-stage template matching has been mentioned a number of times in this book as a means whereby the normally slow and computationally intensive process of template matching may be speeded up. In general it involves looking for easily distinguishable subfeatures, so that locating the features that are ultimately being sought involves only the minor problem of eliminating false alarms. The reason this strategy is useful is that the first stage eliminates the bulk of the raw image data, so that only a relatively trivial testing process remains. This latter process can then be made as rigorous as necessary. In contrast, the first "skimming" stage can be relatively crude (see for example the lateral histogram approach of Chapter 12), the main criterion being that it must not eliminate any of the desired features: false positives are permitted but not false negatives. Needless to say, however, the efficiency of the overall two-stage process is limited by the number of false alarms thrown up by the first stage.

Suppose that the first stage is subject to a threshold h_1 and the second stage to a threshold h_2. If h_1 is set very low, then the process reverts to the normal template matching situation, since the first stage does not eliminate any part of the image. In fact, setting $h_1 = 0$ initially is useful so that h_2 may be adjusted to its normal working value. Then h_1 can be increased to improve efficiency (reduce overall computation); a natural limit arises when false negatives start to occur—i.e. some of the desired features are not being located. Further increases in h_1 now have the effect of cutting down available signal, although speed continues to increase. This clearly gives a tradeoff between signal-to-noise ratio, and hence accuracy of location, and speed.

In a particular application in which objects were being located by the HT, the numbers of edge points located were reduced as h_1 increased, so accuracy of object location was reduced (Davies, 1988i). A criterion function approach was then used to determine an optimum working condition. A suitable criterion function turned out to be $C = T/A$, where T is the total execution time and A the achievable accuracy. Although this approach gave a useful optimum, the optimum can be improved further if a mix of normal two-stage template matching and random sampling is used. This turns the problem into a 2-D optimization problem with adjustable parameters h_1 and u (the random sampling coefficient, equal to $1/\alpha$). However, in reality these types of problem are even more complex than indicated so far: in general, this is a 3-D optimization problem, the relevant parameters being h_1, h_2 and u, although in fact a good approximation to the global optimum may be obtained by the procedure of adjusting h_2 first, and then optimizing h_1 and u together—or even of adjusting h_2 first, then h_1 and then u (Davies, 1988i). Further details are beyond the scope of the present discussion.

25.4 Future directions

It has been indicated once or twice that the constraints and tradeoffs limiting algorithms are sometimes not accidental but rather the result of underlying technological or natural constraints. If so, it is important to determine this in as many cases as possible; otherwise, workers may spend much time on algorithm development only to find their efforts repeatedly being thwarted. Usually this is more easily said than done, but it underlines the necessity for scientific analysis of fundamentals. At this point it is worth considering technological constraints, to see what possibilities exist for advancing machine vision dramatically in the future.

The well-known law due to Moore (Noyce, 1977) relating to computer hardware states that the number of components that can be incorporated onto a single integrated circuit increases by a factor of about two per year. Certainly, this was so for the 20 years following 1959, although the rate subsequently decreased somewhat (not enough, however, to prevent the growth from remaining approximately exponential). It is not the purpose of this chapter to speculate on the accuracy of Moore's law. However, it is useful to suppose that computer memory and power will grow by a factor approaching two per year in the foreseeable future. Similarly, computer speeds may also grow at roughly this rate in the foreseeable future.

Unfortunately, many vision processes such as search are inherently NP-complete and hence demand computation that grows exponentially with some internal parameter such as the number of nodes in a match graph. This means that the advance of technology is able to give only a roughly linear improvement in this internal parameter (e.g. something like one extra node in a match graph every two years): it is therefore not solving the major search and other problems but only easing them.

While individual processor and memory circuits offer only these limited improvements, the possibilities to be obtained by combining the computational powers of many such processing elements in large parallel systems are significant. For the same reasons as outlined above, we probably want exponential growth in the numbers of processing elements that can be combined together into such systems. As seen in Chapter 24, data bottlenecks limit the overall power attainable by these means. Certainly the efficiency of each processing element seems always doomed to drop as more such elements are added to the system. Yet there is the hope of achieving a breakthrough in the architecture of massive computers. In particular, suitable means are required for achieving functional breakdowns of general vision algorithms so that data can flow naturally from one processing area to another—as no doubt occurs in biological processing systems. These ideas are consistent with the notion that once the processing system reaches a certain critical size, communication between one major fundamental part and another need not be too much of a limiting factor. Present-day multiprocessors are probably some way from achieving this critical size but there is some hope that with the advent of wafer-scale integration, we will be able to achieve increasingly impressive real-time vision capabilities.

25.5 Hardware, algorithms and processes

The previous section expressed the hope that improvements in hardware systems will provide the key to the development of impressive vision capabilities. However, it seems likely that breakthroughs in vision algorithms will also be required before this can come about. My belief is that until robots can play with objects and materials in the way that tiny children do they will not be able to build up sufficient information and the necessary databases for handling the complexities of real vision. The real world is too complex for all the rules to be written down overtly: these rules have to be internalized by training each brain individually. In some ways this approach is better, since it is more flexible and adaptable and at the same time more likely to be able to correct for the errors that would arise in direct transference of huge databases or programs. Nor should it be forgotten that it is the underlying processes of vision and intelligence that are important: hardware merely provides a means of implementation. If an idea is devised for a hardware solution to a visual problem, it reflects an underlying algorithmic process that either is or is not effective. Once it is known to be effective, then the hardware implementation can be analysed to confirm its utility. Of course, we must not segregate algorithms too much from hardware design. In the end it is necessary to optimize the whole system, which means considering both together. However, the underlying processes should be considered first, before a hardware solution is frozen in. Hardware should not be the driving force since there is a danger that some type of hardware implementation (especially one that is temporarily new and promising) will take over and make workers blind to underlying processes. Hardware should not be the tail that wags the vision dog.

25.6 A retrospective view

This book has progressed steadily from low-level ideas, through intermediate-level methods to high-level processing, covering 3-D image analysis, the necessary technology and so on—admittedly with its own type of detailed examples and emphasis. Many ideas have been covered and many strategies described. But where have we got to, and to what extent have we solved the problems of vision referred to in Chapter 1?

Perhaps the worst of all the problems of vision is that of minimizing the amount of processing required to achieve particular image recognition and measurement tasks. Not only do images contain huge amounts of data, but often they need to be interpreted in frighteningly small amounts of time and the underlying search and other tasks tend to be subject to combinatorial explosions. Yet, in retrospect, we seem to have remarkably few *general* tools for coping with these problems. Indeed, the truly general tools available[*] appear to be:

(i)   reducing high-dimensional problems to lower-dimensional problems that can be solved in turn;
(ii)   the Hough transform approach;
(iii) location of features that are in some sense sparse, and which can hence help to reduce redundancy quickly (obvious examples of such features are edges and corners);
(iv)   two-stage and multistage template matching;
(v)   random sampling.

These are said to be general tools since they appear in one guise or another in a number of situations, with totally different data. However, it is pertinent to ask to what extent these are genuine tools rather than almost accidental means (or tricks) by which computation may be reduced. Further analysis yields interesting answers to this question, as will now be seen.

First, consider the Hough transform, which takes a variety of forms—the normal parametrization of a line in an abstract parameter space, the GHT which is parametrized in a space congruent to image space, the adaptive thresholding transform (Chapter 4) which is parametrized in an abstract 2-D parameter space, and so on. What is common about these forms is *the choice of a representation in which the data peak naturally at various points*, so that analysis can proceed with improved efficiency. The relation with item (iii) above now becomes clear, making it less likely that either of these procedures is purely accidental in nature.

Next, item (i) appears in many guises—see for example the lateral histogram approach (Chapter 12) and the approaches used to locate ellipses (Chapter 11). Thus item (i) has much in common with item (iv). Note also that item (v) can be considered a special case of item (iv) (random sampling is a form of two-stage template matching with a "null" first stage, capable of eliminating large numbers of input patterns with particularly high efficiency:

---

[*] We here consider only intermediate-level processing, ignoring for example efficient AI tree-search methods relevant for purely abstract high-level processing.

see Davies, 1988i). Finally, note that the example of so-called two-stage template matching covered in Section 25.3.1 was actually part of a larger problem which was really multistage: the edge detector was two-stage but this was incorporated in an HT which was itself two-stage, making the whole problem at least four-stage. It can now be seen that items (i)–(v) are all forms of multistage matching (or sequential pattern recognition) which are potentially more powerful and efficient than a single-stage approach. Similar conclusions are arrived at in Appendix D, which deals with robust statistics and its application to machine vision.

Hence we are coming to a realization that there is just one general tool for increasing efficiency. However, in practical terms this may not be too useful a conclusion, since the subject of image analysis is also concerned with the ways in which this underlying idea may actually be realized—how are complex tasks to be broken down into the most appropriate multistage processes, and how then is the most suitable representation found for sparse feature location? Probably the five-point list given above throws the right sort of light onto this particular problem—although in the long run it will hardly turn out to be the whole truth!

25.7 Just a glimpse of vision?

This book has solved some of the problems it set itself—starting with low-level processing, concentrating on strategies, limitations and optimizations of intermediate-level processing, going some way with higher-level tasks, and attempting to create an awareness of the underlying processes of vision. Yet this programme has taken a lot of space and has thereby had to omit many exciting developments. Notable omissions are advanced work on artificial intelligence and knowledge-based systems and their application in vision; details of much important work on motion and image sequence analysis; details of hardware architectures and techniques including dataflow computers, systolic arrays and VLSI technology; and detailed theoretical studies of neural nets and connection-ism which offer many possibilities for marrying algorithms and architectures. Arguably, separate volumes are available on all these topics but until now a gap appeared to be present which the present book has aimed to fill.

25.8 Bibliographical and historical notes

Much of this chapter has summarized the work of earlier chapters and attempted to give it some perspective so the reader is referred back to appropriate sections for further detail. However, two-stage template matching has been highlighted in the current chapter: the earliest work on this topic was carried out by Rosenfeld and VanderBrug (1977) and VanderBrug and Rosenfeld (1977); later work included that on lateral histograms for corner detection by Wu and Rosenfeld (1983), while the ideas of Section 25.3.2 were developed by Davies (1988i). Two-stage template matching harks back to the spatial matched filtering concept discussed in Chapters 10, 13 and elsewhere. Ultimately this concept is limited by the variability of the objects to be detected. However, it has been shown that some account can be taken of this problem, e.g. in the design of filter masks (see Davies, 1992e).

As for the topics that could not be covered in this book (see previous section), the reader should refer initially to such volumes as Winston (1984) and Ballard and Brown (1982) for AI techniques and their application in vision: to Hildreth (1984) and Nagel (1986) for work on motion and image sequence analysis; and to Hwang and Briggs (1984), Offen (1985) and Fountain (1987), and references contained in these volumes, for hardware architectures and techniques relating to image processing. The subject of artificial neural networks is still a young field and some excellent volumes on it have appeared in the past few years: see for example Hertz *et al.* (1991) and Haykin (1994); see also Ripley (1996).

Appendix A

Programming Notation

A.1 Introduction

This book is concerned with the ideas of machine vision. This subject is still relatively young and the science of how to get computers to see is not fully developed. To a large extent practitioners in this field work intuitively, being guided by the experience of others—not least by the considerable literature that now exists. For this book to achieve its aims of helping workers with their various tasks, it is not sufficient to present ideas: they must be given substance. In a computer-based subject such as machine vision, that substance must involve computer programs. Now the methodology of writing computer programs is a vast subject in its own right and one that involves immense rigour and discipline—either a program works or it doesn't; either it degrades gracefully or it crashes. Hence we are in an area which has exacting requirements of its own. However, the subject of computer programming is too vast for this book to tackle that task as well. Hence we must adopt a stance in which readers will be assumed to be conversant with one computer language and merely need to be shown how to understand in detail the algorithms presented in this volume.

The language chosen for presenting programs in this book is Pascal (to which certain extensions have been made at relevant junctures). Such a choice is not lightly made but is influenced by: (i) how well known the language is; (ii) the availability of suitable compilers; and (iii) factors such as elegance, completeness and suitability for the task. In this respect it has been borne in mind that Pascal compilers are available on machines covering the whole range from supercomputers down to small personal computers. In addition the language is subject to strong "typing", so that variables are declared as being integer, boolean and so on, and then the computer is able to note programming errors by detecting violations relative to the original declarations. This helps with program robustness and reliability. Furthermore the language is "block-structured"—a factor that helps with the writing of correct code, and later with debugging and program modification (even if this is done by a different programmer!). Another important feature of Pascal is that it is "format-free", i.e. blank spaces and new lines in general make no difference to the operation of the code. This is both a convenience and a help since programs can be laid out in a manner that facilitates understanding of program structure and operation.

A.2　The Pascal language

A.2.1　Control structures

To underline the block-structuring characteristics of Pascal, note the following:

- (i)　a program consists of a sequence of instructions;
- (ii)　any instruction may be replaced by a block of instructions: if necessary (see below) these may be grouped together by **begin** and **end** markers;
- (iii)　blocks may be nested inside each other;
- (iv)　blocks must be wholly nested, i.e. one block must be wholly inside another, or wholly outside it.

These characteristics may make the language sound somewhat simplistic. However, it is important that there are several powerful constructs for directing the flow of control around a program: in particular, the **if** statement, the **for** loop, the **while** loop and the **repeat** loop. Of these the **if** statement is perhaps the most basic and permits control to be transferred further on in a program, following a specific test. In fact, an **if** statement tests a condition, and if it is true the **then** statement is executed: if the condition is not true, then the **else** statement (if one is present) is executed. Two examples are:

$$\text{if } u = 2 \text{ then } v := 3; \tag{A.1}$$

and

$$\text{if } u = 2 \text{ then } v := 3 \text{ else } w := 5; \qquad \text{(A.2)}$$

Note the semicolons used to indicate the end of each instruction.

The **for** loop is one in which an instruction is performed a certain number of times; for example, variable u can be incremented five times as follows:

$$\text{for } i := 1 \text{ to } 5 \text{ do } u := u + 1; \qquad \text{(A.3)}$$

Here the constant bounds on the **for** loop can be replaced by formulae from which the bound values may be calculated at run time:

$$\text{for } i := j + 1 \text{ to } k + 3 \text{ do } u := 1; \qquad \text{(A.4)}$$

The **while** and **repeat** loop constructs permit an instruction to be executed a variable number of times, termination being according to an appropriate condition—the meaning in each case being exactly what would be expected in English:

$$\text{while } i < 0 \text{ do } i := i + 1; \qquad \text{(A.5)}$$

$$\text{repeat } i := i + 1 \text{ until } i >= 0; \qquad \text{(A.6)}$$

In either case the action of incrementing i continues until i is at least zero. Hence, if i is initially less than zero, the final result will be $i = 0$. However, the results will differ in the two cases if i is initially zero or more: the **while** loop does not change i at all in this case but the **repeat** loop acts once and the final result is one greater than the initial value.

The constructs described above may be written in the notional forms:

$$\text{if } \langle \text{condition} \rangle \text{ then } \langle \text{statement} \rangle \text{ else } \langle \text{statement} \rangle; \quad \text{(A.7)}$$

$$\text{for } \langle \text{variable} \rangle := \langle \text{expression} \rangle \text{ to } \langle \text{expression} \rangle \text{ do } \langle \text{statement} \rangle;$$
$$\text{(A.8)}$$

$$\text{while } \langle \text{condition} \rangle \text{ do } \langle \text{statement} \rangle; \qquad \text{(A.9)}$$

$$\text{repeat } \langle \text{statement} \rangle \text{ until } \langle \text{condition} \rangle; \qquad \text{(A.10)}$$

In all four cases the triangular brackets show the *form* of the construct and would be omitted in any actual program. Finally, note that in each case the

⟨statement⟩ may be replaced by a block of statements grouped together by **begin** and **end** markers (although the latter are unnecessary in the **repeat** loop).

A.2.2 Procedures and functions

Two further important programming constructs are of use in this book: these are procedures and functions. A procedure is a short section of code that is likely to be used many times, so that it is useful to give it a name and call it by name instead of including it in its entirety every time it is needed. A function is similar to a procedure in that it may be used many times and is called by name. However, a function provides a value rather than just performing a function; since a function is usually a function of at least one variable, the latter is generally specified in brackets after the function name. A typical function is the sine function, which is called as in the following statement:

$$u := \sin(\text{theta}); \tag{A.11}$$

The following function returns the cube of an integer:

```
function cube (i: integer): integer;
   begin
     cube := i * i * i;                              (A.12)
   end;
```

This function can be used within the following procedure:

```
procedure cubes
   begin
     cube i := 1 to 10 do
        writeln ('The cube of', i:4, 'is', cube (i) :6);
   end;                                              (A.13)
```

If we now introduce the statement *cubes*; at any stage in a program, it will repeatedly call the function *cube* and print out a table of cubes. Note the distinction between the *declarations* (definitions) of the procedures and functions and the *calls* which cause them to be executed.

A.2.3   Other details of Pascal syntax

Now that the main Pascal constructs of use in this book have been described, it remains to explain some of the other notation that is employed within the language. Here *no attempt is made at completeness*—the main aim is to explain the notation employed within the various chapters. For a fuller explanation of Pascal see other texts, such as Rohl and Barrett (1980) and Forsyth (1982). Table A.1 summarizes the relevant notation. Note in particular that *set* and *string* notation is omitted from this table and that input/output procedures have for the most part been disregarded.

A number of further points should be made about Pascal, as follows.

Comments
These may be placed anywhere within a program, within curly brackets, in order to explain functionality: as an alternative, in case curly brackets are not available, "(* ", " *)" brackets may be used.

Assignments
These are made in Pascal using the ": =" notation, the variable on the left of the sign indicating the location where the value on the right is placed: e.g. $u := 3$ means that the number 3 is placed in the location corresponding to variable u (note that, apart from its use in constant and type declarations, "=" is used only to make tests for equality—never for assignments).

Standard variable types
Variables may take one of the standard *types* "integer", "real", "boolean" or "char", where these terms refer respectively to normal integer values, floating-point numbers, boolean variables that are either true or false, and characters (the last of which are not used in this book).

Special types
Variables may also take special forms that are declared after the **type** directive. For example, we may declare a type of the form *digit* to mean the numbers 0–9:

$$\text{type digit} = 0 \ .. \ 9; \tag{A.14}$$

and then declare various numbers of this type:

$$\text{var } c, \ d, \ e: \text{ digit}; \tag{A.15}$$

Table A.1 Summary of relevant Pascal notation.

Reserved words

and	array	begin	div	do	downto
else	end	for	function	if	not
or	procedure	repeat	then	to	type
until	var	while			

Identifiers

boolean	char	false	integer	real	true

Punctuation marks

( ) [ ] (* *) , ; : . '

Operators

:=	assignment
+	plus
–	minus
*	multiply
div	divide (for integer result, rounded down if necessary)
/	divide (for real result)
=	equal
<>	not equal
<	less than
>	greater than
<=	less than or equal
>=	greater than or equal
not	negation
and	conjunction
or	disjunction

Standard functions

abs	absolute value
arctan	inverse tangent
cos	cosine
exp	exponential
max	maximum of the set of arguments[*]
min	minimum of the set of arguments[*]
round	round real number to nearest integer value
sin	sine
sqr	square
sqrt	square root
trunc	round real number *down* to next integer value

Standard procedures

readln	read a line of data
writeln	write a line of data

---

[*] These two functions are not available in all implementations of Pascal.

Type conversions

Pascal has strict "typing" and does not permit a variable of one type to be used as another type. However, routines (procedures) may be declared for converting from one type to another. In particular, the standard functions *round* and *trunc* convert floating-point ("real") numbers (e.g. u, v) to integers (e.g. i, j):

```
i := round(u);
        (* converts u to the nearest integer value *)   (A.16)
j := trunc(v);
        (* rounds v down to the next integer *)          (A.17)
```

Different implementations of Pascal differ in how they permit variables to be converted from integer to floating-point values. In many cases it is permitted to mix integers (e.g. i, j) and floating-point numbers (e.g. f, g) in an expression, so long as the result is assigned to a floating-point number (e.g. q):

$$q := (f*i + g*j)/(i*i + j*j);   \qquad (A.18)$$

However, certain implementations require use of a *float* function:

$$q := (f*\text{float}(i) + g*\text{float}(j))/\text{float}(i*i + j*j);   \qquad (A.19)$$

There is no need to delve further into such matters here.

Arrays

Pascal permits arrays of variables to be used, these being of crucial value in handling image data. To declare an array that might act as a 256×256 pixel image we may first declare type *image*; then we may declare specific image storage areas:

```
type image = array [0 .. 255, 0 .. 255] of integer;
var P, Q, R: image;                                      (A.20)
```

Alternatively P, Q, R may be declared directly as arrays of this size:

```
var P, Q, R: array [0 .. 255, 0 .. 255] of integer;  (A.21)
```

In either case, array elements take the form (for example) P[3, 41]. Hence an image space may be cleared by the double **for** loop:

```
for j := 0 to 255 do
    for i := 0 to 255 do
        P[i, j] := 0;                                    (A.22)
```

A.2.4 The need for special syntax

So far we have outlined the Pascal notation rather briefly and have given a few examples to indicate its nature. This should on the whole be sufficient to understand the program extracts given in the main text of the book. However, two extensions of Pascal are required to complete the task. The first is simple: that in cases where the whole program would be too voluminous and detailed to warrant inclusion in the text, it is valuable to provide a pseudo-Pascal program in which the statements are written in reasonably precise English, using Pascal constructs such as **repeat ... until**, and so on. The second is more complex and is concerned with the nature of images and how they are processed; it is also concerned with the need to provide clear, succinct program statements when the full Pascal version would look rather clumsy—and indeed the reader would have to study the program very carefully in order to disentangle the inner sense from the programming detail. The next section shows the nonstandard notation that has been added to the usual Pascal constructs to arrive at the programs in the main text.

A.3 Special syntax embedded in Pascal

A.3.1 Image handling notation

Image processing and image analysis are concerned very much with the need to run systematically over all the pixels in an image, as in the double **for** loop example given earlier. In this book a double square bracket notation is used to indicate the double **for** loop, and the double subscript array is replaced by a single variable: e.g. P0 is used to replace $P[i, j]$ as the general pixel in an image in P-space. Thus the example given earlier for clearing an image in P-space now takes the much more succinct form:

$$[[\ P0 \ := \ 0 \ ]]; \qquad\qquad\qquad (A.23)$$

We now extend this notation by employing a linear array structure to refer to any pixel in the neighbourhood of the general pixel P0; normally, the pixels adjacent to P0 are denoted by P1–P8 as follows:

P4	P3	P2
P5	P0	P1
P6	P7	P8

although if an array notation for these 9 pixels is needed, the notation used is P[*i*], the numbering being such that for example P[5] means the same as P5. It turns out that these types of notation shorten and clarify programs very significantly, as will be seen on referring to the text (especially Chapters 2, 3 and 6). More is said in Section A.4 on the double square bracket notation, which we specialize to denote parallel processing, modified notation being available to denote sequential processing (see Table A.2).

Table A.2 Special syntax.

~	equivalent to not
&	equivalent to and
A0 ... A8 ⎫	
B0 ... B8 ⎬	bit image variables
C0 ... C8 ⎭	
P0 ... P8 ⎫	
Q0 ... Q8 ⎬	byte image variables
R0 ... R8 ⎭	
P[0],...	equivalent to P0,...
[[...]]	double for loop over image (parallel processing)
[[+...+]]	double for loop for forward sequential raster scan
[[-...-]]	double for loop for reverse sequential raster scan
x, y	local coordinates for random access to image
input(P)	grab a new image into P-space
xxxx:	equivalent to procedure xxxx;

A.3.2 Other succinct notation

Another simplifying notation is to use "&" for **and** and "~" for **not** in certain cases where succinctness is particularly important (see below). Further clarification can be achieved if single-bit pixel values can be regarded as both numbers and boolean values. This sort of mixing of types is disallowed in Pascal—indeed, it is anathema to the whole philosophy of Pascal. However, it is adopted here for two reasons: one is the simplicity and succinctness it gives to programs as they appear in text; and the other is the fact that speed of processing often demands such programs (or at least specific parts of them) to be coded optimally in machine code—at which stage niceties of typing disappear. From the latter point of view, single-bit integers and booleans can very easily be regarded as identical (and indeed, if special hardware is used to implement the storage locations, the difference between the types becomes even more indistinguishable).

Let us see in detail what types of statement we are attempting to write, and exactly how the types become mixed. Here we take two examples from Chapter 6:

```
sigma := A1 + A2 + A3 + A4 + A5 + A6 + A7 + A8
        + (A1 & ~A2 & A3) + (A3 & ~A4 & A5);
        + (A5 & ~A6 & A7) + (A7 & ~A8 & A1);      (A.24)
```

and:

```
chi := (A1 <> A3) + (A3 <> A5) + (A5 <> A7) + (A7 <> A1)
      + 2*((~A1 & A2 & ~A3) + (~A3 & A4 ~A5);
      + (~A5 & A6 & ~A7) + (~A7 & A8 & ~A1));      (A.25)
```

Both examples involve long sequences of computations, which are apparently simple but are actually not trivial to understand (see Chapter 6). If we insisted on using rigorously typed Pascal, we would have to use special type conversion functions such as the following:

```
(* function to convert booleans to bits *)
function unit (B: boolean): bit;
  begin
    if B then unit :=1 else unit := 0;
  end;                                              (A.26)
```

```
(* function to convert bits to booleans *)
function bool(A: bit): boolean;
  begin
    if A = 1 then bool := true else bool := false;
  end;                                              (A.27)
```

In either case it is assumed that a type *bit* has been declared as follows:

$$\text{type bit} = 0 .. 1;   (A.28)$$

and that the variables A0–A8 have been predeclared as:

```
var A0, A1, A2, A3, A4, A5, A6, A7, A8: bit;      (A.29)
```

With all these prior declarations, the two forms sigma and chi (Chapters 2 and 6) can be re-expressed as their standard Pascal equivalents:

```
sigma := A1 + A2 + A3 + A4 + A5 + A6 + A7 + A8
          + unit(bool(A1) & ~bool(A2) & bool(A3))
          + unit(bool(A3) & ~bool(A4) & bool(A5))
          + unit(bool(A5) & ~bool(A6) & bool(A7))
          + unit(bool(A7) & ~bool(A8) & bool(A1));    (A.30)
```

and

```
chi := unit(bool(A1) <> bool(A3))+unit(bool(A3) <> bool(A5))
        + unit(bool(A5) <> bool(A7))
        + unit(bool(A7)) <> bool(A1))
        + 2 *( unit(~bool(A1) & bool(A2) & ~bool(A3))
          + unit(~bool(A3) & bool(A4) & ~bool(A5))
          + unit(~bool(A5) & bool(A6) & ~bool(A7))
          + unit(~bool(A7) & bool(A8) & ~bool(A1)));
                                                      (A.31)
```

These examples show fully the value of the notation used in the text. A further notation that clarifies the text is omission of the word **procedure** and use merely of a label plus colon followed by a defining block as the declaration. Note that, in general, declarations of variables are not given in the text but should be clear from the context. However, image variables are taken to be predeclared as bit or byte integers (see Table A.2), bit variables being needed to hold shape information in binary images, and byte variables to hold information on up to 256 grey levels in grey-scale images (see Chapter 2).

Finally, clumsy situations that arise with **if** statements, such as the following:

```
if PO > thr then AO := 1 else AO := 0;        (A.32)
```

are made considerably simpler by writing:

```
AO := if PO > thr then 1 else 0;              (A.33)
```

This follows the style of languages such as Algol68.

On the whole, program extracts given in the text should always be in one-to-one correspondence both with an underlying mathematical meaning and with a rigorous Pascal form obtainable via conversion functions such as *bool* and *unit* defined above.

A.4   On the validity of the "repeat until finished" construct

In the text the notation [[...]] is used to represent parallel processing over the pixels in an image. In true parallel processing this would be carried out independently for each pixel by the $N \times N$ processors of a SIMD array (see Chapter 24). However, parallel processing may be simulated by using a double **for** loop (see equations (A.22) and (A.23)). This means that a single processor is actually being applied to perform the same function as a parallel processor. Note that the order in which the pixels are processed is immaterial, since one of the rules in simulating parallel processing is that the information is always placed in a new storage space (see Chapter 2). However, there are problems with constructs such as that used in the algorithm of Table A.3. This is because the "**repeat until** finished" construct is formulated in such a way that parallel processing is simulated, so there only needs to be one variable *finished*. However, in a parallel processor there would have to be many independent variables of this type, and these would then all have to be examined to find whether the process had really terminated. However, a straightforward logical operation could easily tell whether the overall process had terminated. This logical operation would be performed by very simple hardware (little more than one large OR-gate!) linking the processors with the single *finished* control variable.

Table A.3 Parallel algorithm using the "**repeat until** finished" construct.

```
( * mock thinning algorithm * )
repeat
  finished := true;
  if (chi = 2) and (sigma <> 1) then
    begin
      remove pixel;
      finished := false;
    end;
until finished;
```

We do not dwell on this further since this book is concerned more with the processes of image analysis than with the niceties of parallel processing protocols.

Appendix B

Mathematical Morphology

B.1 Introduction

In Chapter 2 we met the operations of erosion and dilation: in Chapter 6 we applied them to the filtering of binary images, and showed that with suitable combinations of these operators it is possible to eliminate certain types of object from images, and also to locate other objects. It turns out that these possibilities are not fortuitous, but on the contrary reflect important mathematical properties of shape which are dealt with in the subject known as mathematical morphology. This subject has grown up over the past 20 or so years, and over the past decade knowledge in this area has become consolidated and is now understood in considerable depth. It is the purpose of this appendix to give some insight into this important area of study. The mathematical nature of the topic mitigated against inserting it in the main text of the book, where it might impede the flow of the text and at the same time put off less mathematically minded readers. However, these considerations must not give the impression that mathematical morphology is a mere sideline: on the contrary, it is highly important as it provides a backbone for the whole study of shape which is capable of unifying techniques as disparate as noise suppression, shape analysis, feature recognition, skeletonization, convex hull formation, and a host of other topics.

B.2   Dilation and erosion in binary images

B.2.1   Dilation and erosion

As we have seen in Chapter 2, dilation expands objects into the background and is able to eliminate "salt" noise within an object. It can also be used to remove cracks in objects which are less than three pixels in width.

In contrast, erosion shrinks binary picture objects, and has the effect of removing "pepper" noise. It also removes thin object "hairs" whose widths are less than three pixels.

As we shall see in more detail below, erosion is strongly related to dilation, in that a dilation acting on the inverted input image acts as an erosion, and vice versa.

B.2.2   Cancellation effects

An obvious question is whether erosions cancel out dilations, or vice versa. We can easily answer this question: for if a dilation has been carried out, salt noise and cracks will have been removed, and once they are gone, erosion cannot bring them back; hence exact cancellation will not in general occur. Thus, for the set S of object pixels in a general image I, we may write:

$$\text{erode}(\text{dilate}(S)) \neq S \tag{B.1}$$

equality only occurring for certain specific types of image (these will lack salt noise, cracks and fine boundary detail). Similarly, pepper noise or hairs that are eliminated by erosion will not in general be restored by dilation:

$$\text{dilate}(\text{erode}(S)) \neq S \tag{B.2}$$

Overall, the most general statements that can be made are:

$$\text{erode}(\text{dilate}(S)) \supseteq S \tag{B.3}$$
$$\text{dilate}(\text{erode}(S)) \subseteq S \tag{B.4}$$

We may note, however, that large objects will be made 1 pixel larger all round by dilation, and will be reduced by 1 pixel all round by erosion, so a considerable amount of cancellation will normally take place when the two operations are applied in sequence. This means that sequences of erosions and dilations provide a good basis for filtering noise and unwanted detail from images.

B.2.3   Modified dilation and erosion operators

It sometimes happens that images contain structures which are aligned more or less along the image axes directions, and in such cases it is useful to be able to process these structures differently. For example, it might be useful to eliminate fine vertical lines, without altering broad horizontal strips. In that case the following "vertical erosion" operator will be useful:

```
ERODE: [[ sigma := A1 + A5;
            BO := if sigma < 2 then 0 else AO ]];        (B.5)
```

though it will be necessary to follow it with a compensating dilation operator[*] so that horizontal strips are not shortened:

```
DILATE: [[ sigma := A1 + A5;
             BO := if sigma > 0 then 1 else AO ]];       (B.6)
```

This example demonstrates some of the potential for constructing more powerful types of image filter. To realize these possibilities we next develop a more general mathematical morphology formalism.

B.3   Mathematical morphology

B.3.1   Generalized morphological dilation

The basis of mathematical morphology is the application of set operations to images and their operators. We start by defining a generalized dilation mask as a set of locations within a 3 × 3 neighbourhood. When referred to the centre of the neighbourhood as origin, each of these locations causes a shift of the image in the direction defined by the vector from the origin to the location. When several shifts are prescribed by a mask, the 1 locations in the various shifted images are combined by a set union operation.

---

[*] Here and elsewhere in this appendix, any operations required to restore the image to the original image space are not considered or included (see Section 2.4).

The simplest example of this type is the identity operation which leaves the image unchanged:

	1	

(Notice that we leave the 0s out of this mask, as we are now focusing on the set of elements at the various locations, and set elements are either present or absent.)

The next operation to consider is:

1		

which is a left shift, equivalent to the one discussed in Section 2.2. Combining the two operations into a single mask:

1	1	

leads to a horizontal thickening of all objects in the image, by combining it with a left-shifted version of itself. An isotropic thickening of all objects is achieved by the operator:

1	1	1
1	1	1
1	1	1

(clearly, this is equivalent to the dilation operator discussed in Sections 2.2 and B.2), while a vertical dilation operation (see Section B.2.3) is achieved by the mask:

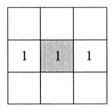

A rule of such operations is that if we want to guarantee that all the original object pixels are included in the output image, then we must include a 1 at the centre (origin) of the mask.

Finally, there is no compulsion for all masks to be 3×3. Indeed, all but one of those listed above are effectively smaller than 3×3, and in more complex cases larger masks could be used. To emphasize this point, and to allow for asymmetrical masks in which the full 3×3 neighbourhood is not given, we shall shade the origin—as shown in the above cases.

B.3.2   Generalized morphological erosion

We now move on to describe erosion in terms of set operations. The definition is somewhat peculiar in that it involves reverse shifts, but the reason for this will become clear as we proceed. Here the masks define directions as before, but in this case we shift the image in the reverse of each of these directions and perform intersection operations to combine the resulting images. For masks with a single element (as for the identity and shift left operators of Section B.3.1) the intersection operation is improper and the final result is as for the corresponding dilation operator, but with a reverse shift. For more complex cases the intersection operation results in objects being reduced in size. Thus the mask:

has the effect of stripping away the left sides of objects (the object is moved right and *and*ed with itself). Similarly, the mask:

1	1	1
1	1	1
1	1	1

results in an isotropic stripping operation, and is hence identical to the erosion operation described in Section B.2.1.

B.3.3   Duality between dilation and erosion

We shall write the dilation and erosion operations formally as $A \oplus B$ and $A \ominus B$ respectively, where A is an image and B is the mask of the relevant operation:

$$A \oplus B = \bigcup_{b \in B} A_b \qquad \text{(B.7)}$$

$$A \ominus B = \bigcap_{b \in B} A_{-b} \qquad \text{(B.8)}$$

In these equations, A_b indicates a basic shift operation in the direction of element b of B and A_{-b} indicates the reverse shift operation.

We next prove an important theorem relating the dilation and erosion operations:

$$(A \ominus B)^c = A^c \oplus B^r \qquad \text{(B.9)}$$

where A^c represents the complement of A, and B^r represents the reflection of B in its origin. We first note that[*]:

$$x \in A^c \Leftrightarrow x \notin A \qquad \text{(B.10)}$$

and

$$b \in B^r \Leftrightarrow -b \in B \qquad \text{(B.11)}$$

---

[*] The sign "$\Leftrightarrow$" means "if and only if", i.e. the statements so connected are equivalent.

We now have[*]:

$$x \in (A \ominus B)^c \Leftrightarrow x \notin A \ominus B$$
$$\Leftrightarrow \exists b \text{ such that } x \notin A_{-b}$$
$$\Leftrightarrow \exists b \text{ such that } x + b \notin A$$
$$\Leftrightarrow \exists b \text{ such that } x + b \in A^c$$
$$\Leftrightarrow \exists b \text{ such that } x \in (A^c)_{-b}$$
$$\Leftrightarrow x \in \cup_{b \in B}(A^c)_{-b}$$
$$\Leftrightarrow x \in \cup_{b \in B^r}(A^c)_b$$
$$\Leftrightarrow x \in A^c \oplus B^r \tag{B.12}$$

This completes the proof. The related theorem:

$$(A \oplus B)^c = A^c \ominus B^r \tag{B.13}$$

is proved similarly.

The fact that there are two such closely related theorems, following the related union and intersection definitions of dilation and erosion given above, indicates an important duality between the two operations. Indeed, as stated earlier, erosion of the objects in an image corresponds to dilation of the background, and *vice versa*. However, the two theorems indicate that this relation is not absolutely trivial, on account of the reflections of the masks required in the two cases. It is perhaps curious that, in contrast with the case of the de Morgan rule for complementation of an intersection:

$$(P \cap Q)^c = P^c \cup Q^c \tag{B.14}$$

the effective complementation of the dilating or eroding mask is its reflection rather than its complement *per se*, while that for the operator is the alternate operator.

A further interesting aspect of the results of the two theorems arises since:

$$A \oplus B = B \oplus A \tag{B.15}$$

This leads to the following similar but more complex relation for the erosion operator:

$$A^c \ominus B^r = B^c \ominus A^r \tag{B.16}$$

---

[*] This proof is based on that of Haralick *et al.* (1987).

Unfortunately, this latter relation is probably of little direct practical value: reflecting small masks is a useful concept, but reflecting whole images is unlikely to be readily visualizable. Its importance lies in its demonstration that, unlike dilation, erosion is noncommutative, so the apparent symmetry between the two operators is more subtle than their simple origins in expanding and shrinking might indicate.

We also find that successive dilations are associative:

$$(A \oplus B) \oplus C = A \oplus (B \oplus C) \tag{B.17}$$

whereas successive erosions are not. In fact the corresponding relation for erosions is:

$$(A \ominus B) \ominus C = A \ominus (B \oplus C) \tag{B.18}$$

It turns out that this relation is of value in showing how a large erosion might be factorized so that it can be implemented more efficiently as two smaller erosions applied in sequence. This may also be achieved by making use of combinations of dilation or erosion operators with set elements, though great care must be exercised to note which particular distribution operations are valid:

$$A \oplus (B \cup C) = (A \oplus B) \cup (A \oplus C) \tag{B.19}$$

$$A \ominus (B \cup C) = (A \ominus B) \cap (A \ominus C) \tag{B.20}$$

$$(A \cap B) \ominus C = (A \ominus C) \cap (B \ominus C) \tag{B.21}$$

In other cases, the strongest statements that can be made are typified by the following:

$$A \ominus (B \cap C) \supseteq (A \ominus B) \cup (A \ominus C) \tag{B.22}$$

Finally, we explore why the morphological definition of erosion involves a reflection. The idea is so that dilation and erosion are able, under the right circumstances, to cancel each other out. Take the left shift dilation operation and the right shift erosion operation. These are both achieved via the mask:

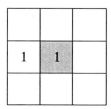

but in the erosion operation it is applied in its reflected form, thereby producing the right shift required to erode the left edge of any objects. This makes it clear why an operation of the type $(A \oplus B) \ominus B$ has a chance of cancelling to give A. More specifically, there must be shifts in opposite directions as well as appropriate subtractions produced by *and*ing instead or *or*ing in order for cancellation to be possible. Of course, in many cases the dilation mask will have 180° rotation symmetry, and then the distinction between B^r and B will be purely academic.

B.3.4 Closing and opening

The discussion at the end of the last subsection leads naturally to the concepts of closing and opening. Closing is the process of dilating an image and then immediately eroding it using the same mask operator:

$$A \bullet B = (A \oplus B) \ominus B \qquad (B.23)$$

As indicated in Section B.2, the purpose of this procedure is to fill in salt noise or cracks or narrow channels, and then to reverse the process so that larger scale image features are recovered as far as possible in their original form. The process can also be used to fill in concavities or holes.

Opening is the process of eroding an image and then immediately dilating it using the same mask operator:

$$A \circ B = (A \ominus B) \oplus B \qquad (B.24)$$

As indicated earlier, this procedure is able to eliminate pepper noise and remove fine hairs while leaving the rest of the image almost unchanged. It can also be used to remove protrusions or convexities.

Closing and opening have the interesting and useful property that they are idempotent, i.e. that repeated application of either operation has no further effect. This property contrasts strongly with the case of dilation (which, if applied repeatedly, will enlarge objects until they fill all or part of the image space) and erosion (which, if applied repeatedly, will empty all or part of the image space). These properties are summed up by the following relations:

$$(A \bullet B) \bullet B = A \bullet B \qquad (B.25)$$

$$(A \circ B) \circ B = A \circ B \qquad (B.26)$$

These properties can be understood by the fact that once a hole or crack has been filled in, it remains filled in and cannot be filled in again. Likewise, once a hair or protrusion has been removed, it remains removed and cannot be recreated without evidence (e.g. without appealing to the original image data).

There are a number of other properties of closing and opening; amongst the most important ones are the following set containment properties:

$$A \oplus B \supseteq A \bullet B \supseteq A \tag{B.27}$$

$$A \ominus B \subseteq A \circ B \subseteq A \tag{B.28}$$

In particular, objects in a closed image will if anything be larger than the original objects, and objects in an opened image will if anything be smaller than the original objects.

B.3.5   Hit-and-miss transform

We now move on to a transform that can be used to search an image for particular instances of a shape or other characteristic image feature. It is defined by the expression:

$$A \otimes (B, C) = (A \ominus B) \cap (A^c \ominus C) \tag{B.29}$$

The fact that this expression contains an *and* (intersection) sign would appear to make the description "hit-and-miss" transform more appropriate than the name "hit-or-miss" transform which is sometimes used.

We first of all consider the image analysing properties of the $A \ominus B$ bracket. Applying B in this way (i.e. using erosion) leads to the location of a set of pixels where the 1s in the image match the 1s in the mask. Now notice that applying C to image A^c reveals a set of pixels where the 1s in C match the 1s in A^c or the 0s in A. However, it is usual to take a C mask which does not intersect with B (i.e. $B \cap C = \varnothing$). This indicates that we are matching a set of 0s in a more general mask D against the 0s in A. Thus D is an augmented form of B containing a specified set of 0s (corresponding to the 1s in C) and a specified set of 1s (corresponding to the 1s in B). Certain other locations in D contain neither 0s nor 1s and can be regarded as "don't care" elements (which we will mark with an "×"). As an example, take the following lower left concave corner locating masks:

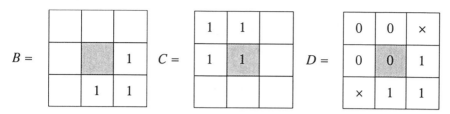

It will be seen that, in reality, this transform is equivalent to a code coincidence detector, with the power to ignore irrelevant pixels. Thus it searches for instances where specified pixels have the required 0 values and other specified pixels have the required 1 values: when it encounters such a pixel, it marks it with a 1 in the new image space, and otherwise returns a 0. This approach is quite general and can be used to detect a wide variety of local image conditions, including particular features, structures, shapes or objects. It is by no means restricted to 3×3 masks as the above example might indicate. In short, it is a general binary template matching operator. Its power might serve to indicate that erosion is ultimately a more important concept than dilation—at least in its discriminatory, pattern recognition properties.

B.3.6 Template matching

As indicated in the previous subsection, the hit-and-miss transform is essentially a template matching operator and can be used to search images for characteristic features or structures. However, it needs to be generalized significantly before it can be used for practical feature detection tasks, because features may appear in a number of different orientations or guises, and any one of them will have to trigger the detector. To achieve this we merely need to take the union of the results obtained from the various pairs of masks:

$$\cup_i A \otimes (B_i, C_i) = \cup_i (A \ominus B_i) \cap (A^c \ominus C_i) \qquad (B.30)$$

or, equivalently, to take the union of the results obtained from the various D-type masks (see previous subsection).

As an example, we consider how the ends of thin lines may be located. To achieve this we need eight masks of the following types:

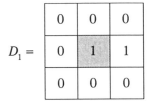

It is of interest that a much simpler algorithm could be used in this case to count the number of neighbours of each pixel and check whether the number is unity. (It is worth bearing in mind that alternative approaches can sometimes be highly beneficial in cutting down on excessive computational loads: on the other hand, morphological processing is of value in this general context because of the extreme simplicity of the masks being used for analysing images.) Further examples of template matching will appear in the following section.

B.4 Connectedness-based analysis of images

In this appendix there is only space for one illustration of the application of morphology to connectedness-based analysis. We have chosen the case of skeletonization and thinning.

B.4.1 Skeletons and thinning

A widely used thinning algorithm uses eight D-type hit-and-miss transform masks (see Section B.3.5):

$$
T_1 = \begin{array}{|c|c|c|}
\hline
0 & \times & 1 \\
\hline
0 & 1 & 1 \\
\hline
0 & \times & 1 \\
\hline
\end{array}
\qquad
T_2 = \begin{array}{|c|c|c|}
\hline
\times & 1 & 1 \\
\hline
0 & 1 & 1 \\
\hline
0 & 0 & \times \\
\hline
\end{array}
$$

$$T_3 = \begin{array}{|c|c|c|} \hline 1 & 1 & 1 \\ \hline \times & 1 & \times \\ \hline 0 & 0 & 0 \\ \hline \end{array} \qquad T_4 = \begin{array}{|c|c|c|} \hline 1 & 1 & \times \\ \hline 1 & 1 & 0 \\ \hline \times & 0 & 0 \\ \hline \end{array}$$

$$T_5 = \begin{array}{|c|c|c|} \hline 1 & \times & 0 \\ \hline 1 & 1 & 0 \\ \hline 1 & \times & 0 \\ \hline \end{array} \qquad T_6 = \begin{array}{|c|c|c|} \hline \times & 0 & 0 \\ \hline 1 & 1 & 0 \\ \hline 1 & 1 & \times \\ \hline \end{array}$$

$$T_7 = \begin{array}{|c|c|c|} \hline 0 & 0 & 0 \\ \hline \times & 1 & \times \\ \hline 1 & 1 & 1 \\ \hline \end{array} \qquad T_8 = \begin{array}{|c|c|c|} \hline 0 & 0 & \times \\ \hline 0 & 1 & 1 \\ \hline \times & 1 & 1 \\ \hline \end{array}$$

Each of these masks recognizes a particular situation where the central pixel is unnecessary for maintaining connectedness and can be eliminated (this assumes that the foreground is regarded as 8-connected). Thus all pixels can be subjected to the T_1 mask and eliminated if the mask matches the neighbourhood; then all pixels can be subjected to the T_2 mask and eliminated if appropriate; and so on until upon repeating with all the masks there is no change in the output image (see Chapter 6 for a full description of the process).

However, although this particular procedure is guaranteed not to disconnect objects, it does not terminate the skeleton location procedure, as there remain a number of locations where pixels can be removed without disconnecting the skeleton. The advantage of the masks given above is that they are very simple and produce fast convergence towards the skeleton. Their disadvantage is that a final stage of processing is required to complete the process (for example, one of the crossing number based methods of Chapter 6 may be used for this purpose, but we do not pursue the matter further here).

B.5   Concluding remarks

Binary images contain all the data needed to analyse the shapes, sizes, positions and orientations of objects in two dimensions, and thereby to recognize them and even to inspect them for defects. Many simple small neighbourhood operations exist for processing binary images and moving towards the goals stated above. At first sight these may appear a somewhat random set, reflecting historical development rather than systematic analytic tools. However, in the past few years, mathematical morphology has emerged as a unifying theory of shape analysis. Even this subject has been subject to a distinct historical development, but key papers over rather less than a decade have built up a solid body of knowledge: we have aimed to give the flavour of the subject in this appendix. Unfortunately, mathematical morphology, as its name suggests, is mathematical in nature, but there are a number of key theorems and results which are worth remembering: a few of these have been considered here and placed in context. For example, generalized dilation and erosion have acquired a central importance, since further vital concepts and constructs are based on them—closing, opening, template matching (via the hit-and-miss transform), to name but a few. And when one moves on to connectedness properties and concepts such as skeletonization, the relevance of mathematical morphology has already been proven (though space has prevented a detailed discussion of its application to these latter topics).

The application of morphology to grey-scale processing involves considerable additional theory and is outside the scope of this appendix. The reader is referred to the literature for further information on this topic (see Section B.6).

B.6   Bibliographical and historical notes

The book by Serra (1982) is an important early landmark in the development of morphology. Many subsequent papers helped to lay the mathematical foundations, perhaps the most important and influential being that by Haralick et al. (1987); see also Zhuang and Haralick (1986) for methods for decomposing morphological operators, and Crimmins and Brown (1985) for more practical aspects of shape recognition. The paper by Dougherty and Giardina (1988) was important in the development of methods for grey-scale morphological processing; Heijmans (1991) and Dougherty and Sinha (1995a, b) have extended this earlier work, while Haralick and Shapiro (1992) provide a useful general introduction to the topic. The work of

Huang and Mitchell (1994) on grey-scale morphology decomposition and that of Jackway and Deriche (1996) on multiscale morphological operators show that the theory of the subject is by no means fully developed, though more applied developments such as morphological template matching (Jones and Svalbe, 1994), edge detection in medical images (Godbole and Amin, 1995) and boundary detection of moving objects (Yang and Li, 1995) demonstrate the potential value of morphology. The reader is referred to the volumes by Dougherty (1992) and Heijmans (1994) for further information on the subject.

An interesting point is that it is by no means obvious how to decide on the sequence of morphological operations that is required in any application. This is an area where genetic algorithms might be expected to contribute to the systematic generation of complete systems and seems likely to be a focus for much future research (see for example Harvey and Marshall, 1994).

Appendix C

Image Transformations and Camera Calibration

C.1 Introduction

When images are obtained from 3-D scenes, the exact position and orientation of the camera sensing device is often unknown and there is a need for it to be related to some global frame of reference. This is especially important if accurate measurements of objects are to be made from their images, e.g. in inspection applications. On the other hand, it may in certain cases be possible to dispense with such detailed information – as in the case of a security system for detecting intruders, or a system for counting cars on a motorway. There are also more complicated cases, such as those in which cameras can be rotated or moved on a robot arm, or the objects being examined can move freely in space. In such cases, camera calibration becomes a central issue. Before we can consider camera calibration, we need to understand in some detail the transformations that can occur between the original world points and the formation of the final image. We attend to these image transformations in the following section.

C.2 Image transformations

First, we consider the rotations and translations of object points relative to a global frame. After a rotation through an angle θ about the Z-axis (Fig. C.1), the coordinates of a general point (X, Y) change to:

$$X' = X \cos \theta - Y \sin \theta \tag{C.1}$$

$$Y' = X \sin \theta + Y \cos \theta \tag{C.2}$$

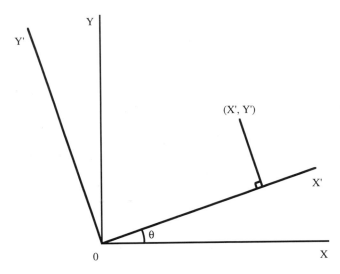

Fig. C.1   Effect of a rotation θ about the origin.

This result is neatly expressed by the matrix equation:

$$\begin{bmatrix} X' \\ Y' \end{bmatrix} = \begin{bmatrix} \cos\theta & -\sin\theta \\ \sin\theta & \cos\theta \end{bmatrix} \begin{bmatrix} X \\ Y \end{bmatrix} \tag{C.3}$$

Clearly, similar rotations are possible about the X- and Y-axes. To satisfactorily express rotations in 3-D we require a more general notation using 3×3 matrices, the matrix for a rotation θ about the Z-axis being:

$$\mathbf{Z}(\theta) = \begin{bmatrix} \cos\theta & -\sin\theta & 0 \\ \sin\theta & \cos\theta & 0 \\ 0 & 0 & 1 \end{bmatrix} \tag{C.4}$$

those for rotations ψ about the X-axis and φ about the Y-axis are:

$$\mathbf{X}(\psi) = \begin{bmatrix} 1 & 0 & 0 \\ 0 & \cos\psi & -\sin\psi \\ 0 & \sin\psi & \cos\psi \end{bmatrix} \tag{C.5}$$

$$\mathbf{Y}(\varphi) = \begin{bmatrix} \cos\varphi & 0 & \sin\varphi \\ 0 & 1 & 0 \\ -\sin\varphi & 0 & \cos\varphi \end{bmatrix} \tag{C.6}$$

We can make up arbitrary rotations in 3-D by applying sequences of such rotations. Similarly, we can express arbitrary rotations as sequences of rotations about the coordinate axes. Thus $\mathbf{R} = \mathbf{X}(\psi)\mathbf{Y}(\varphi)\mathbf{Z}(\theta)$ is a composite rotation in which $\mathbf{Z}(\theta)$ is applied first, then $\mathbf{Y}(\varphi)$, and finally $\mathbf{X}(\psi)$. Rather than multiplying out these matrices, we write down here the general result expressing an arbitrary rotation $\mathbf{R}$:

$$\begin{bmatrix} X' \\ Y' \\ Z' \end{bmatrix} = \begin{bmatrix} R_{11} & R_{12} & R_{13} \\ R_{21} & R_{22} & R_{23} \\ R_{31} & R_{32} & R_{33} \end{bmatrix} \begin{bmatrix} X \\ Y \\ Z \end{bmatrix} \tag{C.7}$$

Note that the matrix $\mathbf{R}$ is not completely general: it is orthogonal and thus has the property that $\mathbf{R}^{-1} = \mathbf{R}^{\mathrm{T}}$. In contrast with rotation, translation through a distance (T_1, T_2, T_3) is given by:

$$X' = X + T_1 \tag{C.8}$$

$$Y' = Y + T_2 \tag{C.9}$$

$$Z' = Z + T_3 \tag{C.10}$$

which is not expressible in terms of a multiplicative 3×3 matrix. However, just as general rotations can be expressed as rotations about various coordinate axes, so general translations and rotations can be expressed as sequences of basic rotations and translations relative to individual coordinate axes. Thus it would be most useful to have a notation which unified the mathematical treatment so that a generalized displacement could be expressed as a product of matrices. This is indeed possible if so-called *homogeneous coordinates* are used. To achieve this the matrices must be augmented to 4×4. A general rotation can then be expressed in the form:

$$\begin{bmatrix} X' \\ Y' \\ Z' \\ 1 \end{bmatrix} = \begin{bmatrix} R_{11} & R_{12} & R_{13} & 0 \\ R_{21} & R_{22} & R_{23} & 0 \\ R_{31} & R_{32} & R_{33} & 0 \\ 0 & 0 & 0 & 1 \end{bmatrix} \begin{bmatrix} X \\ Y \\ Z \\ 1 \end{bmatrix} \tag{C.11}$$

while the general translation matrix becomes:

$$
\begin{bmatrix} X' \\ Y' \\ Z' \\ 1 \end{bmatrix} = \begin{bmatrix} 1 & 0 & 0 & T_1 \\ 0 & 1 & 0 & T_2 \\ 0 & 0 & 1 & T_3 \\ 0 & 0 & 0 & 1 \end{bmatrix} \begin{bmatrix} X \\ Y \\ Z \\ 1 \end{bmatrix}
$$

(C.12)

The generalized displacement (i.e. translation plus rotation) transformation clearly takes the form:

$$
\begin{bmatrix} X' \\ Y' \\ Z' \\ 1 \end{bmatrix} = \begin{bmatrix} R_{11} & R_{12} & R_{13} & T_1 \\ R_{21} & R_{22} & R_{23} & T_2 \\ R_{31} & R_{32} & R_{33} & T_3 \\ 0 & 0 & 0 & 1 \end{bmatrix} \begin{bmatrix} X \\ Y \\ Z \\ 1 \end{bmatrix}
$$

(C.13)

We now have a convenient notation for expressing generalized transformations including operations other than the translations and rotations which account for the normal motions of rigid bodies. First we take a scaling in size of an object, the simplest case being given by the matrix:

$$
\begin{bmatrix} S & 0 & 0 & 0 \\ 0 & S & 0 & 0 \\ 0 & 0 & S & 0 \\ 0 & 0 & 0 & 1 \end{bmatrix}
$$

The more general case:

$$
\begin{bmatrix} S_1 & 0 & 0 & 0 \\ 0 & S_2 & 0 & 0 \\ 0 & 0 & S_3 & 0 \\ 0 & 0 & 0 & 1 \end{bmatrix}
$$

introduces a shear in which an object line λ will be transformed into a line that is not in general parallel to λ. Skewing is another interesting transformation, being given by linear translations varying from the simple case:

$$\begin{bmatrix} 1 & B & 0 & 0 \\ 0 & 1 & 0 & 0 \\ 0 & 0 & 1 & 0 \\ 0 & 0 & 0 & 1 \end{bmatrix}$$

to the general case:

$$\begin{bmatrix} 1 & B & C & 0 \\ D & 1 & F & 0 \\ G & H & 1 & 0 \\ 0 & 0 & 0 & 1 \end{bmatrix}$$

Rotations can be regarded as combinations of scaling and skewing, and are sometimes implemented as such (Weiman, 1976).

The other simple but interesting case is that of reflection, which is typified by:

$$\begin{bmatrix} 0 & 1 & 0 & 0 \\ 1 & 0 & 0 & 0 \\ 0 & 0 & 1 & 0 \\ 0 & 0 & 0 & 1 \end{bmatrix}$$

This generalizes to other cases of improper rotation where the determinant of the top left 3×3 matrix is -1.

In all the cases discussed above it will be observed that the bottom row of the generalized displacement matrix is redundant. In fact, we can put this row to good use in certain other types of transformation. Of particular interest in this context is the case of perspective projection. Following Section 15.3, equation (15.1), the equations for projection of object points into image points are:

$$x = fX/Z \tag{C.14}$$

$$y = fY/Z \tag{C.15}$$

$$z = f \tag{C.16}$$

We next make full use of the bottom row of the transformation matrix by defining the homogeneous coordinates as $(X_h, Y_h, Z_h, h) = (hX, hY, hZ, h)$, where h is a nonzero constant which we can take to be unity. To proceed, we examine the homogeneous transformation:

$$\begin{bmatrix} 1 & 0 & 0 & 0 \\ 0 & 1 & 0 & 0 \\ 0 & 0 & 1 & 0 \\ 0 & 0 & 1/f & 0 \end{bmatrix} \begin{bmatrix} X \\ Y \\ Z \\ 1 \end{bmatrix} = \begin{bmatrix} X \\ Y \\ Z \\ Z/f \end{bmatrix} \tag{C.17}$$

We see that dividing by the fourth coordinate gives the required values of the transformed Cartesian coordinates $(fX/Z, fY/Z, f)$.

Let us now review this result. First, we have found a 4×4 matrix transformation which operates on 4-D homogeneous coordinates. These do not correspond directly to real coordinates, but real 3-D coordinates can be calculated from them by dividing the first three by the fourth homogeneous coordinate. Thus there is an arbitrariness in the homogeneous coordinates in that they can all be multiplied by the same constant factor without producing any change in the final interpretation. Likewise, when deriving homogeneous coordinates from real 3-D coordinates, we can employ any convenient constant multiplicative factor h, though we will normally take h to be unity.

The advantage to be gained from use of homogeneous coordinates is the convenience of having a single multiplicative matrix for any transformation, in spite of the fact that perspective transformations are intrinsically nonlinear: thus a quite complex nonlinear transformation can be reduced to a more straightforward linear transformation. This eases computer calculation of object coordinate transformations, and other computations such as those for camera calibration (see below). We may also note that almost every transformation can be inverted by inverting the corresponding homogeneous transformation matrix. The exception is the perspective transformation, for which the fixed value of z leads merely to Z being unknown, and X, Y only being known relative to the value of Z (hence the need for binocular vision or other means of discerning depth in a scene).

C.3 Camera calibration

The above discussion has shown how homogeneous coordinate systems are used to help provide a convenient linear 4×4 matrix representation for 3-D transformations including rigid body translations and rotations, and nonrigid operations including scaling, skewing and perspective projection. In this last case, it was implicitly assumed that the camera and world coordinate systems are identical, since the image coordinates were expressed in the same frame of reference. However, in general the objects viewed by the camera will have positions which may be known in world coordinates, but which will not *a priori* be known in camera coordinates, since the camera will in general be mounted in a somewhat arbitrary position and will point in a somewhat arbitrary direction. Indeed, it may well be on adjustable gimbals, and may also be motor driven, with no precise calibration system. If the camera is on a robot arm, there are likely to be position sensors which could inform the control system of the camera position and orientation in world coordinates, though the amount of slack may well make the information too imprecise for practical purposes (e.g. to guide the robot towards objects).

These factors mean that the camera system will have to be calibrated very carefully before the images can be used for practical applications such as robot pick-and-place. A useful approach is to assume a general transformation between the world coordinates and the image seen by the camera under perspective projection, and to locate in the image various calibration points which have been placed in known positions in the scene. If enough such points are available, it should be possible to compute the transformation parameters, and then all image points can be interpreted accurately until recalibration becomes necessary.

The general transformation **G** takes the form:

$$\begin{bmatrix} X_H \\ Y_H \\ Z_H \\ H \end{bmatrix} = \begin{bmatrix} G_{11} & G_{12} & G_{13} & G_{14} \\ G_{21} & G_{22} & G_{23} & G_{24} \\ G_{31} & G_{32} & G_{33} & G_{34} \\ G_{41} & G_{42} & G_{43} & G_{44} \end{bmatrix} \begin{bmatrix} X \\ Y \\ Z \\ 1 \end{bmatrix} \tag{C.18}$$

where the final Cartesian coordinates appearing in the image are $(x, y, z) = (x, y, f)$, and these are calculated from the first three homogeneous coordinates by dividing by the fourth:

$$x = X_H/H = (G_{11}X + G_{12}Y + G_{13}Z + G_{14})/(G_{41}X + G_{42}Y + G_{43}Z + G_{44}) \tag{C.19}$$

$$y = Y_H/H = (G_{21}X + G_{22}Y + G_{23}Z + G_{24})/(G_{41}X + G_{42}Y + G_{43}Z + G_{44}) \tag{C.20}$$

$$z = Z_H/H = (G_{31}X + G_{32}Y + G_{33}Z + G_{34})/(G_{41}X + G_{42}Y + G_{43}Z + G_{44}) \tag{C.21}$$

However, as we know z, there is no point in determining parameters G_{31}, G_{32}, G_{33}, G_{34}. Accordingly we proceed to develop the means for finding the other parameters. In fact, because only the ratios of the homogeneous coordinates are meaningful, only the ratios of the G_{ij} values need be computed, and it is usual to take G_{44} as unity: this leaves only 11 parameters to be determined. Multiplying out the first two equations and rearranging gives:

$$G_{11}X + G_{12}Y + G_{13}Z + G_{14} - x(G_{41}X + G_{42}Y + G_{43}Z) = x \qquad \text{(C.22)}$$

$$G_{21}X + G_{22}Y + G_{23}Z + G_{24} - y(G_{41}X + G_{42}Y + G_{43}Z) = y \qquad \text{(C.23)}$$

Noting that a single world point (X, Y, Z) which is known to correspond to image point (x, y) gives us *two* equations of the above form, it requires a minimum of six such points to provide values for all 11 G_{ij} parameters: Fig. C.2 shows a convenient near-minimum case. An important factor is that the world points used for the calculation should lead to independent equations: thus it is important that they should not be coplanar. More precisely, there must be at least six points, no four of which are coplanar. However, further points are useful in that they lead to overdetermination of the parameters and

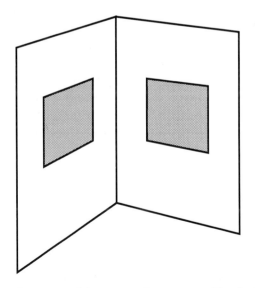

Fig. C.2   A convenient near-minimum case for camera calibration. Here two sets of four coplanar points, each set of four being at the corners of a square, provide more than the absolute minimum number of points required for camera calibration.

increase the accuracy with which the latter can be computed. There is no reason why the additional points should not be coplanar with existing points: indeed, a common arrangement is to set up a cube so that three of its faces are visible, each face having a pattern of squares with 30–40 easily discerned corner features (as for a Rubik cube).

Least squares analysis can be used to perform the computation of the 11 parameters, e.g. via the pseudo-inverse method. First, the $2n$ equations have to be expressed in matrix form:

$$\mathbf{Ag} = \xi \qquad \text{(C.24)}$$

where $\mathbf{A}$ is a $2n \times 11$ matrix of coefficients, which multiplies the $\mathbf{G}$-matrix, now in the form:

$$\mathbf{g} = (G_{11}\ G_{12}\ G_{13}\ G_{14}\ G_{21}\ G_{22}\ G_{23}\ G_{24}\ G_{41}\ G_{42}\ G_{43})^{\mathrm{T}} \quad \text{(C.25)}$$

and ξ is a $2n$-element column vector of image coordinates. The pseudo-inverse solution is:

$$\mathbf{g} = \mathbf{A}^{\dagger} \xi \qquad \text{(C.26)}$$

where:

$$\mathbf{A}^{\dagger} = (\mathbf{A}^{\mathrm{T}}\mathbf{A})^{-1}\mathbf{A}^{\mathrm{T}} \qquad \text{(C.27)}$$

The solution is more complex than might have been expected, since a normal matrix inverse is only defined, and can only be computed, for a square matrix. Note that solutions are only obtainable by this method if the matrix $\mathbf{A}^{\mathrm{T}}\mathbf{A}$ is invertible. For further details of this method, see Golub and van Loan (1983).

C.4 Intrinsic and extrinsic parameters

At this point it is useful to look in more detail at the general transformation leading to camera calibration. When we are calibrating the camera, we are actually trying to bring the camera and world coordinate systems into coincidence. The first step is to move the origin of the world coordinates to the origin of the camera coordinate system. The second step is to rotate the world coordinate system until its axes are coincident with those of the camera coordinate system. The third step is to move the image plane laterally until there is complete agreement between the two coordinate systems (this step is required since it is not known initially which point in the world coordinate system corresponds to the principal point[*] in the image).

There is an important point to be borne in mind during this process. If the camera coordinates are given by C, then the translation T required in the first step will be $-C$. Similarly, the rotations that are required will be the inverses of those which correspond to the actual camera orientations. The reason for these reversals is that (for example) rotating an object (here the camera) forwards gives the same effect as rotating the axes backwards. Thus all operations have to be carried out with the reverse arguments to those indicated above in Section C.1. The complete transformation for camera calibration is hence:

$$G = PLRT$$

$$= \begin{bmatrix} 1 & 0 & 0 & 0 \\ 0 & 1 & 0 & 0 \\ 0 & 0 & 1 & 0 \\ 0 & 0 & 1/f & 0 \end{bmatrix} \begin{bmatrix} 1 & 0 & 0 & t_1 \\ 0 & 1 & 0 & t_2 \\ 0 & 0 & 1 & t_3 \\ 0 & 0 & 0 & 1 \end{bmatrix} \begin{bmatrix} R_{11} & R_{12} & R_{13} & 0 \\ R_{21} & R_{22} & R_{23} & 0 \\ R_{31} & R_{32} & R_{33} & 0 \\ 0 & 0 & 0 & 1 \end{bmatrix} \begin{bmatrix} 1 & 0 & 0 & T_1 \\ 0 & 1 & 0 & T_2 \\ 0 & 0 & 1 & T_3 \\ 0 & 0 & 0 & 1 \end{bmatrix}$$

$$(C.28)$$

where matrix P takes account of the perspective transformation required to form the image. In fact, it is usual to group together the transformations P and L and call them internal camera transformations which include the *intrinsic camera parameters*, while R and T are taken together as external camera transformations corresponding to *extrinsic camera parameters*:

$$G = G_{internal}G_{external} \qquad (C.29)$$

where:

$$G_{internal} = PL = \begin{bmatrix} 1 & 0 & 0 & t_1 \\ 0 & 1 & 0 & t_2 \\ 0 & 0 & 1 & t_3 \\ 0 & 0 & 1/f & t_3/f \end{bmatrix} \longrightarrow \begin{bmatrix} 1 & 0 & t_1 \\ 0 & 1 & t_2 \\ 0 & 0 & 1/f \end{bmatrix} \qquad (C.30)$$

---

[*] The *principal point* is the image point lying on the principal axis of the camera: it is the point in the image which is closest to the centre of projection. Correspondingly, the *principal axis* (or *optical axis*) of the camera is the line through the centre of projection normal to the image plane.

$$G_{external} = RT = \begin{bmatrix} R_1 & R_1 \cdot T \\ R_2 & R_2 \cdot T \\ R_3 & R_3 \cdot T \\ 0 & 1 \end{bmatrix} \qquad (C.31)$$

In the matrix for $G_{internal}$ we have assumed that the initial translation matrix T moves the camera's centre of projection to the correct position, so that the value of t_3 can be made equal to zero: in that case the effect of L will indeed be lateral as indicated above. At that point we can express the (2-D) result in terms of a 3×3 homogeneous coordinate matrix. In the matrix for $G_{external}$ we have expressed the result succinctly in terms of the rows R_1, R_2, R_3 of R, and have taken dot products with $T(T_1, T_2, T_3)$: the (3-D) result is a 4×4 homogenous coordinate matrix.

Although the above treatment gives a good indication of the underlying meaning of G, it is not general because we have not so far included scaling and skew parameters in the internal matrix. In fact the generalized form of $G_{internal}$ is:

$$G_{internal} = \begin{bmatrix} s_1 & b_1 & t_1 \\ b_2 & s_2 & t_2 \\ 0 & 0 & 1/f \end{bmatrix} \qquad (C.32)$$

Potentially, $G_{internal}$ should include the following:

(i)   a transform for correcting scaling errors;
(ii) a transform for correcting translation errors [*];
(iii) a transform for correcting sensor skewing errors (due to non-orthogonality of the sensor axes);
(iv) a transform for correcting sensor shearing errors (due to unequal scaling along the sensor axes);
(v)   a transform for correcting for unknown sensor orientation within the image plane.

---

[*] For this purpose, the origin of the image should be on the principal axis of the camera. Misalignment of the sensor may prevent this point from being at the centre of the image.

Clearly, translation errors (item 2) are corrected by adjusting t_1 and t_2. All the other adjustments are concerned with the values of the 2×2 submatrix:

$$\begin{bmatrix} s_1 & b_1 \\ b_2 & s_2 \end{bmatrix}$$

However, it should be noted that application of this matrix performs rotation within the image plane immediately after rotation has been performed in the world coordinates by $G_{external}$, and it is virtually impossible to separate the two rotations. This explains why we now have a total of six external and six internal parameters totalling 12 rather than the expected 11 parameters (we return to the factor $1/f$ below). As a result it is better to exclude item (v) in the above list of internal transforms and to subsume it into the external parameters[*]. Since the rotational component in $G_{internal}$ has been excluded, b_1 and b_2 must now be equal, and the internal parameters will be: s_1, s_2, b, t_1, t_2. Note that the factor $1/f$ provides a scaling which cannot be separated from the other scaling factors during camera calibration, without specific (i.e. separate) measurement of f. Thus we have a total of six parameters from $G_{external}$ and five parameters from $G_{internal}$: this totals 11 and equals the number cited in the previous section.

At this point it is worth considering the special case where the sensor is known to be Euclidean to a high degree of accuracy. This will mean that $b = b_1 = b_2 = 0$, and $s_1 = s_2$, bringing the number of internal parameters down to three. In addition, if care has been taken over sensor alignment, and there are no other offsets to be allowed for, it may be known that $t_1 = t_2 = 0$. This will bring the total number of internal parameters down to just one, namely $s = s_1 = s_2$, or s/f, if we take proper account of the focal length. In this case there will be a total of seven calibration parameters for the whole camera system, and this may permit it to be set up unambiguously by viewing a known object having four clearly marked features instead of the six that would normally be required (see Section C.2).

---

[*] While doing so may not be ideal, there is no way of separating the two rotational components by purely optical means: only measurements on the internal dimensions of the camera system could determine the internal component, but separation is not likely to be a cogent or even meaningful matter. On the other hand, the internal component is likely to be stable, whereas the external component may be prone to variation if the camera is not mounted securely.

Finally, it should be remarked that nonlinear lens distortions such as pin-cushion and barrel distortion have not been taken into account at all in the above discussion. Typically, these distortions become progressively more severe with increasing radial distance from the principal point. Full discussion of these distortions and their correction is beyond the scope of the present volume: the reader is referred to the *Manual of Photogrammetry* (Slama, 1980) and to Tsai (1986) for further details.

C.5   Concluding remarks

This appendix has discussed the transformations required for camera calibration and has outlined how calibration can be achieved. The camera parameters have been classified as "internal" and "external", thereby simplifying the conceptual problem and throwing light on the origins of errors in the system. It has been shown that a minimum of six points is required to perform calibration in the general case where 11 transformation parameters are involved; however, the number of points required might be reduced somewhat in special cases, e.g. where the sensor is known to be Euclidean. Nevertheless, it is normally more important to increase the number of points used for calibration than to attempt to reduce it, since substantial gains in accuracy can be obtained via the resulting averaging process.

C.6   Bibliographical and historical notes

One of the first to use the various transformations described in this chapter was Roberts (1965). Two important references for camera calibration are the *Manual of Photogrammetry* (Slama, 1980) and Tsai (1986). Tsai's paper is especially useful in that he provides an extended and highly effective treatment which copes with non-linear lens distortions. More recent papers on this topic include Haralick (1989), Crowley *et al.* (1993), Cumani and Guiducci (1995), and Robert (1996): see also Zhang (1995). Note that parametrized plane curves can be used instead of points for the purpose of camera calibration (Haralick and Chu, 1984). Faugeras (1993) provides further useful background relating to the problem.

Appendix D

Robust Statistics

D.1 Introduction

We have found many times in this volume that noise can interfere with image signals and result in inaccurate measurements—for example, of object shapes, sizes and positions. Perhaps more important, however, is the fact that signals other than the particular one being focused upon can lead to gross shape distortions and can thus prevent an object from being recognized or even being discerned at all. In many cases, this will render some obvious interpretation algorithm useless, though algorithms with intrinsic "intelligence" may be able to save the day. For this reason the Hough transform has achieved some prominence: indeed, this approach to image interpretation has frequently been described as "robust", though no rigorous definition of robustness has been ventured so far in this volume. This appendix aims to throw further light on the problem.

Research into robustness did not originate in machine vision but evolved as the specialist area of statistics now known as robust statistics. Perhaps the paradigm problem in this area is that of fitting a straight line to a set of points. In the physics laboratory, least-squares analysis is commonly used to tackle the task. Figure D.1(a) shows a straightforward situation where all the data points can be fitted with a reasonably uniform degree of exactness, in the sense that the residual errors[*] approximate to the expected Gaussian distribution. Figure D.1(b) shows a less straightforward case, where a

---

[*] The residual errors or "residuals" are the deviations between the observed values and the theoretical predictions of the current model or current iteration of that model.

(*a*)

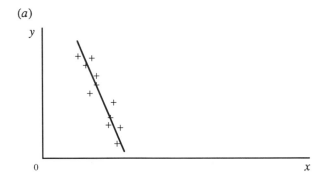

(*b*)

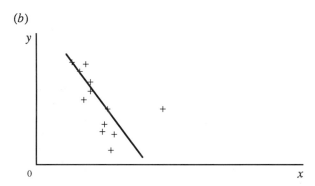

Fig. D.1 Fitting of data points to straight lines. (a) A straightforward situation where all the data points can be fitted with reasonable precision; (b) a less straightforward case, where a particular data point seems not to fall within a Gaussian distribution; (c) a situation where the correct solution has been ignored by the numerical analysis procedure; (d) a situation where there are many rogue points, and it is not clear which points lie on the straight line and which do not: in such cases it may not be known whether there are several, or any, lines to be fitted.

particular data point seems not to fall within a Gaussian distribution. Intuition says that this particular point represents data which has become corrupted in some way, e.g. by misreading an instrument or through a transcription error. Although the wings of a Gaussian distribution stretch out to infinity, the probability that a point will be more than five standard deviations from the centre of the distribution is very small, and indeed, $\pm 3\sigma$ limits are commonly taken as demarcating practical limits of correctness: it is taken as reasonable to disregard data points lying outside this range.

(c)

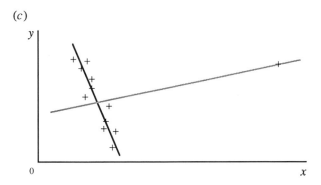

(d)

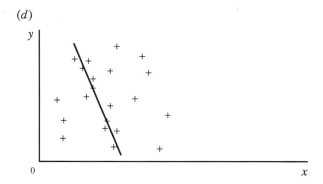

Fig. D.1 (*continued*).

Unfortunately, the situation can be much worse than this simple example suggests. For suppose that there is a rogue data point which is a very long way off. In least-squares analysis it will have such a large leverage that the correct solution may not be found. And if the correct solution is not found, there will be no basis for excluding the rogue data point. This situation is illustrated in Fig. D.1(c), where the obviously correct solution has been ignored by the numerical analysis procedure.

A worse case of line fitting occurs when there are many rogue points, and it is not clear which points lie on the straight line and which do not (Fig. D.1(d)). In fact, it may not be known whether there are several lines to be fitted, or whether there are *any* lines to be fitted. Whereas this circumstance would appear not to occur while data points are being plotted in physics experiments, it can arise when high-energy particles are being tracked; it also occurs frequently in images of indoor and outdoor scenes where a myriad of straight lines of various lengths can appear in a great many orientations and positions. Thus it is a real problem for which answers are required. An attempt at a full statement for this type of problem might be: devise a means for finding all the straight lines—of whatever length—in a generalized[*] image, so as to obtain the best overall fit to the dataset. Unfortunately, there are likely to be many solutions to any line fitting task, particularly if the data points are not especially accurate (if they are highly accurate then the number of solutions will be small, and it should be easy to decide intuitively or automatically what the best solution is). In fact, a rigorous answer to the question of which solution provides the best fit requires the definition of a criterion function which in some way takes account of the number of lines and the *a priori* length distribution. We shall not pursue this line of attack here, as the purpose of this appendix is to give a basic account of the subject of robust statistics, not one that is tied to a particular task. Hence we shall return below to the simpler case where there is only one line present in the generalized image, and there are a substantial number of rogue data points or "outliers" present.

D.2 Preliminary definitions and analysis

In the previous section we saw that robustness is an important factor in deciding on a scheme for fitting experimental data to numerical models. It is clearly important to have an exact measure of robustness, and the concept of a "breakdown point" long ago emerged as such a measure. The breakdown point ε of a regression scheme is defined as the smallest proportion of outlier contamination which may force the value of the estimate to exceed an arbitrary range. As we have seen, even a single outlier in a set of plots can cause least-squares regression to give completely erroneous results. However, a much simpler example is to hand, namely a 1-D distribution for which the mean is computed: here again, a single outlier can cause the mean to exceed any stated bound. This means that the breakdown point for the mean must be zero. On the other hand, the median of a distribution is well known to be highly robust to outliers, and remains unchanged if nearly half the data is corrupted. Specifically, for a set of n data points, the median will remain unchanged if the lowest $\lfloor n/2 \rfloor$

---

[*] i.e. an image which might correspond to off-camera images, or to situations such as data points being plotted on a graph.

points[*] are moved to arbitrary lower values, or the highest $\lfloor n/2 \rfloor$ points are moved to arbitrary higher values, but in either case the median value will be changed to an arbitrary value if $\lfloor n/2 \rfloor + 1$ points are so moved. By definition (see above), this means that the breakdown point of the median is $(\lfloor n/2 \rfloor + 1)/n$; this value should be compared with the value $1/n$ for the mean. In the case of the median the breakdown point approaches 0.5 as n tends to infinity (see Table D.1). Thus the median attains the apparently maximum achievable breakdown point of 0.5, and is therefore optimal—at least in the 1-D case described in this paragraph.

Table D.1 Breakdown points for means and medians.

n	Mean	Median
1	1	1
3	1/3	2/3
5	1/5	3/5
11	1/11	6/11
∞	0	0.5

This table shows how the respective breakdown points for the mean and median approach 0 and 0.5 as n tends to infinity, in the case of 1-D data.

In fact, the breakdown point is not the only relevant parameter for characterizing regression schemes. For example, the "relative efficiency" is also important, and is defined as the ratio between the lowest achievable variance and the actual variance achieved by the regression method. It turns out that the relative efficiency depends on the particular noise distribution that the data is subject to. It can be shown that the mean is optimal for elimination of Gaussian noise, having a relative efficiency of unity, while the median has a relative efficiency of $2/\pi = 0.637$. However, when dealing with impulse noise the median has a higher relative efficiency than the mean, the exact values depending on the nature of the noise. This point will be discussed in more detail below.

Time complexity is a further parameter which is needed for characterising regression methods. We shall not pursue this aspect further here, beyond

---

[*] The function $\lfloor \cdot \rfloor$ denotes the "floor" (rounding down) operation, and indicates the largest integer less than or equal to the enclosed value. In the present case we have $\lfloor n/2 \rfloor \leq n/2 < \lfloor n/2 \rfloor + 1$.

making the observation that the time complexity of the mean is $O(n)$, while that for the median varies with the method of computation (e.g. $O(n)$ for the histogram approach of Section 3.3 and $O(n^2)$ when using a bubble sort): in any case the absolute time for computing a median normally far exceeds that for the mean.

Of the parameters referred to above, the breakdown point has been at the forefront of workers' minds when devising new regression schemes. While it might appear that the median already provides an optimal approach for robust regression, its breakdown value of 0.5 only applies to 1-D data. It is therefore worth considering what breakdown point could be achieved for tasks such as line fitting, bearing in mind the poor performance of least-squares regression. Let us take the method of Theil (1950) in which the slope of each pair of a set of n data points is computed, and the median of the resulting set of

$$\binom{n}{2} = \frac{1}{2} n(n-1)$$

values is taken as the final slope; in fact the intercept can be determined more simply because the problem has at that stage been reduced to one dimension. As the median is used in this procedure, at least half the slopes have to be correct in order to obtain a correct estimate of the actual slope. If we assume that the proportion of outliers in the data is η, the proportion of inliers[*] will be $1 - \eta$, and the proportion of correct slopes will be $(1 - \eta)^2$, and this has to be at least 0.5. This means that η has to lie in the range:

$$\eta \leqslant 1 - 1/\sqrt{2} = 1 - 0.707 = 0.293 \qquad\qquad (D.1)$$

Thus the breakdown point for this approach to linear regression is less than 0.3. In a 3-D data-space where a best-fit plane has to be found, the best breakdown point will be even smaller, with a value $1 - 2^{-1/3} \approx 0.2$. The general formula for p dimensions is:

$$\eta_p \leqslant 1 - 2^{-1/p} \qquad\qquad (D.2)$$

Clearly, there is a need for more robust regression schemes which becomes more urgent for larger values of p.

---

[*] Inliers are normal valid data points: the dataset is to be regarded as composed of inliers and outliers.

The development of robust multidimensional regression schemes took place relatively recently, in the 1970s. The basic estimators which were developed at that time, and classified by Huber in 1981, were the M-, R- and L-estimators. The M-estimator is by far the most widely used, and appears in a variety of forms which encompass median and mean estimators and least-squares regression: we shall study this type of estimator in more detail below. The L-estimators employ linear combinations of order statistics, and include the alpha-trimmed mean, with the median and mean as special cases. However, it will be easier to consider the median and the mean under the heading of M-estimators, and in what follows we concentrate on this approach.

D.3 The M-estimator (influence function) approach

M-estimators operate by minimizing the sum of a suitable function ρ of the residuals r_i. Normally ρ is taken to be a positive definite function, and for least-squares (L_2) regression it is the square of the residuals:

$$\rho(r_i) = r_i^2 \tag{D.3}$$

In general it is necessary to perform the M-estimation minimization operation iteratively until a stable solution is obtained (at each iteration the new set of offsets has to be added to the previous set of parameter values).

To improve upon the poor robustness of L_2 regression, reflected by its zero breakdown point, an improved function ρ must be obtained which is well adapted to the particular noise[*] and outlier content of the data. To understand this process it is easiest to analyse the situation for 1-D datasets, and to consider the influence of each data point. We represent the influence of a data point by an influence function $\psi(r_i)$, where:

$$\psi(r_i) = \frac{d\rho(r_i)}{dr_i} \tag{D.4}$$

---

[*] At this point a certain ambiguity creeps into the discussion. "Noise" tends to originate from electronic processes in the image source, and typically leads to a Gaussian distribution in the pixel intensity values. By the time positions of objects are being measured, it is strictly speaking errors rather than noise that are being considered, and the error distribution is not necessarily identical to the noise distribution that gave rise to it. However, in the later sections of this appendix we usually refer to noise and noise distributions: the term noise will be taken to refer either to the original noise source or to the derived errors, as appropriate to the discussion.

Notice that minimizing $\sum_{i=0}^{n} \rho(r_i)$ is equivalent to reducing $\sum_{i=0}^{n} \psi(r_i)$ to zero, and in the case of L_2 regression:

$$\psi(r_i) = 2r_i \tag{D.5}$$

In one dimension this equation has a simple interpretation—moving the origin of coordinates to a position where $\sum_{i=0}^{n} r_i = 0$, i.e. to the position of the mean. Now that we have shown the equivalence of L_2 regression to simple averaging, the source of the lack of robustness becomes all too clear—however far away from the mean a data point is, it still retains a weight proportional to its residual value r_i. Accordingly, a wide range of possible alternative influence functions has been devised to limit the problem by cutting down the weights of distant points which are potential outliers.

An obvious approach is to limit the influence of a distant point to some maximum value: another is to eliminate its influence altogether once its residual error exceeds a certain limiting value (Fig. D.2). We could achieve this by a variety of schemes, either cutting off the influence suddenly at this limiting distance (as in the case of the $\pm 3\sigma$ points), or letting it approach zero according to a linear profile, or opting for a more mathematically ideal functional form with a smoother profile. In fact, there are other considerations, such as the amount of computation involved in dealing with large numbers of data points taken over a fair number of iterations. Thus it is not surprising that a variety of piecewise linear profiles approximating to the smoother ideal profiles have been devised. In general, however, influence functions are linear near the origin, zero at large distances from the origin, and possess a region of over which they give significant weight to the data points (Fig. D.2).

Prominent among these possibilities are the Hampel 3-part redescending M-estimator, whose influence function is composed simply of convenient linear components, and the Tukey biweight estimator (Beaton and Tukey, 1974) which takes a form similar to that shown in Fig. D.2(e):

$$\psi(r_i) = r_i(\gamma^2 - r_i^2)^2 \qquad |r_i| \leq \gamma$$
$$= 0 \qquad\qquad |r_i| > \gamma \tag{D.6}$$

It was remarked above that the median operation is a special case of the M-estimator: here all data points on one side of the origin have a unit positive weight, and all data points on the other side of the origin have unit

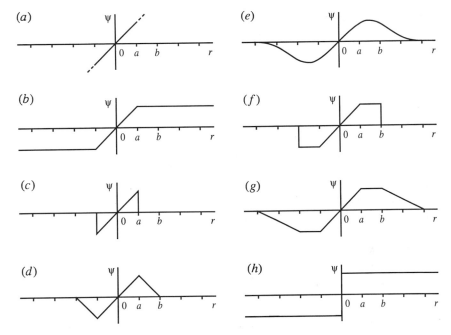

Fig. D.2 Influence functions which limit the effects of outliers. (*a*) The case where no limit is placed on the influence of distant points; (*b*) the influence is limited to some maximum value; (*c*) the influence is eliminated altogether once the residual exceeds a certain maximum value; (*d*) a piecewise-linear profile which gives a less abrupt variation; (*e*) a mathematically more well-behaved influence function; (*f*) another possible piecewise-linear case; (*g*) a Hampel 3-part redescending M-estimator which approximates the mathematically ideal case (*e*) with reasonable accuracy; (*h*) the situation for a median estimator.

negative weight:

$$\psi(r_i) = \text{sign } (r_i) \tag{D.7}$$

Thus if more data points are on one side than the other, the solution will be pulled in that direction, iteration proceeding until the median is at the origin.

It is important to appreciate that while the median has exceptionally useful outlier suppression characteristics, it actually gives outliers significant weight: in fact, the median clearly ignores how far away an outlier is, but it still counts up how many outliers there are on either side of the current origin. As a result, the median is liable to produce a biased estimate. This is good reason for considering other types of influence function for analysing data. Finally, we remark that the median influence function leads to the value of ρ for L_1 regression:

$$\rho(r_i) = |r_i| \tag{D.8}$$

When selecting an influence function, it is important not only that the function must be appropriate but also that its scale must match that of the data. If the width of the influence function is too great, too few outliers will be rejected; it the width is too small, the estimator may be surrounded by a rather homogeneous sea of data points with no guarantee that it will do more than find a locally optimal fit to the data. These factors mean that preliminary measurements must be made to determine the optimal form of the influence function for any application.

At this point it seems clear that there ought to be a more scientific approach which would permit the influence function to be calculated from the noise characteristics. This is indeed so, and if the expected noise distribution is given by $f(r_i)$, the optimal form of the influence function (Huber, 1964) is:

$$\psi(r_i) = -f'(r_i)/f(r_i) = -\frac{d}{dr_i} \ln [f(r_i)] \tag{D.9}$$

The existence of the logarithmic form of this solution is interesting and useful, as it simplifies exponential-based noise distributions such as the Gaussian and double exponential functions. For the former, $\exp(-r_i^2/2\sigma^2)$, we find:

$$\psi(r_i) = r_i/\sigma^2 \tag{D.10}$$

and for the latter, $\exp(-|r_i|/s)$:

$$\psi(r_i) = \operatorname{sign}(r_i)/s \tag{D.11}$$

Since the constant multipliers may be ignored, we conclude that the mean and median are optimal estimators for signals in Gaussian and double exponential noise, respectively.

Gaussian noise may be expected to arise in many situations (most particularly because of the effects of the central limit theorem), demonstrating the intrinsic value of employing the mean or L_2 regression. On the other hand the double exponential distribution has no obvious justification in practical situations. However, it represents situations where the wings of the noise distribution stretch out rather widely, and it is good to see under what conditions the widely used median would in fact be optimal. Nevertheless, our purpose in wanting an explicit mathematical form for the influence function was to optimize the detection of signals in arbitrary noise conditions and specifically those where outliers may be present.

Let us suppose that the noise is basically Gaussian, but that outliers may also be present and that these would be drawn approximately from a uniform distribution: there might for example be a uniform (but low-level) distribution of outlier values over a limited range. An overall distribution of this type is shown in Fig. D.3. Near $r_i = 0$, the uniform distribution of outliers will have relatively little effect and $\psi(r_i)$ will approximate to r_i. For large $|r_i|$, the value of f' will be due mainly to the Gaussian noise contribution, whereas the value of f will arise mainly from the uniform distribution f_u, and the result will be:

$$\psi(r_i) \simeq \frac{r_i}{\sigma^2 f_u} \exp\left(-r_i^2/2\sigma^2\right) \tag{D.12}$$

a function which peaks at an intermediate value of r_i. This essentially proves that the form shown in Fig. D.2(e) is reasonable. However, there is a severe problem in that outliers are by definition unusual and rare, so it is almost impossible in most cases to be able to produce on optimum form of $\psi(r_i)$ as suggested above. Unfortunately, the situation is even worse than this discussion might indicate. Redescending M-estimators are even more limited in that they are sensitive to local densities of data points, and are therefore prone to finding false solutions—unique solutions are *not* guaranteed. Non-redescending M-estimators are guaranteed to arrive at unique solutions, though the accuracy of the latter depend on the accuracy of the preliminary scale estimate. In addition, the quality of the initial approximation tends to be of very great importance for M-estimators, particularly for redescending M-estimators.

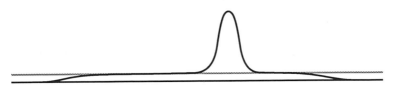

Fig. D.3 Distribution resulting from Gaussian noise and outliers. The usual Gaussian noise contribution is augmented by a distribution of outliers which is nearly uniform over a limited range.

Finally, we should point out that the above analysis has concentrated on optimization of accuracy and is ultimately based on maximum likelihood strategies (Huber, 1964). It is really concerned with maximizing relative efficiency on the assumption that the underlying distribution is known. Robustness measured according to the breakdown point criterion is not optimized, and this factor will be of vital importance in any situation where the outliers form part of a totally unexpected distribution, or do not form part of a predictable distribution[*]. Clearly, methods must be engineered which are intrinsically highly robust according to the breakdown point criterion. This is what motivated the development of the least median of squares approach to regression during the 1980s.

D.4   The least median of squares approach to regression

Above we have seen that a variety of estimators exist which can be used to suppress noise from numerical data, and to optimize the robustness and accuracy of the final result. The M-estimator (or influence function) approach is extremely widely used and is successful in eliminating the main problems associated with the use of least-squares regression (including, in 1-D, use of the mean). However, it does not in general achieve the ideal breakdown value of 0.5, and requires careful setting up to give optimal matching to the scale of the variation in the data. Accordingly, much attention has been devoted to a newer approach—least median of squares regression.

The aim of least median of squares (LMedS) regression is to capitalize on the known robustness of the median in a totally different way—by replacing the mean of the least (mean) squares averaging technique by the far more robust median. The effect of this is to ignore errors from the distant parts of the distribution and also from the central parts where the peak is often noisy and ill defined, and to focus on the parts about half-way up and on either side of the distribution. Minimization then balances the contributions from the two sides of the distribution, thereby sensitively estimating the mode position, though clearly this is achieved rather indirectly. Perhaps the simplest view of the technique is that it determines the location of the narrowest width region which includes half the population of the distribution. In a 2-D straight-line location application, this interpretation

---

[*] It is perhaps a philosophical question whether an outlier distribution does not exist, cannot exist, or cannot be determined by any known experimental means, e.g. because of rarity.

amounts to locating the narrowest parallel-sided strip which includes half the population of the distribution (Fig. D.4). In principle, in such cases the method operates just as effectively if the distribution is sparsely populated—as happens where the best-fit straight line for a set of experimental plots has to be determined.

The LMedS technique involves minimizing the median of the squares of the residuals r_j for all possible positions in the distribution which are potentially mode positions, i.e. it is the position x_i which minimizes $M = \mathrm{med}_j \, (r_j^2)$. While it might be thought that minimizing M is equivalent to minimizing $\mathcal{M} = \mathrm{med}_j \, (|r_j|)$, this is not so if there are two adjacent central positions giving equal responses (as in Fig. D.5(a–c)); however, the form of M guarantees that a position midway between these two will give an appropriate minimum. For clarity we shall temporarily ignore this technicality and concentrate on $\mathcal{M}$: the reason for doing this is to take advantage of piecewise-linear responses which considerably simplify theoretical analysis.

Figure D.5(a) shows the response $\mathcal{M}$ when the original distribution is approximately Gaussian. There is a clear minimum of $\mathcal{M}$ at the mode position, and the method works perfectly. Figure D.5(b) shows a case where there is a very untidy distribution, and there is a minimum of $\mathcal{M}$ at an appropriate position. Figure D.5(c) shows a more extreme situation in which there are two peaks, and again the response $\mathcal{M}$ is appropriate, except that it is now clear that the technique can only focus on one peak at a time. Nevertheless, it gets an appropriate and robust answer for the case it is focusing on. If

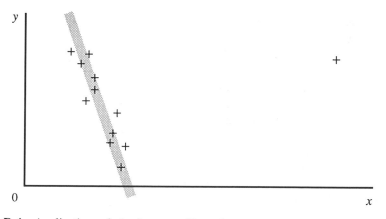

Fig. D.4 Application of the least median of squares technique. The narrowest parallel-sided strip is found which includes half the population of the distribution, in an attempt to determine the best-fit line. Notice the effortless superiority in performance when compared with the situation in Fig. D.1(c).

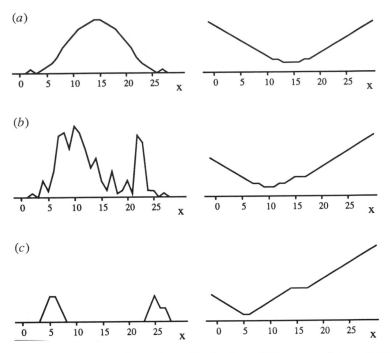

Fig. D.5   Minimizing $\mathcal{M}$ for various distributions. This figure shows (left) the original distributions and (right) the resulting response functions $\mathcal{M}$, in the following cases: (a) an approximately Gaussian distribution, (b) an "untidy" distribution, (c) a distribution with two peaks.

the two peaks are identical the method will still work, but will clearly not give a unique solution.

Although the LMedS approach to regression appeared only a decade ago (Rousseeuw, 1984), it has acquired considerable support, since it has the maximum possible breakdown point of 0.5. In particular, the LMedS approach has been used for pattern recognition and image analysis applications (see for example Kim *et al.*, 1989). In these areas, the method is useful for (i) location of straight lines in digital images, (ii) location of Hough transform peaks in parameter space, and (iii) location of clusters of points in feature space.

Unfortunately, the LMedS approach is liable to give a biased estimate of the modes if two distributions overlap, and in any case focuses on the main mode of a multimodal distribution. Thus the LMedS technique has to be applied several times, alternating with necessary truncation processes, to find all the cluster centres, while weighted least-squares fitting is required to optimize accuracy. The result is a procedure of some complexity and considerable computational load. Indeed, the load is in general so large that it is normally approximated by taking subsets of the data points, though this aspect cannot be examined in detail here (see for example Kim *et al.*, 1989). Once this has been carried out, the method can give quite impressive results.

Ultimately, the value of the LMedS approach lies in its increased breakdown point in situations of multidimensional data. If we have n data points in p dimensions, the LMedS breakdown point is:

$$\varepsilon_{\text{LMedS}} = (\lfloor n/2 \rfloor - p + 2)/n \qquad \text{(D.13)}$$

which tends to 0.5 as n approaches infinity (Rousseeuw, 1984). This value must be compared with a maximum of

$$\varepsilon = 1/(p+1) \qquad \text{(D.14)}$$

for standard methods of robust regression such as the M-, R- and L-estimators discussed earlier (Kim *et al.*, 1989). (Equation D.2 represents the suboptimal solution achieved by the Theil approach to line estimation.) Thus in these latter cases, 0.33 is the best breakdown point that can be achieved for $p = 2$, while the LMedS approach offers 0.5. However, the relative efficiency of LMedS is relatively low (ultimately because it is a median-based estimator); as stated above, this means that it has to be used with the weighted least-squares technique. We should also point out that the LMedS technique is intrinsically 1-D, so it has to be used in a "projection pursuit" manner (Huber, 1985), concentrating on one dimension at a time. For implementation details the reader is referred to the literature (see Section D.7).

D.5   Overview of the robustness problem

For greatest success in solving the robustness and accuracy problems— represented respectively by the breakdown point and relative efficiency criteria—it has been found in the foregoing sections that the LMedS technique should be used for finding signals (whether peaks, clusters, lines or hyperplanes, etc.), and weighted least-squares regression should be used for refining accuracy, the whole process being iterated until satisfactory results are achieved. This is a complex and computation intensive process, but reflects an overall strategy which has been outlined several times in earlier chapters—namely, search for an approximate solution, and then refinement to optimize location accuracy. The major question to be considered at this stage is: what is the best method for performing an efficient and effective initial search? In fact, there is a further question which is of especial relevance: is there any means of achieving a breakdown point of greater than 0.5?

It is worth considering the extent to which the Hough transform tackles and solves these problems. First, it is a highly effective search procedure, though in some contexts its computational efficiency has been called into question (however, in the present context it must be remembered that the LMedS technique is especially computation intensive). Second, it seems able to yield breakdown points far higher than 0.5 and even approaching unity. Consider a parameter space where there are many peaks and also a considerable number of randomly placed votes. Then any individual peak is likely to include only a small fraction of the votes, and the peak location proceeds without difficulty in spite of the presence of 90–99% contamination by outliers (the latter arising from noise and clutter). Thus the strategy of searching for peaks appears to offer significant success at avoiding outliers. Yet this does not mean that the LMedS technique is valueless, since subsequent application of LMedS is essentially able to *verify* the identification of a peak, to locate it more accurately via its greater relative efficiency, and thus to feed reliable information to a subsequent least-squares regression stage. Overall, we can see that a staged progression is taking place from a high breakdown point, low relative efficiency procedure, to a procedure of intermediate breakdown point and moderate relative efficiency, and finally to a procedure of low breakdown point and high relative

Table D.2   Breakdown points and efficiency values for peak finding.

	HT	LMedS	LS	Overall
ε	0.98	0.50	0.2	0.98
η	0.2	0.4	0.95	0.95

Possible breakdown points ε and relative efficiency values η for peak finding. A Hough transform is used to perform an initial search for peaks; then the LMedS technique is employed for validating the peaks and eliminating outliers; finally, least-squares regression is used to optimize location accuracy. The result is far higher overall effectiveness than that obtainable by any of the techniques applied alone; however, computational load is not taken into account, and is likely to be a major consideration.

efficiency. We summarize the progression by giving possible figures for the relevant quantities in Table D.2.

D.6 Concluding remarks

This appendix has aimed to place the discussion of robustness on a sounder basis than might have been thought possible in the earlier chapters of the book, where a more intuitive approach was presented. It has been necessary to delve quite deeply into the maturing and highly mathematical subject of robust statistics, and there are certain important lessons to be learnt. In particular, three relevant parameters have been found to form the basis for study in this area. The first is the breakdown point of an estimator, which shows the latter's resistance to outliers and provides the core meaning of robustness. The second is the relative efficiency of an estimator, which provides a measure of how efficiently it will use the inlier data at its disposal to arrive at accurate estimates. The third is the time complexity of the estimator when it is implemented as a computer algorithm. While this last parameter is a vital consideration in practical situations, available space has not permitted it to be covered in any depth here, though it is clear that the most robust techniques (especially LMedS) tend to be highly computation intensive. It is also found that there is a definite tradeoff between the other two parameters—techniques which have high breakdown points have low relative efficiencies and *vice versa*[*]. These factors make it reasonable, and desirable, to use several techniques in sequence, or iteratively in cycle, in order to obtain the best overall performance. Thus LMedS is frequently used in conjunction with least-squares regression (see for example Kim *et al.*, 1989).

Finally, it is worth pointing out that the basis of robust statistics is that of statistical analysis of the available data: there is thus a tendency to presume that outliers are rare events due typically to erroneous readings or transcriptions. Yet in vision the most difficult problems tend to arise from the clutter of irrelevant objects in the background, and only a tiny fraction of the incoming data may constitute the relevant inlier portion. This makes the problem of robustness all the more serious, and in principle *could* mean that until a whole image has been interpreted satisfactorily no single object can

---

[*] The reason for this may be summarized as the aim of achieving high robustness requiring considerable potentially outlier data to be discarded, even when this could be accurate data which would contribute to the overall accuracy of the estimate.

finally be identified and its position and orientation measured accurately. While this view may be pessimistic, it does indicate that the conventional approach to robust statistics may have significant limitations in its application to machine vision.

D.7 Bibliographical and historical notes

This appendix has given a basic introduction to the rapidly maturing subject of robust statistics which has made a substantial impact on machine vision over the past decade. The most popular and successful approach to robust statistics must still be seen as the M-estimator (influence function) approach, though in many-dimensional spaces its robustness is called to question, and it is here that the newer LMedS approach has gathered a firm following. Quite recently, the value of using a sequence of estimators which can optimize the overall breakdown point and relative efficiency has been pointed out (Kim *et al.*, 1989): in particular, the right combination of Hough transform (or other relevant technique), LMedS and weighted least-squares regression would seem especially powerful.

Robust statistics has been applied in a number of areas of machine vision, including robust window operators (Besl *et al.*, 1989), pose estimation (Haralick and Joo, 1988), motion studies (Bober and Kittler, 1993), camera location and calibration (Kumar and Hanson, 1989), and surface defect inspection (Koivo and Kim, 1989), to name but a few.

The original papers by Huber (1964) and Rousseeuw (1984) are still worth reading, and the books by Huber (1981), Hampel *et al.* (1986), and Rousseeuw and Leroy (1987) are valuable references, containing much insight and readable material. On the application of the LMedS technique, and for a more recent review of robust regression in machine vision, see Meer *et al.* (1990, 1991).

It should be pointed out that the RANSAC technique (Fischler and Bolles, 1981) was introduced before LMedS and presaged its possibilities: RANSAC was thus historically of great importance. The work of Siegel (1982) was also important historically in providing the background from which LMedS could take off, while the work of Steele and Steiger (1986) showed how LMedS might be implemented with attainable levels of computation.

References

Abdou, I.E. and Pratt, W.K. (1979). Quantitative design and evaluation of enhancement/threshold edge detectors. *Proc. IEEE* **67**, 753–763.

Abend, K., Harley, T. and Kanal, L.N. (1965). Classification of binary random patterns. *IEEE Trans. Inf. Theory* **11**, 538–544.

Abutaleb, A.S. (1989). Automatic thresholding of gray-level pictures using two-dimensional entropy. *Comput. Vision Graph. Image Process.* **47**, 22–32.

Ade, F. (1983). Characterization of texture by "eigenfilters". *Signal Process.* **5**, 451–457.

Agin, G.J. and Binford, T.O. (1973). Computer description of curved objects. *Proc. 3rd Int. Joint Conf. on Artif. Intell., Stanford, California*, pp. 629–640.

Agin, G.J. and Binford, T.O. (1976). Computer description of curved objects. *IEEE Trans. Comput.* **25**, 439–449.

Akey, M.L. and Mitchell, O.R. (1984). Detection and sub-pixel location of objects in digitized aerial imagery. *Proc. 7th Int. Conf. on Pattern Recogn., Montreal* (30 July–2 August), pp. 411–414.

Aleksander, I., Thomas, W.V. and Bowden, P.A. (1984). WISARD: A radical step forward in image recognition. *Sensor Rev.* **4**, 120–124.

Ali, S.M. and Burge, R.E. (1988). A new algorithm for extracting the interior of bounded regions based on chain coding. *Comput. Vision Graph. Image Process.* **43**, 256–264.

Aloimonos, J. (1988). Shape from texture. *Biol. Cybern.* **58**, 345–360.

Aloimonos, J. and Swain, M.J. (1985). Shape from texture. *Proc. 9th Int. Joint Conf. on Artif. Intell.* **2**, 926–931.

Alter, T.D. (1994). 3-D pose from 3 points using weak-perspective. *IEEE Trans. Pattern Anal. Mach. Intell.* **16**, 802–808.

Ambler, A.P., Barrow, H.G., Brown, C.M., Burstall, R.M. and Popplestone, R.J. (1975). A versatile system for computer-controlled assembly. *Artif. Intell.* **6**, 129–156.

Ankenbrandt, C.A., Buckles, B.P. and Petry, F.E. (1990). Scene recognition using genetic algorithms with semantic nets. *Pattern Recogn. Lett.* **11**, 285–293.

Arcelli, C. and di Baja, G.S. (1985). A width-independent fast-thinning algorithm. *IEEE Trans. Pattern Anal. Mach. Intell.* **7**, 463–474.

Arcelli, C. and Ramella, G. (1995). Finding grey-skeletons by iterated pixel removal. *Image Vision Comput.* **13**, no. 3, 159–167.

Arcelli, C., Cordella, L.P. and Levialdi, S. (1975). Parallel thinning of binary pictures. *Electron. Lett.* **11**, 148–149.

Arcelli, C., Cordella, L.P. and Levialdi, S. (1981). From local maxima to connected skeletons. *IEEE Trans. Pattern Anal. Mach. Intell.* **3**, 134–143.

Arnold, R.D. (1978). Local context in matching edges for stereo vision. *Proc. Image Understanding Workshop, Cambridge, Massachusetts*, pp. 65–72.

Aström, K. (1995). Fundamental limitations on projective invariants of planar curves. *IEEE Trans. Pattern Anal. Mach. Intell.* **17**, 77–81.

Ataman, E., Aatre, V.K. and Wong, K.M. (1980). A fast method for real time median filtering. *IEEE Trans. Acoust. Speech Signal Process.* **28**, 415–420.

Atiquzzaman, M. (1994). Pipelined implementation of the multiresolution Hough transform in a pyramid multiprocessor. *Pattern Recogn. Lett.* **15**, 841–851.

Atiquzzaman, M. and Akhtar, M.W. (1994). Complete line segment description using the Hough transform. *Image Vision Comput.* **12**, no. 5, 267–273.

Babaud, J., Witkin, A.P., Baudin, M. and Duda, R.O. (1986). Uniqueness of the Gaussian kernel for scale-space filtering. *IEEE Trans. Pattern Anal. Mach. Intell.* **8**, 26–33.

Backer, E. (ed.) (1992). Special Issue on Artificial Neural Networks. *Pattern Recogn. Lett.* **13**, no. 5.

Bajcsy, R. (1973). Computer identification of visual surface. *Comput. Graph. Image Process.* **2**, 118–130.

Bajcsy, R. and Liebermann, L. (1976). Texture gradient as a depth cue. *Comput. Graph. Image Process.* **5**, 52–67.

Baldi, P. and Hornik, K. (1989). Neural networks and principal component analysis: learning from examples without local minima. *Neural Networks* **2**, 53–58.

Ball, G.H. and Hall, D.J. (1966). ISODATA, an iterative method of multivariate data analysis and pattern classification. *IEEE Int. Communications Conf., Philadelphia, Digest of Techn. Papers* II, pp. 116–117.

Ballard, D.H. (1981). Generalizing the Hough transform to detect arbitrary shapes. *Pattern Recogn.* **13**, 111–122.

Ballard, D.H. and Brown, C.M. (1982). *Computer Vision.* Prentice-Hall, Englewood Cliffs, NJ.

Ballard, D.H. and Sabbah, D. (1983). Viewer independent shape recognition. *IEEE Trans. Pattern Anal. Mach. Intell.* **5**, 653–660.

Barbe, D.F. (1975). Imaging devices using the charge-coupled principle. *Proc. IEEE* **63**, 38–66.

Barker, A.J. and Brook, R.A. (1978). A design study of an automatic system for on-line detection and classification of surface defects on cold-rolled steel strip. *Optica Acta* **25**, 1187–1196.

Barnard, S.T. and Thompson, W.B. (1980). Disparity analysis of images. *IEEE Trans. Pattern Anal. Mach. Intell.* **2**, no. 4, 333–340.

Barnea, D.I. and Silverman, H.F. (1972). A class of algorithms for fast digital image registration. *IEEE Trans. Comput.* **21**, 179–186.

Barnes, G.H., Brown, R.M.. Kato, M.. Kuck, D.J. Slotnick, D.L and Stokes, R.A. (1968). The ILLIAC IV computer. *IEEE Trans. Comput.* **17**, 746–757.

Barrow, H.G. and Popplestone, R.J. (1971). Relational descriptions in picture processing. In: Meltzer, B. and Michie, D. (eds) *Machine Intelligence 6.* Edinburgh University Press, Edinburgh, pp. 377–396.

Barrow, H.G. and Tenenbaum, J.M. (1981). Computational vision. *Proc. IEEE* **69**, 572–595.

Barrow, H.G., Ambler, A.P. and Burstall, R.M. (1972). Some techniques for recognising structures in pictures. In: Watanabe, S. (ed.) *Frontiers of Pattern Recognition.* Academic Press, New York, pp. 1–29.

Bartz, M.R. (1968). The IBM 1975 optical page reader. *IBM J. Res. Dev.* **12**, 354–363.

Bascle, B., Bouthemy, P., Deriche, R. and Meyer, F. (1994). Tracking complex primitives in an image sequence. *Proc. 12th Int. Conf. on Pattern Recogn., Jerusalem, Israel* (9–13 October), Vol. A, pp. 426–431.

Batchelor, B.G. (1974). *Practical Approach to Pattern Classification*. Plenum Press, London.

Batchelor, B.G. (1979). Using concavity trees for shape description. *Comput. Digital Techniques* **2**, 157–165.

Batchelor, B.G., Hill, D.A. and Hodgson, D.C. (1985). *Automated Visual Inspection*. IFS (Publications) Ltd, Bedford, UK/North-Holland, Amsterdam.

Beaton, A.E. and Tukey, J.W. (1974). The fitting of power series, meaning polynomials, illustrated on band-spectroscopic data. *Technometrics* **16**, no. 2, 147–185.

Beaudet, P.R. (1978). Rotationally invariant image operators. *Proc. 4th Int. Conf. on Pattern Recogn., Kyoto*, pp. 579–583.

Beckers, A.L.D. and Smeulders, A.W.M. (1989). A comment on "A note on 'Distance transformations in digital images'". *Comput. Vision Graph. Image Process.* **47**, 89–91.

Benke, K.K. and Skinner, D.R. (1987). Segmentation of visually similar textures by convolution filtering. *Australian Comput. J.* **19**, no. 3, 134–139.

Bergholm, F. (1986). Edge focusing. *Proc. 8th Int. Conf. on Pattern Recogn., Paris* (27–31 October), pp. 597–600.

Berman, S., Parikh, P. and Lee, C.S.G. (1985). Computer recognition of two overlapping parts using a single camera. *IEEE Comput.* (Mar.) 70–80.

Besl, P.J. and Jain, R.C. (1984). Three-dimensional object recognition. *Comput. Surveys* **17**, 75–145.

Besl, P.J., Delp, E.J. and Jain, R. (1985). Automatic visual solder joint inspection. *IEEE J. Robot. Automation* **1**, 42–56.

Besl, P.J., Birch, J.B. and Watson, L.T. (1989). Robust window operators. *Machine Vision and Applications* **2**, 179–191.

Beun, M. (1973). A flexible method for automatic reading of handwritten numerals. *Philips Techn. Rev.* **33**, 89–101; 130–137.

Beynon, J.D.E. and Lamb, D.R. (eds) (1980). *Charge-coupled Devices and their Applications*. McGraw-Hill, London.

Bhanu, B. and Faugeras, O.D. (1982). Segmentation of images having unimodal distributions. *IEEE Trans. Pattern Anal. Mach. Intell.* **4**, 408–419.

Bhattacharya, A.K. and Roysam, B. (1994). Joint solution of low, intermediate and high-level vision tasks by evolutionary optimization: application to computer vision at low SNR. *IEEE Trans. Neural Networks* **5**, no. 1, 83–95.

Biberman, L.M. and Nudelman, S. (eds) (1971). *Photoelectronic Imaging Devices. Vol. 2. Devices and their Evaluation*. Plenum Press, New York.

Billingsley, J. (ed.) (1985). *Robots and Automated Manufacture*. IEE Control Engineering Series 28. Peter Peregrinus Ltd, London.

Blake, A., Zisserman, A. and Knowles, G. (1985). Surface descriptions from stereo and shading. *Image Vision Comput.* **3**, 183–191.

Bledsoe, W.W. and Browning, I. (1959). Pattern recognition and reading by machine. *Proc. Eastern Joint Comput. Conf.*, pp. 225–232.

Blostein, D. and Ahuja, N. (1989). Shape from texture: integrating texture element extraction and surface estimation. *IEEE Trans. Pattern Anal. Mach. Intell.* **11**, no. 12, 1233–1251.

Blum, H. (1967). A transformation for extracting new descriptors of shape. In: Wathen-Dunn, W. (ed.) *Models for the Perception of Speech and Visual Form*. MIT Press, Cambridge, MA, pp. 362–380.

Blum, H. and Nagel, R.N. (1978). Shape description using weighted symmetric axis features. *Pattern Recogn.* **10**, 167–180.

Bober, M. and Kittler, J. (1993). Estimation of complex multimodal motion: an approach based on robust statistics and Hough transform. *Proc. 4th British Machine Vision Assoc. Conf., Univ. of Surrey* (21–23 September), Vol. 1, pp. 239–248.

Boerner, H. and Strecker, H. (1988). Automated X-ray inspection of aluminium castings. *IEEE Trans. Pattern Anal. Mach. Intell.* **10**, no. 1, 79–91.

Bolles, R.C. (1979). Robust feature matching via maximal cliques. *SPIE, 182. Proc. Technical Symposium on Imaging Applications for Automated Industrial Inspection and Assembly, Washington D.C.* (April), pp. 140–149.

Bolles, R.C. and Cain, R.A. (1982). Recognizing and locating partially visible objects: the local-feature-focus method. *Int. J. Robot. Res.* **1**, 57–82.

Bolles, R.C. and Horaud, R. (1986). 3DPO: a three-dimensional part orientation system. *Int. J. Robot. Res.* **5**(3), 3–26.

Bourland, H. and Kamp, Y. (1988). Auto-association by multilayer perceptrons and singular value decomposition. *Biol. Cybern.* **59**, 291–294.

Bovik, A.C., Huang, T.S. and Munson, D.C. (1987). The effect of median filtering on edge estimation and detection. *IEEE Trans. Pattern Anal. Mach. Intell.* **9**, 181–194.

Brady, J.M. and Wang, H. (1992). Vision for mobile robots. *Phil. Trans. R. Soc. (London).* **B337**, 341–350.

Brady, J.M. and Yuille, A. (1984). An extremum principle for shape from contour. *IEEE Trans. Pattern Anal. Mach. Intell.* **6**, 288–301.

Brady, M. (1982). Computational approaches to image understanding. *Comput. Surveys* **14**, 3–71.

Braggins, D. and Hollingham, J. (1986). *The Machine Vision Sourcebook.* IFS (Publications) Ltd, Bedford, UK/Springer-Verlag, Berlin.

Bretschi, J. (1981). *Automated Inspection Systems for Industry.* IFS Publications Ltd, Bedford, UK.

Bron, C. and Kerbosch, J. (1973). Algorithm 457: finding all cliques in an undirected graph [H]. *Comm. ACM* **16**, 575–577.

Brooks, M.J. (1976). Locating intensity changes in digitised visual scenes. Computer Science Memo-15 (from MSc thesis), University of Essex.

Brooks, M.J. (1978). Rationalising edge detectors. *Comput. Graph. Image Process.* **8**, 277–285.

Brown, C.M. (1983). Inherent bias and noise in the Hough transform. *IEEE Trans. Pattern Anal. Mach. Intell.* **5**, 493–505.

Brown, C.M. (1984). Peak-finding with limited hierarchical memory. *Proc. 7th Int. Conf. on Pattern Recogn., Montreal* (30 July–2 August), pp. 246–249.

Brown, J. and Glazier, E.V.D. (1974). *Telecommunications* (2nd edn). Chapman and Hall, London.

Browne, A. and Norton-Wayne, L. (1986). *Vision and Information Processing for Automation.* Plenum Press, New York.

Bruckstein, A.M. (1988). On shape from shading. *Comput. Vision Graph. Image Process.* **44**, 139–154.

Bryant, D.J. and Bouldin, D.W. (1979). Evaluation of edge operators using relative and absolute grading. *Proc. IEEE Comput. Soc. Conf. on Pattern Recogn. and Image Process., Chicago* (31 May–2 June), pp. 138–145.

Burr, D.J. and Chien, R.T. (1977). A system for stereo computer vision with geometric models. *Proc. 5th Int. Joint Conf. on Artif. Intell., Boston*, p. 583.

Califano, A. and Mohan, R. (1994). Multidimensional indexing for recognizing visual shapes. *IEEE Trans. Pattern Anal. Mach. Intell.* **16**, no. 4, 373–392.

Canny, J. (1986). A computational approach to edge detection. *IEEE Trans. Pattern Anal. Mach. Intell.* **8**, 679–698.

Cantoni, V. and Levialdi, S. (1983). Matching the task to an image processing architecture. *Comput. Vision Graph. Image Process.* **22**, 301–309.

Castleman, K.R. (1979). *Digital Image Processing.* Prentice-Hall, Englewood Cliffs, NJ.

Chakravarty, I. and Freeman, H. (1982). Characteristic views as a basis for three-dimensional object recognition. *Proc. Soc. Photo-opt. Instrum. Eng. Conf. Robot Vision* **336**, 37–45.

Chan, J.P., Batchelor, B.G., Harris, I.P. and Perry, S.J. (1990). Intelligent visual inspection of food products. *Proc. SPIE Conf. on Machine Vision Systems in Industry,* Vol. 1386, pp. 171–179.

Chang, S., Davis, L.S., Dunn, S.M., Eklundh, J.-O. and Rosenfeld, A. (1987). Texture discrimination by projective invariants. *Pattern Recogn. Lett.* **5**, no. 5, 337–342.

Charniak, E. and McDermott, D. (1985). *Introduction to Artificial Intelligence.* Addison Wesley, Reading, MA.

Chasles, M. (1855). Question no. 296. *Nouv. Ann. Math.* **14**, 50.

Chauduri, B.B. (1994). Dynamic clustering for time incremental data. *Pattern Recogn. Lett.* **15**, no. 1, 27–34.

Chen, T.C. (1971). Parallelism, pipelining and computer efficiency. *Comput. Design* (Jan.), 69–74.

Chiang, Y.P. and Fu, K.-S. (1983). Matching parallel algorithm and architecture. In: *Proceedings of the International Conference on Parallel Processing,* Computer Society Press, pp. 374–380.

Chin, R.T. (1988). Automated visual inspection: 1981 to 1987. *Comput. Vision Graph Image Process.* **41**, 346–381.

Chin, R.T. and Harlow, C.A. (1982). Automated visual inspection: a survey. *IEE Trans. Pattern Anal. Mach. Intell.* **4**, 557–573.

Chittineni, C.B. (1980). Efficient feature subset selection with probabilistic distance criteria. *Inf. Sci.* **22**, 19–35.

Chow, C.K. and Kaneko, T. (1972). Automatic boundary detection of the left ventricle from cineangiograms. *Comput. Biomed. Res.* **5**, 388–410.

Choy, S.S.O., Choy, C.S.-T. and Siu, W.-C. (1995). New single-pass algorithm for parallel thinning. *Comput. Vision Image Understanding* **62**, no. 1, 69–77.

Clocksin, W.F. and Mellish, C.S. (1984). *Programming in Prolog.* Springer-Verlag, New York.

Cohen, F.S. and Cooper, D.B. (1987). Simple parallel hierarchical and relaxation algorithms for segmenting noncausal Markovian random fields. *IEEE Trans. Pattern Anal. Mach. Intell.* **9**, no. 2, 195–219.

Coleman, G.B. and Andrews, H.C. (1979). Image segmentation by clustering. *Proc. IEEE* **67**, 773–785.

Conners, R.W. and Harlow, C.A. (1980a). A theoretical comparison of texture algorithms. *IEEE Trans. Pattern Anal. Mach. Intell.* **2**, no. 3, 204–222.

Conners, R.W. and Harlow, C.A. (1980b). Toward a structural textural analyzer based on statistical methods. *Comput. Graph. Image Process* **12**, 224–256.

Cook, R.L. and Torrance, K.E. (1982). A reflectance model for computer graphics. *ACM Trans. Graphics* **1**, 7–24.

Corneil, D.G. and Gotlieb, C.C. (1970). An efficient algorithm for graph isomorphism. *J. ACM* **17**, 51–64.

Cottrell, G.W., Munro, P. and Zipser, D. (1987). Learning internal representations from grey-scale images: an example of extensional programming. *Proc. 9th Annual Conf. of the Cognitive Science Soc., Seattle.* Erlbaum, Hillsdale, pp. 462–473.

Crimmins, T.R. and Brown, W.R. (1985). Image algebra and automatic shape recognition. *IEEE Trans. Aerospace and Electronic Systems* **21**, 60–69.

Cross, G.R. and Jain, A.K. (1983). Markov random field texture models. *IEEE Trans. Pattern Anal. Mach. Intell.* **5**, no. 1, 25–39.

Crowley, J.L., Bobet, P. and Schmid, C. (1993). Auto-calibration by direct observation of objects. *Image Vision Comput.* **11**, no. 2, 67–81.

Cumani, A. and Guiducci, A. (1995). Geometric camera calibration: the virtual camera approach. *Machine Vision and Applications* **8**, no. 6, 375–384.

Cybenko (1988). *Continuous Valued Neural Networks with Two Hidden Layers are Sufficient.* Techn. Report, Dept. of Comput. Sci., Tufts Univ., Medford, MA.

Cybenko, G. (1989). Approximation by superpositions of a sigmoidal function. *Mathematics of Control, Signals and Systems* **2**, no. 4, 303–314.

Danielsson, P.-E. (1981). Getting the median faster. *Comput. Graph. Image Process.* **17**, 71–78.

Danielsson, P.-E. and Levialdi, S. (1981). Computer architectures for pictorial information systems. *IEEE Comput.* (Nov.), **14**, 53–67.

Davies, E.R. (1982). Image processing. In: Sumner, F.H. (ed.) *State of the Art Report: Supercomputer Systems Technology.* Pergamon Infotech, Maidenhead, pp. 223–244.

Davies, E.R. (1983). Image processing—its milieu, its nature and constraints on the design of special architectures for its implementation. In: Duff, M.J.B. (ed.) *Computing Structures for Image Processing,* Academic Press, London, 57–76.

Davies, E.R. (1984a). The median filter: an appraisal and a new truncated version. *Proc. 7th Int. Conf. on Pattern Recogn., Montreal* (30 July–2 August), pp. 590–592.

Davies, E.R. (1984b). Circularity—a new principle underlying the design of accurate edge orientation operators. *Image Vision Comput.* **2**, 134–142.

Davies, E.R. (1984c). Design of cost-effective systems for the inspection of certain food products during manufacture. In: Pugh, A. (ed.) *Proc. 4th Int. Conf. on Robot Vision and Sensory Controls, London* (9–11 October). IFS (Publications) Ltd, Bedford and North-Holland, Amsterdam, pp. 437–446.

Davies, E.R. (1984d). A glance at image analysis—how the robot sees. *Chartered Mech. Eng.* (Dec.), 32–35.

Davies, E.R. (1985a). A comparison of methods for the rapid location of products and their features and defects. In: McKeown, P.A. (ed.) *Proc. 7th Int. Conf. on Automated Inspection and Product Control, Birmingham* (26–28 March), pp. 111–120.

Davies, E.R. (1985b). Radial histograms as an aid in the inspection of circular objects. *IEE Proc. D* **132** (*4, Special Issue on Robotics*), 158–163.

Davies, E.R. (1985c). Precise measurement of radial dimensions in automatic visual inspection and quality control—a new approach. In: Billingsley, J. (ed.) *Robots and Automated Manufacture.* IEE Control Engineering Series 28. Peter Peregrinus Ltd, London, pp. 157–171.

Davies, E.R. (1986a). Constraints on the design of template masks for edge detection. *Pattern Recogn. Lett.* **4**, 111–120.

Davies, E.R. (1986b). Corner detection using the generalised Hough transform. *Proc. 2nd Int. Conf. on Image Processing and Its Applications* (24–26 June). *IEE Conf. Publ.* **265**, 175–179.

Davies, E.R. (1986c). Image space transforms for detecting straight edges in industrial images. *Pattern Recogn. Lett.* **4**, 185–192.

Davies, E.R. (1986d). Reduced parameter spaces for polygon detection using the generalised Hough transform. *Proc. 8th Int. Conf. on Pattern Recogn., Paris* (27–31 October), pp. 495–497.

Davies, E.R. (1987a). Methods for the rapid inspection of food products and small parts. In: McGeough, J.A. (ed.) *Proceedings of the 2nd International Conference on Computer-Aided Production Engineering, Edinburgh* (13–15 April), pp. 105–110.

Davies, E.R. (1987b). Visual inspection, automatic (robotics). In: Meyers, R.A. (ed.) *Encyclopedia of Physical Science and Technology, Vol. 14.* Academic Press, San Diego, pp. 360–377.

Davies, E.R. (1987c). A new framework for analysing the properties of the generalised Hough transform. *Pattern Recogn. Lett.* **6**, 1–7.

Davies, E.R. (1987d). A new parametrisation of the straight line and its application for the optimal detection of objects with straight edges. *Pattern Recogn. Lett.* **6**, 9–14.

Davies, E.R. (1987e). Design of optimal Gaussian operators in small neighbourhoods. *Image Vision Comput.* **5**, 199–205.

Davies, E.R. (1987f). Design of robust algorithms for automated visual inspection. *Proc. R. Swedish Acad. Eng. Sci. Symp. on "Machine Vision-the Eyes of Automation", Stockholm* (27 August), *IVA Rapp.* **336**, 55–81.

Davies, E.R. (1987g). Lateral histograms for efficient object location: speed versus ambiguity. *Pattern Recogn. Lett.* **6**, 189–198.

Davies, E.R. (1987h). The performance of the generalised Hough transform: concavities, ambiguities and positional accuracy. *Proc. 3rd Alvey Vision Conf., Cambridge* (15–17 September), pp. 327–333.

Davies, E.R. (1987i). Industrial vision systems: segmentation of gray-scale images. In: *Encyclopedia of Systems and Control*, Vol. 4. Pergamon, Oxford, pp. 2479–2484.

Davies, E.R. (1987j). Improved localisation in a generalised Hough scheme for the detection of straight edges. *Image Vision Comput.* **5**, 279–286.

Davies, E.R. (1987k). The effect of noise on edge orientation computations. *Pattern Recogn. Lett.* **6**, 315–322.

Davies, E.R. (1987l). A high speed algorithm for circular object location. *Pattern Recogn. Lett.* **6**, 323–333.

Davies, E.R. (1988a). Application of the generalised Hough transform to corner detection. *IEE Proc. E* **135**, 49–54.

Davies, E.R. (1988b). A modified Hough scheme for general circle location. *Pattern Recogn. Lett.* **7**, 37–43.

Davies, E.R. (1988c). On the noise suppression and image enhancement characteristics of the median, truncated median and mode filters. *Pattern Recogn. Lett.* **7**, 87–97.

Davies, E.R. (1988d). Median-based methods of corner detection. In: Kittler, J. (ed.) *Proceedings of the 4th BPRA International Conference on Pattern Recognition, Cambridge* (28–30 March). Lecture Notes in Computer Science, Vol. 301. Springer-Verlag, Heidelberg, pp. 360–369.

Davies, E.R. (1988e). A hybrid sequential-parallel approach to accurate circle centre location. *Pattern Recogn. Lett.* **7**, 279–290.

Davies, E.R. (1988f). Training sets and *a priori* probabilities with the nearest neighbour method of pattern recognition. *Pattern Recogn. Lett.* **8**, 11–13.

Davies, E.R. (1988g). An alternative to graph matching for locating objects from their salient features. *Proc. 4th Alvey Vision Conf., Manchester* (31 August–2 September), pp. 281–286.

Davies, E.R. (1988h). Efficient image analysis techniques for automated visual inspection. *Proc. Unicom Semin. on Comput. Vision and Image Process., London* (29 November–1 December), pp. 1–19.

Davies, E.R. (1988i). Tradeoffs between speed and accuracy in two-stage template matching. *Signal Process.* **15**, 351–363.

Davies, E.R. (1989a). Finding ellipses using the generalised Hough transform. *Pattern Recogn. Lett.* **9**, 87–96.

Davies, E.R. (1989b). Edge location shifts produced by median filters: theoretical bounds and experimental results. *Signal Process.* **16**, 83–96.

Davies, E.R. (1989c). Minimising the search space for polygon detection using the generalised Hough transform. *Pattern Recogn. Lett.* **9**, 181–192.

Davies, E.R. (1989d). Occlusion analysis for object detection using the generalised Hough transform. *Signal Process.* **16**, 267–277.

Davies, E.R. (1990) *Machine Vision: Theory, Algorithms, Practicalities*, Academic Press, London.

Davies, E.R. (1991a). The minimal match graph and its use to speed identification of maximal cliques. *Signal Process* **22**, no. 3, 329–343.

Davies, E.R. (1991b). Median and mean filters produce similar shifts on curved boundaries. *Electronics Lett.* **27**, no. 10, 826–828.

Davies, E.R. (1991c). Insight into operation of Kulpa boundary distance measure. *Electronics Lett.* **27**, no. 13, 1178–1180.

Davies, E.R. (1992a). Simple fast median filtering algorithm, with application to corner detection. *Electronics Lett.* **28**, no. 2, 199–201.

Davies, E.R. (1992b). Modelling peak shapes obtained by Hough transform. *IEE Proc. E.* **139**, no. 1, 9–12.

Davies, E.R. (1992c). A skimming technique for fast accurate edge detection. *Signal Process* **26**, no. 1, 1–16.

Davies, E.R. (1992d). Locating objects from their point features using an optimised Hough-like accumulation technique. *Pattern Recogn. Lett.* **13**, no. 2, 113–121.

Davies, E.R. (1992e). Procedure for generating template masks for detecting variable signals. *Image Vision Comput.* **10**, no. 4, 241–249.

Davies, E.R. (1992f). Accurate filter for removing impulse noise from one- or two-dimensional signals. *IEE Proc. E.* **139**, no. 2, 111–116.

Davies, E.R. (1992g). Simple two-stage method for the accurate location of Hough transform peaks. *IEE Proc. E.* **139**, no. 3, 242–248.

Davies, E.R. (1992h). A framework for designing optimal Hough transform implementations. *Proc. 11th IAPR Int. Conf. on Pattern Recogn., The Hague* (30 August–3 September), Vol. III, pp. 509–512.

Davies, E R. (1993a). *Electronics, Noise and Signal Recovery*, Academic Press, London.

Davies, E.R. (1993b). Computationally efficient Hough transform for 2-D object location. *Proc. 4th British Machine Vision Assoc. Conf., Univ. of Surrey* (21–23 September), Vol. 1, pp. 259–268.

Davies, E.R. (1995). Machine vision in manufacturing—what are the real problems? *Proc. 2nd Int. Conf. on Mechatronics and Machine Vision in Practice, Hong Kong* (12–14 September), pp. 15–24.

Davies, E.R. and Barker, S.P. (1990). An analysis of hole detection schemes. *Proc. British Machine Vision Assoc. Conf., Oxford* (24–27 September), pp. 285–290.

Davies, E.R. and Celano, D. (1993a). Orientation accuracy of edge detection operators acting on binary and saturated grey-scale images. *Electronics Lett.* **29**, no. 7, 603–604.

Davies, E R. and Celano, D. (1993b). Analysis of skeleton junctions in 3×3 windows. *Electronics Lett.* **29**, no. 16, 1440–1441.

Davies, E.R. and Johnstone, A.I.C. (1986). Engineering trade-offs in the design of a real-time system for the visual inspection of small products. *Proc. 4th Conf. on UK Research in*

Advanced Manufacture (10–11 December). IMechE Conference Publications, London, pp. 15–22.

Davies, E.R. and Johnstone, A.I.C. (1989). Methodology for optimising cost/speed tradeoffs in real-time inspection hardware. *IEE Proc. E* **136**, 62–69.

Davies, E.R. and Plummer, A.P.N. (1980). A new method for the compression of binary picture data. *Proc. 5th Int. Conf. on Pattern Recogn., Miami Beach, Florida (IEEE Comput. Soc.)*, pp. 1150–1152.

Davies, E.R. and Plummer, A.P.N. (1981). Thinning algorithms: a critique and a new methodology. *Pattern Recogn.* **14**, 53–63.

Davies, E.R., Patel, D. and Johnstone, A.I.C. (1995). Crucial issues in the design of a real-time contaminant detection system for food products. *Real-Time Imaging* **1**, no. 6, 397–407.

Davis, L.S. (1979). Computing the spatial structure of cellular textures. *Comput. Graph. Image Process* **11**, no. 2, 111–122.

Deans, S.R. (1981). Hough transform from the Radon transform. *IEEE Trans. Pattern Anal. Mach. Intell.* **3**, 185–188.

Delagnes, P., Benois, J. and Barba, D. (1995). Active contours approach to object tracking in image sequences with complex background. *Pattern Recogn. Lett.* **16**, no. 2, 171–178.

Deravi, F. and Pal, S.K. (1983). Grey level thresholding using second-order statistics. *Pattern Recogn. Lett.* **1**, pp. 417–422.

Derin, H. and Elliott, H. (1987). Modelling and segmentation of noisy and textured images using Gibbs random fields. *IEEE Trans. Pattern Anal. Mach. Intell.* **9**, no. 1, 39–55.

Devijver, P.A. and Kittler, J. (1980). On the edited nearest neighbour rule. *Proc. 5th Int. Conf. on Pattern Recogn., Miami Beach, Florida (IEEE Comput. Soc.)*, pp. 72–80.

Devijver, P.A. and Kittler, J. (1982). *Pattern Recognition: a Statistical Approach*. Prentice-Hall, Englewood Cliffs, NJ.

Dewaele, P., Van Gool, L., Wambacq, P. and Oosterlinck, A. (1988). Texture inspection with self-adaptive convolution filters. *Proc. 9th Int. Conf. on Pattern Recogn.* pp. 56–60.

Dhome, M., Rives, G. and Richetin, M. (1983). Sequential piecewise linear segmentation of binary contours. *Pattern Recogn. Lett.* **2**, 101–107.

Dickmanns, E.D. and Mysliwetz, B.D. (1992) Recursive 3-D road and relative ego-state recognition. *IEEE Trans. Pattern Anal. Mach. Intell.* **14**, no. 2, 199–213.

Dodd, G.G. and Rossol, L. (eds) (1979). *Computer Vision and Sensor-Based Robots*. Plenum Press, New York.

Dorst, L. and Smeulders, A.W.M. (1987). Length estimators for digitized contours. *Comput. Vision Graph. Image Process* **40**, 311–333.

Dougherty, E.R. (1992) *An Introduction to Morphological Image Processing*. SPIE Press, Bellingham, MA.

Dougherty, E.R. and Giardina, C.R. (1988). Morphology on umbra matrices. *Int. J. Pattern Recogn. Artif. Intell.* **2**, 367–385.

Dougherty, E.R. and Sinha, D. (1995a). Computational gray-scale mathematical morphology on lattices (a comparator-based image algebra) Part I: Architecture. *Real-Time Imaging* **1**, no. 1, 69–85.

Dougherty, E.R. and Sinha, D. (1995b). Computational gray-scale mathematical morphology on lattices (a comparator-based image algebra) Part II: Image operators. *Real-Time Imaging* **1**, no. 4, 283–295.

Doyle, W. (1962). Operations useful for similarity-invariant pattern recognition. *J. ACM* **9**, 259–267.

Dreschler, L. and Nagel, H.-H. (1981). Volumetric model and 3D-trajectory of a moving car derived from monocular TV-frame sequences of a street scene. *Proc. Int. Joint Conf. on Artif. Intell.* pp. 692–697.

Du Buf, J.M.H., Kardan, M. and Spann, M. (1990). Texture feature performance for image segmentation. *Pattern Recogn.* **23**, 291–309.

Duda, R.O. and Hart, P.E. (1972). Use of the Hough transformation to detect lines and curves in pictures. *Comm. ACM* **15**, 11–15.

Duda, R.O. and Hart, P.E. (1973). *Pattern Classification and Scene Analysis.* Wiley, New York.

Dudani, S.A. and Luk, A.L. (1978). Locating straight-line edge segments on outdoor scenes. *Pattern Recogn.* **10**, 145–157.

Dudani, S.A., Breeding, K.J. and McGhee, R.B. (1977). Aircraft identification by moment invariants. *IEEE Trans. Comput.* **26**, 39–46.

Duff, M.J.B., Watson, D.M., Fountain, T.J. and Shaw, G.K. (1973). A cellular logic array for image processing. *Pattern Recogn.* **5**, 229–247.

Duin, R.P.W., Haringa, H. and Zeelen, R. (1986). Fast percentile filtering. *Pattern Recogn. Lett.* **4**, 269–272.

Dyer, C.R. and Rosenfeld, A. (1976). Fourier texture features: suppression of aperture effects. *IEEE Trans. Systems Man Cybern.* **6**, no. 10, 703–705.

Edmonds, J.M. and Davies, E.R. (1991). High-speed processor for realtime visual inspection. *Microprocess. Microsys.* **15**, 11–19.

Eichmann, G. and Kasparis, T. (1988). Topologically invariant texture descriptors. *Comput. Vision Graph. Image Process* **41**, 267–281.

Eisberg, R.M. (1961) *Fundamentals of Modern Physics.* Wiley, New York.

Ellis, T.J., Abbood, A. and Brillault, B. (1992). Ellipse detection and matching with uncertainty. *Image Vision Comput.* **10**, no. 5, 271–276.

Fathy, M. and Siyal, M.Y. (1995). Real-time image processing approach to measure traffic queue parameters. *IEE Proc. Vision Image Signal Process* **142**, no. 5, 297–303.

Faugeras, O.D. (1978). Texture analysis and classification using a human visual model. *Proc. 4th Int. Joint Conf. on Pattern Recogn., Kyoto* (7–10 November), pp. 549–552.

Faugeras, O.D. (1993) *Three-Dimensional Computer Vision—a Geometric Viewpoint,* MIT Press, Cambridge, MA.

Faugeras, O.D. and Hebert, M. (1983). A 3-D recognition and positioning algorithm using geometrical matching between primitive surfaces. *Proc. 8th Int. Joint Conf. on Artif. Intell.* pp. 996–1002.

Feng, T.-Y. (1981). A survey of interconnection networks. *IEEE Comput.* **14**, 12–17.

Ferrie, F.P. and Levine, M.D. (1989). Where and why local shading analysis works. *IEEE Trans. Pattern Anal. Mach. Intell.* **11**, 198–206.

Fesenkov, V.P. (1929). Photometric investigations of the lunar surface. *Astronomochhesk. Zh.* **5**, 219–234.

Fischler, M.A. and Bolles, R.C. (1981). Random sample consensus: a paradigm for model fitting with applications to image analysis and automated cartography. *Comm. ACM.* **24**, no. 6, 381–395.

Fitch, J.P., Coyle, E.J. and Gallagher, N.C. (1985). Root properties and convergence rates of median filters. *IEEE Trans. Acoust. Speech Signal Process* **33**, 230–239.

Flynn, M.J. (1972). Some computer organizations and their effectiveness. *IEEE Trans. Comput.* **21**, 948–960.

Föglein, J. (1983). On edge gradient approximations. *Pattern Recogn. Lett.* **1**, 429–434.

Forgy, E.W. (1965). Cluster analysis of multivariate data: efficiency versus interpretability of classification. *Biometrics.* **21**, 768–769.

Forsyth, D.A., Mundy, J.L., Zisserman, A., Coelho, C., Heller, A. and Rothwell, C.A. (1991). Invariant descriptors for 3-D object recognition and pose. *IEEE Trans. Pattern Anal. Mach. Intell.* **13**, no. 10, 971–991.

Forsyth, R.S. (1982). *Pascal at Work and Play*. Chapman & Hall, London.

Fountain, T. (1987). *Processor Arrays: Architectures and Applications*. Academic Press, London.

Freeman, H. (1961). On the encoding of arbitrary geometric configurations. *IEEE Trans. Electron. Comput.* **10**, 260–268.

Freeman, H. (1974). Computer processing of line drawing images. *Comput. Surveys* **6**, 57–97.

Freeman, H. (1978). Shape description via the use of critical points. *Pattern Recogn.* **10**, 159–166.

Frei, W. and Chen. C.-C. (1977). Fast boundary detection: a generalization and a new algorithm. *IEEE Trans. Comput.* **26**, 988–998.

Fu, K.-S. and Mui, J.K. (1981). A survey on image segmentation. *Pattern Recogn.* **13**, 3–16.

Gallagher, N.C. and Wise, G.L (1981). A theoretical analysis of the properties of median filters. *IEEE Trans. Acoust. Speech Signal Process* **29**, 1136–1141.

Gårding, J. (1988). Properties of fractal intensity surfaces. *Pattern Recogn. Lett.* **8**, no. 5, 319–324.

Gavrila, D.M. and Groen, F.C.A. (1992). 3D object recognition from 2D images using geometric hashing. *Pattern Recogn. Lett.* **13**, no. 4, 263–278.

Gelsema, E.S. (ed.) (1994). Special Issue on Genetic Algorithms. *Pattern Recogn. Lett.* **16**, no. 8.

Geman, D. (1987). Stochastic model for boundary detection. *Image Vision Comput.* **5**, no. 2, 61–65.

Geman, S. and Geman, D. (1984). Stochastic relaxation, Gibbs distributions, and the Bayesian restoration of images. *IEEE Trans. Pattern Anal. Mach. Intell.* **6**, no. 6, 721–741.

Gerig, G. and Klein, F. (1986). Fast contour identification through efficient Hough transform and simplified interpretation strategy. *Proc. 8th Int. Conf. on Pattern Recogn., Paris* (27–31 October), pp. 498–500.

Ghosh, A. and Pal, S.K. (1992). Neural network, self-organisation and object extraction. *Pattern Recogn. Lett.* **13**, no. 5, 387–397.

Gibbons, A. (1985). *Algorithmic Graph Theory*. Cambridge University Press, Cambridge.

Gibson, J.J. (1950). *The Perception of the Visual World*. Houghton Mifflin, Boston, MA.

Godbole, S. and Amin, A. (1995). Mathematical morphology for edge and overlap detection for medical images. *Real-Time Imaging* **1**, no. 3, 191–201.

Golub, G.H. and van Loan, C.F. (1983). *Matrix Computations*. North Oxford, Oxford, UK.

Gonnet, G.H. (1984). *Handbook of Algorithms and Data Structures*. Addison-Wesley, London.

Gonzalez, R.C. and Wintz, P. (1987). *Digital Image Processing* (2nd edn). Addison-Wesley, Reading, MA.

Granlund, G.H. (1980). Description of texture using the general operator approach. *Proc. 5th Int. Conf. on Pattern Recogn., Miami Beach, Florida* (1–4 December), pp. 776–779.

Graves, M., Batchelor, B.G. and Palmer, S. (1994). 3D X-ray inspection of food products. *Proc. SPIE Conf. on Applications of Digital Image Process 17*, Vol. 2298, pp. 248–259.

Greenhill, D. and Davies, E.R. (1993). Texture analysis using neural networks and mode filters. *Proc. 4th British Machine Vision Assoc. Conf., Univ. of Surrey* (21–23 September), Vol. 2, pp. 509–518.

Greenhill, D. and Davies, E.R. (1994a). Relative effectiveness of neural networks for image noise suppression. In: Gelsema, E.S. and Kanal, L.N. (eds.) *Pattern Recognition in Practice IV*. Elsevier Science B.V., pp. 367–378.

Greenhill, D. and Davies, E.R. (1994b). How do neural networks compare with standard filters for image noise suppression? IEE Digest no. 1994/248, Colloquium on *Applications of Neural Networks to Signal Processing, IEE* (15 December), pp. 3/1–4.

Greenhill, D. and Davies, E.R. (1995). A new approach to the determination of unbiased thresholds for image segmentation. *5th Int. IEE Conf. on Image Processing and its Applications, Heriot-Watt University* (3–6 July). IEE Conf. Publication no. 410, pp. 519–523.

Gregory, J. and McReynolds, R. (1963). The SOLOMON computer. *IEEE Trans. Electron. Comput.* **12**, 774–781.

Gregory, R.L. (1971). *The Intelligent Eye*. Weidenfeld and Nicolson, London.

Gregory, R.L. (1972). *Eye and Brain* (2nd edn). Weidenfeld and Nicolson, London.

Grimson, W.E.L. and Huttenlocher, D.P. (1990). On the sensitivity of the Hough transform for object recognition. *IEEE Trans. Pattern Anal. Mach. Intell.* **12**, no. 3, 255–274.

Grimson. W.E.L. and Lozano-Perez, T. (1984). Model-based recognition and localisation from sparce range or tactile data. *Int. J. Robot. Res.* **3**(3), 3–35.

Hall, E.L (1979). *Computer Image Processing and Recognition*. Academic Press, New York.

Hall, E.L, Tio, J.B.K., McPherson, C.A. and Sadjadi, F.A. (1982). Measuring curved surfaces for robot vision. *IEEE Comput.* **15**(12), 42–54.

Hampel, F.R., Ronchetti, E.M., Rousseeuw, P.J. and Stahel, W.A. (1986). *Robust statistics, The Approach Based on Influence Functions*. Wiley, New York.

Hannah, I., Patel, D. and Davies, E.R. (1995). The use of variance and entropic thresholding methods for image segmentation. *Pattern Recognition* **28**, no. 8. 1135–1143.

Hansen, F.R. and Elliott, H. (1982). Image segmentation using simple Markov field models. *Comput. Graph. Image Process* **20**, 101–132.

Haralick, R.M. (1979). Statistical and structural approaches to texture. *Proc. IEEE* **67**, 786–804.

Haralick, R.M. (1980). Edge and region analysis for digital image data. *Comput. Graph. Image Process* **12**, 60–73.

Haralick, R.M. (1984). Digital step edges from zero crossing of second directional derivatives. *IEEE Trans. Pattern Anal. Mach. Intell.* **6**, 58–68.

Haralick, R.M. (1989). Determining camera parameters from the perspective projection of a rectangle. *Pattern Recogn.* **22**, 225–230.

Haralick, R.M. and Chu, Y.H. (1984). Solving camera parameters from the perspective projection of a parameterized curve. *Pattern Recogn.* **17**, no. 6, 637–645.

Haralick, R.M. and Joo, H. (1988). 2D–3D pose estimation. *Proc. 9th Int. Conf. on Pattern Recogn., Rome, Italy* (14–17 November), pp. 385–391.

Haralick, R.M. and Shapiro, L.G. (1985). Image segmentation techniques. *Comput. Vision Graph. Image Process* **29**, 100–132.

Haralick, R.M. and Shapiro, L.G. (1992). *Computer and Robot Vision*, Vol. 1. Addison Wesley, Reading, Mass.

Haralick, R.M., Shanmugam, R. and Dinstein, I. (1973). Textural features for image classification. *IEEE Trans. Systems Man Cybern.* **3**, 610–621.

Haralick, R.M., Chu, Y.H., Watson, L.T. and Shapiro, L.G. (1984). Matching wire frame objects from their two dimensional perspective projections. *Pattern Recogn.* **17**, no. 6, 607–619.

Haralick, R.M., Sternberg, S.R. and Zhuang, X. (1987). Image analysis using mathematical morphology. *IEEE Trans. Pattern Anal. Mach. Intell.* **9**, 532–550.

Hart, P.E. (1968). The condensed nearest neighbour rule. *IEEE Trans. Inf. Theory* **14**, 515–516.

Harvey, N.R. and Marshall, S. (1994). Using genetic algorithms in the design of morphological filters. *IEE Colloquium on Genetic Algorithms in Image Processing and Vision, IEE* (20 October). IEE Digest no. 1994/193, pp. 6/1–5.

Harwood, D., Subbarao, M. and Davis, L.S. (1985). Texture classification by local rank correlation. *Comput. Vision Graph. Image Process* **32**, 404–411.

Haykin, S. (1994). *Neural Networks*. Macmillan College Pub. Co., New York.

Hebb, D.O. (1949). *The Organization of Behavior*. Wiley, New York.

Heijmans, H. (1991). Theoretical aspects of gray-level morphology. *IEEE Trans. Pattern Anal. Mach. Intell.* **13**, 568–582.

Heijmans, H. (1994). *Morphological Operators*. Academic Press, New York.

Heikkonen, J. (1995). Recovering 3-D motion parameters from optical flow field using randomized Hough transform. *Pattern Recogn. Lett.* **16**, no. 9, 971–978.

Heinonen, P. and Neuvo, Y. (1987). FIR-median hybrid filters. *IEEE Trans. Acoust. Speech Signal Process* **35**, 832–838.

Henderson, T.C. (1984). A note on discrete relaxation. *Comput. Vision Graph. Image Process* **28**, 384–388.

Hertz, J., Krogh, A. and Palmer, R.G. (1991). *Introduction to the Theory of Neural Computation*. Addison-Wesley, Reading, MA.

Hildreth, E.C. (1984). *Measurement of Visual Motion*. MIT Press, Cambridge, MA.

Hill, A. and Taylor, C.J. (1992). Model-based image interpretation using genetic algorithms. *Image Vision Comput.* **10**, no. 5, 295–300.

Hockney, R.W. and Jesshope, C. R. (1981). *Parallel Computers*. Adam Hilger Ltd, Bristol.

Hodgson, R.M., Bailey, D.G., Naylor, M.J., Ng, A.L.M. and McNeill, S.J. (1985). Properties, implementations, and applications of rank filters. *Image Vision Comput.* **3**, 3–14.

Holland, J.H. (1975). *Adaptation in Natural and Artificial Systems*. Univ. of Michigan Press, Ann Arbor, MI.

Hollinghurst, N. and Cipolla, R. (1994). Uncalibrated stereo hand-eye coordination. *Image Vision Comput.* **12**, no. 3, 187–192.

Horaud, R. (1987). New methods for matching 3-D objects with single perspective views. *IEEE Trans. Pattern Anal. Mach. Intell.* **9**, 401–412.

Horaud, R. and Brady, M. (1988). On the geometric interpretation of image contours. *Artif. Intell.* **37**, 333–353.

Horaud, R., Conio, B., Leboulleux, O. and Lacolle, B. (1989). An analytic solution for the perspective 4-point problem. *Comput. Vision Graph. Image Process* **47**, 33–44.

Horn, B.K.P. (1975). Obtaining shape from shading information. In: Winston, P.H. (ed.) *The Psychology of Computer Vision*. McGraw-Hill, New York, pp. 115–155.

Horn, B.K.P. (1977). Understanding image intensities. *Artif. Intell.* **8**, 201–231.

Horn, B.K.P. (1986). *Robot Vision*. MIT Press, Cambridge, MA.

Horn, B.K.P. and Brooks, M.J. (1986). The variational approach to shape from shading. *Comput. Vision Graph. Image Process* **33**, 174–208.

Horn, B.K.P. and Brooks, M.J. (eds) (1989). *Shape from Shading*. MIT Press, Cambridge, MA.

Horn, B.K.P. and Schunck, B.G. (1981). Determining optical flow. *Artif. Intell.* **17**, nos. 1–3, 185–203.

Hornik, K., Stinchcombe, M. and White, H. (1989). Multilayer feedforward networks are universal approximators. *Neural Networks* **2**, 359–366.

Horowitz, S.L. and Pavlidis, T. (1974). Picture segmentation by a directed split-and-merge procedure. *Proc. 2nd Int. Joint Conf. on Pattern Recogn.*, pp. 424–433.

Hough, P.V.C. (1962). Method and means for recognising complex patterns. *US Patent 3069654*.

Hsiao, J.Y. and Sawchuk, A.A. (1989). Supervised textured image segmentation using feature smoothing and probabilistic relaxation techniques. *IEEE Trans. Pattern Anal. Mach. Intell.* **11**, no. 12, 1279–1292.

Hsiao, J.Y. and Sawchuk, A.A. (1990). Unsupervised textured image segmentation using feature smoothing and probabilistic relaxation techniques. *Comput. Vision Graph. Image Process* **48**, 1–21.

Hu, M.K. (1961). Pattern recognition by moment invariants. *Proc. IEEE* **49**, 1428.

Hu, M.K. (1962). Visual pattern recognition by moment invariants. *IRE Trans. Inf. Theory* **8**, 179–187.

Huang, C.T. and Mitchell, O.R. (1994). A Euclidean distance transform using grey-scale morphology decomposition. *IEEE Trans. Pattern Anal. Mach. Intell.* **16**, no. 4, 443–448.

Huang, T.S. (ed.) (1983). *Image Sequence Processing and Dynamic Scene Analysis*. Springer-Verlag, New York.

Huang, T.S., Yang, G.J. and Tang, G.Y. (1979). A fast two-dimensional median filtering algorithm. *IEEE Trans. Acoust. Speech Signal Process* **27**, 13–18.

Huang, T.S., Bruckstein, A.M., Holt, R.J. and Netravali, A.N. (1995). Uniqueness of 3D pose under weak perspective: a geometrical proof. *IEEE Trans. Pattern Anal. Mach. Intell.* **17**, no. 12, 1220–1221.

Huber, P.J. (1964). Robust estimation of a location parameter. *Annals Math. Statist.* **35**, 73–101.

Huber, P.J. (1981). *Robust Statistics*. Wiley, New York.

Huber, P.J. (1985). Projection pursuit. *Annals Statist.* **13**, no. 2, 435–475.

Hueckel, M.F. (1971). An operator which locates edges in digitised pictures. *J. ACM* **18**, 113–125.

Hueckel, M.F. (1973). A local visual operator which recognises edges and lines. *J. ACM* **20**, 634–647.

Hughes, G.F. (1968). On the mean accuracy of statistical pattern recognisers. *IEEE Trans. Inf. Theory* **14**, 55–63.

Hummel, R.A. and Zucker, S.W. (1983). On the foundations of relaxation labelling processes. *IEEE Trans. Pattern Anal. Mach. Intell.* **5**, 267–287.

Hunt, D.J. (1981). The ICL DAP and its application to image processing. In: Duff, M.J.B. and Levialdi, S. (eds) *Languages and Architectures for Image Processing*. Academic Press, London, pp. 275–282.

Hush, D.R. and Horne, B.G. (1993). Progress in supervised neural networks. *IEEE Signal Processing Magazine* **10**, no. 1, 8–39.

Hussain, Z. (1988). A fast approximation to a convex hull. *Pattern Recogn. Lett.* **8**, 289–294.

Huttenlocher, D.P., Klanderman, G.A. and Rucklidge, W.J. (1993). Comparing images using the Hausdorff distance. *IEEE Trans. Pattern Anal. Mach. Intell.* **15**, no. 9, 850–863.

Hwang, K. and Briggs, F.A. (1984). *Computer Architecture and Parallel Processing*. McGraw-Hill, New York.

Hwang, V.S., Davis, L.S. and Matsuyama, T. (1986). Hypothesis integration in image understanding systems. *Comput. Vision Graph. Image Process* **36**, 321–371.

Iannino, A. and Shapiro, S.D. (1979). An iterative generalisation of the Sobel edge detection operator. *Proc. IEEE Comput. Soc. Conf. on Pattern Recogn. and Image Process., Chicago* (31 May–2 June), pp. 130–137.

IBM (1986). Cluster estimation in Hough space for image analysis. *IBM Techn. Disclosure Bull* **28**, 3667–3668.

IEE (1994). *IEE Colloquium on Genetic Algorithms in Image Processing and Vision, IEE* (20 October). IEE Digest no. 1994/193.

Ikeuchi, K. and Horn, B.K.P. (1981). Numerical shape from shading and occluding boundaries. *Artif. Intell.* **17**, 141–184.

Illingworth, J. and Kittler, J. (1987). The adaptive Hough transform. *IEEE Trans. Pattern Anal. Mach. Intell.* **9**, 690–698.

Illingworth, J. and Kittler, J. (1988). A survey of the Hough transform. *Comput. Vision Graph. Image Process* **44**, 87–116.

Inmos Ltd (1987). *The Transputer Family 1987*. Inmos Ltd, Bristol.

Inmos Ltd (1988). *OCCAM 2 Reference Manual*. Prentice-Hall, London.

Ising, E. (1925). *Z. Physik* **31**, 253.

Ito, M. and Ishii, A. (1986). Three-view stereo analysis. *IEEE Trans. Pattern Anal. Mach. Intell.* **8**, no. 4, 524–532.

Jackway, P.T. and Deriche, M. (1996). Scale-space properties of the multiscale morphological dilation-erosion. *IEEE Trans. Pattern Anal. Mach. Intell.* **18**, no. 1, 38–51.

Jain, A.K. and Dubes, R.C. (1988). *Algorithms for Clustering Data*. Prentice-Hall, Englewood Cliffs, NJ.

Jain, A.K. and Farrokhnia, F. (1991). Unsupervised texture segmentation using Gabor filters. *Pattern Recogn.* **24**, 1167–1186.

Jain, R. (1983). Direct computation of the focus of expansion. *IEEE Trans. Pattern Anal. Mach. Intell.* **5**, 58–63.

Jolion, J.-M. and Rosenfeld, A. (1989). Cluster detection in background noise. *Pattern Recogn.* **22**, no. 5, 603–607.

Jones, R. and Svalbe, I. (1994). Morphological filtering as template matching. *IEEE Trans. Pattern Anal. Mach. Intell.* **16**, no. 4, 438–443.

Juan, A. and Vidal, E. (1994). Fast K-means-like clustering in metric spaces. *Pattern Recogn. Lett.* **15**, no. 1, 19–25.

Kaizer, H. (1955). A Quantification of Textures on Aerial Photographs. MS thesis, Boston Univ.

Kamel, M.S., Shen, H.C., Wong, A.K.C., Hong, T.M. and Campeanu, R.I. (1994). Face recognition using perspective invariant features. *Pattern Recogn. Lett.* **15**, no. 9, 877–883.

Kapur, J.N., Sahoo, P.K. and Wong, A.K.C. (1985). A new method for gray-level picture thresholding using the entropy of the histogram. *Comput. Vision Graph. Image Process* **29**, 273–285.

Kasif, S., Kitchen, L and Rosenfeld, A. (1983). A Hough transform technique for subgraph isomorphism. *Pattern Recogn. Lett.* **2**, 83–88.

Kass, M. and Witkin, A. (1987). Analyzing oriented patterns. *Comput. Vision Graph. Image Process* **37**, no. 3, 362–385.

Kass, M., Witkin, A. and Terzopoulos, D. (1988). Snakes: active contour models. *Int. J. Comput. Vision 1*, 321–331.

Katatani, K. and Chou, T. (1989). Shape from texture: general principle. *Artif. Intell.* **38**, no. 1, 1–49.

Katz, A.J. and Thrift, P.R. (1994). Generating image filters for target recognition by genetic learning. *IEEE Trans. Pattern Anal. Mach. Intell.* **16**, no. 9, 906–910.

Kehtarnavaz, N. and Mohan, S. (1989). A framework for estimation of motion parameters from range images. *Comput. Vision Graph. Image Process* **45**, 88–105.

Keller, J.M., Chen, S. and Crownover, R.M. (1989). Texture description and segmentation through fractal geometry. *Comput. Vision Graph. Image Process* **45**, 150–166.

Kelley, R.B. and Gouin, P. (1984). Heuristic vision hole finder. In: Pugh, A. (ed.) *Proc. 4th Int. Conf. on Robot Vision and Sensory Controls, London* (9–11 October). IFS (Publications) Ltd, Bedford and North-Holland, Amsterdam, pp. 341–350.

Kender, J.R. (1981). Shape from texture. *Carnegie-Mellon University, Comput. Sci. Techn. Rep.* CMU-CS-81-102.

Kim, D.Y., Kim, J.J., Meer, P., Mintz, D. and Rosenfeld, A. (1989). Robust computer vision: a least median of squares based approach. *Proc. DARPA Image Understanding Workshop, Palo Alto, CA* (23–26 May), pp. 1117–1134.

Kim, H.-B. and Park, R.-H. (1990). Extraction of periodicity vectors from structural textures using projection information. *Pattern Recogn. Lett.* **11**, no. 9, 625–630.

Kimme, C., Ballard, D. and Sklansky, J. (1975). Finding circles by an array of accumulators. *Comm. ACM* **18**, 120–122.

King, T.G. and Tao, L.G. (1995). An incremental real-time pattern tracking algorithm for line-scan camera applications. *Mechatronics* **4**, no. 5, 503–516.

Kirsch, R.A. (1971). Computer determination of the constituent structure of biological images. *Comput. Biomed Res.* **4**, 315–328.

Kiryati, N. and Bruckstein, A.M. (1991). Antialiasing the Hough transform. *Comput. Vision Graph. Image Process.: Graph. Models Image Process* **53**, 213–222.

Kitchen, L. and Rosenfeld, A. (1979). Discrete relaxation for matching relational structures. *IEEE Trans. Systems Man Cybern.* **9**, 869–874.

Kitchen, L. and Rosenfeld, A. (1982). Gray-level corner detection. *Pattern Recogn. Lett.* **1**, 95–102.

Kittler, J. (1983). On the accuracy of the Sobel edge detector. *Image Vision Comput.* **1**, 37–42.

Kittler, J. and Duff, M.J.B. (eds) (1985). *Image Processing System Architectures*. Research Studies Press Ltd/Wiley, New York.

Kittler, J., Illingworth, J., Föglein, J. and Paler, K. (1984). An automatic thresholding algorithm and its performance. *Proc. 7th Int. Conf. on Pattern Recogn., Montreal* (30 July–2 August), pp. 287–289.

Kittler, J., Illingworth, J. and Föglein, J. (1985). Threshold selection based on a simple image statistic. *Comput. Vision Graph. Image Process* **30**, 125–147.

Koenderink, J.J. and van Doorn, A.J. (1979). The internal representation of solid shape with respect to vision. *Biol. Cybern.* **32**, 211–216.

Koivo, A.J. and Kim, C.W. (1989). Robust image modelling for classification of surface defects on wood boards. *IEEE Trans. Systems Man Cybern.* **19**, no. 6, 1659–1666.

Koller, D., Weber, J., Huang, T., Malik, J., Ogasawara, G., Rao, B. and Russell, S. (1994). Towards robust automatic traffic scene analysis in real-time. *Proc. 12th Int. Conf. on Pattern Recogn., Jerusalem, Israel* (9–13 October), pp. 126–131.

Koplowitz, J. and Bruckstein, A.M. (1989). Design of perimeter estimators for digitized planar shapes. *IEEE Trans. Pattern Anal. Mach. Intell.* **11**, 611–622.

Kruger, R.P. and Thompson, W.B. (1981). A technical and economic assessment of computer vision for industrial inspection and robotic assembly. *Proc. IEEE* **69**, 1524–1538.

Kuehnle, A. (1991). Symmetry-based recognition of vehicle rears. *Pattern Recogn. Lett.* **12**, 249–258.

Kulpa, Z. (1977). Area and perimeter measurement of blobs in discrete binary pictures. *Comput. Graph. Image Process* **6**, 434–451.

Kumar, R. and Hanson, A.R. (1989). Robust estimation of camera location and orientation from noisy data having outliers. *Proc. Workshop on Interpretation of 3D Scenes, Austin, TX* (27–29 November), pp. 52–60.

Kung, H.T. (1980). The structure of parallel algorithms. *Adv. Comput.* **19**, 69–112.

Kung, H.T. (1984). Systolic algorithms for the CMU Warp processor. *Proc. 7th Int. Conf. on Pattern Recogn., Montreal* (30 July–2 August), pp. 570–577.

Kung, H.T. and Leiserson, C.E (1979). Systolic arrays (for VLSI). In: Duff, I.S. and Stewart, G.W. (eds) *Sparce Matrix Proceedings 1978.* SIAM, Philadelphia, pp. 256–282.

Kwok, P.C.K. (1989). Customising thinning algorithms. *Proceedings of the 3rd International Conference on Image Processing and its Applications, Warwick* (18–20 July), IEE Conf. Publ. **307**, 633–637.

Lacroix, V. (1988). A three-module strategy for edge detection. *IEEE Trans. Pattern Anal. Mach. Intell.* **10**, 803–810.

Lam, S.W.C. and Ip, H.H.S. (1994). Structural texture segmentation using irregular pyramid. *Pattern Recogn. Lett.* **15**, no. 7, 691–698.

Landgrebe, D.A. (1981). Analysis technology for land remote sensing. *Proc. IEEE* **69**, 628–642.

Lane, R.A., Thacker, N.A. and Seed, N.L. (1994). Stretch-correlation as a real-time alternative to feature-based stereo matching algorithms. *Image Vision Comput.* **12**, no. 4, 203–212.

Laws, K.I. (1979). Texture energy measures. *Proc. Image Understanding Workshop,* November, pp. 47–51.

Laws, K.I. (1980a). Rapid texture identification. *Proc. SPIE Conf. on Image Processing for Missile Guidance, San Diego, Calif.* (28 July–1 August), **238**, pp. 376–380.

Laws, K.I. (1980b). *Textured Image Segmentation.* PhD thesis, Univ. of Southern California, Los Angeles.

Leavers, V.F. (1992). *Shape Detection in Computer Vision Using the Hough Transform.* Springer-Verlag, Berlin.

Leavers, V.F. (1993). Which Hough transform?. *CVGIP: Image Understanding* **58**, no. 2, 250–264.

Leavers, V.F. and Boyce, J.F. (1987). The Radon transform and its application to shape parametrization in machine vision. *Image Vision Comput.* **5**, 161–166.

Lev, A., Zucker, S.W. and Rosenfeld, A. (1977). Iterative enhancement of noisy images. *IEEE Trans. Systems Man Cybern.* **7**, 435–442.

Levine, M.D. (1985). *Vision in Man and Machine.* McGraw-Hill, New York.

Li, H. and Lavin, M.A. (1986). Fast Hough transform based on bintree data structure. *Proc. Conf. Comput. Vision and Pattern Recogn., Miami Beach, Florida,* pp. 640–642.

Li, H., Lavin, M.A. and LeMaster, R.J. (1985). Fast Hough transform. *Proc. 3rd Workshop on Comput. Vision: Representation and Control, Bellair,* pp. 75–83.

Lin, C.C. and Chellappa, R. (1987). Classification of partial 2-D shapes using Fourier descriptors. *IEEE Trans. Pattern Anal. Mach. Intell.* **9**, 686–690.

Linsker, R. (1986). From basic network principles to neural architecture. *Proc. National Acad. Sciences, USA.* **83**, 7508–7512, 8390–8394, 8779–8783.

Linsker, R. (1988). Self-organization in a perceptual network. *IEEE Computer* (March). 105–117.

Lippmann, R.P. (1987). An introduction to computing with neural nets. *IEEE ASSP Mag,* **4**, 4–22.

Longuet-Higgins, H.C. (1981). A computer algorithm for reconstructing a scene from two projections. *Nature* **293**, 133–135.

Longuet-Higgins, H.C. (1984). The visual ambiguity of a moving plane. *Proc. R. Soc. (London)* **B233**, 165–175.

Longuet-Higgins, H.C. and Prazdny, K. (1980). The interpretation of a moving retinal image. *Proc. R. Soc. (London)* **B208**, 385–397.

Lowe, D. (ed.) (1994). Special Issue on Applications of Artificial Neural Networks. *IEE Proc. Vision Image Signal Process* **141**, no. 4.

Lowe, D.G. (1987). Three-dimensional object recognition from single two-dimensional images. *Artif. Intell.* **31**, 355–395.

Lu, S. and Szeto, A. (1991). Improving edge measurements on noisy images by hierarchical neural networks. *Pattern Recogn. Lett.* **12**, 155–164.

Lutton, E. and Martinez, P. (1994). A genetic algorithm for the detection of 2D geometric primitives in images, *Proc. 12th Int. Conf. on Pattern Recogn., Jerusalem* (9–13 October), Vol. **1**, pp. 526–528.

Lutton, E., Maître, H. and Lopez-Krahe, J. (1994). Contribution to the determination of vanishing points using Hough transform. *IEEE Trans. Pattern Anal. Mach. Intell.* **16**, no. 4, 430–438.

Lyvers, E.P. and Mitchell, O.R. (1988). Precision edge contrast and orientation estimation. *IEEE Trans. Pattern Anal. Mach. Intell.* **10**, 927–937.

Mackeown, W.P.J., Greenway, P., Thomas, B.T. and Wright, W.A. (1994). Contextual image labelling with a neural network. *IEE Proc. Vision Image Signal Process* **141**, no. 4, 238–244.

MacQueen, J.B. (1967). Some methods for classification and analysis of multivariate observations. *Proc. 5th Berkeley Symp. on Math. Stat. and Prob.*, **I**, pp. 281–297.

Malik, J., Weber, J., Luong, Q.-T. and Koller, D. (1995). Smart cars and smart roads. *Proc. 6th British Machine Vision Assoc. Conf., Birmingham* (11–14 September), pp. 367–381.

Mandelbrot, B.B. (1982). *The Fractal Geometry of Nature*. Freeman, San Fransisco, CA.

Marchant, J.A. and Brivot, R. (1995). Real-time tracking of plant rows using a Hough transform. *Real-Time Imaging* **1**, no. 5, 363–371.

Marr, D. (1976). Early processing of visual information. *Phil. Trans. R. Soc. (London)* **B275**, 483–524.

Marr, D. and Hildreth, E. (1980). Theory of edge detection. *Proc. R. Soc. (London)* **B207**, 187–217.

Marr, D. and Poggio, T. (1979). A computational theory of human stereo vision. *Proc. R. Soc. (London)* **B204**, 301–328.

Marslin, R.F., Sullivan, G.D. and Baker, K.D. (1991). Kalman filters in constrained model based tracking. *Proc. 2nd British Machine Vision Assoc. Conf., Glasgow* (23–26 September), pp. 371–374.

Matsuyama, T., Saburi, K. and Nagao, M. (1982). A structural analyzer for regularly arranged textures. *Comput. Graph. Image Process* **18**, no. 3, 259–278.

Maybank, S. (1992). *Theory of Reconstruction from Image Motion*. Springer-Verlag, Berlin, Heidelberg.

Maybank, S.J. (1986). Algorithm for analysing optical flow based on the least squares method. *Image Vision Comput.* **4**, 38–42.

McIvor, A.M. (1988). Edge detection in dynamic vision. *Proc. 4th Alvey Vision Conf., Manchester* (31 August–2 September), pp. 141–145.

Mead, C.A. and Conway, L.A. (1980). *Introduction to VLSISystems*. Addison-Wesley, Reading, MA.

Meer, P., Mintz, D. and Rosenfeld, A. (1990). Least median of squares based robust analysis of image structure. *Proc. DARPA Image Understanding Workshop, Pittsburgh, Pennsylvania* (11–13 September), pp. 231–254.

Meer, P., Mintz, D., Rosenfeld, A. and Kim, D.Y. (1991). Robust regression methods for computer vision: a review. *Int. J. Comput. Vision.* **6**, no. 1, 59–70.

Merlin, P.M. and Farber, D.J. (1975). A parallel mechanism for detecting curves in pictures. *IEEE Trans. Comput.* **28**, 96–98.

Mérö, L. and Vassy, Z. (1975). A simplified and fast version of the Hueckel operator for finding optimal edges in pictures. *Proc. 4th Int. Joint Conf. on Artif. Intell., Tbilisi, Georgia, USSR,* pp. 650–655.

Michalewicz, Z. (1992). *Genetic Algorithms + Data Structures = Evolution Programs.* Springer-Verlag, Berlin, Heidelberg.

Milgram, D. (1979). Region extraction using convergent evidence. *Comput. Graph. Image Process* **11**, 1–12.

Minsky, M.L and Papert, S.A. (1969). *Perceptrons.* MIT Press, Cambridge, MA.

Minsky, M. and Papert, S. (1971). On some associative, parallel and analog computations. In: Jacks, E.J. (ed.) *Associative Information Techniques.* Elsevier, New York, pp. 27–47.

Moganti, M., Ercal, F., Dagli, C.H. and Tsunekawa, S. (1996). Automatic PCB (98 inspection algorithms: a survey. *Comput. Vision Image Understanding* **63**, no. 2, 287–313.

Moore, G.A. (1968). Automatic scanning and computer processes for the quantitative analysis of micrographs and equivalent subjects. In: Cheng, G.C. *et al.* (eds) *Pictorial Pattern Recognition.* Thompson, Washington, DC, pp. 236–275.

Moravec, H.P. (1980). Obstacle avoidance and navigation in the real world by a seeing robot rover. *Stanford Artif. Intell. Lab. Memo AIM-340.*

Môtus, L. and Rodd, M.G. (1994). *Timing Analysis of Real-Time Software.* Elsevier, Oxford.

Mundy, J.L and Zisserman, A. (eds) (1992a). *Geometric Invariance in Computer Vision.* MIT Press, Cambridge, MA.

Mundy, J.L and Zisserman, A. (1992b). Appendix—projective geometry for machine vision. In Mundy, J.L. and Zisserman, A. (eds) *Geometric Invariance in Computer Vision.* MIT Press, Cambridge, MA, pp. 463–519.

Murray, D.W., Kashko, A. and Buxton, H. (1986). A parallel approach to the picture restoration algorithm of Geman and Geman on an SIMD machine. *Image Vision Comput.* **4**, no. 3, 133–142.

Nagao, M. and Matsuyama, T. (1979). Edge preserving smoothing. *Comput. Graph. Image Process* **9**, 394–407.

Nagel, H.-H. (1983). Displacement vectors derived from second-order intensity variations in image sequences. *Comput. Vision Graph. Image Process* **21**, 85–117.

Nagel, H.-H. (1986). Image sequences-ten (octal) years-from phenomenology towards a theoretical foundation. *Proc. 8th Int. Conf. on Pattern Recogn., Paris* (27–31 October), pp. 1174–1185.

Nagel, R.N. and Rosenfeld, A. (1972). Ordered search techniques in template matching. *Proc. IEEE* **60**, 242–244.

Nakagawa, Y. (1982). Automatic visual inspection of solder joints on printed circuit boards. *Proc. SPIE, Robot Vision* **336**, 121–127.

Nakagawa, Y. and Rosenfeld, A. (1979). Some experiments on variable thresholding. *Pattern Recogn.* **11**, 191–204.

Narendra, P.M. (1978). A separable median filter for image noise smoothing. *Proc. IEEE Comput. Soc. Conf. on Pattern Recogn. and Image Process, Chicago* (31 May-2 June), pp. 137–141.

Newman, T.S. and Jain, A.K. (1995). A survey of automated visual inspection. *Comput. Vision Image Understanding* **61**, no. 2, 231–262.

Niblack, W. (1985). *An Introduction to Digital Image Processing.* Strandberg, Birkeroed, Denmark.

Nieminen, A., Heinonen, P. and Neuvo, Y. (1987). A new class of detail preserving filters for image processing. *IEEE Trans. Pattern Anal. Mach. Intell.* **9**, 74–90.

Nightingale, C. and Hutchinson, R.A. (1990). Artificial neural nets and their applications to image processing. *Brit. Telecom Technol. J.* **8**, no. 3, 81–93.

Nilsson, N.J. (1965). *Learning Machines-Foundations of Trainable Pattern Classifying Systems.* McGraw-Hill, New York.

Nitzan, D., Brain, A.E. and Duda, R.O. (1977). The measurement and use of registered reflectance and range data in scene analysis. *Proc. IEEE* **65**, 206–220.

Nixon, M. (1985). Application of the Hough transform to correct for linear variation of background illumination in images. *Pattern Recogn. Lett.* **3**, 191–194.

Noble, A., Hartley, R., Mundy, J. and Farley, J. (1994). X-ray metrology for quality assurance. *Proc. IEEE Int. Conf. Robotics and Automation, San Diego, CA* (May), Vol. 2, pp.1113–1119.

Noble, J.A. (1988). Finding corners. *Image Vision Comput.* **6**, 121–128.

Noble, J.A. (1995). From inspection to process understanding and monitoring: a view on computer vision in manufacturing. *Image Vision Comput.* **13**, no. 3, 197–214.

North, D.O. (1943). An analysis of the factors which determine signal/noise discrimination in pulsed-carrier systems. *RCA Lab., Princeton, NJ, Rep. PTR-6C*; reprinted in *Proc. IEEE* **51**, 1016–1027 (1963).

Noyce, R.N. (1977) Microelectronics. *Scientific Amer.* **237** (September), 62–69.

Gorman, F. (1978). Edge detection using Walsh functions. *Artif. Intell.* **10**, 215–223.

Gorman, F. and Clowes, M.B. (1976). Finding picture edges through collinearity of feature points. *IEEE Trans. Comput.* **25**, 449–456.

Offen, R.J. (ed.) (1985). *VLSI Image Processing.* Collins, London.

Oflazer, K. (1983). Design and implementation of a single chip ld median filter. *IEEE Trans. Acoust. Speech Signal Process* **31**, 1164–1168.

Ogawa, H. and Taniguchi, K. (1979). Preprocessing for Chinese character recognition and global classification of handwritten Chinese characters. *Pattern Recogn.* **11**, 1–7.

Ohanian, P.P. and Dubes, R.C. (1992). Performance evaluation for four classes of textural features. *Pattern Recogn.* **25**, 819–833.

Ohta, Y., Maenobu, K. and Sakai, T. (1981). Obtaining surface orientation from texels under perspective projection. *Proc. 7th Int. Joint Conf. on Artif. Intell., Vancouver*, pp. 746–751.

Oja, E. (1982). A simplified neuron model as a principal component analyzer. *Int. J. Neural Systems* **1**, 61–68.

Oja, E. and Karhunen, J. (1993). *Nonlinear PCA: Algorithms and Applications.* Report A18, Helsinki Univ. of Technol., Espoo, Finland.

Osteen, R.E. and Tou, J.T. (1973). A clique-detection algorithm based on neighbourhoods in graphs. *Int. J. Comput. Inf. Sci.* **2**, 257–268.

Overington, I. and Greenway, P. (1987). Practical first-difference edge detection with subpixel accuracy. *Image Vision Comput.* **5**, 217–224.

Pal, N.R. and Pal, S.K. (1989). Object-background segmentation using new definitions of entropy. *IEE Proc. E***136**, no. 4, 284–295.

Pal, S.K., King, R.A. and Hashim, A.A. (1983). Automatic grey level thresholding through index of fuzziness and entropy. *Pattern Recogn. Lett.* **1**,141–146.

Pal, S.K., Bhandari, D. and Kundu, M.K. (1994). Genetic algorithms for optimal image enhancement. *Pattern Recogn. Lett.* **15**, no. 3, 261–271.

Paler, K. and Kittler, J. (1983). Greylevel edge thinning: a new method. *Pattern Recogn. Lett.* **1**, 409–416.

Paler, K., Föglein, J., Illingworth, J. and Kittler, J. (1984). Local ordered grey levels as an aid to corner detection. *Pattern Recogn.* **17**, 535–543.

Pan, X.D., Ellis, T.J. and Clarke, T.A. (1995). Robust tracking of circular features. *Proc. 6th British Machine Vision Assoc. Conf., Birmingham* (11–14 September), pp. 553–562

Panda, D.P. and Rosenfeld, A. (1978). Image segmentation by pixel classification in (gray level, edge value) space. *IEEE Trans. Comput.* **27**, 875–879.

Parker, D.B. (1985). *Learning-Logic: Casting the Cortex of the Human Brain in Silicon.* Technical Report TR-47, Center for Comput. Res. in Economics and Management Sci., MIT, Cambridge, MA.

Patel, D. and Davies, E.R. (1995). Low contrast object detection using a MLP network designed by node creation. *Proc. IEEE Int. Conf. on Neural Networks, Perth* (27 November–1 December), Vol. 2, pp. 1155–1159.

Patel, D., Hannah, I. and Davies, E.R. (1994). Texture analysis for foreign object detection using a single layer neural network. *Proc. IEEE Int. Conf. on Neural Networks, Florida* (28 June–2 July), Vol. VII, pp. 4265–4268.

Patel, D., Davies, E.R. and Hannah, I. (1995). Towards a breakthrough in the detection of contaminants in food products. *Sensor Review* **15**, no. 2, 27–28.

Pavlidis, T. (1968). Computer recognition of figures through decomposition. *Inf. Control* **12**, 526–537.

Pavlidis, T. (1977). *Structural Pattern Recognition.* Springer-Verlag, New York.

Pavlidis, T. (1978). A review of algorithms for shape analysis. *Comput. Graph. Image Process* **7**, 243-258.

Pavlidis, T. (1980). Algorithms for shape analysis of contours and waveforms. *IEEE Trans. Pattern Anal. Mach. Intell.* **2**, 301–312.

Pearl, J. (1988). *Probabilistic Reasoning in Intelligent Systems: Networks of Plausible Inference.* Morgan Kaufmann, San Mateo, CA.

Penman, D., Olsson, O. and Beach, D. (1992). Automatic X-ray inspection of canned products for foreign material. *Machine Vision Applications, Architectures and Systems Integration, SPIE,* Vol. **1823**, pp. 342–347.

Pentland, A.P. (1984). Fractal-based description of natural scenes. *IEEE Trans. Pattern Anal. Mach. Intell.* **6**, no. 6, 661–674.

Persoon, E. and Fu, K.-S. (1977). Shape discrimination using Fourier descriptors. *IEEE Trans. Systems Man Cybern* **7**, 170–179.

Petrou, M. and Kittler, J. (1988). On the optimal edge detector. *Proc. 4th Alvey Vision Conf., Manchester* (31 August–2 September), pp. 191–196.

Pfaltz, J.L. and Rosenfeld, A. (1967). Computer representation of planar regions by their skeletons. *Comm. ACM* **10**, 119–125.

Pham, D.T. and Bayro-Corrochano, E.J. (1992) RNeural computing for noise filtering, edge detection and signature extraction. *J. Systems Eng.* **1**, 13–23

Phong, B.-T. (1975). Illumination for computer-generated pictures. *Comm. ACM* **18**, 311–317.

Pickett, R.M. (1970). Visual analysis of texture in the detection and recognition of objects. In: Lipkin, B.S. and Rosenfeld, A. (eds) *Picture Processing and Psychopictorics.* Academic Press, New York, pp. 289–308.

Pietikäinen, M., Rosenfeld, A. and Davis, L.S. (1983). Experiments with texture classification using averages of local pattern matches. *IEEE Trans. Systems Man Cybern.* **13**, no. 3, 421–426.

Plessey Semiconductors (1986). *PDSP16401 2-dimensional edge detector.*

Plummer, A.P.N. and Dale, F. (1984). *The Picture Processing Language Compiler Manual.* National Physical Laboratory, Teddington.

Pollard, S.B., Porrill, J., Mayhew, J.E.W. and Frisby, J.P. (1987). Matching geometrical descriptions in three-space. *Image Vision Comput.* **5**, no. 2, 73–78.

Postaire, J.G. and Touzani, A. (1989). Mode boundary detection by relaxation for cluster analysis. *Pattern Recogn.* **22**, no. 5, 477–489.

Pountain, D. and May, D. (1987). *A Tutorial Introduction to OCCAM Programming*. BSP Professional Books, Oxford.

Pratt, W.K. (1978). *Digital Image Processing*. Wiley, New York.

Prewitt, J.M.S. (1970). Object enhancement and extraction. In: Lipkin, B.S. and Rosenfeld, A. (eds) *Picture Processing and Psychopictorics*. Academic Press, New York, pp. 75–149.

Princen, J., Illingworth, J. and Kittler, J. (1989a). A hierarchical approach to line extraction. *Proc. IEEE Comput. Vision and Pattern Recogn. Conf., San Diego*, pp. 92–97.

Princen, J., Yuen, H.K., Illingworth, J. and Kittler, J. (1989b). Properties of the adaptive Hough transform. *Proc. 6th Scand. Conf. on Image Analysis, Oulu, Finland* (19–22 June), pp. 613–620.

Pringle, K.K. (1969). Visual perception by a computer. In: Grasselli, A. (ed.) *Automatic Interpretation and Classification of Images*. Academic Press, New York, pp. 277–284.

Pugh, A. (ed.) (1983). *Robot Vision*. IFS (Publications) Ltd., Bedford, UK/Springer-Verlag, Berlin.

Pun, T. (1981). Entropic thresholding, a new approach. *Comput. Graph. Image Process* **16**, 210–239.

Ramesh, N., Yoo, J.-H. and Sethi, I.K. (1995). Thresholding based on histogram approximation. *IEE Proc. Vision Image Signal Process* **142**, no. 5, 271–279.

Reed, T.R. and Du Buf, J.M.H. (1993). A review of recent texture segmentation and feature extraction techniques. *Comput. Vision Graph. Image Process. Image Understanding* **57**, 359–372.

Reeves, A.P. (1984). Parallel computer architectures for image processing. *Comput. Vision Graph. Image Process* **25**, 68–88.

Reeves, A.P., Akey, M.L. and Mitchell, O.R. (1983). A moment-based two-dimensional edge operator. *Proc. IEEE Comput. Soc. Conf. on Comput. Vision and Pattern Recogn.* (19–23 June), pp. 312–317.

Rindfleisch, T. (1966). Photometric method for lunar topography. *Photogrammetric Eng.* **32**, 262–276.

Ripley, B.D. (1996). *Pattern Recognition and Neural Networks*. Cambridge Univ. Press, Cambridge, UK.

Rives, G., Dhome, M., Lapreste, J.T. and Richetin, M. (1985). Detection of patterns in images from piecewise linear contours. *Pattern Recogn. Lett.* **3**, 99–104.

Robert, L. (1996). Camera calibration without feature extraction. *Comput. Vision Image Understanding* **63**, no. 2, 314–325.

Roberts, L.G. (1965). Machine perception of three-dimensional solids. In: Tippett, J. *et al.* (eds) *Optical and Electro-optical Information Processing*. MIT Press, Cambridge, MA, pp. 159–197.

Robinson, G.S. (1977). Edge detection by compass gradient masks. *Comput. Graph. Image Process* **6**, 492–501.

Rogers, D.F. (1985). *Procedural Elements for Computer Graphics*. McGraw-Hill, New York.

Rohl, J.S. and Barrett, H.J. (1980). *Programming via Pascal*. Cambridge University Press, Cambridge.

Rosenblatt, F. (1958). The perceptron: a probabilistic model for information storage and organisation in the brain. *Psychol. Review* **65**, 386–408.

Rosenblatt, F. (1962). *Principles of Neurodynamics*. Spartan, New York.

Rosenfeld, A. (1969). *Picture Processing by Computer*. Academic Press, New York.

Rosenfeld, A. (1970). Connectivity in digital pictures. *J. ACM* **17**, 146–160.

Rosenfeld, A. (1979). *Picture Languages: Formal Models for Picture Recognition*. Academic Press, New York.

Rosenfeld, A. and Kak, A.C. (1981). *Digital Picture Processing* (2nd edn). Academic Press, New York.

Rosenfeld, A. and Pfaltz, J.L. (1966). Sequential operations in digital picture processing. *J. ACM* **13**, 471–494.

Rosenfeld, A. and Pfaltz, J.L. (1968). Distance functions on digital pictures. *Pattern Recogn.* **1**, 33–61.

Rosenfeld, A. and Troy, E.B. (1970a). Visual texture analysis. Computer Science Center, Univ. of Maryland Techn. Report TR-116.

Rosenfeld, A. and Troy, E.B. (1970b). Visual texture analysis. *Conf. Record for Symposium on Feature Extraction and Selection in Pattern Recogn.* IEEE Publication 70C-51C, Argonne, Ill., Oct. 115–124.

Rosenfeld, A. and VanderBrug, G.J. (1977). Coarse-fine template matching. *IEEE Trans. Systems Man Cybern.* **7**, 104–107.

Rosenfeld, A., Hummel, R.A. and Zucker, S.W. (1976). Scene labelling by relaxation operations. *IEEE Trans. Systems Man Cybern.* **6**, 420–433.

Rosie, A.M. (1966). *Information and Communication Theory*. Blackie, London.

Rosin, P.L and West, G.A.W. (1995). Curve segmentation and representation by superellipses. *IEE Proc. Vision Image Signal Process* **142**, no. 5, 280–288.

Roth, G. and Levine, M.D. (1994) Geometric primitive extraction using a genetic algorithm. *IEEE Trans. Pattern Anal. Mach. Intell.* **16**, no. 9, pp. 901–905.

Rothwell, C.A., Zisserman, A., Forsyth, D.A. and Mundy, J.L. (1992a). Canonical frames for planar object recognition. *Proc 2nd European Conf. on Computer Vision, Santa Margherita Ligure, Italy* (19–22 May), pp. 757–772.

Rothwell, C.A., Zisserman, A., Marinos, C.I., Forsyth, D.A. and Mundy, J.L. (1992b). Relative motion and pose from arbitrary plane curves. *Image Vision Comput.* **10**, no. 4, 250–262.

Rousseeuw, P.J. (1984). Least median of squares regression. *J. Amer. Statist. Assoc.* **79**, no. 388, 871–880.

Rousseeuw, P.J. and Leroy, A.M. (1987). *Robust Regression and Outlier Detection*. Wiley, New York.

Ruff, B.P.D. (1987). A pipelined architecture for the Canny edge detector. *Proc. 3rd Alvey Vision Conf., Cambridge* (15–17 September), pp. 147–150.

Rumelhart, D.E., Hinton, G.E. and Williams, R.J. (1986). Learning internal representations by error propagation. In: Rumelhart, D.E. and McClelland, J.L. (eds), *Parallel Distributed Processing: Explorations in the Microstructure of Cognition*. MIT Press, Cambridge, MA, pp. 318–362.

Rummel, P. and Beutel, W. (1984). Workpiece recognition and inspection by a model-based scene analysis system. *Pattern Recogn.* **17**, 141–148.

Rutkowski, W. and Benton, R. (1984). *Determination of Object Pose by Fitting a Model to Sparce Range Data*. Interim Techn. Rep. Intelligent Task Automation Program, Honeywell, Inc., pp. 6-62–6-98.

Rutovitz, D. (1970). Centromere finding: some shape descriptors for small chromosome outlines. In: Meltzer, B. and Michie, D. (eds) *Machine Intelligence 5*. Edinburgh University Press, Edinburgh, pp. 435–462.

Rutovitz, D. (1975). An algorithm for in-line generation of a convex cover. *Comput Graph. Image Process* **4**, 74–78.

Sahoo, P.K., Soltani, S., Wong, A.K.C. and Chen, Y.C. (1988). A survey of thresholding techniques. *Comput. Vision Graph. Image Process* **41**, 233–260.

Sanger, T.D. (1989). Optimal unsupervised learning in a single-layer linear feedforward neural network. *Neural Networks* **2**, 459–473.

Sanz, J.L.C. (ed.) (1988). Special Issue on Industrial Machine Vision and Computer Vision Technology. *IEEE Trans. Pattern Anal. Mach. Intell.* **10**, np. 1.

Schneiderman, H., Nashman, M., Wavering, A.J. and Lumia, R. (1995) "Vision-based robotic convoy driving", Machine Vision and Applications, **8**, no. 6, pp. 359–364.

Scott, G.L. (1988). *Local and Global Intepretation of Moving Images.* Pitman, London and Morgan Kaufmann, San Mateo, CA.

Seeger, U. and Seeger, R. (1994). Fast corner detection in grey-level images. *Pattern Recogn. Lett.* **15**, no. 7, 669–675.

Ser, P.-K. and Siu, W.-C. (1995). Novel detection of conics using 2-D Hough planes. *IEE Proc. Vision Image Signal Process* **142**, no. 5, 262–270.

Serra, J. (1982). *Image Analysis and Mathematical Morphology.* Academic Press, New York.

Serra, J. (1986). Introduction to mathematical morphology. *Comput. Vision Graph. Image Process* **35**, 283–305.

Shah, M.A. and Jain, R. (1984). Detecting time-varying corners. *Comput. Vision Graph. Image Process* **28**, 345–355.

Shaming, W.B. (1974). *Digital Image Transform Encoding.* RCA Corp. paper no. PE-622.

Shirai, Y. (1972). Recognition of polyhedra with a range finder. *Pattern Recogn.* **4**, 243–250.

Shirai, Y. (1975). Analyzing intensity arrays using knowledge about scenes. In: Winston, P.H. (ed.) *The Psychology of Computer Vision.* McGraw-Hill, New York, pp. 93–113.

Shirai, Y. (1987). *Three-dimensional Computer Vision.* Springer-Verlag, Berlin.

Siedlecki, W. and Sklansky, J. (1989). A note on genetic algorithms for large-scale feature selection. *Pattern Recogn. Lett.* **10**, no. 5, 335–347.

Siegel, A.F. (1982). Robust regression using repeated medians. *Biometrika.* **69**, no. 1, 242–244.

Silberberg, T.M., Davis, L. and Harwood, D. (1984). An iterative Hough procedure for three-dimensional object recognition. *Pattern Recogn.* **17**, 621–629.

Singh, R.K. and Ramakrishna, R.S. (1990). Shadows and texture in computer vision. *Pattern Recogn. Lett.* **11**, no. 2, 133–141.

Sjoberg, F. and Bergholm, F. (1988). Extraction of diffuse edges by edge focussing. *Pattern Recogn. Lett.* **7**, 181–190.

Skillicorn, D.B. (1988). A taxonomy for computer architectures. *IEEE Comput.* (Nov.), **21**, 46–57.

Sklansky, J. (1970). Recognition of convex blobs. *Pattern Recogn.* **2**, 3–10.

Sklansky, J. (1978). On the Hough technique for curve detection. *IEEE Trans. Comput.* **27**, 923–926.

Sklansky, J., Cordella, L.P. and Levialdi, S. (1976). Parallel detection of concavities in cellular blobs. *IEEE Trans. Comput.* **25**, 187–196.

Slama, C.C. (ed.) (1980). *Manual of Photogrammetry* (4th edn). Amer. Soc. of Photogrammetry, Falls Church, VA.

Snyder, W.E. (ed.) (1981). *Special Issue on Computer Analysis of Time-Varying Images. IEEE Comput. (Aug.)* **14**.

Sobey, P.J.M. (1989). The Automated Visual Inspection and Grading of Timber. PhD Thesis, University of Adelaide.

Spirkovska, L. and Reid, M.B. (1992). Robust position, scale, and rotation invariant object recognition using higher-order neural networks. *Pattern Recogn.* **25**, no. 9, 975–985.

Srinivas, M. and Patnaik, L.M. (1994). Genetic algorithms: a survey. *IEEE Computer.* **27**, no. 6, 17–26.

Steele, J.M. and Steiger, W.L. (1986). Algorithms and complexity for least median of squares regression. *Discrete Applied Math.* **14**, 93–100.

Stella, E., Lovergine, F.P., D'Orazio, T. and Distante, A. (1995). A visual tracking technique suitable for control of convoys. *Pattern Recogn. Lett.* **16**, no. 9, 925–932.

Stephens, R.S. (1991). Probabilistic approach to the Hough transform. *Image Vision Comput.* **9**, no. 1, 66–71.

Stevens, K. (1980). Surface perception from local analysis of texture and contour. *MIT Artif. Intell. Lab. Memo AI-TR-512.*

Stockman, G.C. and Agrawala, A.K. (1977). Equivalence of Hough curve detection to template matching. *Comm. ACM* **20**, 820–822.

Stone, J.V. (1990). Shape from texture: textural invariance and the problem of scale in perspective images of textured surfaces. *Proc. BMVA Conf., Oxford* (24–27 September), pp. 181–186.

Straforini, M., Coelho, C. and Campani, M. (1993). Extraction of vanishing points from images of indoor and outdoor scenes. *Image Vision Comput.* **11**, no. 2, 91–99.

Sullivan, G.D. (1992). Visual interpretation of known objects in constrained scenes. *Phil. Trans. R. Soc.* (London). **B337**, 361–370.

Tabandeh, A.S. (1988). Artificial Intelligence Techniques and Concepts for Integrating a Robot Vision System with a Solid Modeller. PhD Thesis, Cambridge University.

Tabandeh, A.S. and Fallside, F. (1986). Artificial intelligence techniques and concepts for the integration of robot vision and 3D solid modellers. *Proc. Int. Conf. on Intell. Autonomous Systems, Amsterdam* (18–11 December).

Tan, T.N. (1995). Structure, pose and motion of bilateral symmetric objects. *Proc. 6th British Machine Vision Assoc. Conf., Birmingham* (11–14 September), pp. 473–482.

Tan, T.N., Sullivan, G.D. and Baker, K.D. (1994). Recognizing objects on the ground-plane. *Image Vision Comput.* **12**, no. 3, 164–172.

Taylor, J. and S. Coombes (1993). Learning higher order correlations. *Neural Networks* **6**, 423–427.

Theil, H. (1950). A rank-invariant method of linear and polynomial regression analysis (parts 1–3). *Nederlandsche Akad. Wetenschappen Proc.* **A53**, 386–392, 521–525 and 1397–1412.

Thomas, A. D. H. and Rodd, M.G. (1994). Knowledge-based inspection of electric lamp caps. *Engineering Applications of Artificial Intelligence* **7**, no. 1, 31–37.

Thomas, A.D.H., Rodd, M.G., Holt, J.D. and Neill, C.J. (1995). Real-time industrial visual inspection: a review. *Real-Time Imaging* **1**, no. 2, 139–158.

Tou, J.T. and Gonzalez, R.C. (1974). *Pattern Recognition Principles.* Addison Wesley, Reading, MA.

Toulson, D.L. and Boyce, J.F. (1992). Segmentation of MR images using neural nets. *Image Vision Comput.* **10**, no. 5, 324–328.

Tsai, F.C.D. (1996). A probabilistic approach to geometric hashing using line features. *Comput. Vision Image Understanding* **63**, no. 1, 182–195.

Tsai, R.Y. (1986). An efficient and accurate camera calibration technique for 3D machine vision, *Proc. Conf. on Comput. Vision Pattern Recogn., Miami, FL,* pp. 364–374.

Tsuji, S. and Matsumoto, F. (1978). Detection of ellipses by a modified Hough transform. *IEEE Trans. Comput.* **27**, 777–781.

Tsukune, H. and Goto, K. (1983). Extracting elliptical figures from an edge vector field. *Proc. IEEE Conf. on Computer Vision and Pattern Recogn., Washington,* pp. 138–141.

Tuckey, C.O. and Armistead, W. (1953). *Coordinate Geometry*. Longmans, Green & Co Ltd, London.

Turin, G.L. (1960). An introduction to matched filters. *IRE Trans. Inf. Theory* **6**, 311–329.

Turney, J.L., Mudge, T.N. and Volz, R.A. (1985). Recognizing partially occluded parts. *IEEE Trans. Pattern Anal. Mach. Intell.* **7**, 410–421.

Uhr, L. (1978). Recognition cones and some test results; the imminent arrival of well-structured parallel-serial computers; positions, and positions on positions. In: Hanson, A.R. and Riseman, E.M. (eds) *Computer Vision Systems*. Academic Press, New York, pp. 363–377.

Uhr, L. (1980). Psychological motivation and underlying concepts. In: Tanimoto, S. and Klinger, A. (eds) *Structured Computer Vision-Machine Perception through Hierarchical Computation Structures*. Academic Press, London, pp. 1–30.

Ullman, S. (1979). *The Interpretation of Visual Motion*. MIT Press, Cambridge, MA.

Ullmann, J.R. (1969). Experiments with the n-tuple method of pattern recognition. *IEEE Trans. Comput.* **18**, 1135–1137.

Ullmann, J.R. (1973). *Pattern Recognition Techniques*. Butterworth, London.

Ullmann, J.R. (1974). Binarisation using associative addressing. *Pattern Recogn.* **6**, 127–135.

Ullmann, J.R. (1976). An algorithm for subgraph isomorphism. *J. ACM* **23**, 31–42.

Ullmann, J.R. (1981). Video-rate digital image analysis equipment. *Pattern Recogn.* **14**, 305–318.

Unger, S.H. (1958). A computer orientated towards spatial problems. *Proc. IRE* **46**, 1714–1750.

Unser, M. (1986). Local linear transforms for texture measurements. *Signal Process* **11**, 61–79.

Unser, M. and Ade, F. (1984). Feature extraction and decision procedure for automated inspection of textured materials. *Pattern Recogn. Lett.* **2**, no. 3, 185–191.

Unser, M. and Eden, M. (1989). Multiresolution feature extraction and selection for texture segmentation. *IEEE Trans. Pattern Anal. Mach. Intell.* **11**, no. 7, 717–728.

Unser, M. and Eden, M. (1990). Nonlinear operators for improving texture segmentation based on features extracted by spatial filtering. *IEEE Trans. Systems Man Cybern.* **20**, no. 4, 804–815.

Vaillant, R., Monrocq, C. and Le Cun, Y. (1994). Original approach for the localisation of objects in images. *IEE Proc. Vision Image Signal Process* **141**, no. 4, 245–250.

Van Gool, L., Dewaele, P. and Oosterlinck, A. (1985). Survey: Texture analysis anno 1983. *Comput. Vision Graph. Image Process* **29**, 336–357.

van Digellen, J. (1951). Photometric investigations of the slopes and heights of the ranges of hills in the Maria of the moon. *Bull. Astron. Inst. Netherlands* **11**, 283–289.

VanderBrug, G.J. and Rosenfeld, A. (1977). Two-stage template matching. *IEEE Trans. Comput.* **26**, 384–393.

Vilnrotter, F.M., Nevatia, R. and Price, K.E. (1986). Structural analysis of natural textures. *IEEE Trans. Pattern Anal. Mach. Intell.* **8**, no. 1, 76–89.

Vistnes, R. (1989). Texture models and image measures for texture discrimination. *Int. J. Comput. Vision* **3**, 313–336.

von Helmholtz, H. (1925). *Treatise on Physiological Optics*, Vols 1–3 (translated by Southall, J.P.C.) Dover, New York.

Voss, R. (1986). Random fractals: characterization and measurement. In: Pynn, R. and Skjeltorp, A. (eds) *Scaling Phenomena in Disordered Systems*. Plenum, New York.

Wagner, G.G. (1988). Combining X-ray imaging and machine vision. *Proc. SPIE*. Vol. **850**, pp. 43–53.

Wallace, A.M. (1988). Industrial applications of computer vision since 1982. *IEE Proc. E* **135**, 117–136.

Waltz, D. (1975). Understanding line drawings of scenes with shadows. In: Winston, P.H. (ed.) *The Psychology of Computer Vision*. McGraw-Hill, New York, pp. 19–91.

Wang, C., Sun, H., Yada, S. and Rosenfeld, A. (1983). Some experiments in relaxation image matching using corner features. *Pattern Recogn.* **16**, 167.

Wang, S. and Haralick, R.M. (1984). Automatic multithreshold selection. *Comput. Vision Graph. Image Process* **25**, 46–67.

Wang, T., Zhuang, X. and Xing, X. (1992). Robust segmentation of noisy images using a neural network model. *Image Vision Comput.* **10**, no. 4, 233–240.

Weiman, C.F.R. (1976) Highly parallel digitised geometric transformations without matrix multiplication. *Proc. Int. Joint Conf. on Parallel Processing.* pp. 1–10.

Weimer, P.K. (1975). From camera tubes to solid-state sensors. *RCA Rev. 36*, 385 405.

Werbos, P.J. (1974). Beyond Regression: New Tools for Prediction and Analysis in the Behavioral Sciences. PhD Thesis, Harvard Univ., Cambridge, MA.

Wermser, D. and Liedtke, C.-E. (1982). Texture analysis using a model of the visual system. *Proc. 6th Int. Conf. on Pattern Recogn., Munich* (19–22 October), pp. 1078–1081

Wermser, D., Haussmann, G. and Liedtke, C.-E. (1984). Segmentation of blood smears by hierarchical thresholding. *Comput. Vision Graph. Image Process* **25**, 151–168.

Weska, J.S. (1978). A survey of threshold selection techniques. *Comput. Graph. Image Process* **7**, 259–265.

Weska, J.S. and Rosenfeld, A. (1976). An application of texture analysis to materials inspection. *Pattern Recogn.* **8**, 195–199.

Weska, J.S. and Rosenfeld, A. (1979). Histogram modification for threshold selection. *IEEE Trans. Systems Man Cybern.* **9**, 38–52.

Weska, J.S., Nagel, R.N. and Rosenfeld, A. (1974). A threshold selection technique. *IEEE Trans. Comput.* **23**, 1322–1326.

Weszka, J.S., Dyer, C.R. and Rosenfeld, A. (1976). A comparative study of texture measures for terrain classification. *IEEE Trans. Systems Man Cybern.* **6**, no. 4, 269–285.

White, J.M. and Rohrer, G.D. (1983). Image thresholding for optical character recognition and other applications requiring character image extraction. *IBMJ. Res. Dev.* **27**, 400–411.

Wiejak, J.S., Buxton, H. and Buxton, B.F. (1985). Convolution with separable masks for early image processing. *Comput. Vision Craph. Image Process* **32**, 279–290.

Will, P.M. and Pennington, K.S. (1971). Grid coding: a preprocessing technique for robot and machine vision. *Artif. Intell.* **2**, 319–329.

Wilson, H.R. and Giese, S.C. (1977). Threshold visibility of frequency gradient patterns. *Vision Res.* **17**, 1177–1190.

Winston, P.H. (ed.) (1975). *The Psychology of Computer Vision*. McGraw-Hill, New York.

Winston, P.H. (1984). *Artificial Intelligence* (2nd edn). Addison-Wesley, Reading, MA.

Witkin, A.P. (1981). Recovering surface shape and orientation from texture, *Artif. Intell.* **17**, 17–45.

Witkin, A.P. (1983). Scale-space filtering. *Proc. 4th Int. Joint. Conf. on Artif. Intell., Tbilisi, Georgi, USSR*, pp. 1019–1022.

Wong, RY. and Hall, E.L. (1978). Scene matching with invariant moments. *Comput. Graph. Image Process* **8**, 16–24.

Woodham, R.J. (1978). Reflectance map techniques for analysing surface defects in metal castings. *MIT Artif. Intell. Lab. Memo AI-TR-457*.

Woodham, R.J. (1980). Photometric method for determining surface orientation from multiple images. *Opt. Eng.* **19**, 139–144.

Woodham, R.J. (1981). Analysing images of curved surfaces. *Artif. Intell.* **17**, 117–140.

Woods, J.W. (1972). Two-dimensional discrete Markovian fields. *IEEE Trans. Inf. Theory*, **18**, 232–240.

Worrall, A.D., Baker, K.D. and Sullivan, G.D. (1989). Model based perspective inversion. *Image Vision Comput.* **7**, 17–23.

Wu, A.Y., Hong, T.-H. and Rosenfeld, A. (1982). Threshold selection using quadtrees. *IEEE Trans. Pattern Anal. Mach. Intell.* **4**, 90–94.

Wu, Z.-Q. and Rosenfeld, A. (1983). Filtered projections as an aid in corner detection. *Pattern Recogn.* **16**, 31–38.

Xu, L. and Oja, E. (1993). Randomized Hough transform (RHT): basic mechanisms, algorithms, and computational complexities. *Comput. Vision Graph. Image Process: Image Understanding* **57**, 131–154.

Yang, G.J. and Huang, T.S. (1981). The effect of median filtering on edge location estimation. *Comput. Graph. Image Process* **15**, 224–245.

Yang, J. and Li, X. (1995). Boundary detection using mathematical morphology. *Pattern Recogn. Lett.* **16**, no. 12, 1277–1286.

Yang, J.-D., Chen, Y.-S. and Hsu, W.-H. (1994). Adaptive thresholding algorithm and its hardware implementation. *Pattern Recogn. Lett.* **15**, no. 2, 141–150

Yin, L., Astola, J. and Neuvo, Y. (1993). A new class of nonlinear filters—neural filters. *IEEE Trans. Signal Process.* **41**, no. 3, 1201–1222.

Yuen, H.K., Illingworth, J. and Kittler, J. (1988). Ellipse detection using the Hough transform. *Proc. 4th Alvey Vision Conf., Manchester* (31 August–2 September), pp. 265–271.

Yuen, H.K., Princen, J., Illingworth, J. and Kittler, J. (1989). A comparative study of Hough transform methods for circle finding. *Proc. 5th Alvey Vision Conf., Reading* (31 August–2 September), pp. 169–174.

Yuille, A. and Poggio, T.A. (1986). Scaling theorems for zero crossings. *IEEE Trans. Pattern Anal. Mach. Intell.* **8**, 15–25.

Yuille, A.L, Kammen, D.M. and Cohen, D.S. (1989). Quadrature and the development of orientation selective cortical cells by Hebb rules. *Biol. Cybern.* **61**, 183–194.

Zahn, C.T. and Roskies, R.Z. (1972). Fourier descriptors for plane closed curves. *IEEE Trans. Comput.* **21**, 269–281.

Zhang, J. and Modestino, J.W. (1990). A model-fitting approach to cluster validation with application to stochastic model-based image segmentation. *IEEE Trans. Pattern Anal. Mach. Intell.* **12**, no. 10, 1009–1017.

Zhang, Z. (1995). Motion and structure of four points from one motion of a stereo rig with unknown extrinsic parameters. *IEEE Trans. Pattern Anal. Mach. Intell.* **17**, no. 12, 1222–1227.

Zhuang, X. and Haralick, R.M. (1986). Morphological structuring element decomposition. *Comput. Vision Graph Image Process.* **35**, 370–382.

Zielke, T., Braukermann, M. and von Seelen, W. (1993). Intensity and edge-based symmetry detection with an application to car-following. *Comput. Vision Graph. Image Process.: Image Understanding* **58**, 177–190.

Zisserman, A., Marinos, C., Forsyth, D.A., Mundy, J.L. and Rothwell, C.A. (1990). Relative motion and pose from invariants. *Proc. 1st British Machine Vision Assoc. Conf., Oxford* (24–27 September), pp. 7–12.

Zucker, S.W. (1976a). Toward a model of texture. *Comput. Graph. Image Process.* **5**, 190–202.

Zucker, S.W. (1976b). Region growing: childhood and adolescence. *Comput. Graph. Image Process.* **5**, 382–399.

Zuniga, O.A. and Haralick, R.M. (1983). Corner detection using the facet model. *Proc. IEEE Comput. Vision Pattern Recogn. Conf.*, pp. 30–37.

Zuniga, O.A. and Haralick, R.M. (1987). Integrated directional derivative gradient operator. *IEEE Trans. Pattern Anal. Mach. Intell.* **17**, 508–517.

Subject Index

Author Index